Group Theoretical
Methods in Physics

Organizing Committee

P. WINTERNITZ (Chairman), *Université de Montréal*
A. DAIGNEAULT, *Université de Montréal*
H.-F. GAUTRIN, *Université de Montréal*
W. MILLER, JR., *University of Minnesota*
J. PATERA, *Université de Montréal*
R. T. SHARP, *McGill University*
H. ZASSENHAUS, *Ohio State University*

Advisory Committee

W. OPECHOWSKI (Chairman), *University of British Columbia*
H. BACRY, *Université de Marseille*
J. BECKERS, *Université de Liège*
L. C. BIEDENHARN, *Duke University*
F. GURSEY, *Yale University*
S. HELGASON, *Massachusetts Institute of Technology*
A. JANNER, *University of Nijmegen*
A. JOFFE, *Université de Montréal*
B. R. JUDD, *Johns Hopkins University*
B. KOLMAN, *Drexel University*
P. KRAMER, *University of Tübingen*
L. MICHEL, *Institut des Hautes Etudes Scientifiques*
M. MOSHINSKY, *Université de Mexico*
L. O'RAIFEARTAIGH, *Dublin Institute of Advanced Studies*
G. PAQUETTE, *Université de Montréal*
P. ROMAN, *Boston University*

Group Theoretical Methods in Physics

Proceedings of the Fifth International Colloquium
Université de Montréal—July 1976

Edited by

Robert T. Sharp
McGill University, Montreal

Bernard Kolman
Drexel University, Philadelphia

ACADEMIC PRESS New York San Francisco London 1977
A Subsidiary of Harcourt Brace Jovanovich, Publishers

Academic Press Rapid Manuscript Reproduction

ACADEMIC PRESS, INC.
111 Fifth Avenue, New York, New York 10003

United Kingdom Edition published by
ACADEMIC PRESS, INC. (LONDON) LTD.
24/28 Oval Road, London NW1

Library of Congress Cataloging in Publication Data

International Colloquium on Group Theoretical Methods in
Physics, 5th, Université de Montréal, 1976.
Group theoretical methods in physics.

English or French.
Bibliography: p.
Includes index.
1. Groups, Theory of–Congresses. 2. Mathematical physics–Congresses. I. Sharp, Robert T. Date II. Kolman, Bernard, III. Title.
QC20.7.G76I57 1976 530.1'5'222 77-22167
ISBN 0-12-637650-6

PRINTED IN THE UNITED STATES OF AMERICA

Contents

Part V. Mathematical Physics

Part VI. Representation Theory

Contributors

*Agyei, Alfred K. — *Department of Physics, University of Cape Coast, Cape Coast, Ghana*

Aragone, C. — *Departamento de Fisica, Universidad Simon Bolivar, Caracas, Venezuela*

*Backhouse, Nigel — *Department of Applied Mathematics and Theoretical Physics, University of Liverpool, Liverpool, United Kingdom*

*Bacry, H. — *Centre de Physique Théorique, C.N.R.S., Marseille, France*

*Beckers, J. — *Institut de Physique, Service de Physique Théorique et Mathématique, Université de Liège, Liège, Belgique*

*Berenson, Rhoda — *Physics Department, Nassau Community College, Garden City, New York*

*Biedenharn, L. C. — *Institut für Theoretische Physik, Universität Frankfurt am Main, Frankfurt, West Germany*

Birman, Joseph L. — *Physics Department, City College, City University of New York, New York, New York*

*Boyer, Charles P. — *I.I.M.A.S., Universidad Nacional Autonoma de Mexico, Mexico, D.F.*

*Bracken, A. J. — *Mathematics Department, University of Queensland, St. Lucia, Queensland, Australia*

Burdet, G. — *Département de Mathématiques, Université de Montréal, Montréal, P.Q., Canada*

Chalbaud, E. — *Departamento de Fisica, Universidad Simon Bolivar, Caracas, Venezuela*

Cizek, J. — *Department of Applied Mathematics, University of Waterloo, Waterloo, Ontario, Canada*

*Combe, Ph. — *Centre de Physique Théorique, C.N.R.S., Marseille, France*

*Dirl, R. — *Institut für Theoretische Physik der Technische Universität Wien, Vienna, Austria*

Gal-Ezer, E. — *Department of Physics and Astronomy, Tel-Aviv University, Ramat-Aviv, Israel*

*Gazeau, J. P. — *Laboratoire de Chimie Physique, Université Pierre et Marie Curie, Paris, France*

Gulshani, P. — *Department of Physics, University of Toronto, Toronto, Ontario, Canada*

*Gürsey, Feza — *Physics Department, Yale University, New Haven, Connecticut*

Haba, Zbigniew — *Institute of Theoretical Physics, University of Wroclaw, Wroclaw, Poland*

*Harkavy, Allan A. — *Department of Physics, State University College, New Paltz, New York*

*Harnad, J. P. — *Centre de Recherches Mathématiques, Université de Montréal, Montréal, P.Q., Canada*

Hässelbarth, W. — *Institute of Quantum Chemistry, Free University Berlin, Germany*

*Helgason, Sigurdur — *Department of Mathematics, Massachusetts Institute of Technology, Cambridge, Massachusetts*

*Hongoh, M. — *Département de Physique, Université de Montréal, Montréal, P.Q., Canada*

Horwitz, L. P. — *Department of Physics and Astronomy, Tel-Aviv University, Ramat-Aviv, Israel*

*Ilamed, Yehiel — *Israel Atomic Energy Commission, SOREQ Nuclear Research Center, Yavne, Israel*

*Jakobsen, H. P. — *Department of Mathematics, M.I.T., Cambridge, Massachusetts*

*Janner, A. — *Institute for Theoretical Physics, University of Nijmegen, Nijmegen, The Netherlands*

Janssen, T. — *Institute for Theoretical Physics, University of Nijmegen, Nijmegen, The Netherlands*

Jaspers, M. — *Institut de Physique, Service de Physique Théorique et Mathématique, Université de Liège, Liège, Belgique*

*Kaiser, Gerald — *Mathematics Department, University of Toronto, Toronto, Ontario, Canada*

*Kerner, Richard — *Département de Mécanique, Université Pierre et Marie Curie, Paris, France*

*Kibler, Maurice R. — *Institut de Physique Nucléaire, Villeurbanne, France*

Klein, D. J. — *Department of Physics, University of Texas, Austin, Texas*

*Kramer, P. — *Institut für Theoretische Physik, Universität Tübingen, Tübingen, Germany*

*Ktorides, C. N. — *Department of Physics, University of Athens, Athens, Greece*

*Kumei, Sukeyuki — *Department of Physics, University of the Pacific, Stockton, California*

*Laskar, W. — *U.E.R. de Physique, Université de Nantes, Nantes, France*

Lasocka, Maria — *Pedagogical Academy, Krakow, Poland*

*Lax, Peter D. — *Courant Institute of Mathematical Sciences, New York University, New York, New York*

*Litvin, D. B. — *Department of Physics, University of British Columbia, Vancouver, British Columbia, Canada*

*Marathe, Kishore B. — *Department of Mathematics, Brooklyn College of the City University of New York, Brooklyn, New York*

*Marmo, G. — *Instituto di Fisica Teorica, Universitá di Napoli, Napoli, Italia*

*Michel, Louis — *Institut des Hautes Etudes Scientifiques, Bures-sur-Yvette, France*

*Miller, Willard Jr., — *School of Mathematics, University of Minnesota, Minneapolis, Minnesota*

*Mirman, R. — *155 East 34th Street, New York, New York*

Mishra, A. K. — *Department of Mechanical Engineering, The University of Calgary, Calgary, Alberta, Canada*

*Niederer, U. — *Institut für Theoretische Physik der Universität Zürich, Zürich, Switzerland*

Olszewski, Jan — *Jagellonian University, Krakow, Poland*

*Opechowski, W. — *Department of Physics, The University of British Columbia, Vancouver, British Columbia, Canada*

*Perrin, M. — *Centre de Recherches Mathématiques, Université de Montréal, Montréal, P.Q., Canada*

*Perroud, M. — *Centre de Recherches Mathématiques, Université de Montréal, Montréal, P.Q., Canada*

Pettitt, R. B. — *Department of Mathematics, University of Ottawa, Ottawa, Ontario, Canada*

†*Plebanski, Jerzy F. — *Centro de Investigacion y Estudios Avanzados, Mexico, D.F.*

*Pommaret, J. F. — *Département de Physique, Collège de France, Berthelot, Paris, France*

Rabin, Y. — *Department of Physics and Astronomy, Tel-Aviv University, Ramat-Aviv, Israel*

*Rao, Srinivasa, K. — *MATSCIENCE, The Institute of Mathematical Sciences, Adyar, Madras, India*

Robinson, Ivor — *Institute for Mathematical Sciences, The University of Texas at Dallas, Richardson, Texas*

Rodriguez, R. — *Centre de Physique Théorique, C.N.R.S., Marseille, France*

*Rosensteel, G. — *Department of Physics, McMaster University, Hamilton, Ontario, Canada*

*Rowe, D. J. — *Department of Physics, University of Toronto, Toronto, Ontario, Canada*

*Roy, Ghislain — *Département de Mathématiques, Université Laval, Québec, P.Q., Canada*

†On leave of absence from: *University of Warsaw, Warsaw, Poland*

Ruch, E. — *Institute of Quantum Chemistry, Free University, Berlin, Germany*
*Salamó, S. — *Departamento de Fisica, Universidad Simon Bolivar, Caracas, Venezuela*
*Saletan, E. J. — *Physics Department, Northeastern University, Boston, Massachusetts*
*Sattinger, D. H. — *School of Mathematics, University of Minnesota, Minneapolis, Minnesota*
Schindler, Susan — *Department of Mathematics, Baruch College, City University of New York, New York, New York*
*Segal, I. E. — *Department of Mathematics, M.I.T., Cambridge, Massachusetts*
*Seligman, T. H. — *Departamento de Fisica Teorica, Universidad Nacional Autonoma de Mexico, Mexico, D.F.*
*Singh, M. C. — *Department of Mechanical Engineering, The University of Calgary, Calgary, Alberta, Canada*
Sirugue, M. — *Centre de Physique Théorique, C.N.R.S., Marseille, France*
Sirugue-Collin, M. — *Centre de Physique Théorique, C.N.R.S., Marseille, France*
*Solomon, A. I. — *Faculty of Mathematics, The Open University, Milton Keynes, United Kingdom*
*Sorba, Paul — *Fermi National Accelerator Laboratory, Batavia, Illinois*
*Strasburger, Aleksander — *Katedra Metod Matematycznych Fizyki, Uniwersytet Warszawski, Warszawa, Poland*
Szczyrba, Igor — *Katedra Metod Matematycznych Fizyki, Uniwersytet Warszawski, Warszawa, Poland*
van Dam, H. — *Department of Physics, University of North Carolina, Chapel Hill, North Carolina*
Venkatesh, K. — *MATSCIENCE, The Institute of Mathematical Sciences, Adyar, Madras, India*
*Villarroel, D. — *Departamento de Fisica, Universidad de Chile, Santiago, Chile*
*Vrscay, E. R. — *Department of Applied Mathematics, University of Waterloo, Waterloo, Ontario, Canada*
*Winternitz, P. — *Centre de Recherches Mathématiques, Université de Montréal, Montréal, P.Q., Canada*
*Wolf, K. B. — *I.I.M.A.S., Universidad Nacional Autonoma de Mexico, Mexico, D.F.*
*Yadegar, J. — *Department of Applied Mathematics, Queen Mary College, London, England*

*An asterisk denotes an author who presented a paper.

Preface

The Fifth International Colloquium on Group Theoretical Methods in Physics was held at the Université de Montréal, July 5-9, 1976; it was preceded by a four-week joint session of the Séminaire de Mathématiques Supérieures and the Canadian Association of Physicists Summer School, on the same subject. Both events were hosted by the Centre de Recherches Mathématiques and the Département de Mathématiques of the Université de Montréal.

The earlier colloquia in the series took place at the Centre de Physique Théorique of the C.N.R.S. in Marseille, France, in 1972 and 1974 and the University of Nijmegen, Nijmegen, Holland, in 1973-1975. The sixth colloquium will be in Tübingen, Germany, in 1977. Meetings in 1978 and 1979 are planned for Austin, Texas, and Tel-Aviv, Israel.

This volume contains most of the invited and contributed papers presented at the Fifth Colloquium in Montreal. A talk by M. Moshinsky is omitted at the speaker's request because the material is available elsewhere, in particular in the proceedings of the Summer School on Group Theoretical Methods in Physics, which preceded the colloquium. There were other scientific events that cannot be included here—a free-ranging discussion of the "missing label" problem, poster sessions that paralleled the talks, etc.

The colloquium and these proceedings were made possible by the help of a large number of people. The members of the Advisory and Organization Committees are listed elsewhere. The participants all appreciated the efficiency of the secretariat—Michèle Hubert, Micheline Marano, Johanne Marcoux, and Wendy McKay. The lecture sessions were organized by Mrs. McKay. Miss Jacqueline Reggiori looked after monetary arrangements.

Mrs. Marano and Miss Marcoux were in charge of typing the manuscripts and deserve credit for the relatively early appearance of this volume.

We thank the National Research Council of Canada, the Atomic Energy of Canada Ltd, the Institute of Particle Physics, the Ministère de l'Education du Québec, and the Université de Montréal for financial support.

The proceedings of the first colloquium in the series were published as a joint report of the University of Provence, the University of Aix-Marseille, and the C.N.R.S. The second and third proceedings were printed by the Faculty of Science, University of Nijmegen. The fourth were published by Springer-Verlag, Berlin, Heidelberg, New York. The proceedings of the Séminaire de Mathématiques Supérieures and the Canadian Association of Physicists Summer School are published by Les Presses de l'Université de Montréal.

Part I

Nuclei, Atoms, Solids

CANONICAL TRANSFORMATIONS AND SPECTRUM GENERATING ALGEBRAS IN THE THEORY OF NUCLEAR COLLECTIVE MOTION

P. Gulshani, G. Rosensteel and D.J. Rowe

Theories of nuclear quadrupole dynamics fall essentially into two classes: phenomenological models, which are expressed in terms of ad hoc collective coordinates, and microscopic theories which attempt to explain collectivity in terms of coherent motions of particles [1]. With the algebraic models came the means of relating the two. Thus, for example, in the $[R^5]so(3)$ model [2], one can identify the abelian subalgebra R^5 with the 5 components of the traceless mass quadrupole tensor and so(3) with the rotational angular momentum. In this way one obtains the phenomenological rotational model. But at the same time the algebra $[R^5]so(3)$ has a well-defined action on particle coordinates and so one has the beginnings of a microscopic theory.

In addition to describing quadrupole dynamics, one would like a theory which would also predict, or at least provide the means to observe, what goes on inside a rotational nucleus. For that we need to learn what are the relevant quantum mechanical observables that could, for example, distinguish between some of the possible flow patterns illustrated in Fig. 1. We pursue these questions and the relationships between the phenomenological and microscopic collective models by making a linear canonical point transformation from particle coordinates to centre-of-mass, collective and intrinsic coordinates.

The method is described in detail in Ref. [3]. A canonical transformation is made in two steps.

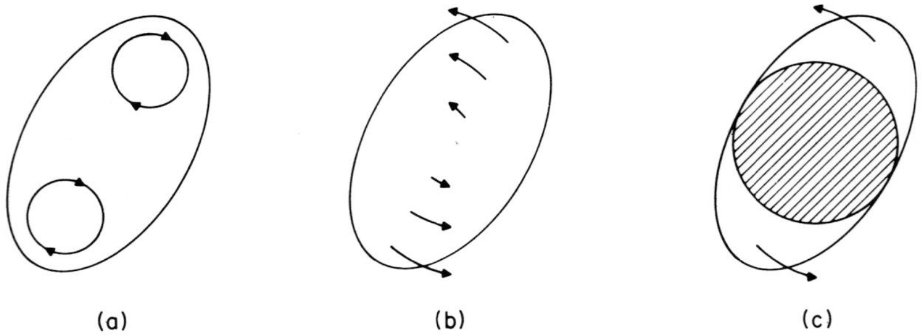

Fig. 1. Possible nuclear flow patterns (a) irrotational flow (b) rigid flow (c) two-fluid flow.

$$\vec{x}_n = \vec{X} + \vec{x}'_n \qquad n = 1,\ldots,N$$

$$x'_{ni} = g_{i\alpha}(\theta,\lambda,\psi)\, x''_{n\alpha}(\xi) \qquad i = 1,2,3, \quad \alpha = 1,2,3$$

where $\vec{X}$ is the c.m. coordinate, $g \in GL(3,R)$ is a function of 9 collective coordinates θ, λ, ψ and $x''_{n\alpha}$ is a function of 3N-12 intrinsic coordinates ξ. When the corresponding canonical transformation is applied to the momentum coordinates, one obtains a separation of the Hamiltonian

$$H = H_{cm} + H_{rel}$$

$$H_{rel} = H_{coll} + H_{intr} + H_{coup}.$$

The c.m. coordinates $\vec{X}$ are defined in the usual manner so that there is no term in the Hamiltonian coupling the relative and c.m. degrees of freedom. We therefore define the collective coordinates by the parallel criterion of minimizing H_{coup}.

Consider the Cartan decomposition of g

$$g = R(\theta)S_o(\lambda)R(\psi)$$

where $R(\theta)$ and $R(\psi)$ are rotation matrices and S_o is diagonal. We choose the 3 Euler angles θ such that $R(\theta)$ effects a rotation from space fixed axes to the principal axes of the quadrupole mass tensor Q. Defining $\lambda_1, \lambda_2, \lambda_3$ to be the principal values of Q, we then choose $S_{oA} = \sqrt{\lambda_A/M}$, where M is the mass of a particle, such that there is no quadrupole deformation in the intrinsic system. Thus S_o is a scale transformation on each of the three principal coordinate axes. Finally $R(\psi)$ is chosen such that there is zero angular momentum of the system relative to the intrinsic coordinate axes. As a consequence of this latter choice, we find that the Coriolis force vanishes and that the kinetic energy becomes

$$K.E. = T_{cm} + T_{vib} + T_{rot} + T_{intr}$$

with

$$T_{vib} = -2\hbar^2 \sum_A \lambda_A \left[\frac{\partial^2}{\partial\lambda_A^2} + \left(\frac{N-3}{2\lambda_A} + \sum_{B\neq A} \frac{1}{\lambda_A-\lambda_B}\right)\frac{\partial}{\partial\lambda_A}\right]$$

$$T_{rot} = \frac{1}{2}\sum_{A<B}\left[\frac{\lambda_A+\lambda_B}{(\lambda_A-\lambda_B)^2}(L_{AB}^2+\mathcal{L}_{AB}^2) - \frac{4\sqrt{\lambda_A\lambda_B}}{(\lambda_A-\lambda_B)^2}L_{AB}\mathcal{L}_{AB}\right].$$

Inspection of these expressions reveals that $T_{coll} = T_{vib} + T_{rot}$ is a function of four objects: the principal moments λ_A of the quadrupole tensor, the corresponding vibrational momenta $\partial/\partial\lambda_A$, the angular momenta L_{AB} which act on θ and the vorticity operators $\mathcal{L}_{AB}$, which are the angular momenta acting on ψ. These quantities are all in the enveloping algebra of cm(3) [4]. This is a highly significant result. For one thing it strongly supports the cm(3) model's status as the relevant algebraic model for quadrupole collective motions. For another, it supplies the appropriate kinetic energy for the cm(3) model.

Recall that the cm(3) algebra contains the quadrupole mass

tensor, the angular momenta which generate rigid rotations, and the shear operators which generate irrotational flow vibrational and rotational displacements. Thus the cm(3) model contains the potentiality for describing rigid flow, irrotational flow and all possible linear combinations of the two. It is of interest therefore to determine what flow patterns correspond to the various irreducible representations of cm(3).

The transformed expression for the momentum of a particle has three components

$$\vec{p}_n = \frac{1}{N}\vec{P} + \vec{p}_n(\text{intr.}) + \vec{p}_n(\text{coll.})$$

where $\vec{P}$ is the c.m. momentum and $\vec{p}_n(\text{intr})$ and $\vec{p}_n(\text{coll})$ are respectively the components associated with the intrinsic and collective degrees of freedom. Now it turns out that $\vec{p}_n(\text{coll})$ is a function of the collective momenta and angular momenta in cm(3) and that it is a linear function of the particle's position coordinates. Thus in the classical situation, we can define the collective velocity $\vec{v}_n$ of particle n at a point $\vec{x}$ by

$$v_{ni} = \frac{1}{M} p_{ni}(\text{coll}) = \chi_{ij} x_j ,$$

and, since the right hand side does not depend on n, it follows that this expression defines a velocity field $\vec{v}(\vec{x})$.

We can therefore investigate the circumstances under which $\vec{v}(\vec{x})$ is irrotational; i.e., when $\vec{\nabla}\times\vec{v} = 0$. The answer turns out to be, when $\vec{\mathcal{L}} = 0$. In quantum mechanics $\vec{v}(\vec{x})$ and $\vec{\mathcal{L}}$ are operators. We can nevertheless extend the analysis to quantum mechanics in the natural way and say that a representation of cm(3) describes irrotational flow if

$$\mathcal{L}^2\Psi = 0$$

for all states Ψ in the carrier space of the irreducible representation. Since $\mathcal{L}^2$ is one of the quadratic invariants of cm(3)

[3,4,5] there is a large class of representations for which this condition holds.

In a similar way we may enquire when the collective velocity field is of the form $\vec{v}(\vec{x}) = \vec{\omega}\times\vec{x}$ corresponding to rigid flow. In the classical situation, we find that $\vec{L}'$ should vanish, where

$$L'_{AB} = \frac{4\lambda_A\lambda_B}{(\lambda_A-\lambda_B)^2}\left[\frac{\lambda_A+\lambda_B}{2\sqrt{\lambda_A\lambda_B}}\mathcal{L}_{AB}-L_{AB}\right].$$

In quantum mechanics then we say that a state Ψ describes rigid collective flow if

$$L'^2\Psi = 0.$$

Physically, $\vec{L}'$ is the angular momentum of the system relative to the principal axes of the mass tensor. However, the components of $\vec{L}'$ do not by themselves form a closed algebra and thus it is not, in general, possible to satisfy the conditions for rigid flow. Thus it appears that rigid flow can occur only in exceptional circumstances or under certain limiting conditions, e.g., a classical limit, which remain to be investigated.

To summarize, we have found that the method of canonical transformations provides valuable physical insights into the interpretation of the algebraic cm(3) model and into its relationship with the phenomenological models of quadrupole collective motions. It provides the kinetic energy component of the cm(3) Hamiltonian and observables $\mathcal{L}^2$ and L'^2 which can measure the extent to which a given state, or set of states, describes irrotational- or rigid-flow. Finally it raises some fundamental questions regarding the nature of collective motions; namely, the impossibility of pure rigid collective flow and the impossibility of expressing a many-particle wave function in terms of the above collective and intrinsic coordinates. The latter conclusion emerges as follows: it can readily be shown that the three Euler angles, denoted by the symbol ψ, are non-integrable as functions of the original space-fixed particle coordinates $\vec{x}_n$. This is not a problem for the canonical transformation of the kinetic energy, since the coordinates ψ are cyclic (i.e., they do not appear explicitly in the kinetic energy). Furthermore, all the variables in the kinetic energy are well-defined and have a well-defined action on many-particle Hilbert space. However, it does mean that a given wave-function, expressed in terms of particle coordinates, cannot be re-expressed in terms of the above c.m., collective and intrinsic coordinates. The full implications of this observation for the microscopic theory of collective motion remain to be investigated.

REFERENCES

1. D.J. ROWE, "Nuclear Collective Motion; Models and Theory", Methuen (London) 1970.
2. H. UI, Prog. Theor. Phys. 44 (1970) 153; O.L. WEAVER, L.C. BIEDENHARN and R.Y. CUSSON, Ann. Phys. (New York) 77 (1973) 250.
3. P. GULSHANI and D.J. ROWE, Can. Journ. Phys. 54 (1976) 970.
4. G. ROSENSTEEL and D.J. ROWE, Ann. Phys. (New York) 96 (1976) 1.
5. O.L. WEAVER, "Factorization of the Invariant Operators of CM(3)", contributed paper, session 1A.

MODULATED SPACE GROUPS

A. Janner

1. INTRODUCTION

Consider a crystal with a superstructure. Its elementary supercell has a finite volume. If this supercell is of macroscopic size, one gets a modulated crystal structure. Mathematically speaking, however, a modulated crystal has an infinite elementary cell, which can be considered as the limiting case of a larger and larger supercell. Indeed by *modulation* we mean a periodic deviation from a basic periodic pattern, both periodicities being incommensurate. In the case of modulated crystals the Euclidean symmetry of the corresponding basic pattern is that of a space group of the same dimension as the crystal. The modulation destroys this property. Furthermore the remaining Euclidean symmetry (if any) does not explain systematic extinctions observed in the diffraction pattern of modulated crystals [see e.g. ref. 1 and 2]. One can of course make use of the space group symmetry of the basic structure and treat the modulation as a perturbation. This is the conventional approach in the determination of superstructures[3] which is insofar not a satisfactory one, as the basic structure does not always have a physical meaning and, in general, is not uniquely determined.

In an alternative approach, the modulated crystal is considered as section of a higher-dimensional periodic structure. Under certain assumptions (of continuity e.g.) the imbedding of the modulated crystal in the higher dimensional Euclidean space is essentially unique and the symmetry of the imbedding crystal (which is a higher dimensional space group) does explain systematic diffractive extinctions [see ref. 4 and 5 for more details]. Going further along this line, it will be shown here, that nevertheless, the symmetry of the modulated crystal can equally well be described by a space group of the same dimension as that of the crystal: This group, called *modulated space group*, however, is *not* a Euclidean symmetry group. The situation is very much the same as that of the point symmetry of a normal crystal, if its space group is non-symmorphic, because then the point group is not the orthogonal symmetry group of such a crystal. This comparison will be discussed in more details in the last section.

Let us here recall some basic properties of normal and of modulated crystals fixing at the same time the notation. A general treatment makes the formal structure more transparent; we therefore consider a n-dimensional crystal (the n=3 case being the most important one from the physical point of view) described by a scalar function:

$$\rho(\vec{r}) = \sum_{\vec{k}\in Sp} \hat{\rho}(\vec{k}) e^{i\vec{k}\vec{r}} \qquad (1)$$

where $\vec{r}, \vec{k} \in V_n$; V_n being a n-dimensional Euclidean space (the metric allowing the identification between the direct and the reciprocal space). By Sp we denote the spectrum of ρ, i.e. the set of $\vec{k}$ vectors such that $\hat{\rho}(\vec{k}) \neq 0$. Here Sp generates V_n.

A. In a *normal crystal* the elements of S can be written as:

$$\vec{k} = \sum_{h=1}^{n} z_h \vec{a}^*_h, \quad z_h \in Z, \quad h = 1,\dots,n \qquad (2)$$

where the vectors $\vec{a}^*_1,\ldots,\vec{a}^*_n$ form a basis B^*_n of a n-dimensional lattice Λ^*_n. The Euclidean symmetry group G of ρ (where $(g\rho)(\vec{r}) = \rho(g^{-1}\vec{r})$ for $g \in E(n)$) is then a n-dimensional space group[6], which is defined by the following properties:

$$\begin{aligned} &\text{(i)} \quad G \subset E(n) \\ &\text{(ii)} \quad G \cap T^n = \Lambda_n \simeq Z^n \\ &\text{(iii)} \quad \{\Lambda_n\} = V_n \end{aligned} \tag{3}$$

where E(n) is the n-dim. Euclidean group, T^n its subgroup of translations, and the group Λ_n of lattice translations (identified after choice of an origin with the lattice of points equivalent with the chosen origin) generates the whole space V_n. Using Seitz's notation[7], the elements of G can be written as:

$$g = \{R|\vec{a}+\vec{v}(R)\} \tag{4}$$

with action on $\vec{x} \in V_n$ given by:

$$g\vec{x} = R\vec{x} + \vec{a} + \vec{v}(R),$$

where $R \in K$ the point group $K \subset O(n)$; $\vec{a} \in \Lambda_n$ the group of lattice (or primitive) translations; $\vec{v}(R) \in T^n$ but $\vec{v}(R) \notin \Lambda_n$ if $\vec{v}(R) \neq 0$, so that $\vec{v}(R)$ is called a non-primitive translation associated to R and $\vec{v}$ a system of non-primitive translations, whose properties are:

$$\begin{aligned} &\vec{v}(1) = 0, \\ &\vec{v}(R) + R\vec{v}(R') - \vec{v}(RR') = \vec{m}(R,R') \in \Lambda_n, \end{aligned} \tag{5}$$

with $\vec{m}$ defining a factor set. (Note that $\vec{v}(R)$ depends on the choice of the origin and is determined only modulo primitive translations.)

Accordingly, the space group G is well defined once Λ_n, K, and $\vec{v}$ or $\vec{m}$ are given. (See ref. 8 for more details.)

B. In a *modulated crystal* the elements of Sp can be written as:

$$\vec{k} = \sum_{h=1}^{n} z_h \vec{a}^*_n + \sum_{j=1}^{d} z_{n+j} \vec{b}^*_j, \qquad z_m \in Z, \qquad m = 1,\ldots,n+d \tag{6}$$

where $\vec{a}^*_1,\ldots,\vec{a}^*_n$ form a basis B^*_n of Λ^*_n, as above and $\vec{b}^*_1,\ldots,\vec{b}^*_d$ a basis B^*_d of a d-dimensional lattice D^*_d. (Confusion, say in the case n = d = 3, can be avoided by writing $B^*_{n=3}$ and $B^*_{d=3}$.)

The periodicity of the modulation implies $1 \leq d \leq n$, and its incommensurability with respect to Λ^*_n can be expressed (without restriction of generality) by the condition:

$$\Lambda^*_n \cap D^*_d = 0, \tag{7}$$

that we take as definition of two incommensurable lattices Λ_n and D_d. Diffraction spots associated with $\vec{k}$ vectors such that $z_{n+1} =\ldots= z_{n+d} = 0$ are called *main reflections;* those for which this is not the case are called *satellite reflections.* Note that the main reflections occur at the lattice points of Λ^*_n, whereas the satellite reflections occur at the points of lattices $D^*_d(\Lambda^*_n)$ generated by D^*_d from each point of Λ^*_n.

Clearly the description of Sp in terms of the bases B^*_n and B^*_d as in (6) implies a choice. The validity of the essential properties derived hereafter is independent of this choice. To get this non trivial result requires a careful inspection of the consequences of this arbitrariness. A first discussion on it can be found elsewhere in these same proceedings[9]. A more complete analysis will be published elsewhere.

By the following proposition the imbedding idea mentioned above becomes natural.

Proposition 1. The incommensurability condition $\Lambda^*_n \cap D^*_d = 0$ implies that the abelian group (Λ^*_n, D^*_d) freely generated by the bases B^*_n and B^*_d is isomorphic to Z^{n+d}:

$$(\Lambda_n^*, D_d^*) \simeq Z^{n+d}. \tag{8}$$

This proposition, whose proof is elementary, means that the set of all possible reflections can be seen as occurring at the projection of points of a (n+d)-dimensional lattice Σ_{n+d}^*, the incommensurability ensuring a unique lifting of the elements of Sp on lattice points of Σ_{n+d}^*, and correspondingly an imbedding of $\rho(\vec{r})$ in a (n+d)-dim. space.

2. IMBEDDING

The lattice Λ_n^* generates V_n (as Euclidean space) and D_d^* generates the (non trivial) Euclidean d-dimensional subspace $V_d \subseteq V_n$. We now consider the (n+d)-dim. Euclidean space V_S given by:

$$V_S = V_E \oplus V_I. \tag{9}$$

We call V_E (isomorphic with V_n) the *external space*, V_I (isomorphic with V_d) the *internal space* and $V_{n+d} = V_S$ the *superspace*.

The metric g_{ij}^E in V_E is the same as that in V_n whereas the metric g_{ij}^I in V_I need not to be that of V_d †) and depends on the choice of the lattice D_d^* among all possible $D_d^*(\Lambda_n^*)$. In the cases where it is desirable to take this explicitly into account we will denote the scalar product in V_E by a dot and that in V_I by an open circle. Usually, however, these symbols will be omitted.

We adopt the notation:

$$r = (\vec{r}_E, \vec{r}_I) \in V_S, \quad \vec{r}_E \in V_E, \quad \vec{r}_I \in V_I$$

$$k = (\vec{k}_E, \vec{k}_I) \in V_S, \quad \vec{k}_E \in V_E, \quad \vec{k}_I \in V_I,$$

and scalar product

---

†) I thank Dr. T. Janssen for drawing my attention to this fact.

$$kr = (\vec{k}_E,\vec{k}_I)(\vec{r}_E,\vec{r}_I) \overset{\mathrm{Def}}{=} \vec{k}_E \cdot \vec{r}_E + \vec{k}_I \circ \vec{r}_I. \tag{10}$$

(Note that in ref. 5 the minus sign was adopted because it was more convenient in that context.)

As (standard) basis B^*_{n+d} for V_S we define:

$$\begin{aligned} a^*_k &= (\vec{a}^*_k, 0) \qquad k = 1,\ldots,n \\ a^*_{n+j} &= (\vec{b}^*_j, \vec{b}^*_j) \quad j = 1,\ldots,d. \end{aligned} \tag{11}$$

This ensures that the projection π_E of the abelian group Σ^*_{n+d}, freely generated by B^*_{n+d}, on its external components is (Λ^*_n, D^*_d), and that V_I is a copy of V_d. From proposition 1 we know that π_E on Σ^*_{n+d} is an isomorphism. Its inverse defines the imbedding ι: $(\Lambda^*_n, D^*_d) \to \Sigma^*_{n+d}$ we are looking for. In components, one has for $\vec{k}$ given as in (6):

$$\iota\vec{k} = \sum_{h=1}^{n+d} z_h a^*_h = k \in \Sigma^*_{n+d}, \tag{12}$$

so that the function ρ describing the modulated crystal in V_n can be extended to one in V_S:

$$\tilde{\rho}(r) = \sum_k \hat{\rho}(k) e^{ikr}, \tag{13a}$$

be defining:

$$\hat{\rho}(k) \overset{\mathrm{Def}}{=} \hat{\rho}(\pi_E k) = \hat{\rho}(\vec{k}), \tag{13b}$$

so that

$$\rho(\vec{r}) = \tilde{\rho}((\vec{r},0)). \tag{14}$$

3. SUPERSPACE GROUP G

Considering transformations $g \in E(n) \times E(d)$, $\tilde{\rho}$ also behaves as a scalar function. Thus $(g\tilde{\rho})(r) = \tilde{\rho}(g^{-1}r)$.

Accordingly its symmetry group G is given by:

$$G = \{g \in E(n) \times E(d) \,|\, g\tilde{\rho} = \tilde{\rho}\}. \tag{15}$$

We adopt the notation:

$$\begin{aligned} g &= \{R|t(R)\} = (g_E, g_I) \in G \subset E(n) \times E(d) \\ g_E &= \{R_E|\vec{t}_E(R)\} \in \pi_E G = G_E \subset E(n) \\ g_I &= \{R_I|\vec{t}_I(R)\} \in \pi_I G = G_I \subset E(d). \end{aligned} \tag{16}$$

Then the symmetry condition expressed in terms of the Fourier components becomes:

$$(g\hat{\rho})(k) = \hat{\rho}(Rk)e^{iRk\cdot t(R)} = \hat{\rho}(k), \tag{17}$$

where dot indicates here the scalar product in V_S.

Proposition 2. G is a (n+d)-dimensional space group of the superspace V_S.

Proof. $a \in G \cap T^{n+d}$ implies $ak \equiv 0 \pmod{2\pi}$ for all $k \in \Sigma^*_{n+d}$. As Σ^*_{n+d} generates V_S, $a \in \Sigma_{n+d}$, where Σ_{n+d} is the group of lattice translations generated by the basis $B_{n+d} = (a_1,\ldots,a_{n+d})$ reciprocal to B^*_{n+d}. Therefore $\Sigma_{n+d} \simeq Z^{n+d}$ and Σ_{n+d} generates V_S. We therefore call G a *superspace group*.

Knowing that G is a space group, we analyse further its group of lattice translations, its point group K and its system of non-primitive translations v.

Considering the bases $B_n = (\vec{a}_1,\ldots,\vec{a}_n)$ and $B_d = (\vec{b}_1,\ldots,\vec{b}_d)$ reciprocal to B^*_n and to B^*_d in V_E and V_I respectively, one finds for B_{n+d}, the basis reciprocal in V_S to B^*_{n+d}:

$$\begin{aligned} a_k &= (\vec{a}_k, -\Delta\vec{a}_k) \qquad & k &= 1,\ldots,n \\ a_{n+j} &= (0, \vec{b}_j) \qquad & j &= 1,\ldots,d \end{aligned} \tag{18}$$

where the condition:

$$\vec{a}_k \cdot \vec{b}^*_j = \Delta\vec{a}_k \circ \vec{b}^*_j \tag{19}$$

implies

$$\Delta\vec{a}_k = \frac{1}{2\pi}\sum_{j=1}^{d}(\vec{a}_k\cdot\vec{b}_j^*)\vec{b}_j \overset{\text{Def}}{=} \sum_{j=1}^{d}\vec{b}_j\sigma_{jk}. \tag{20}$$

Through the isomorphism $V_I \simeq V_d$, the basis B_d of V_I can be identified with the corresponding basis of V_d and so also B_d^*. Even if B_d and B_d^* are no more dual in V_d, (20) says that $\Delta\vec{a}_k$ is the orthogonal projection of $\vec{a}_k$ on the subspace V_d. The incommensurability condition (7) implies that the d×n real matrix σ defined in (20) has irrational row vectors (i.e. $\in \tilde{R}^n - \tilde{Q}^n$), and is of rank d (Tilde means transposed).

Identifying groups of lattice translations with the corresponding lattices (after choice of a fixed origin) and denoting by Λ_n, D_d the lattices reciprocal to Λ_n^* and D_d^* in V_E and V_I, respectively, by considering (11) and (18) one immediately gets:

$$(0,D_d) \subset \Sigma_{n+d}, \qquad \Sigma_{n+d} \cap V_I = (0,D_d) \tag{21}$$

$$(\Lambda_n^*,0) \subset \Sigma_{n+d}^*, \qquad \Sigma_{n+d}^* \cap V_E = (\Lambda_n^*,0). \tag{22}$$

Furthermore identifying $(0,D_d)$ with D_d and $(\Lambda_n^*,0)$ with Λ_n^* we also have:

$$\pi_E\Sigma_{n+d} = \Lambda_n, \qquad \pi_I\Sigma_{n+d}^* = D_d^*, \tag{23}$$

π_E and π_I denoting, when acting on V_S, the orthogonal projection on V_E and V_I respectively, and when acting on group elements of G, the corresponding projection on their external and internal components.

Proposition 3. $G \cap T_I^d = D_d$, where T_I^d denotes the group of internal translations.

Proof. $G \cap T_I^d = (G \cap T^{n+d}) \cap V_I = \Sigma_{n+d} \cap V_I$ by proposition 2. The result then follows from (21).

The point group K of the space group G is defined as the

group of the homogeneous parts of the elements of G:

$$K = \{R|\{t(R)\} \in G\}.$$

According to the general theory on space groups one then knows that K leaves Σ_{n+d} and Σ^*_{n+d} invariant.

Proposition 4.

(a) $R_E\Lambda_n = \Lambda_n$ and

(b) $R_I D_d = D_d$ for any $R \in K$.

Proof. Use (21) and (22) together with $R\,\Sigma_{n+d} = \Sigma_{n+d}$.

Corollary 4. The groups $\pi_E K = K_E$ and $\pi_I K = K_I$ are crystallographic point groups of dimension n and d, respectively.

Proposition 5.

(a) $R_E\vec{b}^* \equiv R_I\vec{b}^* \pmod{\Lambda^*_n}$ for any $\vec{b}^* \in D^*_d$;

(b) $R_I\Delta\vec{a} \equiv \Delta R_E\vec{a} \pmod{D_d}$ for any $\vec{a} \in \Lambda_n$.

Proof.

(a) Calculate Rb^* for $b^* = (\vec{b}^*,\vec{b}^*) \in \Sigma^*_{n+d}$.

(b) In analogous way calculate Ra for $a = (\vec{a},-\Delta\vec{a}) \in \Sigma_{n+d}$.

Proposition 5 expresses the metrical compatibility conditions between V_E and V_I. Indeed any $R_E \in K_E$ leaves the spectrum Sp invariant: $R_E\vec{k} \in Sp$ for any $\vec{k} \in Sp$, so that if applied to the elements of the chosen basis B^*_d one gets

$$R_E\vec{b}^*_j = \sum_{h=1}^{d} \vec{b}^*_h\Gamma^*_I(R)_{hj} + \sum_{k=1}^{n} \vec{a}^*_k\Gamma^*_M(R)_{kj} \tag{24}$$

which defines d×d and n×d integral matrices $\Gamma^*_I(R)$ and $\Gamma^*_M(R)$, (M stands for "mixed"). From (24) and proposition 4 follows that Γ^*_I is a representation of K with carrier space V_I. The group K being finite $\Gamma^*_I(K)$ is equivalent to an orthogonal representation of K. In other words, there exists a positive definite metric tensor $g^I_{jk} = \vec{b}_j \circ \vec{b}_k$ for V_I such that the transformation R_I associated with R_E and defined by:

$$R_I \vec{b}^*_j = \sum_{h=1}^{d} \vec{b}^*_h \Gamma^*_I(R)_{hj} \tag{25}$$

is an orthogonal one: $R_I \in O(d)$. Again the freedom in the metric g^I_{jk} compatible with the given g^E_{jk} does not invalidate the generality of the results derived[9].

Corollary 5. Consider $\vec{a}_k \in B_n$, $\vec{b}^*_j \in B^*_d$ and $R \in K$. Then:

$$\text{(a)} \qquad \Delta R_E \vec{a}_k = R_I \Delta \vec{a}_k + \sum_{j=1}^{d} \vec{b}_j \Gamma_M(R)_{jk}$$

$$\text{(b)} \qquad R_E \vec{b}^*_j = R_I \vec{b}^*_j + \sum_{k=1}^{n} \vec{a}^*_k \Gamma^*_M(R)_{kj} \tag{26}$$

with $\Gamma_M(R)$ the same (d×n) integral matrix as in (24) and $\Gamma^*_M(R) = \tilde{\Gamma}_M(R^{-1})$, tilde meaning transposed.

Proposition 6. D_d is a normal subgroup of G.

Proof. We recall the identification $D_d \equiv (0, D_d)$; with 0 the unit element of G_E and $\vec{b} \in D_d$ one has for $g \in G$:

$$gbg^{-1} = (0, g_I \vec{b} g_I^{-1}) = (0, R_I \vec{b}) \in (0, D_d). \tag{27}$$

Proposition 7. $R_E = 1$ implies $R=1$, thus also $R_I = 1$.

Proof. Consider any $k \in \Sigma^*_{n+d}$. Then

$$Rk = (R_E \vec{k}_E, R_I \vec{k}_I) = k' = (\vec{k}'_E, \vec{k}'_I).$$

As $R_E = 1$, $\vec{k}'_E = \vec{k}_E$, but the projection of k on its external component is an isomorphism by (12); thus $k=k'$. As Σ^*_{n+d} generates V_S then $R=1$.

Corollary 7. $\pi_E: K \to K_E$ is an isomorphism.

We can now split the translational component t(R) of (16) in its primitive and its non-primitive part a and v(R) respectively:

$$t(R) = a+v(R), \qquad a \in \Sigma_{n+d}. \tag{28}$$

From the general theory we know that v defines a system of non-primitive translations associated with a factor set m:

$$v(R) + Rv(R') - v(RR') = m(R,R'); \qquad v(1) = 0. \tag{29}$$

Proposition 8. The group $G_E = \pi_E G$ is a n-dimensional space group whose lattice translation group is Λ_n, the point group K_E and $\pi_E v = \vec{v}_E$ a system of non-primitive translations.

Proof. Consider $g_E \in G_E \cap T^n_E$; then $R_E = 1$ thus R=1 and $g \in G \cap T^{n+d} = \Sigma_{n+d}$. As $\pi_E \Sigma_{n+d} = \Lambda_n$, this implies $g_E = \vec{a} \in \Lambda_n$; therefore $G_E \cap T^n_E = \Lambda_n$ which generates V_n. Furthermore $G_E \subset E(n)$ and the conditions (3) are satisfied. The homogeneous parts R_E of g_E form the point group K_E, which of course leaves Λ_n invariant. The projection π_E of (29) gives

$$\vec{v}_E(R) + R_E\vec{v}_E(R') - \vec{v}_E(RR') = \vec{m}_E(R,R'),$$

and because of corollary 7 defining $\vec{v}_E(R_E) = \vec{v}_E(R)$ with $\vec{v}_E(1) = 0$, $\vec{v}_E$ is a system of non-primitive translations for G_E.

We call G_E the space group of G (or also the external space group).

4. MODULATED SPACE GROUP M

Because of proposition 6 we are now able to define the modulated space group M:

$$M \overset{\text{Def}}{=} G/D_d, \tag{30}$$

whose structure is clarified by the following proposition.

Proposition 9. The group M defined above is isomorphic to the n-dimensional space group G_E (this justifies the name).

Proof. Consider a section $r_o: M \to G$, with $r_o(m)$ representatives of the cosets of $D_d \lhd G$, $(m \in G/D_d)$. Then the elements of G can uniquely be written as: $g = br_o(m)$ with $b = (0,\vec{b}) \in D_d$ and also uniquely as $g = \{R|a+v(R)\}$ implying:

$$r_o(m) = (g_E,\{R_I|-\Delta\vec{a}_E + \vec{v}_I(R)\}). \tag{31}$$

Now g_E determines R_I, $\Delta\vec{a}_E$ and for given G also $\vec{v}_I(R)$. Thus writing:

$$r_o(m) \overset{\text{Def}}{=} r(g_E) \tag{32}$$

one gets a section $r: G_E \to G$, and by $r^{-1}r_o$ an isomorphic mapping of M onto G_E.

Proposition 10. The superspace group G appears in an extension:

$$0 \to D_d \to G \to G_E \to 1 \quad (\phi,f) \tag{33}$$

with $\phi\ (G_E) = K_I \subset \text{Aut}\ D_d$ and factor set f given by:

$$f(g_E,g_E') = (0,-R_I\Delta\vec{a}_E'+\Delta R_E\vec{a}_E'+\vec{m}_I(R,R')+\Delta\vec{m}_E(R,R')). \tag{34}$$

Proof. (33) follows from porpositions 6 and 9, which also give ϕ and f.

In order to establish a relation between the two approaches mentioned in the introduction, the perturbative and the present one, but on the basis of the diffraction pattern, let us split the crystal function $\rho(\vec{r})$ in a ρ_o due to the main reflections only and a ρ_Δ due to remaining satellites. Accordingly we have:

$$\rho(\vec{r}) = \rho_o(\vec{r}) + \rho_\Delta(\vec{r})$$

with

$$\rho_o(\vec{r}) = \sum_{\vec{k}\in\Lambda_n^*} \hat{\rho}(\vec{k})e^{i\vec{k}\vec{r}}. \qquad (35)$$

Note that $\rho_o(\vec{r})$ describes a kind of averaged structure.

Denote by G_o the Euclidean symmetry group of ρ_o, which is a n-dimensional space group having Λ_n as group of lattice translations. In simple cases this group coincides with that G_B of a basic structure which by modulation yields the crystal structure, as one can verify in particular models. In general, one can say that G_B is a n-dimensional space group having the same group of lattice translations Λ_n as G_o.

Proposition 11. The external space group G_E is an equi-translational subgroup of G_o, which is the *average space group*.

Proof. For main reflections $\vec{k}_I = 0$ and $\vec{k} \in \Lambda_n^*$. Thus $g \in G$ implies for such $\vec{k}$'s according to (17)

$$\hat{\rho}(\vec{k}) = \hat{\rho}(R_E\vec{k})e^{iR_E\vec{k}\cdot\vec{t}_E(R)} = (g_E\hat{\rho})(\vec{k})$$

thus $g_E \in G_o$.

5. EXAMPLES

The theory developed above is here applied to two examples of modulated crystal structures, one (γ-Na_2CO_3) an ionic crystal showing a one-dimensional displacive modulation[1], the other (1T-TaS_2) a conducting crystal whose displacive modulation is 3-dimensional and coupled to charge density waves due to instabilities of the Fermi surface[2].

A. Sodium carbonate

According to the investigations of P.M. de Wolff and collaborators the average structure of Na_2CO_3 in the γ-phase is monoclinic with G_o = C2/m; this fixes Λ_3. Choosing its basis

$B_{n=3} = (\vec{a}_1, \vec{a}_2, \vec{a}_3)$ as shown in Fig. 1, the modulation is generated by (expressed in the reciprocal basis $B^*_{n=3}$):

$$\vec{b}^*_1 \equiv \vec{b}^* = \gamma_1(\vec{a}^*_1 + \vec{a}^*_2) + \gamma_3\vec{a}^*_3 \tag{36}$$

with γ_1, $\gamma_3 \in$ R-Q, temperature dependent (for T = 300°C, $\gamma_1 = 0.154$ and $\gamma_3 = 0.286$), and thus in general irrationals: we get d=1. Therefore $\sigma = (\gamma_1\gamma_1\gamma_3)$ and using (20) one gets:

$$\Delta\vec{a}_1 = \gamma_1\vec{b}, \quad \Delta\vec{a}_2 = \gamma_1\vec{b}, \quad \Delta\vec{a}_3 = \gamma_3\vec{b}.$$

Furthermore $D_1 = \{\vec{b}\}$.

As $K_E \subseteq = K_o = 2/m$, one verifies which elements of K_o satisfy the relations of corollary 5. One finds all of them: thus K has two generators R_1 and R_2:

$$\begin{array}{lll} (R_1)_E = 2_y & (R_1)_I = -1 & \Gamma_M(R_1) = 0 \\ (R_2)_E = m_y & (R_2)_I = 1 & \Gamma_M(R_2) = 0. \end{array} \tag{37}$$

Therefore one can take in V_I the same metric tensor as in V_d. Using the "magnetic" notation of primed elements if $R_I = -1$ and unprimed if $R_I = 1$ one gets:

$$K = 2'/m' \simeq K_E = K_o = 2/m. \tag{38}$$

As G_E is an equitranslational subgroup of G_o (prop. 11), here with the same point group, it follows

$$G_E = G_o = C2/m, \tag{39}$$

and $\vec{v}_E = 0$, thus $\vec{m}_E = 0$. Furthermore from (37) $\Gamma_M(R) = 0$, therefore according to proposition 10 the factor set f is given by:

$$f(g_E, g'_E) = (0, \vec{m}_I(R,R')). \tag{40}$$

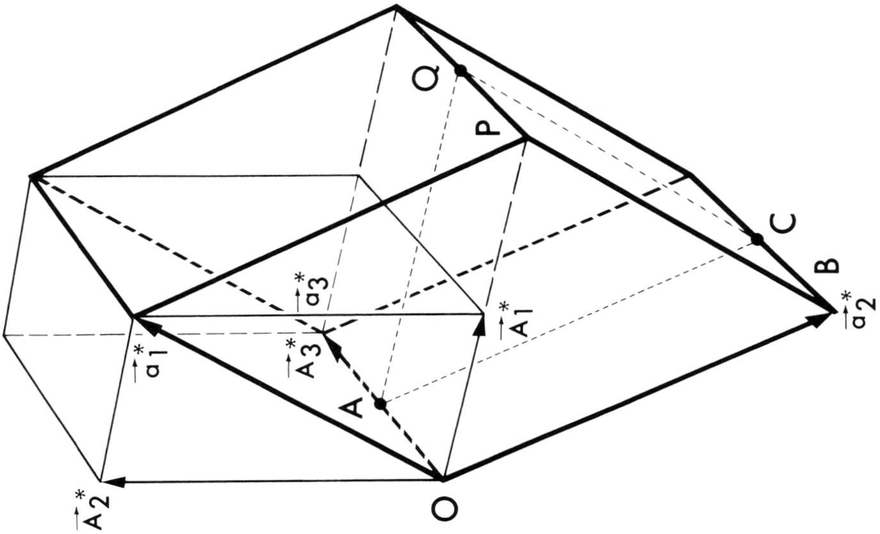
$\vec{A}_2^*$
O
$\vec{a}_1^*$
A
$\vec{A}_3^*$
$\vec{a}_3^*$
$\vec{A}_1^*$
P
Q
C
B
$\vec{a}_2^*$

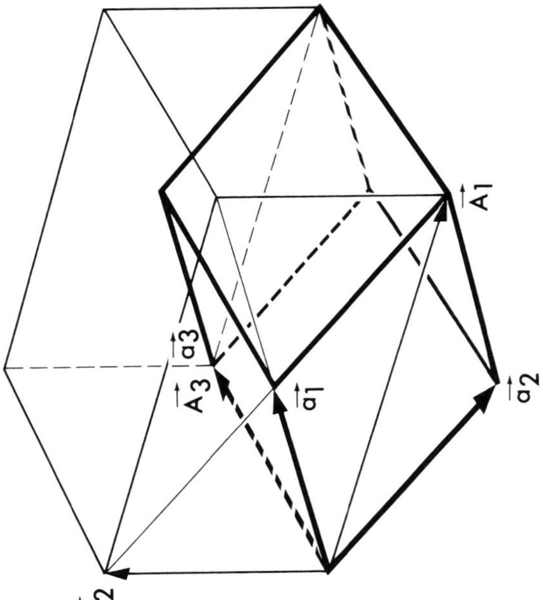
$\vec{A}_2$
$\vec{A}_3$
$\vec{a}_3$
$\vec{a}_1$
$\vec{a}_2$
$\vec{A}_1$

Extinctions in the observed satellite reflections yield the following non-primitive translations:

$$\vec{v}_I(R_1) = \tfrac{1}{2}\vec{b}; \quad \vec{v}_I(R_2) = -\tfrac{1}{2}\vec{b} \quad \text{and} \quad \vec{v}_I(R_1R_2) = 0. \tag{41}$$

The only non-vanishing elements of the factor set $\vec{m}_I$ are therefore (using (29)):

$$\vec{m}_I(R_2,R_1) = \vec{m}_I(R_2,R_2) = \vec{b}.$$

As G_E is generated by $\vec{a}_1$, $\vec{a}_2$, $\vec{a}_3$, 2_y and m_y the modulated space group M is generated by (using (31) and (32)):

$$\begin{aligned} m_1 &= r_o^{-1}r(\vec{a}_1) = (\vec{a}_1,-\gamma_1\vec{b}) \bmod (0,\vec{b}) \\ m_2 &= r_o^{-1}r(\vec{a}_2) = (\vec{a}_2,-\gamma_1\vec{b}) \bmod (0,\vec{b}) \\ m_3 &= r_o^{-1}r(\vec{a}_3) = (\vec{a}_3,-\gamma_3\vec{b}) \bmod (0,\vec{b}) \\ m_4 &= r_o^{-1}r(2_y) = (2_y,\{\bar{1}\,|\,\tfrac{1}{2}\vec{b}\}) \bmod (0,\vec{b}) \\ m_5 &= r_o^{-1}r(m_y) = (m_y,\tfrac{\vec{b}}{2}) \bmod (0,\vec{b}), \end{aligned} \tag{42}$$

and indeed $M \simeq C2/m$.

B. Tantalum disulphide

This example is based on the results reported in ref. 2 for the crystal structure of 1T-TaS_2. The space group of the average structure is trigonal: $G_o = P\bar{3}m\,1$. The 3-dim. modulation expressed with respect to the reciprocal hexagonal basis $B^*_{n=3} = (\vec{a}^*_1,\vec{a}^*_2,\vec{a}^*_3)$ (see Fig. 2) is generated by:

$$\vec{b}^*_1 = \gamma_1\vec{a}^*_1 - \frac{1}{3}\,\vec{a}^*_3; \quad \vec{b}^*_2 = \gamma_1(-\vec{a}^*_1+\vec{a}^*_2) - \frac{1}{3}\,\vec{a}^*_3 \tag{43}$$

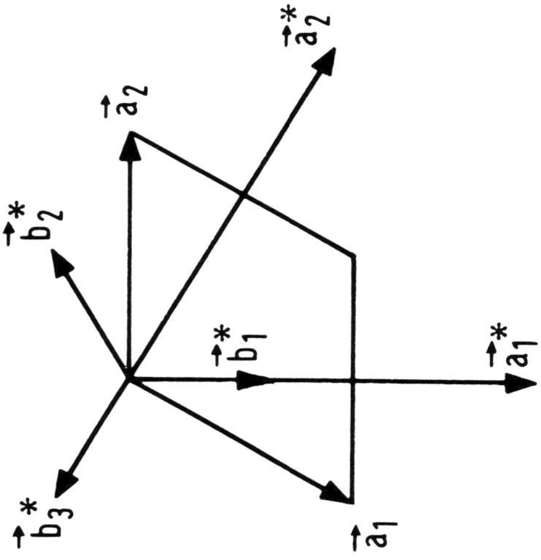

$\vec{a}_2^*$
$\vec{a}_2$
$\vec{b}_2^*$
$\vec{b}_1^*$
$\vec{a}_1^*$
$\vec{b}_3^*$
$\vec{a}_1$

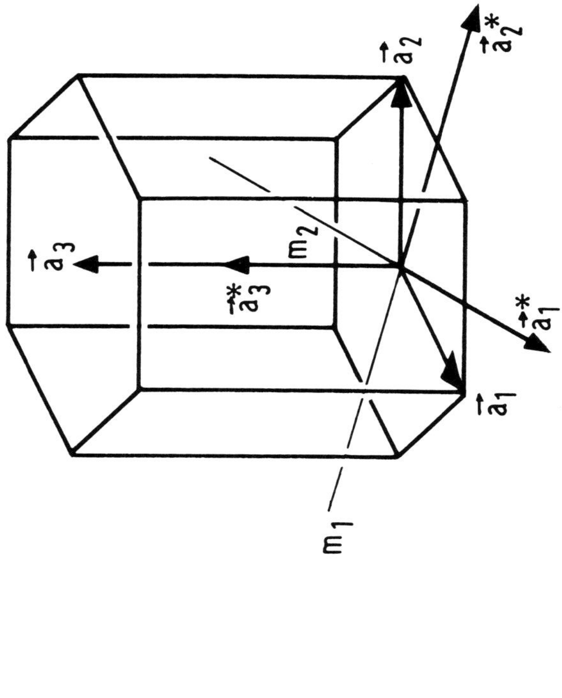

$\vec{a}_3$
$\vec{a}_3^*$
m_2
$\vec{a}_2$
$\vec{a}_2^*$
$\vec{a}_1^*$
$\vec{a}_1$
m_1

$$\vec{b}_3^* = -\gamma_1 \vec{a}_2^* - \frac{1}{3}\vec{a}_3^*, \text{ with } \gamma_1 \text{ irrational } (\gamma_1 \simeq 0{,}283). \tag{43}$$

According to (20) and because $V_{d=3} = V_{n=3}$ (present case) one gets the relations:

$$\begin{aligned} \vec{a}_1 &= \Delta\vec{a}_1 = \gamma_1(\vec{b}_1 - \vec{b}_2) \\ \vec{a}_2 &= \Delta\vec{a}_2 = \gamma_1(\vec{b}_2 - \vec{b}_3) \\ \vec{a}_3 &= \Delta\vec{a}_3 = -\frac{1}{3}(\vec{b}_1 + \vec{b}_2 + \vec{b}_3). \end{aligned} \tag{44}$$

The point group K_o of G_o is generated by two mirrors denoted by

$$m_{\vec{a}_1} \quad \text{and} \quad m_{\vec{a}_2}$$

with mirror plane through $\vec{a}_1^*, \vec{a}_3^*$ and $\vec{a}_2^*, \vec{a}_3^*$, respectively, (Fig. 2)

$$K_o = \bar{3}m\,1 = \{m_{\vec{a}_1}, m_{\vec{a}_2}\}. \tag{45}$$

Again using corollary 5 one finds that $K_E = K_o$, $\Gamma_M(K) = 0$ and the following relations for the generators R_1 and R_2 of K:

$$(R_1)_E = (R_1)_I = m_{\vec{a}_1} \; ; \quad (R_2)_E = (R_2)_I = m_{\vec{a}_2} \tag{46}$$

so that here also the metric of V_I coincides with that of $V_{d=3}$ which is that of $V_{n=3}$.

Using these results and proposition 11 one finds

$$G_E = G_o = P\bar{3}m\,1, \tag{47}$$

so that $\vec{v}_E = \vec{m}_E = 0$. From ref. 2 no systematic satellite extinctions are quoted, so that we deduce $\vec{m}_I = 0$ implying $f(g_E, g_E') = 0$ and $v(R) = 0$.

G_E is generated by the (hexagonal) lattice translations $\vec{a}_1$ $\vec{a}_2$ $\vec{a}_3$ (see Fig. 2) and the point groups elements. The corresponding generators of the modulated space group $M \simeq P\bar{3}m\,1$ are:

$$m_1 = r_o^{-1} r(\vec{a}_1) = (\vec{a}_1, -\gamma_1(\vec{b}_1 - \vec{b}_2)) \bmod (0, D_3)$$

$$m_2 = r_o^{-1} r(\vec{a}_2) = (\vec{a}_2, -\gamma_1(\vec{b}_2 - \vec{b}_3)) \bmod (0, D_3)$$

$$m_3 = r_o^{-1} (\vec{a}_3) = (\vec{a}_3, \tfrac{1}{3}(\vec{b}_1 + \vec{b}_2 + \vec{b}_3)) \bmod (0, D_3) \tag{48}$$

$$m_4 = r_o^{-1} r(m_{\vec{a}_1}) = (m_{\vec{a}_1}, m_{\vec{a}_1}) \bmod (0, D_3)$$

$$m_5 = r_o^{-1} r(m_{\vec{a}_2}) = (m_{\vec{a}_2}, m_{\vec{a}_2}) \bmod (0, D_3),$$

where D_3 is the abelian group freely generated by $\vec{b}_1$ $\vec{b}_2$ and $\vec{b}_3$ (see Fig. 2).

6. THE ANALOGY WITH THE RELATION MACROSCOPIC/MICROSCOPIC CRYSTAL SYMMETRY

The description of a crystal in terms of a superstructure (or also of a modulated structure) represents a refinement of an averaged crystal structure determination.

The conceptual consequences of this development are very similar to those related with the historical evolution which started considering the symmetries of a macroscopic crystal and arrived then at those of its microscopic structure. Indeed on the mathematical level also, one finds a strong analogy between the two cases, which is very deep allowing one to get a better understanding of the meaning of the present approach to crystal symmetry. Let us therefore stress the common conceptual development in a number of characteristic steps indicating in a right and in a left column the corresponding group theoretical situations.

CRYSTAL SYMMETRY

A. From a macroscopic to a microscopic description	B. From an averaged to a refined microscopic description

(i) Basic symmetry (approximated)

macroscopic crystal form	main reflexions only
point group $K_o \subset O(3)$	(or basic structure)
	average space group $G_o \subset E(3)$

(ii) New admitted transformations

space translations: T^3	"internal" translations T_I^d
requiring the extension to:	requiring the extension to:
$E(3) = T^3 \wedge O(3)$	$E(3) \times E(d)$

(iii) Pure new symmetries (associated with the identity of the old ones)

group of "primitive" (or lattice) translations Λ_3	group of "internal" lattice translations D_d

(iv) New symmetry group G

microscopic symmetry group:	refined symmetry group
3-dim space group	(3+d)dim superspace group

(v) Mathematical structure of G (group extension (in general non trivial))

$0 \to \Lambda_3 \to G \to K \to 1$		$0 \to D_d \to G \to G_E \to 1$
or $\Lambda_3 \lhd G;\ G/\Lambda \simeq K$		$D_d \lhd G;\ G/D_d \equiv M \simeq G_E$
non-symmorphic case	$\leftarrow$	modulated case

(vi) Relation with the basic (approximated) symmetry

$K \subseteq K_o$	$G_E \subseteq G_o$
K: point group of	G_E: space group of
G = "homogeneous" parts of	G = "external" parts of the
the space group G	superspace group G

(vii) Physical properties (like systematic extinctions) (in general)

- explained by G but not by K only
- K_o symmetry elements easily recognized

- explained by G but not by G_E only
- G_o symmetry elements easily recognized.

In this comparison superstructures have been treated on the same foot as the modulated ones. There are good reasons for believing this; there is, however, at the moment no theoretical treatment justifying it. The present one is essentially based on the incommensurability condition (7). The correspondence underlines the analogies; there are important differences also: e.g. in the extensions (v) $|K| < \infty$ whereas $|G_E| = \infty$.

7. CONCLUDING REMARKS

The name "modulated space group" has been introduced after considering the analogy with the already introduced nomenclature for "magnetic space groups". Let us briefly recall that situation (referring to 10) for details).

Consider a n-dim space group F with a subgroup D of index two, and write

$$F = D + g_o D. \tag{45}$$

Denote by Θ the time-reversal transformation. Then by:

$$M = D + \Theta g_o D \tag{46}$$

one gets a *magnetic space group*, and all such groups can be obtained in this way. Note that

$$M \simeq F \subset E(n) \text{ but } M \not\subset E(n). \tag{47}$$

If now M is a *modulated space group*, then it also satisfies a relation analogous to (47).

The role of magnetic (and more generally of so called colored groups) in the classification of modulated space groups has been recognized and discussed for the case n=3 and d=1 by Professor P.M. de Wolff[11]. His paper was very inspiring to the author for getting the results presented here. Further developments (concerning Bravais classes in particular) can be found elsewhere in these proceedings[9].

REFERENCES

1. W. VAN AALST, J. DEN HOLLANDER, W.J.A.M. PETERSE and P.M. DE WOLFF, Acta Cryst. B32 (1976), 47.
2. J.A. WILSON, F.J. DI SALVO and S. MAHAJAN, Adv. Phys. 24 (1975), 117.
3. M. KOREKAWA, Theorie der Satellitenreflexe (1967), Habilitationschrift, Universität München.
4. P.M. DE WOLFF, Acta Cryst. A30 (1974), 777 (see also ref. 1).
5. A. JANNER and T. JANSSEN, Symmetry of periodically distorted crystals (to appear in Phys. Rev. B).
6. L. BIEBERBACH, Math. Annalen 70 (1911), 297 and 72 (1912), 400.
7. F. SEITZ, Z. Krist. 88 (1934), 433; 90 (1935), 289; 91 (1935) 336 and 94 (1936), 100.
8. E. ASCHER and A. JANNER, Helv. Phys. Acta 38 (1965), 551 and Commun. Math. Phys. 11 (1968), 138.
9. A. JANNER and T. JANSSEN, Properties of lattices associated with a modulated crystal, these proceedings.
10. W. OPECHOWSKI and R. GUCCIONE, Magnetic symmetry in Magnetism, ed. by G.T. Rado and H. Shul, Academic Press 1965, Vol. IIA, p. 105.
11. P.M. DE WOLFF, Symmetry operations for displacively modulated structures (to appear in Acta Cryst.).

FIGURES CAPTIONS

Fig. 1. Direct and reciprocal cell for γ-Na_2CO_3.

Fig. 2. Modulation wave vectors, direct and reciprocal cell for 1T-TaS_2.

PROPERTIES OF LATTICES ASSOCIATED
WITH A MODULATED CRYSTAL

A. Janner and T. Janssen

1. INTRODUCTION

Concepts and notations appearing here are based on those introduced and discussed in the invited paper appearing elsewhere in these proceedings[1] (denoted here by I). There the appearence of different lattices in the diffraction pattern of a modulated crystal has been discussed: Λ_n^* is the n-dim. lattice of the main reflections (in V_n), D_d^* the d-dim. lattice (in V_d), which if centered at the lattice points of Λ_n^*, describes the positions of the satellite reflections and the (n+d)-dim. lattice Σ_{n+d}^* in $V_S = V_E \oplus V_I$ (where $V_E = V_n$ and $V_I \simeq V_d$) has a projection onto V_E which gives the positions of all reflections. These reflections have been described in I by choosing three lattice bases B_n^*, B_d^* and B_{n+d}^* of Λ_n^*, D_d^* and Σ_{n+d}^* respectively given by:

$$B_n^* = \{\vec{a}_1^*, \vec{a}_2^*, \ldots, \vec{a}_n^*\}, \quad B_d^* = \{\vec{b}_1^*, \vec{b}_2^*, \ldots, \vec{b}_d^*\}$$

$$B_{n+d}^* = \{a_1^*, a_2^*, \ldots, a_{n+d}^*\} \tag{1}$$

and related by (I11), which corresponds to a description of the spectrum Sp of the crystal function $\rho(\vec{r})$ in terms of the elements of Z^{n+d}.

2. REPRESENTATIONS OF THE POINT GROUP K

We denote by B_n and B_d lattice bases of Λ_n and D_d, respectively, and by B'_{n+d} the (split) basis of $V_S = V_E \oplus V_I$ given by $\{(B_n,0),(0,B_d)\}$.

Proposition 1. V_E and V_I are representation spaces for the point group K, and B_n, B_d are bases for integral representations.

Proof. Take $B_n = \{\vec{a}_1,\ldots,\vec{a}_n\}$. Define Γ_E by:

$$R\vec{a}_k = \sum_{h=1}^{n} \vec{a}_h \Gamma_E(R)_{hk}, \qquad R \in K \tag{2}$$

(after identification of $(\vec{a}_k,0)$ with $\vec{a}_k$). Then Γ_E is a representation of K and $\Gamma_E(R) \in Gl(n,Z)$. This last because $R\vec{a}_k = R_E\vec{a}_k$ and $R_E\Lambda_n = \Lambda_n$. In the same way one gets the corresponding result for V_I.

Corollary 1. The action of K on a split basis B'_{n+d} of V_S defines a reduced integral representation $\Gamma' = \Gamma_E \oplus \Gamma_I$ of K.

Proposition 2. Γ' is equivalent to an integral representation Γ of K having the form:

$$\Gamma(R) = \begin{pmatrix} \Gamma_E(R) & 0 \\ \Gamma_M(R) & \Gamma_I(R) \end{pmatrix}, \qquad R \in K \tag{3}$$

with the d×n matrix $\Gamma_M(R)$ defined by (I27):

$$\Delta R_E\vec{a}_k = R_I\Delta\vec{a}_k + \sum_{j=1}^{d} \vec{b}_j\Gamma_M(R)_{jk}.$$

Proof.

$$\Delta R_E\vec{a}_k - R_I\Delta\vec{a}_k = \sum_{j=1}^{d} \vec{b}_j[\sigma\Gamma_E(R)-\Gamma_I(R)\sigma]_{jk},$$

which implies:

$$\sigma\Gamma_E(R) - \Gamma_I(R)\sigma = \Gamma_M(R). \tag{4}$$

The change of basis from the split basis B'_{n+d} to the standard one B_{n+d} by $B_{n+d} = B'_{n+d}S$ with $S = \begin{pmatrix} 1 & 0 \\ -\sigma & 1 \end{pmatrix}$ gives with (4) and $\Gamma(R) = S^{-1}\Gamma'(R)S$ the desired result.

Proposition 3. There exists a primitive or a centered basis $\bar{B}_{n+d}$ of the lattice Σ_{n+d} which reduces the representation Γ to $\Gamma' = \Gamma_E \oplus \Gamma_I$.

Proof. The point group K is of finite order. According to the theorem of Maschke the matrix S which decomposes Γ is given by $S = \begin{pmatrix} 1 & 0 \\ X & 1 \end{pmatrix}$, where X is a d×n matrix with rational coefficients. Indeed one finds[2]:

$$X = \sigma - \frac{1}{|K|} \sum_{R\in K} \Gamma_I(R)\sigma\Gamma_E(R^{-1}). \tag{5}$$

Proposition 4. The real vector space W_{dn} spanned by the real d×n matrices μ is a representation space for K according to:

$$\phi(R)\mu \stackrel{\text{def}}{=} \Gamma_I(R)\mu\Gamma_E(R^{-1}), \qquad R \in K. \tag{6}$$

Proof. $\phi(R)$ is linear and $\phi(R)\phi(R') = \phi(RR')$. Note that the representation ϕ is equivalent to the tensor representation $\Gamma_I \otimes \Gamma_E^*$, where star means adjoint.

Proposition 5. W_{dn} is decomposable into a subspace L_K carrying all trivial representations of K contained in ϕ, and its orthogonal complement L_K^c. Thus:

$$W_{dn} = L_K \oplus L_K^c$$

where

$$L_K \stackrel{\text{def}}{=} \{\mu \in W_{dn} \mid \phi(K)\mu = \mu\}. \tag{7}$$

Proof. $|K| < \infty$. Thus ϕ is fully reducible. One gets L_K by:

$$L_K = \frac{1}{|K|} \sum_{R\in K} \phi(R)W_{dn}. \tag{8}$$

Proposition 6. Consider the unique decomposition of σ, defined by $\sigma_{jk} = \frac{1}{2\pi}\vec{b}_j^*\cdot\vec{a}_k$ (scalar product in V_E), into:

$$\sigma = \sigma^{(i)} + \sigma^{(r)} \text{ with } \sigma^{(i)} \in L_K \text{ and } \sigma^{(r)} \in L_K^c. \tag{9}$$

Then $\sigma^{(i)}$ has irrational row vectors ($\in \tilde{R}^n - \tilde{Q}^n$) and $\sigma^{(r)}$ is a rational matrix.

Proof.

$$\sigma^{(i)} = \frac{1}{|K|}\sum_{R\in K}\Gamma_I(R)\sigma\Gamma_E(R^{-1}). \tag{10}$$

Since $\sigma^{(r)} = \sigma-\sigma^{(i)} = X$ as in eq. (5), $\sigma^{(r)}$ is rational, and as σ has irrational rows so also $\sigma^{(i)}$.

Proposition 7. The elements $\vec{b}^* \in D_d^*$ can be decomposed uniquely into a component $\vec{b}_c^*$ commensurate with Λ_n^* and an incommensurate component $\vec{b}_{in}^*$:

$$\vec{b}^* = \vec{b}_c^* + \vec{b}_{in}^*, \text{ with } |K|\vec{b}_c^* \in \Lambda_n^* \text{ and } \{\vec{b}_{in}^*\} \cap \Lambda_n^* = 0. \tag{11}$$

Proof.

$$\vec{b}^* = \sum_{j=1}^{d} z_j\vec{b}_j^*, \qquad z_j \in Z.$$

Then

$$\vec{b}_c^* \equiv \sum_{j=1}^{d}\sum_{k=1}^{n} z_j\sigma_{jk}^{(r)}\vec{a}_k^*;\quad \vec{b}_{in}^* \equiv \sum_{j=1}^{d}\sum_{k=1}^{n} z_j\sigma_{jk}^{(i)}\vec{a}_k^*. \tag{12}$$

Corollary 7. The matrix $\sigma^{(r)}$ is rational and $|K|\sigma^{(r)}$ has integral entries. Furthermore:

$$\Gamma_M(R) = \sigma^{(r)}\Gamma_E(R) - \Gamma_I(R)\sigma^{(r)}. \tag{13}$$

3. ARITHMETIC EQUIVALENCE

The physical (and geometrical) meaning of the matrices occurring in the representation Γ of K defined in (3) puts restrictions on the corresponding lattice basis $B = (a_1,\dots,a_{n+d})$ of Σ_{n+d} (and B^* of Σ^*_{n+d}). $\Gamma(K)$ has the form (3) if and only if the last d basis vectors a_{n+j} lie in V_I. We therefore extend the concept of *standard basis* to bases with this property. ($a_{n+j} = (0,\vec{b}_j)$, $j=1,\dots,d$, for chosen D_d). Accordingly we restrict the lattice basis transformations of Σ_{n+d} to those transforming standard bases into standard bases. Defining $B \to \bar{B} = BS$ by

$$\bar{a}_k = \sum_{h=1}^{n+d} a_h S_{hk}$$

it follows that $S \in Gl(n+d,Z)$ has the form:

$$S = \begin{pmatrix} S_E & 0 \\ S_M & S_I \end{pmatrix} \quad \text{with} \quad \begin{cases} S_E \in Gl(n,Z) \\ S_I \in Gl(d,Z) \end{cases}. \tag{14}$$

All such matrices form a subgroup of $Gl(n+d,Z)$ that we denote by $Gl(n,d,Z)$. Note that $\Gamma(K) \subset Gl(n,d,Z)$ and that conjugation by S gives conjugated subgroups $\bar{\Gamma}(K)$ and $\Gamma(K)$ in $Gl(n,d,Z)$. If this is the case, we call $\bar{\Gamma}$ and Γ *arithmetically equivalent*. For the corresponding submatrices we have:

$$\begin{aligned} \bar{\Gamma}_E(R) &= S_E^{-1}\Gamma_E(R)S_E, \quad \bar{\Gamma}_I(R) = S_I^{-1}\Gamma_I(R)S_I \\ \bar{\Gamma}_M(R) &= S_I^{-1}\Gamma_M(R)S_E - S_I^{-1}S_M\bar{\Gamma}_E(R) + \bar{\Gamma}_I(R)S_I^{-1}S_M. \end{aligned} \tag{15}$$

A given $\Gamma_M(K)$ determines σ mod L_K, thus according to (13) the rational component $\sigma^{(r)}$. As the irrational one $\sigma^{(i)}$ transforms as:

$$\bar{\sigma}^{(i)} = S_I^{-1}\sigma^{(i)}S_E \tag{16}$$

one gets for $\sigma^{(r)}$

$$\sigma^{(r)} = S_I^{-1}\sigma^{(r)}S_E - S_I^{-1}S_M. \tag{17}$$

4. BRAVAIS CLASSES

By $\Sigma_{n+d}(\Lambda_n, D_d)$ we denote a set of lattices of dim. n+d, n and d, respectively, lying in $V_S = V_E \oplus V_I$ such that one has (I7) (I21), (I23):

$$(0, D_d) \subset \Sigma_{n+d}, \quad \pi_E \Sigma_{n+d} = \Lambda_n, \quad \Lambda_n^* \cap D_d^* = 0. \tag{18}$$

Denote by H the holohedry of $\Sigma_{n+d}(\Lambda_n, D_d)$, which is the largest point group of Σ_{n+d} satisfying Proposition 1. A given $\Sigma_{n+d}(\Lambda_n, D_d)$ determines a class $[\Gamma(H)]$ of arithmetically equivalent representations, i.e. conjugated subgroups in Gl(n,d,Z). Conversely, as in the case of space groups, or in that of magnetic space groups[3], this arithmetic equivalence gives rise to an equivalence relation among sets of lattices as above, leading to a generalization of the Bravais classes.

Definition. Two sets of lattices $\Sigma_{n+d}(\Lambda_n, D_d)$ and $\bar{\Sigma}_{n+d}(\bar{\Lambda}_n, \bar{D}_d)$ belong to the same Bravais class if and only if their holohedries H and $\bar{H}$ with respect to standard lattice bases B and $\bar{B}$, respectively, are conjugated subgroups of Gl(n,d,Z). Thus if:

$$\bar{\Gamma}(\bar{H}) = S^{-1}\Gamma(H)S,$$

$$S \in Gl(n,d,Z) \text{ and } \Gamma(H) \subset Gl(n,d,Z). \tag{19}$$

We now are able to show that the arbitrariness in the choice of the lattice D_d^* among all the possible ones $D_d^*(\Lambda_n^*)$ is irrelevant in the sense that one remains within the same Bravais class. A classification of these classes for n and d up to 3 is in progress.

Proposition 8. Two sets of lattices $\Sigma_{n+d}(\Lambda_n, D_d)$ and $\bar{\Sigma}_{n+d}(\Lambda_n, \bar{D}_d)$ such that $\bar{D}_d^* \equiv D_d^* \pmod{\Lambda_n^*}$ belong to the same Bravais class.

Proof. Consider a basis $B_d^* = (\vec{b}_1^*, \ldots, \vec{b}_d^*)$ of D_d^*. Then a corresponding basis $\bar{B}^*_d$ of $\bar{D}^*_d$ can be written as:

$$\vec{\bar{b}}^*_j = \vec{b}^*_j + \sum_{k=1}^{n} \vec{a}^*_k S^*_{Mkj}$$

with S^*_M a n×d integral matrix. Note that $\Sigma_{n+d}(\Lambda_n,D_d)$ and $\bar{\Sigma}_{n+d}(\Lambda_n,\bar{D}_d)$ have the same holohedry H. For the corresponding representations $\Gamma(H)$ and $\bar{\Gamma}(H)$ obtained by referring H to bases B and $\bar{B}$, respectively, given by (1) (which are standard) one verifies using (I27) that $\Gamma(H)$ and $\bar{\Gamma}(H)$ are conjugated subgroups of Gl(n,d,Z).

More generally the arbitrariness in the choice of bases (1) is as follows:

(i) One may change the description of the main reflections by choosing another basis of Λ^*_n:

$$\vec{a}^*_k \to \sum_{h=1}^{n} \vec{a}^*_h S^*_{Ehk}, \quad S^*_E \in Gl(n,Z). \tag{20}$$

(ii) One may change the description of the satellite reflections:

a) by choosing another basis of D^*_d. Thus

$$\vec{b}^*_j \to \sum_{k=1}^{d} \vec{b}^*_k S^*_{Ikj}, \quad S^*_I \in Gl(d,Z), \tag{21}$$

b) by considering another lattice $\bar{D}^*_d$

$$\vec{b}^*_j \to \vec{b}^*_j + \sum_{k=1}^{n} \vec{a}^*_k S^*_{Mkj}, \quad S^*_{Mkj} \in Z. \tag{22}$$

By an immediate generalization of the proof of Proposition 8 one sees that the corresponding representations Γ and $\bar{\Gamma}$ defined by (3) for K=H, are arithmetically equivalent. Analogously, one defines equivalence of superspace groups G and modulated space groups M in such a way that different choices (20-22) give equivalent groups.

REFERENCES

1. A. JANNER, Modulated space groups, these proceedings.
2. H. BOERNER, Darstellungen von Gruppen, Springer, Berlin 1955.
3. A. JANNER, Helv. Phys. Acta 39 (1966), 665.
4. A. JANNER and T. JANSSEN, in preparation.

A HAMILTONIAN APPROACH TO THE KdV AND OTHER EQUATIONS

Peter D. Lax

1. INTRODUCTION

A recent series of investigations of nonlinear wave motion, commencing with Kruskal and Zabusky's paper[29], have led to the unexpected discovery that an astonishingly large number of important differential equations of mathematical physics are completely integrable Hamiltonian systems. Included among these are the Korteweg-de Vries (KdV) and Boussinesq equations for waves in shallow water, the equations governing self-induced transparency, and self-focusing and self-modulating waves in optics, the vibrations of the Toda lattice, the motion of particles under an inverse square potential, and some others.

In this talk we describe the Hamiltonian formalism, and how it is applied to the KdV equation, as well as to a slight variant of the KdV equation.

2. HAMILTONIAN MECHANICS

In this section we review briefly classical Hamiltonian Mechanics, and put it in a form suitable for infinite dimensional phase space.

The phase space of the classical theory is 2N-dimensional

space, or some portion of it; the coordinates in this space are denoted by p_j, q_j, j=1,...,N. A Hamiltonian is any sufficiently differentiable function H in phase space; the Hamiltonian form of the equations of motion is

$$(2.1) \qquad \frac{d}{dt} p_j = -H_{q_j}, \quad \frac{d}{dt} q_j = H_{p_j}, \quad j=1,\dots,N,$$

where the subscripts of H indicate partial derivatives.

Let F and K be any two functions in phase space; their *Poisson bracket* is defined to be

$$(2.2) \qquad [F,K] = \sum_1^N (F_{q_j} K_{p_j} - F_{p_j} K_{q_j}).$$

It is convenient to rewrite both (2.1) and (2.2) in block vector and matrix notation. We write

$$(2.3) \qquad \begin{pmatrix} p \\ q \end{pmatrix} = u, \qquad \begin{pmatrix} H_p \\ H_q \end{pmatrix} = H_u.$$

Then we can write (2.1) as

$$(2.4) \qquad \frac{d}{dt} u = J_0 H_u,$$

where

$$(2.5) \qquad J_0 = \begin{pmatrix} 0 & -I \\ I & 0 \end{pmatrix},$$

I denoting the N×N unit matrix, 0 the N×N zero matrix. The Poisson bracket (2.2) can be written as

$$(2.6) \qquad [F,K] = (F_u, J_0 K_u),$$

where the parentheses (,) are defined by

$$(2.7) \qquad (u,u') = \sum (p_j p'_j + q_j q'_j).$$

In terms of (,) the gradient H_u of H can be defined by

(2.8)
$$\frac{d}{d\varepsilon} H(u+\varepsilon w)\,\Big|_{\varepsilon=0} = (H_u, w).$$

We ask (and answer) the following entirely elementary question:

When can a system of equations of the form

(2.9)
$$\frac{d}{dt} v = JK_v,$$

J some given constant matrix, K some function in phase space, be reduced to Hamiltonian form by a linear change of variables

(2.10)
$$u = Tv,$$

T a constant matrix independent of t? We define the transformed Hamiltonian H by

$$H(u) = H(Tv) = K(v).$$

Using (2.8) we get

(2.11)
$$T^*H_u = K_v,$$

where T* is the transpose of T. Multiplying (2.9) by T and using (2.10), (2.11) we get

$$\frac{d}{dt} v = TJT^*H_u.$$

This is of Hamiltonian form iff

(2.12)
$$TJT^* = J_0.$$

Given a *real* matrix J, this relation can be satisfied by some real T iff J satisfies

(2.13)
i) $J^* = -J$,
ii) J is nonsingular.

The conditions are obviously necessary; to show their sufficiency we note that by antisymmetry and reality, the eigenvalues of J are purely imaginary; since J is nonsingular the eigenvalues are nonzero, and the eigenvectors come in conjugate pairs:

$$Jf_k = i\lambda_k f_k, \quad \lambda_k > 0, \quad k=1,\dots,N. \tag{2.14}$$

Writing

$$f = g+ih$$

we can rewrite (2.14) as

$$Jg = -\lambda h, \quad Jh = \lambda g.$$

Clearly, if we set

$$T_1^* = (g_1,\dots,g_N, h_1,\dots,h_N),$$

then

$$T_1 J T_1^* = \begin{pmatrix} 0 & -\Lambda \\ \Lambda & 0 \end{pmatrix}$$

where Λ is the diagonal matrix

$$\Lambda = \begin{pmatrix} \lambda_1 & & \\ & \ddots & \\ & & \lambda_N \end{pmatrix}.$$

Define T_2 by

$$T_2 = \begin{pmatrix} \Lambda^{-\frac{1}{2}} & 0 \\ 0 & \Lambda^{-\frac{1}{2}} \end{pmatrix}.$$

Clearly $T = T_2T$, satisfies (2.12).

Given a real inner product space, a real valued function H defined on that space is called differentiable if the directional derivatives (2.8) exist for all directions w, depend linearly on w, and can be represented in form (2.8), with H_u an element of the space. H is called twice differentiable if the directional derivatives of H_u exist in every direction v, and can be represented in the form

(2.15) $$\frac{d}{d\varepsilon} H_u(u+\varepsilon v) = H_{uu} v,$$

where H_{uu} is a linear operator mapping the space into itself. H_{uu}, which is an operator valued function of u, is called the *second derivative* of H; H_{uu} is a *symmetric operator*:

$$(H_{uu} v, w) = (v, H_{uu} w).$$

We shall call an equation of form

$(2.16)_H$ $$u_t = JH_u$$

Hamiltonian if J is an antisymmetric operator, independent of u:

(2.17) $$J^* = -J.$$

Since we have dropped the condition (2.13)ii) requiring nondegeneracy, equation $(2.16)_H$ cannot quite be put into genuine Hamiltonian form; thus even in the finite dimensional case we have a miniscule generalization. The important fact is that equations of form (2.16) share all important properties of Hamiltonian equations, and find applications in the examples presented in Section 3.

Properties of Hamiltonian motion are most conveniently expressed in terms of the *Poisson bracket*, defined by the analogue of formula (2.6) for any pair of differentiable functions F and K:

(2.18) $$[F,K] = (F_u, JK_u).$$

This Poisson bracket has the usual properties:

a) $[F,K]$ is a bilinear function of F and K.

b) $[F,K] = -[K,F]$.

c) The Jacobi identity

$$[[F,H],K] + [[H,K],F] + [[K,F],H] = 0 \tag{2.19}$$

holds.

Part a) is an immediate consequence of the definition (2.18), and b) follows from the antisymmetry of J. To prove c) we have to calculate the gradient of [F,H]. Using the definition of gradient given in (2.8), the antisymmetry of J, its independence of u, and the symmetry of the second derivatives F_{uu} and H_{uu} we get, after a brief calculation, that

$$[F,H]_u = F_{uu}H_u - H_{uu}F_u. \tag{2.20}$$

From this (2.19) is easily deduced, using once more the symmetry of F_{uu}, H_{uu} and K_{uu}.

Suppose that u(t) satisfies (2.16); for any function F(u)

$$\frac{d}{dt} F = (F_u, \frac{d}{dt} u) = (F_u, JH_u) = [F,H].$$

It follows from this that F *is constant along trajectories of the Hamiltonian flow* $(2.16)_H$ iff [F,H] = 0.

Since [F,H] = 0 is a symmetrical relation between F and H, it also follows from [F,H] = 0 that H is constant along the trajectories of the Hamiltonian flow $(2.16)_F$. Furthermore the following classical result holds:

If [F,H] = 0, *then the Hamiltonian flows* $(2.16)_H$ *and* $(2.16)_F$ *commute.* This is a consequence of the Jacobi identity.

The following concept is classical in 2N-dimensional space:

Given an antisymmetric J that does not depend on u, a collection of N functions $H_1, \ldots, H_N$ is called a *completely integrable Hamiltonian system* if

a) $[H_i, H_j] = 0$ for all i,j.

b) The functions H_i are independent in the sense that their gradients are linearly independent, except on lower dimensional sets.

A single Hamiltonian equation $(2.16)_H$ is called completely integrable if H is a member of a completely integrable set $H_1, \ldots, H_N$. In this case, each H_i is constant along trajectories of $(2.16)_H$, so that these trajectories lie on N-dimensional submanifolds. On the other hand, each trajectory $(2.16)_{H_i}$ lies on this manifold. It is not hard to deduce from this that the N-dimensional submanifolds in question are products of N lines or circles. If the sets H_i = const., i=1,...,N, are *compact*, it follows that they are N-dimensional tori. In this case each Hamiltonian motion is quasi-periodic or periodic.

3. INFINITE DIMENSIONAL HAMILTONIAN SYSTEMS

In this section phase space consists of all real valued C^∞ functions which are periodic, say with period 1; another equally interesting example are the functions of Schwartz class $\mathcal{L}$ on the entire real line $\mathbb{R}$. The inner product is taken to be

$$(u,v) = \int u(x)v(x)\,dx, \tag{3.1}$$

where the integration is over a single period in the first instance, and all of $\mathbb{R}$ in the second.

Below we list some functions of u and their gradients defined in this phase space:

$$(3.2)_1 \qquad F_1(u) = \int \tfrac{1}{2} u^2 dx, \qquad F_{1_u} = u.$$

$$(3.2)_2 \qquad F_2(u) = \int (\tfrac{1}{6} u^3 - \tfrac{1}{2} u_x^2)\,dx, \qquad F_{2_u} = \tfrac{1}{2} u^2 + u_{xx}.$$

$$(3.3)_1 \qquad H_1(u) = \int (\tfrac{1}{2} u^2 + \tfrac{1}{2} u_x^2)\,dx, \qquad H_{1_u} = u - u_{xx}.$$

$$(3.3)_2 \qquad H_2(u) = \int \left(\frac{1}{6}u^3 + \frac{1}{2}u^2\right)dx, \qquad H_{2_u} = \frac{1}{2}u^2 + u.$$

A more exotic class of functions of u are the eigenvalues $\lambda = \lambda_j$ of the Schrödinger operator

$$(3.4) \qquad L = D^2+u, \qquad D = \frac{d}{dx},$$

given by

$$(3.5) \qquad Lw = \lambda w.$$

The gradient of $\lambda(u)$ is easily determined by standard perturbation calculation. Replace u by $u+\varepsilon v$ and differentiate (3.5) with respect to ε. Denoting differentiation with respect to ε by a dot, we get

$$L\dot{w} + vw = \lambda\dot{w} + \dot{\lambda}w.$$

Multiplying by w and integrating we get, after integration by parts and using (3.5) that

$$\int vw^2dx = \dot{\lambda} \int w^2dx.$$

If w is normalized, we get

$$\dot{\lambda} = (\lambda_u, v) = \int vw^2dx = (w^2, v).$$

This proves that

$$(3.6) \qquad \lambda_u = w^2.$$

To build Hamiltonian systems we need to choose some operator J. The choice

$$(3.7) \qquad J = D$$

was suggested by Clifford Gardner[7]. Note that this J is antisymmetric and independent of u.

For this choice of J, the Hamiltonian equations (2.16) associated with the Hamiltonians F_1 and F_2 given in $(3.2)_1$ and $(3.2)_2$, respectively, are

$$(3.8)_1 \qquad u_t = JF_{1_u} = Du = u_x$$

$$(3.8)_2 \qquad u_t = JF_{2_u} = D(\frac{1}{2}u^2+u_{xx}) = uu_x+u_{xxx}.$$

Equation $(3.8)_1$ is linear and easily solved: $u = u(x+t)$. Equation $(3.8)_2$ on the other hand is nonlinear and far from being easy to solve; it is the celebrated Korteweg-de Vries (KdV) equation.

We turn now to the exotic function $\lambda(u)$ defined as the eigenvalue of the Schrödinger operator L which differs from (3.4) by a factor 6; i.e., we take

$$(3.9) \qquad L = D^2 + \frac{1}{6}u.$$

The eigenvalue equation is

$$(3.10) \qquad w_{xx} + \frac{1}{6}uw = \lambda w$$

and the gradient of λ is, analogously to (3.6),

$$(3.11) \qquad \lambda_u = \frac{1}{6}w^2.$$

We shall calculate the Poisson bracket of F_2 and λ; using the definition (2.6) we have

$$(3.12) \qquad [F_2,\lambda] = (F_{2u},J\lambda_u) = \int (\frac{1}{2}u^2+u_{xx})D(\frac{1}{6}w^2)\,dx$$
$$= \frac{1}{6}\int (u^2ww_x+2u_{xx}ww_x)\,dx,$$

where we have used the values of F_{2u}, λ_u and J as given above. In the second term on the right in (3.12) we integrate by parts, obtaining

$$- \frac{2}{6} \int u_x(ww_{xx}+w_x^2)\,dx.$$

In the second term above we integrate by parts once more, obtaining

$$- \frac{2}{6} \int (u_x ww_{xx} - 2uw_x w_{xx})\,dx$$

$$= - \frac{1}{3} \int (u_x w - 2uw_x) w_{xx}\,dx.$$

Now we replace w_{xx} from equation (3.10), obtaining

$$- \frac{1}{3} \int (u_x w - 2uw_x)(\lambda w - \frac{1}{6} uw)\,dx.$$

Two further integrations by parts eliminate u_x; we get

$$\frac{1}{3} \int [4\lambda uww_x - \frac{u^2 ww_x}{6} - \frac{2u^2 ww_x}{6}]$$

$$= \frac{4\lambda}{3} \int uww_x\,dx - \frac{1}{6} \int u^2 ww_x\,dx.$$

The second term above cancels out exactly the first term on the right of (3.12), leaving us with

$$\frac{4\lambda}{3} \int uww_x\,dx. \tag{3.13}$$

To evaluate this integral we multiply (3.10) by w_x and integrate. The first term on the left and the term on the right are perfect x-derivatives; therefore their integral is zero. But then so is the integral of the remaining term; since that is a constant multiple of (3.13), we conclude that (3.13) is zero. This proves that (3.12) is zero, i.e. that

$$[F_2, \lambda] = 0. \tag{3.14}$$

According to Hamiltonian theory, it follows from this that λ is constant along the trajectories of the Hamiltonian flow for the Hamiltonian F_2. We saw earlier that the latter is furnished by solutions of the KdV equation. So we conclude:

If u changes subject to the KdV equation, the eigenvalues of the operator L given by (3.9) remain constant, a result originally derived another way by Gardner, Green, Kruskal and Miura[8].

Let λ and μ be two distinct eigenvalues of L given by (3.9); let's compute their Poisson bracket. Using the definition (2.6), and the gradients as given by formula (3.11), we get

$$[\lambda,\mu] = (\lambda_u, D\mu_u) = \frac{2}{6}\int w^2 z z_x dx,$$

where z is the eigenfunction corresponding to the eigenvalue μ. Integrating by parts, and adding the resulting integral to the original one we get

$$(3.15)\qquad \begin{aligned}[\lambda,\mu] &= \frac{1}{6}\int wz(wz_x - zw_x)dx \\ &= \frac{1}{6}\int wzWdx,\end{aligned}$$

where W is the Wronskien

$$W = wz_x - zw_x.$$

A standard calculation, using the eigenvalue equation (3.10) and the analogous one for μ shows that

$$W_x = (\mu-\lambda)wz.$$

Substituting this into (3.15) shows that

$$(3.16)\qquad [\lambda,\mu] = \frac{1}{6}\int \frac{W_x}{\mu-\lambda} Wdx = 0;$$

thus the Poisson bracket of two distinct eigenvalues in zero. The above elegant derivation whose idea goes back to Borg was shown to the author by Eugene Trubowitz; other derivations were given by Zakharov, and by the author[13].

The above results show that the KdV equation is a Hamiltonian system with an infinite number of conserved functions $\lambda_j(u)$, $j=1,2,\ldots$ such that the Poisson bracket of any two conserved functions is zero.

We saw in the finite dimensional case, discussed at the end of Section 2, that the existence of N such conserved functions in 2N-dimensional phase space implies that each trajectory is restricted to an N-dimensional submanifold, and behaves in a very regular fashion. In the present case where $N=\infty$ it is not quite clear whether there are sufficiently many conserved functionals. However Faddeev and Zakharov have shown in [4], on the basis of [8], that the KdV equation can be regarded as a completely integrable Hamiltonian system on the entire line $\mathbb{R}$, when the phase space consists of C^∞ functions which tend to 0 sufficiently rapidly when $x \to \pm\infty$. In this case there exist for any integer $N=1,2,\ldots$ N-dimensional manifolds of solutions. A trajectory $U(t)$ on such an N-dimensional manifold is called an *N-soliton solution*, and has the property that as $t \to \infty$ or to $-\infty$, $U(t)$ appears as a superposition of N solitons proceeding at N different speeds c_k, $k=1,\ldots,N$. A *soliton* is a steady progressing wave solution of the KdV equation, i.e., of the form

$$u(x,t) = s(x-ct). \tag{3.17}$$

The solitons satisfy an ordinary differential equation, obtained by substituting (3.17) into $(3.8)_2$; this ODE can be solved explicitly in terms of exponential functions.

In the periodic case Novikov[17] and the author[12] have succeeded in constructing N-dimensional manifolds of solutions, N arbitrary, on which trajectories behave in a quasi-periodic fashion. Then McKean and Trubowitz have shown in [15], using results from [14], that the space-periodic KdV equation is completely integrable, and that each trajectory is almost periodic in time.

We conclude this section by replacing J, until now given by (3.7), with the following one:

$$J = D(1-D^2)^{-1}. \tag{3.18}$$

Note that J is antisymmetric and independent of u.

We write down the Hamiltonian equations (2.16) for the Hamiltonians H_1 and H_2 given by formulas $(3.3)_1$ and $(3.3)_2$, and J given by (3.18). We get

$$u_t = JH_{1_u} = D(1-D^2)^{-1}(u-u_{xx}) = u_x, \tag{$(3.19)_1$}$$

and

$$u_t = JH_{2_u} = D(1-D^2)^{-1}(\tfrac{1}{2}u^2+u). \tag{$(3.19)_2$}$$

Multiplying both sides of $(3.19)_2$ by $(1-D^2)$ we get

$$u_t - u_{txx} = D(\tfrac{1}{2}u^2+u) = uu_x + u_x,$$

or

$$u_t = uu_x + u_x + u_{txx}. \tag{3.20}$$

This equation is called the *regularized long wave equation*, first considered by Peregrine. Note that, apart of the extra u_x term, (3.20) is very much like the KdV equation $(3.8)_2$, except that the term u_{xxx} is replaced by u_{txx}. In fact, Benjamin[2], has championed (3.20) as preferable to the KdV equation for the description of long waves. The present analysis shows that equation (3.20), too, is Hamiltonian. Furthermore H_1 is a conserved quantity for the Hamiltonian flow (3.20) induced by the Hamiltonian H_2. For clearly, the Hamiltonian flow induced by H_1, satisfying equation $(3.19)_1$, is just translation along the x-axis, and so H_2, which is independent of x, is conserved under this flow.

This shows that $[H_1, H_2] = 0$, from which the former statement follows. A further conserved quantity under the flow (3.20) is

$$H_0(u) = \int u dx.$$

But no further conserved quantities are known for equation (3.20).

Equation (3.20) has solitons, i.e. solutions of form (3.17); in fact these are identical with the solitons of the KdV equation, except that they are parametrized differently by the speed c. Recently Eilbeck[3] has undertaken to study the possible existence of N-soliton solutions by means of numerical calculations. He solved numerically the initial value problem for (3.20), starting with an initial configuration

$$u(x,0) = s_1(x) + s_2(x-\ell),$$

where s_1 and s_2 are solutions that would propagate with speed c_1 and c_2, respectively, with $c_1 > c_2$. The initial separation ℓ is taken to be so large that the effective overlap between the two solitons was negligible; this is easy to achieve since solitons tend to zero as $x \to \infty$ at an exponential rate. After the lapse of sufficient time T, the numerically computed solution was found to have, within very narrow margin of error, the form

$$u(x,T) \simeq s_1(x-c_1T-\theta_1) + s_2(x-c_2T-\theta_2).$$

If the difference between the two sides were to tend to zero as ℓ and $T \to \infty$, this would prove the existence of 2-soliton solutions, which is strong evidence for the complete integrability of (3.20). However calculations performed in the Soviet Union[1] indicate a small deviation, about .3%, which persists, no matter how large T is taken. So the matter is in the hands of computers.

4. THE COMMUTATOR EQUATION FOR ISOSPECTRAL DEFORMATION

In the last section we have shown that the eigenvalues of the Schrödinger operator do not change if the potential u is deformed subject to the KdV equation; the proof was a tour de force. In this section we describe a systematic method developed by the author[11], for constructing nonlinear equations whose solutions leave the eigenvalues of an associated linear operator unchanged, and apply the method to the Schrödinger operator. In the hands of others[18,5,16], this method has produced further interesting examples.

Let L(t) be a one-parameter family of operators all of which are similar to each other. That is, we assume that each L(t) can be mapped by a similarity transformation into L(0):

(4.1) $$U(t)^{-1}L(t)U(t) = L(0).$$

We assume that both L and U depend differentiably on t, and we introduce the notation

(4.2) $$U_tU^{-1} = B(t),$$

from which we deduce

(4.3) $$U_t = BU.$$

Differentiate (4.1) with respect to t; using

$$\frac{d}{dt} U^{-1} = -U^{-1}U_tU^{-1}$$

and (4.3) we get

(4.4) $$-U^{-1}BLU + U^{-1}L_tU + U^{-1}LBU = 0$$

which implies

(4.5) $$L_t = BL-LB.$$

Conversely, suppose (4.4) is satisfied and suppose the initial value problem for the differential equation

$$v_t = B(t)v \tag{4.6}$$

can be solved for a sufficiently wide class of initial values $v(0)$. Then the operator

$$U(t) : v(0) \to v(t)$$

satisfies (4.1).

Similar operators L have the same spectrum; so it follows from (4.1) that *the eigenvalues* $\{\lambda_j\}$ *of* $L(t)$ *are independent of* t.

In any concrete representation the operator L appears as an integral or differential operator, described in terms of coefficients. Relation (4.5) is a nonlinear differential equation for these coefficients. The eigenvalues of L are functionals of the coefficients; being independent of t, they constitute the sought-after integrals.

If the operators L are symmetric or hermitean symmetric then similarity implies unitary equivalence. In fact if the spectrum of L is simple then the operator U appearing in (4.1) must be unitary.

If $U(t)$ is unitary

$$UU^* = I,$$

differentiating with respect to t, we get

$$U_t U^* + UU_t^* = 0.$$

The meaning of this equation is that $U_t U^*$ is antisymmetric. Since U is unitary, $U_t U^* = U_t U^{-1}$, the operator denoted in (4.2) as B. So we conclude:

For L hermitean symmetric, B should be chosen antisymmetric:

$$B^* = -B.$$

We choose now L to be the Schrödinger operator (3.9):

$$L = D^2 + \frac{1}{6}u, \qquad D = \frac{d}{dx}. \tag{4.7}$$

For L given by (4.7), $L_t = u_t$ is multiplication by u_t; therefore in order to satisfy Eq. (4.5) we need operators B_j whose commutator with L is multiplication. In [11] the author has shown how to construct a sequence B_j of such operators; these operators have these properties:

(i) B_j is a differential operator of order 2j-1.

(ii) B_j is antisymmetric.

(iii) $B_jL - LB_j$ is multiplication by $K_j(u)$; $K_j(u)$ depends in a nonlinear fashion on u and its derivatives up to order 2j-1.

Following (4.5) we consider the equations

$$u_t = B_jL - LB_j = K_j(u); \tag{4.8}$$

these equations have the property that for their solutions the spectrum of L defined by (4.1) is independent of t.

The first two of these operators are

$$(4.9)_1 \qquad B_1 = 6D, \qquad (B_1L-LB_1)u = K_1(u) = u_x$$

and

$$(4.9)_2 \qquad B_2 = 24D^3 + 36uD + 18u_x, \qquad (B_2L-LB_2)u = K_2(u) = u_{xxx} + uu_x.$$

Gardner has shown that each K_j is of the form

$$K_j = DF_{j_u}.$$

Furthermore Gardner has shown that these functions F_j satisfy

$$[F_j, F_k] = 0, \tag{4.10}$$

where the Poisson bracket is defined with respect to J given in (3.7). Another proof is given in [14].

Thus the F_j constitute yet another infinite sequence of functions, of which the first two are given by (3.2), that are in involution, i.e. which satisfy (4.10). We remark that these functions are *not* independent of the sequence λ_j formed by the eigenvalues of L; as shown by Kruskal and Zabusky[10], each F_j can be expressed as a function of all the λ_j.

BIBLIOGRAPHY

1. Kh.O. ABDULLOEV, One More Example of Inelastic Soliton Interaction, Physics Letters, vol. 56A, no. 6, p. 427, May, 1976.
2. T.B. BENJAMIN, Lectures on nonlinear wave motion, Lectures in Applied Math., vol. 15, American Mathematical Society, Providence, R.I., 1974.
3. J.C. EILBECK, Numerical Study of the Regularized Long-Wave Equation 1: Numerical Methods, J. Comp. Phys., vol. 19, no. 1, pp. 43-57, Sept., 1975.
4. L. FADDEEV and V.E. ZAKHAROV, Korteweg-de Vries equation as completely integrable Hamiltonian system, Funk. Anal. Priloz. 5, pp. 18-27 (1971) (in Russian).
5. H. FLASCHKA, Integrability of the Toda lattice, Phys. Rev. B, 703 (1974).
6. H. FLASCHKA, On the Toda lattice, II. Inverse scattering solution, Phys. Rev. B9, 1924 (1974).
7. C.S. GARDNER, Korteweg-de Vries equation and generalizations, IV. The Korteweg-de Vries equation as a Hamiltonian system, J. Math. Phys. 12, 1548-1551 (1971).
8. C.S. GARDNER, J.M. GREENE, M.D. KRUSKAL and M. MIURA, Method for solving the Korteweg-de Vries equation, Phys. Rev. Lett. 19, 1095-1097 (1967).
9. C.S. GARDNER, M.D. KRUSKAL and R.M. MIURA, Korteweg- de Vries equation and generalizations, II. Existence of conservation laws and constants of motion, J. Mathematical Phys., 9, 1204-1209 (1968).

10. M.D. KRUSKAL and N.J. ZABUSKY, Interaction of "solitons" in a collisionless plasma and the recurrence of initial states, Phys. Rev. Letters, 15, 240-243 (1965).

11. P.D. LAX, Integrals of nonlinear equations of evolution and solitary waves, Comm. Pure Appl. Math. 21, 467-490 (1968).

12. P.D. LAX, Periodic solutions of the KdV equation, Comm. Pure Appl. Math. 28 (1975).

13. P.D. LAX, Almost Periodic Solutions of the KdV Equation, SIAM Review, vol. 18, no. 3, p. 351, July 1976.

14. H. McKEAN and P. VAN MOERBEKE, The spectrum of Hill's equation, Invenciones Mat. 30, 217-274 (1975).

15. H.P. McKEAN and E. TRUBOWITZ, Hill's operator and hyperelliptic function theory in the presence of infinitely many branch points, CPAM, vol. XXIX, no. 2, p. 143, March 1976.

16. J. MOSER, Three integrable Hamiltonian systems connected with isospectral deformations, Advances in Math., vol. 16, no. 2, p. 197, May 1975.

17. S.P. NOVIKOV, The periodic problem for the Korteweg-de Vries equation, I. Funk. Anal. Priloz, 8, no. 3, 54-66 (1974) (in Russian).

18. V.E. ZAKHAROV and A.B. SHABAT, Soviet Phys. JETP 34, 62 (1972).

USE OF AN ELEMENTARY GROUP THEORETICAL METHOD IN DETERMINING THE STRUCTURE OF A BIOLOGICAL CRYSTAL FROM ITS PATTERSON FUNCTION

D.B. Litvin

Crystallographers have been using X-rays to investigate the structure of biologically important macromolecules for over forty years[1]. One type of these are the so-called spheroidal or globular macromolecules, such as myoglobin, which is responsible for the storage of oxygen in muscle tissue, and hemoglobin, which is responsible for transporting oxygen in the blood stream. There are other types, such as the fibrous, fiber-like macromolecules found in for example hair and skin, but we will limit our interest to the globular type. To have an idea of the size, weight, and number of atoms in such macromolecules, in Table 1 we compare these characteristics of some typical globular macromolecules with those of some familiar molecules.

Most globular macromolecules can be crystallized. In forming a crystal these macromolecules are not to any large extent distorted, there are in general only a few molecules in the unit cell of the crystal, and the identity of each molecule is preserved. The term "biological crystal" in the title of this talk refers to such a crystal.

The object of investigating the structure of such biological crystals is to determine the structure of the macromolecules. To determine the structure of the crystal one attempts to calculate

TABLE 1

Molecule	Molecular weight	Size	Number of atoms
water H_2O	18	4Å across	3
benzine H_6C_6	78	6Å diameter	12
myaglobin	17,000	spheroid 44Å × 44Å × 25Å	~2,600
hemoglobin	64,000	spheroid 65Å × 55Å × 50Å	~10,000
disk of TMV protein coat	600,000	160Å in diameter	~100,000

the electron density $\rho(r)$ of the crystal via the Fourier expansion.

$$\rho(r) = \frac{1}{V}\sum_{k} F(k)e^{-2\pi i k\cdot r}$$

where the Fourier coefficients $F(k) = |F(k)|e^{i\alpha(k)}$ are called the complex structure factors, $\alpha(k)$ the phase of the structure factor, and V is the volume of the unit cell of the crystal. Because of the translational symmetry of the crystal, there is a summation over all vectors k of the reciprocal lattice associated with the translation group of the crystal. One then interprets this electron density in terms of the atomic structure of the macromolecule.

From experimental X-ray diffraction data one obtains the magnitude of the structure factors $|F(k)|$. (In practice one does not obtain $|F(k)|$ for all values of k, and in fact uses only an approximation for the above expression for the electron density of the crystal. However, we will not discuss this complication.) Knowing only the magnitude of the structure factors one cannot determine the electron density; one needs also the phases.

To determine the phases the following method has been used: the X-ray data from the crystal containing the molecules whose structure we want to determine, is compared with the X-ray data from the same crystal where in each molecule a small number of atoms have been replaced by a second set of atoms which contains a heavy atom. This method is called the "isomorphous replacement method". The differences in the X-ray data are used to determine the phases. Using this method of determining the phases, the electron density maps of myoglobin and hemoglobin were determined in 1959.

To investigate the structure of these biological macromolecules and bypass the need for artificially altering the structure of the macromolecules, in the past fifteen years the so-called "molecular replacement" method has been developed[2]. This method, which we will discuss here, uses only the magnitude of the structure factors $|F(k)|$ which are obtained from the X-ray data. This method consists of three steps.

1) The rotation problem: determine the point group of the molecules in the crystal, and their orientation with respect to the translational symmetry vectors of the crystal.

2) The translation problem: determine the vectors between the molecular positions in the crystal. These vectors are called translation vectors.

3) The phase problem: using the information obtained in the first two steps, determine the phases of the structure factors.

With respect to the third step, I will only comment that there has been so far only limited success in determining phases using this method[9]. What has been successful recently is the use of the information obtained in the first two steps in combination with the isomorphous replacement method[10]. I will concentrate on the first two steps. Procedures were introduced over ten years ago to generate data from which one is able to solve the rotation and translation problems. However during this time no general method was derived to analyze this data systematically, and consequently, as I will show, misinterpretations of data have occurred. I will show that a general method to systematically analyze this data can be formulated, and that to do so one needs to introduce only truly elementary group theoretical arguments. It is surprising that these elementary group theoretical arguments were not applied to this problem until very recently.

The solutions of the rotation and translation problems are based on the analysis of a function which can be calculated from the magnitude of the structure factors $|F(k)|$. This is the Patterson function introduced in 1934.

$$P(r) = \frac{1}{V} \sum_{k} |F(k)|^2 e^{2\pi i k \cdot r}.$$

This function is the self-convolution (self-correlation) of the electron density of the crystal.

$$P(r) = \int \rho(r')\rho(r'+r)dr'.$$

Because in the biological crystal the macromolecules preserve their identity, the electron density of the crystal can be written as a sum of the electron densities of the molecules, and the Patterson function can be written as

$$P(r) = \sum_{j,k,t} P_{jkt}(r)$$

where j and k index the molecules in the unit cell of the crystal, t the translations of the crystal, and

$$P_{jkt}(r) = \int \rho_{j0}(r')\rho_{kt}(r'+r)dr'$$

the convolution of the electron density of the jth molecule in the unit cell with the kth molecule in the tth unit cell. The convolution function $P_{jkt}(r)$, in general, is called a cross-Patterson function, and in the special case where $j = k$ and $t = 0$, $P_{jj0}(r)$ is called a self-Patterson.

The two problems:

1. The rotation problem

Here one wants to determine the point group and orientation of the molecules in a biological crystal. The crystals which we consider are assumed to be made up of only one kind of biological macromolecule.

To determine this, one looks for relations between the point group and orientation of the molecules in the crystal and properties of the Patterson function. Consider the electron density $\rho_{j0}(r)$ of the jth molecule in the unit cell. This electron density is localized because of the finite dimension of the molecule, about the center of mass r_j of the jth molecule. The symmetry point group of the molecule is the set of all proper rotation matrices P such that

$$\rho_{j0}(r_j+Pu) = \rho_{j0}(r_j+u).$$

(Because the biological macromolecules are made up of "left-handed" amino acids, the symmetry point group of the molecule consists only of proper rotation matrices.) We choose a coordinate system in the crystal; then the group of matrices P, which is the symmetry point group of the jth molecule, is defined with respect to this coordinate system. The orientation of the molecule is the orientation of the rotation axes of the rotations which are represented by these matrices. Another molecule in the crystal has as its symmetry point group a group of matrices P' also defined with respect to this coordinate system. The two groups of matrices are in general different but equivalent, i.e. they both belong to the same class of point groups and are denoted by the same symbol in, e.g., international notation. When referring to the point group of a molecule we will sometimes mean a specific group of matrices defined in the chosen coordinate system, and at other times, the class of point groups to which this group of matrices belong. The meaning should be clear from the context.

It follows that if P is the symmetry point group of the jth molecule, then the self-convolution of the electron density of the jth molecule, i.e. the jth self-Patterson function

$$P_{jj0}(r) = \int \rho_{j0}(r')\rho_{j0}(r'+r)dr$$

is invariant under all rotations of P, i.e., $P_{jj0}(r) = P_{jj0}(Pr)$. $P_{jj0}(r)$ is also invariant under inversion, and consequently $P \times \bar{1}$ is an invariance point group of $P_{jj0}(r)$. We shall assume that $P \times \bar{1}$ is the symmetry point group of $P_{jj}(r)$.

As this self-Patterson is localized in a volume about the

origin of the Patterson function, Rossmann and Blow[3], in order to determine from the Patterson function the point group and orientation of the molecules, introduced in 1962 the rotation function R(A):

$$R(A) = \int_U P(r)P(Ar)\,dr$$

where A is a proper rotation, and the integration is over a volume about the origin of the Patterson function. This is an overlap integral of a volume about the origin of the Patterson function with a rotation image of the same volume. Relative maxima of this rotation function, as a function of A, are called peaks of the rotation function. Obviously if $P_{jj0}(r)$ is invariant under a rotation P then there will be a peak in the rotation function at A = P. Consequently by determining the peaks of this rotation function, one can obtain information on the point group of the molecules in the crystal. While this rotation function has been successfully used, a general method has been available for less than a year to determine systematically the point group symmetry of the molecules from information provided by the rotation function[4]. To derive this method, one needs to apply only very elementary group theoretical arguments.

Consider a biological crystal consisting of identical molecules generated by a space group G from a single molecule at position r_1. Let T denote the translational subgroup of G, and $(R_j|\tau_j)$ j=1,...,n, the coset representatives of T in G. We will consider the case where r_1 is a general position, i.e. the n vectors $r_j = (R_j|\tau_j)r_1$ are distinct. We then have n molecules in the unit cell, and the electron density of the molecule at r_j is related to the electron density of the molecule at r_1 by

$$\rho_{j0}(r_j+u) = \rho_{10}(r_1+R_ju)$$

where $r_j = (R_j|\tau_j)r_1$. That is, we have n identical molecules in different orientations in the unit cell, and their mutual orientation is determined by the rotations of the space group of the crystal. It also follows that the self-Pattersons $P_{jj0}(r)$, $j=1,\ldots,n$, which are all localized about the origin of the Patterson, are identical, in different orientation, and their mutual orientations are also determined by the rotations of the space group of the crystal.

All peaks of the rotation function correspond to rotations which

1) Leave a self-Patterson $P_{jj0}(r)$, for some j, invariant; or

2) Rotate a self-Patterson $P_{jj0}(r)$ into the orientation of a self-Patterson $P_{kk0}(r)$, where $j \neq k$.

Let P denote the symmetry point group of the molecule at r_1:

1) The group of rotations $\{R(jj)\} \equiv \{R_j P R_j^{-1}\}$ is the symmetry point group of the self-Patterson $P_{jj0}(r)$; and

2) The set of rotations $\{R(jk)\} = \{R_k P R_j^{-1}\}$ is the set of all rotations which rotates $P_{jj0}(r)$ into the orientation of $P_{kk0}(r)$.

Therefore, all peaks of the rotation function R(A) correspond to all the rotations contained in the set of rotations[4]:

$$[\{R(jk)\} \mid j,k=1,\ldots,n]. \tag{1}$$

One now has a systematic method to analyze the data obtained from the rotation function.

1) From the rotation function calculate all rotations which correspond to all peaks.

2) Determine the point groups such that the set of distinct rotations in (1) is identical with the set of rotations corresponding to peaks of the rotation function. If P is such a point group, the conjugate point groups $R_j P R_j^{-1}$, $j=1,\ldots,n$, where R_j is a rotation of the space group of the crystal, are also such point groups. If there is only one set of such conjugate subgroups $R_j P R_j^{-1}$, $j=1,\ldots,n$, we will say that P is the symmetry point group

of the molecules. The orientation of the rotation axes of these conjugate subgroups, with respect to the translational vectors of the crystal, determine the orientation of the molecules in the crystal. If there is more than one set of such conjugate subgroups, the solution of the rotation function problem is not uniquely determined by this method. However, in practice it is impracticable to search for all peaks of the rotation function, i.e. to calculate R(A) for all possible rotations A. One must then use an alternate method:

1) From the rotation function calculate the rotations corresponding to some of the peaks.

2) Determine the point groups such that the set of distinct rotations of (1) includes all those rotations found in step 1 from the rotation function.

3) Determine if there are peaks of the rotation function corresponding to the additional rotations of (1).

Example. Satellite Tobacco Necrosis Virus (STNV). A few years ago there was a debate as to whether the protein coat of this "spherical" virus was of cubic O(432) or icosahedral J(532) point group symmetry. A rotation function study was then made of a crystal containing two STNV molecules in the unit cell[5]. The crystal was monoclinic of space group symmetry C_2^3(C2), and the orientations of the two STNV molecules were related by a rotation of 180°. It was felt that a rotation function study of this crystal would easily determine the point group since there are 4-fold rotations contained in the cubic point group and none in the icosahedral, and 5-fold rotations in the icosahedral and none in the cubic.

A set of strong peaks were found with corresponding rotations which were exactly those proper rotations of a cube, and this was interpreted as meaning that the STNV molecules were of cubic symmetry. There were peaks corresponding to 5-fold rotations, which are characteristic of icosahedral symmetry, but these peaks were much lower than the cubic peaks.

This interpretation was immediately challenged and it was shown that all the peaks of the rotation function corresponding to 5-fold rotations could be interpreted as two molecules of icosahedral symmetry in two different orientations related by the rotation of 180° of the space group of the crystal[6]. There is general agreement that this is the correct interpretation and that the STNV molecules do have icosahedral symmetry. However in this reinterpretation the stronger cubic peaks were explained away in an argument that approximated this monoclinic crystal as being cubic!

All these peaks can be explained using the above formalism and taking the point group of the STNV molecules as being icosahedral[4]: If one calculates the set of rotations in (1) taking j,k=1,2, R_2 the rotation of 180° of the space group, and P as the icosahedral symmetry point group of one of the molecules, one finds 240 rotations. These include the icosahedral rotations of both molecules, and a set of cubic rotations, exactly that set of cubic rotations determined from the rotation function. Each of these cubic rotations either leaves both molecules invariant or interexchanges the two orientations, explaining the high corresponding peaks, since all other rotations either leave only one molecule invariant, or rotate one molecule into the orientation of the other. One finds also that this set (1) of rotations contains additional rotations which have not yet been determined. However, even without determining peaks of the rotation function corresponding to these rotations, it does seem that the symmetry point group of the STNV molecules is icosahedral.

2. The translation function

Information on the point group and orientation of the molecules in a biological crystal is found, using the rotation function, by considering that volume of the Patterson about the origin of the Patterson function. Information on the translation vectors between molecules is found using a similar method, but considering other parts of the Patterson. One uses a so-called translation function $T(x,A)$:[7]

$$T(x,A) = \int_U P(x+r)P(x+Ar)dr.$$

The translation function like the rotation function is an overlap integral of a volume U of the Patterson function with a rotated image of the same volume, but unlike the rotation function, the centre of the volume is now a variable, and not restricted to be at the origin of the Patterson function.

This translation function is non-zero when

1) The volume U intersects with a cross-Patterson $P_{jkt}(r)$ and the intersection is left invariant by the rotation A about x; or

2) The volume U intersects two cross-Patterson $P_{jkt}(r)$ and $P_{j'k't'}(r)$ and one is transformed into the other by the rotation A about x. The relative maxima of a translation function $T(x,A)$ as a function of x, for constant A, are called the peaks of the translation function, and the positions of these peaks are related to the translation vectors between molecules in the crystal. In Figure 1 we have drawn a model crystal, in Figure 2 the corresponding Patterson function, and in Table 2, the positions of the peaks of the corresponding translation function $T(x,A)$. The group theoretical arguments which enter into determining the relationship between the peaks of the translation function and the translation vectors between the molecules are similar to those used in predicting the peaks of the rotation function:

Necessary and sufficient conditions that the translation function $T(x,A)$ has non-zero values associated with the transformation of $P_{jkt}(r)$ into $P_{j'k't'}(r)$ are:

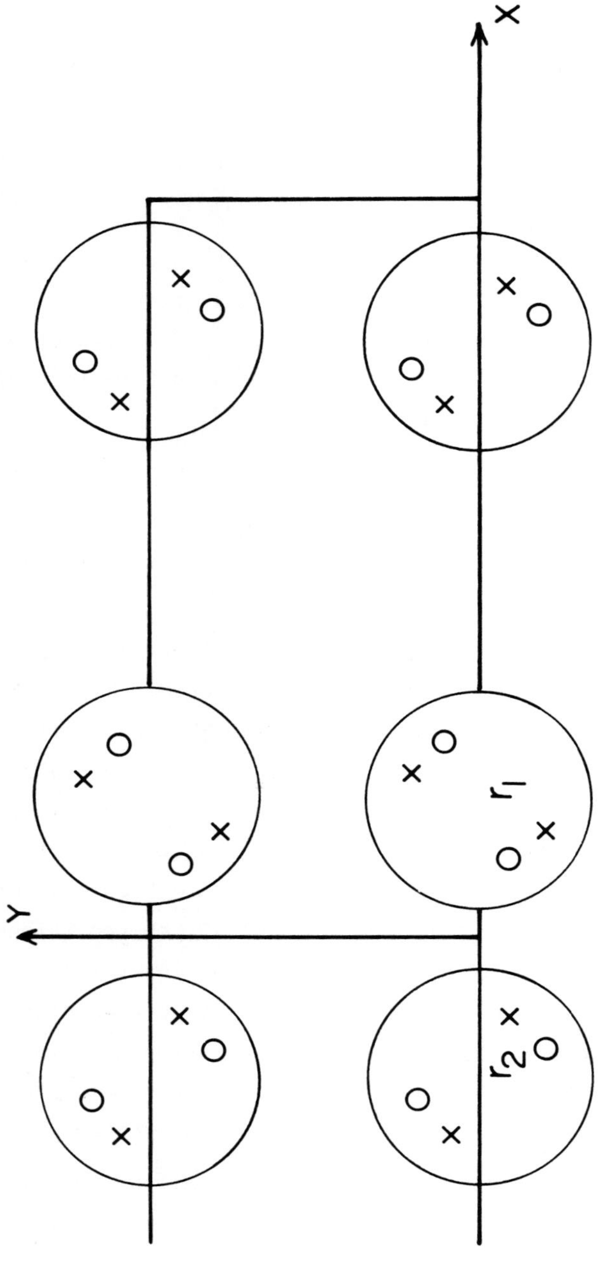
X
Y
r_1
r_2

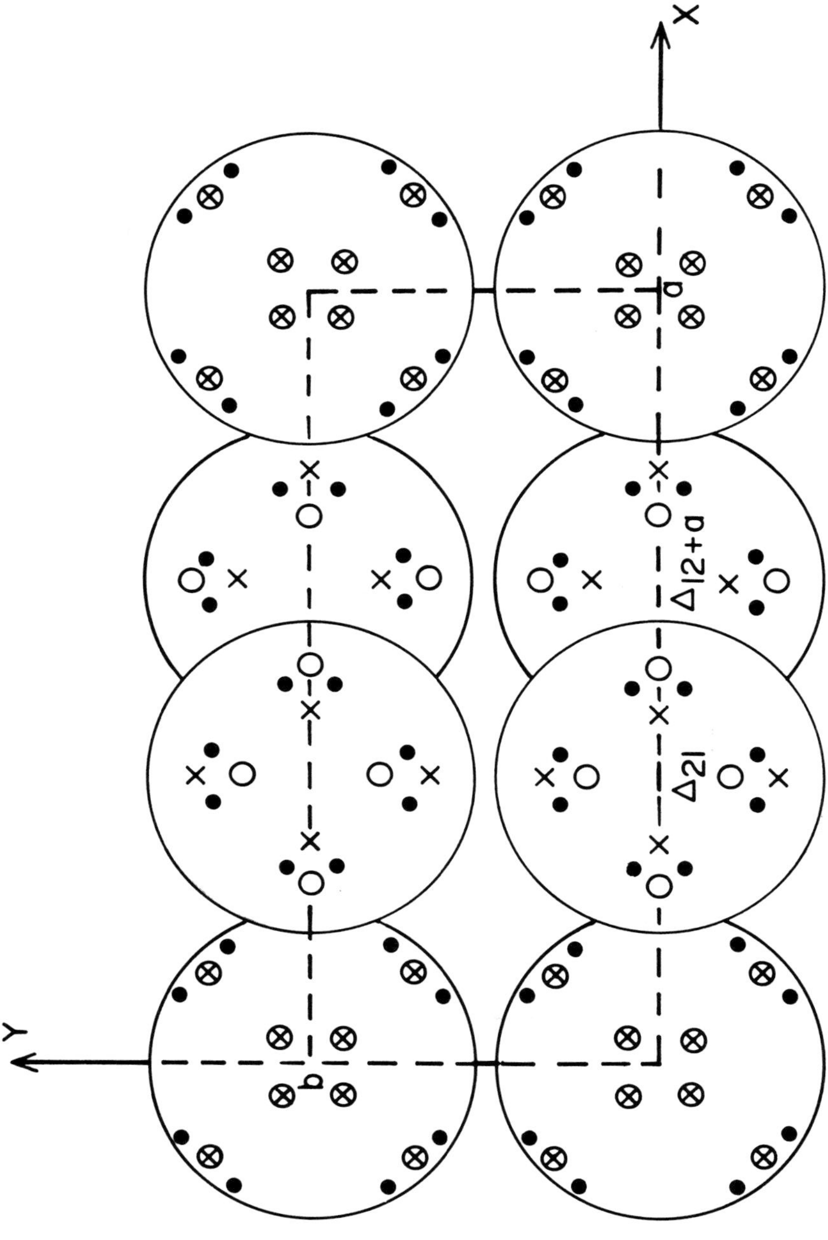
X
Y
a
b
Δ_{12+a}
Δ_{21}

TABLE 2

A	x
2_z	Δ_{21}, $\Delta_{21} + t_x$, $\frac{1}{2}t_y$, $\Delta_{21} + \frac{1}{2}t_y$, $\Delta_{12} + t_x + \frac{1}{2}t_y$
2_{xy}	Δ_{21}, $\Delta_{12} + t_x$
$2_{\bar{x}y}$	Δ_{21}, $\Delta_{12} + t_x$
2_x	$\frac{1}{2}t_y$
4_z	$\frac{b}{2}(1,1,0)$, $\frac{b}{2}(-1,1,0) + t_x$
4_z^3	$\frac{b}{2}(1,1,0)$, $\frac{b}{2}(-1,1,0) + t_x$
m_x	Δ_{21}, $\Delta_{12} + t_x$
m_y	Δ_{21}, $\Delta_{12} + t_x$, $\frac{1}{2}t_y$, $\Delta_{21} + \frac{1}{2}t_y$, $\Delta_{12} + t_x + \frac{1}{2}t_y$
$\bar{1}$	$\frac{1}{2}t_x$, $\frac{1}{2}(t_x+t_y)$
$\bar{4}_z$	Δ_{21}, $\Delta_{12} + t_x$, $\frac{b}{2}(1,1,0)$, $\frac{b}{2}(-1,1,0) + t_x$
	$\Delta_{21} + \frac{b}{2}(1,1,0)$, $\Delta_{21} + \frac{b}{2}(-1,1,0)$
	$\Delta_{12} + t_x + \frac{b}{2}(1,1,0)$, $\Delta_{12} + t_x + \frac{b}{2}(-1,1,0)$
$\bar{4}_z^3$	Δ_{21}, $\Delta_{12} + t_x$, $\frac{b}{2}(1,1,0)$, $\frac{b}{2}(11,1,0) + t_x$
	$\Delta_{21} + \frac{b}{2}(1,1,0)$, $\Delta_{21} + \frac{b}{2}(-1,1,0)$
	$\Delta_{12} + t_x + \frac{b}{2}(1,1,0)$, $\Delta_{12} + t_x + \frac{b}{2}(-1,1,0)$

1) the rotation A is such that

$$P_{jkt}(r_k - r_j + t + A^{-1}u) = P_{j'k't'}(r_{k'} - r_{j'} + t' + u);$$

2) $A(r_k - r_j + t - x) = r_{k'} - r_{j'} + t' - x;$

3) the vector $\frac{1}{2}y$, where $-y = r_k - r_j + t - x$ is within the volume U.

The first two conditions demand that $P_{jkt}(r)$ is transformed into $P_{j'k't'}$ by a rotation A about the point x, and the third condition, that the volume U of the Patterson function centered at x intersects both $P_{jkt}(r)$ and $P_{j'k't'}(r)$. Using group theoretical arguments like those used for the rotation function one can show that all rotations which satisfy the first condition are those denoted by $\{A(jkj'k')\}$

$$\{A(jk,j'k')\} = [\{R(jj')\} \cap \{R(kk')\}] + \bar{1}[\{R(jk')\} \cap \{R(kj')\}] \ .$$

If this set of rotations is empty, then $P_{jkt}(r)$ and $P_{j'k't'}(r)$ are not congruent. For rotations A contained in $\{A(jkj'k')\}$, the positions x are calculated from the second condition.

The concept of the translation function T(x,A) was introduced by Rossmann, Blow, Harding and Coller in 1964[8]. However they considered only the case where A = m, and only in this case has the translation function been put to actual use with some success. Why these people use only this special case is unclear; it is probably because no general method was developed to predict the peaks of the translation function. One now has such a general method, which needs only very elementary group theoretical arguments to derive.

REFERENCES

1. Progress in Physics, Biophysics, a reprint series. W.A. Benjamin, New York, 1969 and the reviews therein by L. BRAGG and A.C.T. NORTH.

2. The molecular replacement method, M.G. Rossmann, ed., New York, Gordon & Breach, 1972.

3. M.G. ROSSMANN and D.M. BLOW, Acta Cryst. 15, 24-31 (1962).

4. D.B. LITVIN, Acta Cryst. A31, 407 (1975).

5. K. AKERVALL, B. STRANDBERG, M.G. ROSSMANN, V. BENGTSSON, K. FRIDBORG, H. JOHANNISON, K. KANNAN, S. LOVGREN, G. PETEF, B. OBERG, D. EAKER, S. HJERTEN, L. RYDEN and I. MORING, Cold Spring Harbour Symp. Quant. Biol. 36, 469, 487 (1971).

6. A. KLUG, Cold Spring Harbour Symp. Quant. Biol. 36, 483 (1971).

7. D.B. LITVIN, Acta Cryst. A33 (1977).

8. M.G. ROSSMANN, D.M. BLOW, M.M. HARDING and E. COLLER, Acta Cryst. 17, 338 (1964).

9. A. JACK, Acta Cryst. A29, 545 (1973).

10. J.N. CHAMPNESS, A.C. BLOOMER, G. BRICOGNE, P.J.G. BUTLER and A. KLUG, Nature 259, 20 (1976).

INVARIANTS POLYNOMIAUX DES GROUPES DE SYMETRIE MOLECULAIRE ET CRISTALLOGRAPHIQUE

Louis Michel

Cette conférence est une revue de quelques aspects de ce problème. L'utilité pour la physique des fonctions invariantes par le groupe de symétrie du problème étudié, est évidente. Beaucoup de travaux récents donnent des tables de polynomes invariants (généralement calculés par ordinateurs) des groupes *ponctuels* (groupes de symétrie des molécules, et quotients par les translations des groupes cristallographiques)[1]. Souvent ces travaux s'ignorent les uns les autres et plus généralement ils ignorent les résultats directement utilisables obtenus par les mathématiciens. Les tables de polynomes invariants peuvent être fort utiles puisque leur forme explicite dépend de la base choisie, le choix d'une base étant souvent déterminé par des considérations physiques. Par contre il est aussi fort utile de connaître la *structure* de l'algèbre des polynomes invariants. C'est à ce problème que va être consacrée la majeure partie de cette conférence.

1. CONCEPTS DE BASE

Définissons d'abord les concepts de base afin de pouvoir

[1] Les articles à ce sujet sont si nombreux que j'ai renoncé à préparer une bibliographie.

ensuite préciser notre sujet. Soit G un groupe fini, de $|G|$ éléments, agissant linéairement et effectivement[2] sur un espace vectoriel E sur le corps K, et de dimension finie m. Ces données définissent aussi une action de G sur la n^e puissance tensorielle $\overset{n}{\otimes} E = E^{(n)}$ de E, ainsi que sur les puissances complètement symétriques $E^{\{n\}}$ et complètement antisymétriques $E^{[n]}$, de dimensions respectives

$$s_n = \binom{n+m-1}{n} \quad \text{et} \quad a_n = \binom{m}{n}, \tag{1}$$

ainsi que sur leur somme directe:

$$S = \bigoplus_{n=0}^{\infty} E^{\{n\}}, \qquad A = \bigoplus_{n=0}^{m} E^{[n]}, \tag{2}$$

les espaces vectoriels des algèbres tensorielles symétriques et extérieures sur E ($E = E^{(1)} = E^{\{1\}} = E^{[1]}$ et $E^{(0)} = E^{\{0\}} = E^{[0]} = K$, le corps de définition de E). Dans cette conférence nous nous limitons à la caractéristique zéro pour K, les applications physiques étant pour les complexes et les réels.

Soit R l'espace dual de S. C'est l'espace des polynomes sur E, le dual R_n de $E^{\{n\}}$ étant l'espace des polynomes homogènes de degré n. L'ensemble des polynomes sur E forme, comme S, un anneau et aussi une K-algèbre.

Soit $g \to D(g)$ la représentation linéaire de G sur E dont nous notons $x_1, x_2, \ldots$ les éléments; $D^{\{n\}}(g)$ est la représentation correspondante sur $E^{\{n\}}$ dont nous notons $x^{\{n\}}$ les éléments. Nous dénotons $g \to D_\alpha(g)$ les représentations irréductibles de G à une équivalence près (D_{α_1} et D_{α_2} sont inéquivalentes si $\alpha_1 \neq \alpha_2$) et $\chi_\alpha(g) = \mathrm{tr}\, D_\alpha(g)$ est le caractère de cette représentation. Par la propriété d'orthogonalité des caractères, l'expression

$$c_n(\chi_\alpha) = \frac{1}{|G|} \sum_{g \in G} \chi_\alpha(g^{-1})\, \mathrm{tr}\, D^{\{n\}}(g) \tag{3}$$

[2] Effectivement signifie que seul l'identité du groupe laisse fixe tous les points de E.

donne la multiplicité de la représentation D_α dans la décomposition de la représentation $D^{\{n\}}$ en représentation irréductible. D'autre part, en diagonalisant D(g) on voit que les valeurs propres de $D^{\{n\}}(g)$ sont les produits de n valeurs propres de D(g), donc tr $D^{\{n\}}(g)$ est le coefficient de t^n dans le développement de $[\det(I-tD(g)]^{-1}$. La relation (3) peut donc s'écrire pour tous les n à la fois [M3].

$$\sum_{n=0}^{\infty} c_n(\chi_\alpha)t^n = F_{\chi_\alpha}(t) = \frac{1}{|G|} \sum_{g\in G} \frac{\chi_\alpha(g-1)}{\det[I-tD(g)]}. \tag{4}$$

Nous sommes plus particulièrement intéressés par l'ensemble R^G (respectivement S^G) des polynomes invariants (resp. des vecteurs invariants de S) c'est-à-dire:

$$p(x_1,\ldots,x_m) = p(D(g)x_1,\ldots,D(g)x_m), \tag{5}$$

qui correspondent à la représentation triviale $\chi_0(g) = 1$ et nous écrivons simplement F(t) pour $F_{\chi_0}(t)$. La somme et le produit de polynomes invariants étant encore des polynomes invariants, R^G est un anneau; c'est aussi une K-algèbre. Par le célèbre théorème de Hilbert [H1] nous savons qu'on peut trouver un nombre fini de générateurs pour cette algèbre, un tel système de générateurs étant en général appelé une *base d'intégrité*[3]. Nous donnerons l'énoncé de théorèmes importants ou utiles pour la construction de telles bases. Nous aurons pour cela besoin de la notion d'*invariant relatif*. Pour toute représentation irréductible de G de dimension 1, et de caractère $\chi_A(g)$ (A pour *abelien*), nous désignons par $R^G_{\chi_A}$ l'ensemble des polynomes invariants relatifs de poids χ_A définis par la relation

$$\forall\ g \in G,\quad p(D(g)x_1,\ldots,D(g)x_m) = \chi_A(g)p(x_1,\ldots,x_m). \tag{6}$$

---

[3] Ce résultat n'est pas vrai pour les groupes de Lie de dimension fini comme l'a montré le contre exemple de Nagata [N1].

Le produit par un invariant ne changeant pas la nature d'un invariant relatif, $R^G_{\chi_A}$ est un R^G-module[4].

2. GROUPES G ENGENDRES PAR (PSEUDO-) REFLEXIONS

Une pseudo-réflexion P dans l'espace E_m de dimension m est un opérateur diagonalisable qui a m-1 fois la valeur propre 1 et qui est d'ordre fini, c'est-à-dire il existe un plus petit entier $c > 1$ tel que $P^c = I$. Donc une pseudo-réflexion satisfait:

$$P \text{ diagonalisable}, \quad (P-I)^{m-1}(P-\rho I) = 0, \rho^c = 1$$

$$\text{pour un entier minimum } c > 1. \tag{7}$$

Pour c = 2, P est une réflexion.

On prouve aisément que toute base d'intégrité de R^G sur E_m doit contenir au moins m polynomes. Chevalley [C1] prouva en 1955 que pour les groupes finis engendrés par des réflexions, toute base d'intégrité de R^G (c'est-à-dire tout ensemble minimum de polynomes invariants) a exactement m éléments. Coxeter a déterminé pour toutes les dimensions finies m, tous les groupes engendrés par réflexions: ils sont symbolisés par un diagramme de Dynkin (généralisé) de m points[5]. Ceux qui laissent invariant un réseau de dimension m (= un cristal à m dimension; il existe donc une base de E_m pour laquelle les éléments de matrice de la représentation de G sont entiers) sont les groupes de Weyl d'un groupe de Lie semi-simple de rang m; l'espace E_m est une sous-algèbre de Cartan. Aux groupes de Lie simples correspondent les diagrammes de Dynkin connexes. Le théorème de Chevalley s'applique donc

---

[4] A étant un anneau d'éléments α_1, α_2 on définit sur le groupe abélien E une structure de A-module si pour tout $\alpha \in A$ et tout $x \in E$, αx est défini comme élément de E, cette correspondance satisfaisant les propriétés $\alpha(x_1+x_2) = \alpha x_1 + \alpha x_2$, $(\alpha_1+\alpha_2)x = \alpha_1 x + \alpha_2 x$, $\alpha_1(\alpha_2 x) = (\alpha_1\alpha_2)x$, $1x = x$.

[5] Ils ont été définis dans le cours du professeur Zassenhaus à l'école d'été qui a précédé cette conférence. Voir aussi [B2].

aussi aux invariants des groupes de Lie semi-simples. Shephard et Todd [S2] ont montré que la réciproque du théorème de Chevalley était vraie à condition de l'étendre aux pseudo-réflexions:

Théorème 1. [S2]. Un sous-groupe fini G de GL(m,K) a une base d'intégrité de m polynomes invariants si et seulement si il est engendré par pseudo-réflexions. On peut alors trouver pour la base d'intégrité m polynomes homogènes θ_i (i = 1 à m) de degré d_i avec $d_1 \leq d_2 \leq \ldots \leq d_m$. Le polynome invariant le plus général de R^G est de la forme $q(\theta_1, \theta_2, \ldots, \theta_m)$ où q est un polynome de m variables. Nous pouvons calculer les degrés d_i en remarquant que

$$\sum_{n=0}^{\infty} t^n \dim R_n^G = \prod_{i=1}^{m} \frac{1}{(1-t^{d_i})} \tag{8}$$

ce qui est encore la fonction de Molien F(t):

$$F(t) = \prod_{i=1}^{m} \frac{1}{(1-t^{d_i})} = \frac{1}{|G|} \sum_{g=G} \frac{1}{\det[I-tD(g)]}. \tag{9}$$

Soit G_k le sous-ensemble d'éléments de G dont l'ensemble des points fixes dans E_m est une variété linéaire de dimension m-k; par exemple G_0 ne contient que l'identité du groupe, G_1 contient les r (pseudo-) réflexions, etc. En définissant:

$$P(t) = \prod_{k=1}^{m} P_k(t) \quad \text{et} \quad P_k(t) = 1 + t + t^2 + \ldots + t^{d_k - 1}, \tag{10}$$

et en multipliant les deux membres de (9) par $|G|(1-t)^m$, on obtient

$$\frac{|G|}{P(t)} = 1 + \sum_{k=1}^{m} (1-t)^k Q_k(t), \tag{11}$$

où

$$Q_k(t) = \sum_{g \in G_k} \frac{(1-t)^{m-k}}{\det[I-tD(g)]}; \tag{12}$$

entre autre

$$Q_1(t) = \sum_{g\in G_1} \frac{1}{1-\rho(g)t}, \qquad (12')$$

où $\rho(g)$ est la valeur propre différente de 1 de la pseudo-réflexion $g \in G_1$.

En faisant $t = 1$ dans (11) on trouve

$$\prod_{i=1}^{m} d_i = |G|. \qquad (13)$$

De même en multipliant (11) par P(t) pris en dérivant par rapport à t et faisant $t = 1$ on obtient après division par $|G|$

$$\sum_i \frac{d_i - 1}{2} = \sum_{g\in G_1} \frac{1}{1-\rho(g)},$$

soit encore

$$\sum_i d_i = m + |G_1| = m + r, \qquad (14)$$

où $|G_1| = r$ est le nombre de (pseudo-) réflexions.

Plus généralement Steinberg [S6] a montré l'égalité[6]:

$$\prod_{k=1}^{m} (1+(d_k-1)t) = \sum_{k=0}^{m} |G_k| t^k. \qquad (15)$$

(En faisant $t = 1$ on obtient (13), tandis que (14) est donnée par le coefficient de t.) Remarquons que ces résultats étaient connus depuis plusieurs siècles dans le cas des fonctions symétriques de m variables. En effet le groupe symétrique S_m des permu-

[6] Quand on connait le degré des invariants de la base (voir aussi plus loin (16), (17), (18)) il est assez facile de les construire en moyennant sur les orbites exceptionnelles (c'est-à-dire les plus petites), qui correspondent aux plus grands stabilisateurs. Pour une étude systématique voir [F1].

tations de m objets est engendré par les transpositions qui sont des réflexions sur E_m. Les d_i prennent toutes les valeurs de 1 à m et une base d'intégrité est donnée par

$$\theta_k = \sum_{i=1}^{m} x_i^k .$$

Pour les groupes engendrés par réflexions Coxeter [C2] avait montré que les opérateurs

$$W = \prod_{g \in G_1} D(g)$$

produit de toutes les réflexions et qui dépendent donc de l'ordre dans lequel est effectué ce produit, sont tous conjugués les uns des autres, leurs valeurs propres étant

$$\text{Spectre } W = \{\exp 2\pi i(d_k-1)/h, k=1 \text{ à } m\} \quad \text{et} \quad h = \frac{2r}{m}. \tag{16}$$

Puisque W est un opérateur orthogonal réel, ses valeurs propres sont complexes conjuguées:

$$d_k + d_{m-k} = 2(1+ \frac{r}{m}) \tag{17}$$

qui est plus précis que (14). Notons que, pour m impair,

$$d_{\frac{m+1}{2}} = 1 + \frac{r}{m}. \tag{18}$$

Finalement les nombres de Betti d'un groupe de Lie simple sont égaux à $2d_k-1$, les d_k étant les degrés des polynomes invariants de la base d'intégrité de son groupe de Weyl [S3].

Dans son article de 1954 Chevalley montre un second théorème. La représentation linéaire de G sur l'espace quotient $R/I(R_+^G)$ est la représentation régulière, $I(R_+^G)$ étant l'idéal engendré par les polynomes non constants de R^G.

Revenons au cas général d'un groupe engendré par pseudo-réflexions, l'équation des hyperplans invariants étant

$$\ell_\alpha(x) = \sum_{i=1}^{m} n_i^{(\alpha)} x_i = n^{(\alpha)} \cdot x = 0, \tag{19}$$

(le point $\cdot$ désignant un produit scalaire orthogonal); les différentes pseudo-réflexions d'hyperplan $\ell_\alpha = 0$ engendrent un groupe cyclique d'ordre $c_\alpha > 1$ (pour les réflexions $c_\alpha = 2$). Soit $p_i(x_1,\ldots,x_n)$ un polynome invariant. Si on choisit la $1^{\text{ère}}$ coordonnée dans la direction $n^{(\alpha)}$, l'invariance par rapport à la pseudo-réflexion qui laisse l'hyperplan $\ell_\alpha = 0$ invariant, exige que p_i ne contiennent que des puissances de x_1 multiples de c_α. Les $\partial p_i/\partial x_j$, $j = 1$ à m, se transforment suivant la représentation contragrédiente de E_m, et ils s'annulent tous pour $x_1 = 0$. Si on forme le jacobien de m polynomes invariants linéairement indépendants, ce jacobien doit s'annuler sur chaque hyperplan invariant. Or en comptant les degrés on vérifie que: [B2]

Théorème 2. Si θ_i, $i = 1$ à m, sont les m polynomes d'une base d'intégrité d'un groupe fini G engendré par pseudo-réflexions, on a

$$\det\left(\frac{\partial\theta_i}{\partial x_j}\right) = K \prod_\alpha \ell_\alpha^{c_\alpha - 1}(x), \tag{20}$$

où K est une constante.

Remarquons que $\det(\partial\theta_i/\partial x_j)$ n'est pas un invariant de G. Si G est engendré par réflexions, c'est un invariant relatif de poids det (g). Mais c'est un invariant du sous-groupe invariant H, d'index 2, des éléments de déterminant 1.

3. SOUS-GROUPES INVARIANTS H DES GROUPES G ENGENDRES PAR PSEUDO-REFLEXIONS DONT LE QUOTIENT G/H EST ABELIEN

Soit G' le groupe engendré par les commutateurs d'un groupe G engendré par pseudo-réflexions: G' est sous-groupe invariant de

G, ainsi que tout sous-groupe $H < G$ contenant G'; de plus le groupe quotient G/H est abélien. Un preprint de R.P. Stanley [S5] permet de construire explicitement l'algèbre R^H.

C'est un résultat bien connu de la théorie des groupes que toute représentation commutative (= abélienne) $g \to \chi(g)$ irréductible de G est unidimensionnelle et son noyau contient G', c'est-à-dire:

$$G' < \ker \chi \Leftrightarrow \chi(G') = 1. \tag{21}$$

L'ensemble de ces représentations abéliennes irréductibles forment un groupe abélien A_G, dual de G/G' et qui lui est isomorphe puisque G/G' est fini. Les invariants relatifs de A de poids $\chi_A \in A_G$ tel que $\chi_A(H) = 1$, sont évidemment des invariants de H, le travail récent de Stanley permet de préciser:

Lemme. [S5]. Si $G' < H < G$, engendré par pseudo-réflexions:

$$R^H = \sum_{\substack{\chi_A \in A_G \\ \chi_A(H)=1}} R^G_{\chi_A}, \tag{22}$$

$\sum$ indiquant une somme d'espaces vectoriels; mais Stanley montre ensuite qu'il s'agit même d'une somme de R^H-modules, chaque $R^G_{\chi_A}$ étant un R^H-module de dimension 1 engendré par le polynome f_{χ_A}:

$$f_{\chi_A}(x) = \prod_{\alpha=1}^{r} \ell_\alpha^{s_\alpha}(x), \tag{23}$$

où s_α est le plus petit entier $0 \le s_\alpha < c_\alpha$ tel que $\chi(P_\alpha) = (\det P_\alpha)^{s_\alpha}$ pour la pseudo-réflexion P_α.

Par exemple, si G est engendré par réflexions, $f_{\chi_A}(x)$ est le produit des équations des hyperplans dont les réflexions sont représentées par -1 dans la représentation abélienne χ_A de G; de telles réflexions n'appartiennent pas à H puisque $\chi_A(H) = 1$. Et si H est le sous-groupe d'index 2, noyau de la représentation abélienne, $g \to \det(g)$, f_{det} est le produit de tous les ℓ_α; c'est le jacobien de la formule (20) avec $c_\alpha = 2$.

Pour l'espace E à trois dimensions, les sous-groupes discrets de O(3) qui sont engendrés par réflexions[7] sont

$$C_{nv},\ D_{nh},\ T_d,\ O_h,\ Y_i \tag{24}$$

(les trois derniers étant les groupes de symétrie respectivement du tétraèdre, de l'octaèdre ou du cube, de l'icosaèdre ou du dodécahèdre). Tous les autres groupes discrets de O(3) sont des sous-groupes invariants à quotient abélien d'un des groupes de la liste (24). Excepté pour les sous-groupes S_{2n}, les R^H sont des R^G modules de dimension 2, c'est-à-dire les polynomes invariants de H, sous-groupe discret de O(3) non engendré par réflexion, sont de la forme

$$q_0(\theta_1,\theta_2,\theta_3) + \varphi_1 q_1(\theta_1,\theta_2,\theta_3), \tag{25}$$

où $\theta_1(x)$, $\theta_2(x)$, $\theta_3(x)$ sont 3 polynomes formant une base d'intégrité pour G (de la liste 24), q_0 et q_1 sont des polynomes arbitraires de 3 variables et φ_1 est le polynome calculé en (23). Pour les groupes $C_{n,i}$ il faut remplacer (25) par un R^G module de dimension 4:

$$q_0(\theta_1,\theta_2,\theta_3) + \varphi_1(x)q_1(\theta_1,\theta_2,\theta_3)$$
$$+ \varphi_2(x)q_2(\theta_1,\theta_2,\theta_3) + \varphi_3(x)q_3(\theta_1,\theta_2,\theta_3). \tag{25'}$$

Nous donnons dans la table 1 des bases d'intégrité θ_1, θ_2, θ_3 et éventuellement φ_1, φ_2, φ_3 pour tous les sous-groupes discrets de O(3) ainsi que la fonction génératrice donnant la dimension de R_n^G, i.e. le nombre de polynomes homogènes invariants de degré n

[7] Nous suivons ici les notations et les définitions de Landau et Lifshitz [L1].

linéairement indépendants. Bethe dans son travail de pionnier [B1] en 1929 avait donné les invariants des groupes de symétrie cubique; ceux de l'icosaèdre étaient connus de F. Klein et W. Burnside [B3]. La plus ancienne table publiée équivalente à la table 1, est à ma connaissance celle de Meyer [M2] de 1954, Killingbeck (sans citer Meyer) a publié l'équivalent de la table 2, mais l'équation (1) de la table 1 ne semblait pas être connue dans la littérature physique.

4. LES PROJETS ANNONCES DES PHYSICIENS

Certainement de nombreuses autres tables de polynomes invariants ont été annoncées et paraîtront dans un avenir proche. J'espère que la vulgarisation que je viens de faire de résultats mathématiques établis ces vingt-cinq dernières années, aidera ces projets (cf. [K2], [M1]). Il y a beaucoup à faire si l'on tient compte de toutes les représentations irréductibles, non seulement des groupes ponctuels mais aussi des 230 groupes cristallographiques. Jean Mozrzymas, qui a établi des tables très condensées de ces représentations, et moi avons réfléchi à cette question. Nous n'annonçons aucun projet, mais je vous livre informellement quelques unes de nos réflexions:

i) Les représentations irréductibles projectives des groupes cristallographiques pour lesquelles il serait le plus utile aux physiciens de connaître l'algèbre R^G des polynomes invariants sont celles qui ont pour image un groupe fini[8].

ii) Il n'y a pas de méthodes pour déduire simplement l'algè-

[8] Dans tout groupe cristallographique G les translations forment un sous-groupe invariant $\sim Z^3$, dont le groupe dual $\sim T_3$ (le tore à 3 dimensions) est appelé zone de Brillouin. Le quotient $G/Z^3 = Q$ agit effectivement sur T_3. Les points de T_3 (en dehors de la strate générique ouverte dense) c'est-à-dire les vecteurs d'onde $0 \neq k \in T$, qui ont un stabilisateur (= petit groupe) maximal, fournissent - par le mécanisme des représentations induites - de telles représentations.

bre R^G pour une représentation quelconque de G, en supposant que le problème ait été résolu pour les représentations irréductibles.

iii) Obtient-on R^G pour une représentation isotypique (= factorielle = somme directe de représentations irréductibles équivalentes) par polarisation des polynomes invariants pour la représentation irréductible correspondante: [K2]? Pas toujours.

iv) Au lieu de la représentation la plus générale, on aura donc simplement à considérer les représentations sommes directes de représentations irréductibles toutes inéquivalentes. Pour un groupe fini de n classes de conjugaison (= $|G|$ pour un groupe abélien), cela fait 2^n cas à considérer.

v) Si G est abélien, la méthode de Stanley s'applique à toutes les représentations. Soit m la dimension d'une telle représentation. $G \times G \times \ldots \times G$ m fois, est alors engendré par pseudo-réflexions, le sous-groupe G, identifié au diagonal G^d est sous-groupe invariant à quotient abélien.

vi) La méthode de Stanley n'est pas la panacée, mais d'autres outils comparables peuvent être développés. Par exemple si $G' \leq H \leq G$ mais c'est H qui est engendré par pseudo-réflexions, G agit sur R^H et on peut trouver une base d'intégrité transformée en elle-même. On est alors ramené au problème de la détermination des polynomes invariants par G/H (agissant linéairement sur l'espace à une dimension de la base), que l'on sait résoudre (cas iv).

Autre remarque

Au lieu de considérer l'action d'un sous-groupe discret G de O(3) sur les polynomes définis sur notre espace à 3 dimensions, on peut considérer l'algèbre enveloppante $\mathcal{U}$ de O(3). C'est ce que viennent de faire Patera et Winternitz dans un preprint récent [P1] (un programme plus général avait été esquissé en [J1]). Les générateurs L_x, L_y, L_z de $\mathcal{U}$ étant les composantes d'un pseudo-vecteur, seul le sous-groupe $G \cap SO(3)$ de G agit effectivement.

On est ramené à étudier la famille de groupes C_n, D_n, T, O, Y. Dès qu'on connait trois invariants d'une base, le quatrième est engendré par les relations de commutations dans $\mathcal{U}$; quelle est la généralité de ce phénomène? Cette approche est intéressante pour attaquer le problème de l'*étiquette manquante* dont a parlé R. Sharp le premier jour. Voir aussi la conférence de P. Winternitz le dernier jour.

REMERCIEMENTS

Je remercie spécialement Jan Mozrzymas; nous avons ensemble étudié la littérature mathématique et réfléchi à ce problème. Tous deux nous avons beaucoup bénéficié de conversations à l'IHES avec P. Cartier, P. Deligne, D. Eisenbud - qui nous a aussi communiqué le preprint de Stanley - et D. Mumford. J'ai aussi bénéficié de discussions avec W. Opechowski (Vancouver), J. Patera, P. Winternitz, M. Kibler et R. Sharp (Montréal) qui de plus m'ont donné une bibliographie plus complète que celle qui apparaît ici.

BIBLIOGRAPHIE

B1. H.A. BETHE, Ann. Phys. Leipzig. 3, 133 (1929).

B2. N. BOURBAKI, *Groupes et Algèbres de Lie*, Ch. IV, V, VI, Hermann (Paris).

B3. W. BURNSIDE, *Theory of Groups of Finite Order*, 2nd Ed., Cambridge, 1911; reprint, Dover, 1955.

C1. C. CHEVALLEY, Amer. J. Math. 77, 778 (1955).

C2. H.S.M. COXETER, Duke Mat. J. 18, 765 (1951).

F1. L. FLATTO et M.M. WEINER, Amer. J. Math. 91, 591 (1969).

H1. D. HILBERT, Math. Ann. 36, 473 (1890).

J1. B.R. JUDD, W. MILLER, Jr., J. PATERA et P. WINTERNITZ, J. Math. Phys. 15, 1787 (1974).

K1. J. KILLINGBECK, J. Phys. (Solid State Phys.) C5, 2497 (1972).

K2. V. KOPSKY, J. Phys. (Solid State Phys.) C8, 3251 (1975).

L1. L.D. LANDAU et E.M. LIFSHITZ, *Mécanique Quantique* (Mir, Moscou).

M1. A.G. McLELLAN, J. Phys. (Solid State Phys.) C7, 3326 (1974).

M2. B. MEYER, Can. J. Math. 6, 135 (1953).

M3. J. MOZRZYMAS, Bull. Acad. Polon. Sci. Ser. Math. et Phys. 23, 485, 493, 499 (1975), et preprint "Algebraic description of space groups and their representations", Stony Brook, N.Y.

M4. T. MOLIEN, Sitzungsber. König Preuss Akad. Wiss., 1152 (1897).

N1. M. NAGATA, Amer. J. Math. 81, 766 (1959).

P1. J. PATERA et P. WINTERNITZ, J. Chem. Phys. 65, 2725 (1976).

S1. G.C. SHEPHARD, Enseignement Mathématique 2, 42 (1956).

S2. G.C. SHEPHARD et J.A. TODD, Can. J. Math. 6, 274 (1954).

S3. L. SOLOMON, Nagoya Math. J. 22, 57 (1963).

S4. L. SOLOMON, Trans. Amer. Math. Soc. 113, 274 (1964).

S5. R.P. STANLEY, "Relative invariants of finite groups generated by pseudo-reflexions", preprint, 1976.

S6. R. STEINBERG, Can. J. Math. 12, 616 (1960).

TABLE 1. Base d'intégrité de l'algèbre des polynomes (en x, y, z) invariants par un sous-groupe discret de O(3).

Cette base est donnée par les polynomes θ_1, θ_2, θ_3 et, éventuellement φ_1, φ_2, φ_3, qui sont des polynomes en x, y, z. Le polynome invariant le plus général est de la forme:

$$p_0(\theta_1,\theta_2,\theta_3) + \sum_i \varphi_i p_i(\theta_1,\theta_2,\theta_3), \tag{1}$$

où p_0 et p_i sont des polynomes *arbitraires* de 3 variables et où, suivant les groupes, la somme $\sum_i$ comprend 0, 1 ou 3 termes.

La dimension c_n, de l'espace vectoriel R_n^G des polynomes invariants homogènes de degré n, est donnée par

$$\sum_{n=0}^{\infty} c_n t^n = \frac{N(t)}{(1-t^{d_1})(1-t^{d_2})(1-t^{d_3})}, \quad d_i = \text{degré de } \theta_i,$$

où selon qu'il y a 0, 1 ou 3 polynomes φ_i,

$$N(t) = 1 \quad \text{ou} \quad 1 + t^{\delta_1} \quad \text{ou} \quad 1 + t^{\delta_1} + t^{\delta_2} + t^{\delta_3}, \quad \delta_i = \text{degré de } \varphi_i.$$

G	$\|G\|$	θ_1	θ_2	θ_3	φ_1	Notations
C_n $n>1$	n	z	ρ	γ_n	σ_n	$\rho=x^2+y^2$ $\gamma_n=\mathrm{Re}(x+iy)^n$,
C_{nv}	$2n$	z	ρ	γ_n		$\sigma_n=\mathrm{Im}(x+iy)^n$
C_{nh} $n>1$	$2n$	z^2	ρ	γ_n	σ_n	notons que $\gamma_n^2+\sigma_n^2=\rho^n$
S_{2n}	$2n$	z^2	ρ	γ_{2n}	σ_{2n}	$\varphi_2=z\gamma_n$, $\varphi_3=z\sigma_n$
D_n	$2n$	z^2	ρ	γ_n	$z\sigma_n$	
D_{nd}	$4n$	z^2	ρ	γ_{2n}	$z\sigma_n$	$\tau_n=x^n+y^n+z^n$
D_{nh}	$4n$	z^2	ρ	γ_n		$K=(1+\sqrt{5})/2$
T	12	τ_2	τ_4	β	α_0	$\alpha_m=(K^m x^2-y^2)(K^m y^2-z^2)$
T_h	24	τ_2	τ_4	β^2	α_0	$\times(K^m z^2-x^2)$
T_d	24	τ_2	τ_4	β		$\beta=xyz$
O	24	τ_2	τ_4	β^2	$\beta\alpha_0$	$\chi_1(x,y,z)=K^4x^4+K^{-4}y^4+$
O_h	48	τ_2	τ_4	β^2		$z^4-2x^2y^2-2z^2(Kx^2+K^{-1}y^2)$
Y	60	τ_2	α_2	$\tau_4\alpha_4$	$\beta\chi_1\chi_2\chi_3$	$\chi_2(x,y,z)=\chi_1(z,x,y)$
Y_h	120	τ_2	α_2	$\tau_4\alpha_4$		$\chi_3(x,y,z)=\chi_1(y,z,x)$

La nomenclature et la présentation des groupes sont celles de Landau et Lifschitz, *Mécanique quantique*.

La table 2 donne explicitement ces bases pour les 32 groupes cristallographiques ponctuels.

TABLE 2. Base d'intégrité de l'algèbre des polynomes invariants des 32 groupes ponctuels cristallographiques.

G	$\|G\|$	θ_1	θ_2	θ_3	φ_1
C_1	1	z	x	y	
$C_i = S_2$	2	z^2	x^2	y^2	$\varphi_1, \varphi_2, \varphi_3$
$C_s = C_{1h}$	2	z^2	x	y	
C_2	2	z	x^2	y^2	xy
C_{2h}	4	z^2	x^2	y^2	xy
C_{2v}	4	z	}	}	
D_2	4	z^2	} x^2	} y^2	xyz
D_{2h}	8	z^2	}	}	
C_3	3	z	}	}	$y(3x^2-y^2)$
C_{3v}	6	z	}	} $x(x^2-3y^2)$	
D_3	6	z^2	} x^2+y^2	}	$zy(3x^2-y^2)$
S_6	6	z^2	}	} $x^6-15x^4y^2$	$\varphi_1, \varphi_2, \varphi_3$
D_{3h}	12	z^2	}	} $+15x^2y^4-y^6$	
C_4	4	z	}	}	$xy(x^2-y^2)$
C_{4v}	8	z	}	}	
S_4	4	z^2	}	}	$\varphi_1, \varphi_2, \varphi_3$
C_{4h}	8	z^2	} x^2+y^2	} x^2y^2	$xy(x^2-y^2)$
D_{2d}	8	z^2	}	}	xyz
D_4	8	z^2	}	}	$xyz(x^2-y^2)$
D_{4h}	16	z^2	}	}	

TABLE 2. (suite)

G	$\|G\|$	θ_1	θ_2	θ_3	φ_1
C_{3h}	6	z^2		$x(x^2-3y^2)$	$y(3x^2-y^2)$
C_6	6	z			$xy(3x^2-y^2)(x^2-3y^2)$
D_{3d}	12	z^2			$zy(3x^2-y^2)$
C_{6v}	12	z	x^2+y^2	$x^6-15x^4y^2$	
C_{6h}	12	z^2		$+15x^2y^4-y^6$	$xy(3x^2-y^2)(x^2-3y^2)$
D_6	12	z^2			$zxy(3x^2-y^2)(x^2-3y^2)$
D_{6h}	24	z^2			
T	12			xyz	$(x^2-y^2)(y^2-z^2)(z^2-x^2)$
T_h	24			$x^2y^2z^2$	$(x^2-y^2)(y^2-z^2)(z^2-x^2)$
T_d	24	$x^2+y^2+z^2$	$x^4+y^4+z^4$	xyz	
O	24			$x^2y^2z^2$	$xyz(x^2-y^2)(y^2-z^2)(z^2-x^2)$
O_h	48			$x^2y^2z^2$	

G	φ_1	φ_2	φ_3
$C_i = S_2$	xy	yz	zx
S_4	zxy	$z(x^2-y^2)$	$xy(x^2-y^2)$
$C_{3i} = S_6$	$xz(x^2-3y^2)$	$yz(3x^2-y^2)$	$xy(3x^4-10x^2y^2+3y^4)$

METACRYSTALLOGRAPHIC GROUPS

W. Opechowski

1.1. INTRODUCTION

Any group whose definition must logically be preceded by that of a crystallographic group will in this paper be called a *metacrystallographic group*. This is, however, not a generally accepted term. In fact, there is no generally accepted term for such groups, although several terms such as *cryptosymmetry groups*, *cambiant symmetry groups*, *quasisymmetry groups*, have been proposed for large classes of metacrystallographic groups in the last 20 years, and the terms used for specific classes of metacrystallographic groups form a rather bewildering collection of words.

The only purpose of this paper is to formulate, and compare, the definitions of the various metacrystallographic groups precisely, by using in all cases the same mathematical terminology. The terminology is much less picturesque than that used by most crystallographers, but has the advantage of clearly bringing out the analogies and differences between these groups. The treatment will be purely classical; for lack of space, no implications for a quantum mechanical treatment will be discussed.

The metacrystallographic groups so far considered in the literature are groups acting on certain function spaces, and as such can be regarded as groups of transformations of these function spaces.

1.2. GENERAL REMARKS ON TERMINOLOGY AND NOTATION

If U is any set (elements: $u_a, u_b, u_c, \ldots$), and an action of a group $\tilde{G}$ on this set is defined (to each element G of $\tilde{G}$ there corresponds an operator [G] such that $[G]u_a = u_b$), then a subset V of U is *invariant* under G if $[G]v_\alpha = v_\beta$ for each v_α and some v_β, both in V: G is a *symmetry element* of V. The subgroup $\tilde{H}$ of $\tilde{G}$ which consists of all symmetry elements of V is the *symmetry group* of V, and each subgroup of $\tilde{H}$ is an *invariance group* of V. If the elements of U are functions, and ψ is one of the functions in U, then *the symmetry group of* ψ is the symmetry group of the set V consisting of just the one element ψ. (Tilde above a letter, as in $\tilde{G}$, indicates in this paper that the letter denotes a group.)

A discrete group of isometries $(R|v)$, where $(R|o)$ is an orthogonal transformation and $(E|v)$ a translation of an n-dimensional Euclidean space, is a *space group* if its subgroup of translations contains n independent translations. A group is a *crystallographic group* if it is a subgroup of some space group. In this paper n = 3, 2 or 1. In the case n = 3 the Euclidean space will be identified with the *space* around us.

A *crystal* is a set of points in space whose symmetry group is a space group.

A function ψ from the set X (elements: $x_\alpha, x_\beta, \ldots$) into a set Y (elements: $y_a, y_b, \ldots$) will here be denoted by $\{x;y\}$ or by $\{x_\alpha;y_a\}$ rather than by $\psi(x) = y$; that is, $\psi = \{x;y\}$. The symbol (x,y) or y_x will then denote a specific pair in the set of pairs which constitute ψ; in other words, y is the value of ψ at x. If X is a subspace of an Euclidean space and Y is a vector space, it will be often tacitly assumed that a coordinate system in X, and a basis in Y, have been introduced. A symbol $[A\|B]$ will denote an operator which is a pair of transformations, A of X and B of Y.

An element of the direct product $\tilde{G} \times \tilde{H}$, or of a semidirect product $\tilde{G} \circledS \tilde{H}$, will be denoted by (G,H), and G will be called the G-*part* and H its H-*part*. The following theorem holding for any subgroup $\tilde{J}$ of $\tilde{G} \times \tilde{H}$ will be taken for granted:

(T1) The distinct G-parts of the elements of $\tilde{J}$ constitute a subgroup $\tilde{G}_J$ of $\tilde{G}$, and their distinct H-parts a subgroup $\tilde{H}_J$ of $\tilde{H}$; therefore $\tilde{J}$ is also a subgroup of $\tilde{G}_J \times \tilde{H}_J$.

2.1. COLOUR FUNCTIONS AND COLOUR GROUPS

The set X will always be in this case a discrete set $\tilde{K}r_1$ of points in space, where the discrete group $\tilde{K}$ (elements: $K_\alpha, K_\beta, \ldots$) of isometries is the symmetry group of $\tilde{K}r_1$, and r_1 is a point such that $K_\alpha r_1 \neq K_\beta r_1$ if $K_\alpha \neq K_\beta$; that is, r_1 is in a *general position* rather than in a *special position* ($\tilde{K}r_1$ is thus the orbit of $\tilde{K}$ at r_1). The treatment of the general case, where r_1 is in a special position, does not present any difficulty, but is somewhat lengthy.

The set Y will be a finite set $C = \{c_1, c_2, \ldots, c_d\}$ with no algebraic structure, called the *colour set*, its elements being *colours*. A function $\phi = \{r;c\}$ from $\tilde{K}r_1$ onto C will be called a colour function. If an element c of C is the value of ϕ at r, we shall say that *the point* r *has the colour* c. Correspondingly, the phrase *coloured point set* will be used as a synonym of the phrase *colour function*. Each colour in C will thus appear as the colour of some points in such a coloured point set ϕ.

Instead of defining the action of $\tilde{K}$ on ϕ in the usual way,

$$\text{(A1)} \qquad [K_\alpha]\{r;c\} = \{K_\alpha^{-1} r;c\},$$

we first introduce a transitive group $\tilde{P}$ of permutations, $P_a, P_b, \ldots$, of the set C, and next define the action of the direct product group $\tilde{P} \times \tilde{K}$ on ϕ as follows:

(A2) $$[P_a \| K_\alpha]\{r;c\} = \{K_\alpha^{-1}r;P_a c\}.$$

To each element (P_a, K_α) of $\tilde{P} \times \tilde{K}$, there will thus correspond an operator $[P_a \| K_\alpha]$ in the space of d-colour functions defined on $\tilde{K}r_1$, where d is the degree of the permutations in $\tilde{P}$. (Dealing with intransitive groups of permutations would introduce trivial complications but nothing essentially different.)

A colour function that is not invariant under an element of $\tilde{K}$, in the sense of (A1), may very well be invariant under an element of $\tilde{P} \times \tilde{K}$, in the sense of (A2). For example, if the 2-coloured point set ϕ consisting of the 4 corners of a square, the corners diagonally opposite being of the same colour, then a rotation K through $\pi/2$ in the plane of the square about its centre is not a symmetry element of ϕ, while (K,P), where P is the permutation (1,2) of the 2-colour set C = {1,2}, is.

A group $\tilde{B}$ is called a *colour group* if it satisfies the conditions:

(C1) $\tilde{B}$ is a subgroup of $\tilde{P} \times \tilde{K}$, where $\tilde{K}$ is a discrete group of isometries, and $\tilde{P}$ is a transitive group of permutations of degree d such that $\tilde{P}_B = \tilde{P}$ and $\tilde{K}_B = \tilde{K}$ (see (T1));

(C2) $\tilde{B}$ is an invariance group, in the sense of (A2), of some d-colour function ϕ.

According to whether $\tilde{K}$ is or is not a crystallographic group a colour group is called a *crystallographic* or *non-crystallographic colour group*. In particular if $\tilde{K}$ is a space group, point group, etc. one speaks of a *colour space group*, *colour point group*, etc.

Since a colour set has no algebraic structure, colour groups have little applications in physics, except in such cases as those of a crystal or a molecule whose atoms have d different masses. However, colour groups do play an important role in the classification of all kinds of 1-, 2- and 3-dimensional ornaments.

No colour function is invariant under those elements of $\tilde{P} \times \tilde{K}$ which are of the form (P,E), where $P \neq E$ (the unit element). Therefore, by (C2), no colour group $\tilde{B}$ contains such elements. From (C1) it then follows that each element of $\tilde{K}$ occurs exactly once as the K-part in the elements of $\tilde{B}$. Therefore $\tilde{K}$ is isomorphic onto $\tilde{B}$, and homomorphic onto $\tilde{P}$. Hence $\tilde{K}/\tilde{k}$ is isomorphic onto $\tilde{P}$, where $\tilde{k}$ is the kernel of the homomorphism ρ of $\tilde{K}$ onto $\tilde{P}$ and consists of the K-parts of the elements (E,K) of $\tilde{B}$; these elements constitute a normal subgroup $\tilde{B}_k$ of $\tilde{B}$. It follows that

$$\tilde{B} = \tilde{B}_k + (P_2,K_2)\tilde{B}_k + \ldots + (P_p,K_p)\tilde{B}_k,$$

where K_α in the coset representative (P_α,K_α) of $\tilde{B}_k$ in $\tilde{B}$ is mapped to P_α under ρ; $\alpha = 2,3,\ldots,p$. The symbol $\tilde{B}(\rho:\tilde{K}\rightarrow\tilde{P})$ thus specifies a colour group $\tilde{B}$ completely. If d is the degree of the permutations in $\tilde{P}$, then $d \leq |\tilde{P}| \leq |\tilde{K}|$.

Conversely, any group $\tilde{B}$ that satisfies (C1) with $\tilde{K}$ homomorphic onto $\tilde{P}$ is a colour group. To prove this, one has to show that $\tilde{B}$ also satisfies (C2). This can easily be done by using the results of the theory of representations of groups by the transitive groups of permutations. According to that theory, every homomorphism of $\tilde{K}$ onto $\tilde{P}$ is determined by some subgroup $\tilde{L}$ of $\tilde{K}$. If the decomposition of $\tilde{K}$ into left-cosets of $\tilde{L}$ is

$$\tilde{K} = \tilde{L} + K_2\tilde{L} + \ldots + K_d\tilde{L},$$

where $\{K_1{=}E,K_2,\ldots,K_d\}$ is a set of the left-coset representatives of $\tilde{L}$ in $\tilde{K}$, then by making correspond to the element K_α of $\tilde{K}$ the permutation

$$P = \begin{pmatrix} \tilde{L}, & K_2\tilde{L},\ldots,K_d\tilde{L} \\ K_\alpha\tilde{L}, & K_\alpha K_2\tilde{L},\ldots,K_\alpha K_d\tilde{L} \end{pmatrix}$$

one obtains a homomorphism ρ of $\tilde{K}$ onto the group $\tilde{P}$ of such permutations. Each group $\tilde{B}(\rho:\tilde{K}\to\tilde{P})$ can thus be obtained in this way.

If we decompose the point set $\tilde{K}r_1$ (r_1 is at a general position!) into d disjoint subsets correspondingly,

$$\tilde{K}r_1 = \tilde{L}r_1 + K_2\tilde{L}r_1 +,..+ K_d\tilde{L}r_1,$$

then the d-colour function ϕ defined on $\tilde{K}r_1$ in which all points $K_\alpha\tilde{L}r_1$ with a fixed K_α ($\alpha=1,2,...,d$), have the same colour c_a, will be invariant under the group $\tilde{B}$ for which $\tilde{P}$ is a representation of $\tilde{K}$ determined by $\tilde{L}$, so that $\tilde{B}$ satisfies (C2) and is therefore a colour group.

From this proof it follows that colour groups can be invariance groups of only those coloured point sets in which the number (possibly infinite) of points of one colour is the same for each colour. It also follows that if the colour group $\tilde{B}(\rho:\tilde{K}\to\tilde{P})$ is the symmetry group, in the sense of (A2), of a coloured point set ϕ then $\tilde{k}$ = ker ρ is the symmetry group of ϕ in the sense of (A1).

Two d-colour groups $\tilde{B}(\rho_1:\tilde{K}_1\to\tilde{P}_1)$ and $\tilde{B}(\rho_2:\tilde{K}_2\to\tilde{P}_2)$ are said to belong to the same *class of* d-*colour groups* if there exists a permutation S of the d-colours and a similarity transformation T of $\tilde{K}_2$ onto $\tilde{K}_1$ such that the homomorphism $T^{-1}\tilde{K}_2T \to S^{-1}\tilde{P}_2S$ obtained in this way from ρ_2 is identical with ρ_1.

If we denote a colour group by $\tilde{K}(\tilde{L})$, and if $\tilde{L}'$ is a subgroup of $\tilde{K}$ conjugate to $\tilde{L}$, then $\tilde{K}(\tilde{L})$ and $\tilde{K}(\tilde{L}')$ are in the same class. But even if $\tilde{L}$ and $\tilde{L}'$ are not conjugate, $\tilde{K}(\tilde{L})$ and $\tilde{K}(\tilde{L}')$ may be in the same class, because it may happen that $T^{-1}\tilde{L}T = \tilde{L}'$ and $T^{-1}\tilde{K}T = \tilde{K}$ for some T.

Each one-colour group $\tilde{K}(\tilde{K})$ can of course be identified with the group $\tilde{K}$.

To obtain, in the way indicated above (Method I), all crystallographic colour point groups is a simple matter, since all sets $\{\tilde{L}\}$ of conjugate subgroups for each point group $\tilde{K}$ are known. By taking one group $\tilde{K}$ from each class of crystallographic point groups, and one group $\tilde{L}$ from each $\{\tilde{L}\}$, one obtains a complete list of colour point groups $\tilde{B} = \tilde{K}(\tilde{L})$. However, one could proceed in a different way (Method II). One could, for each fixed $\tilde{K}$ and each fixed normal subgroup $\tilde{k}$ of $\tilde{K}$, determine the transitive groups $\tilde{P}_1, \tilde{P}_2, \ldots, \tilde{P}_m$ of permutations which are isomorphic onto $\tilde{K}/\tilde{k}$, no two of them being conjugate subgroups of a symmetric group. For a fixed $\tilde{P}_n$ (n=1,2,...,m), each automorphism γ of $\tilde{P}_n$ corresponds to replacing one such isomorphism by another one, and, therefore, one homomorphism of $\tilde{K}$ onto $\tilde{P}_n$ by another one; or, in other words, one colour group $\tilde{K}(\tilde{L})$ by another one $\tilde{K}(\tilde{L}')$. The two colour groups then will or will not be in the same class according to whether the groups $\tilde{L}$ and $\tilde{L}'$ are or are not in the same crystallographic class (they always will if γ is an inner automorphism).

There are 58 classes of two-colour crystallographic point groups (or, as they are also called, *black-and-white point groups*), and there are altogether 212 of them, for any $d \geq 2$.

In the case of colour space groups, a complete list of them is known only for $d = 2$, because the task of determining them even in this simplest case is very laborious. It turns out that there are 1191 classes of black-and-white space groups.

2.2. COLOUR FUNCTIONS AND COLOUR W-GROUPS

The action (A2) of a colour group on a colour function $f = \{r;c\}$ has the same effect on a colour c at all points r of the point set on which f is defined. One can remove this restriction by making use of the *wreath product* $\tilde{P} \circledw \tilde{G}$. The wreath product will here be defined for any two groups $\tilde{P}$ and $\tilde{G}$. Later $\tilde{P}$ will be a transitive group of permutations and $\tilde{G}$ will be taken to be a discrete group $\tilde{K}$ of isometries, as in Section 2.1. The elements of $\tilde{G}$ will be labelled by means of Greek letters: $\tilde{G}_\alpha, \tilde{G}_\beta, \ldots$; the elements of $\tilde{P}$ will be labelled by means of Latin letters: $\tilde{P}_a, \tilde{P}_b, \ldots$; or simply $\alpha, \beta, \ldots$, and $a, b, \ldots$.

We first consider the direct product $\tilde{P}^{\tilde{G}}$ of $|\tilde{G}|$ (thus possibly an infinite number of) copies of $\tilde{P}$,

$$\tilde{P}^{\tilde{G}} = \ldots \times \tilde{P}_{G_\alpha} \times \tilde{P}_{G_\beta} \times \tilde{P}_{G_\lambda} \times \ldots ,$$

each factor in $\tilde{P}^{\tilde{G}}$ being labelled by a different element of $\tilde{G}$. Each element of $\tilde{P}^{\tilde{G}}$ is thus a function from $\tilde{G}$ into $\tilde{P}$, which will be denoted by $\{\alpha;a\}$, or more explicitly:

$$\{\alpha;a\} = [\ldots,(\alpha,a),(\beta,b),\ldots,(\lambda,\ell),\ldots]$$

or, for brevity,

$$\{\alpha;a\} = [\ldots,a_\alpha,b_\beta,\ldots,\ell_\lambda,\ldots];$$

the product of two elements of $\tilde{P}^{\tilde{G}}$ is then:

$$\{\alpha;a\}\{\alpha;a'\} = \{\alpha;aa'\}$$

or, more explicitly:

$$[\ldots,a_\alpha,b_\beta,\ldots,\ell_\lambda,\ldots][\ldots,a'_\alpha,b'_\beta,\ldots,\ell'_\lambda,\ldots]$$

$$= [\ldots,a_\alpha a'_\alpha,\ldots,b_\beta b'_\beta,\ldots,\ell_\lambda \ell'_\lambda,\ldots].$$

If $\tilde{P}$ and $\tilde{G}$ are finite, $|\tilde{P}| = m$, $|\tilde{G}| = n$, then the order of $\tilde{P}^{\tilde{G}}$ is m^n. Each element of $\tilde{P}^{\tilde{G}}$ can be thought of as a row of a rectangular array consisting of m^n rows, each row being a sequence of n (not necessarily distinct) elements of $\tilde{P}$, and each column being labelled by a distinct element of $\tilde{G}$.

Next, we make correspond, to each element G_γ of $\tilde{G}$, a permutation Φ_γ of $\tilde{P}^{\tilde{G}}$ as follows: $\Phi_\gamma\{\alpha;a\} = \{\alpha\gamma;a\}$; instead of $\Phi_\gamma\{\alpha;a\}$, we shall often write $\{\alpha;a\}^\gamma$. Such a permutation is always an automorphism of $\tilde{P}^{\tilde{G}}$, since it preserves the product of any two elements of $\widetilde{P}^{\widetilde{G}}$. We shall say that Φ_γ is the automorphism of $\widetilde{P}^{\widetilde{G}}$ *determined* by G_γ. The automorphism of $\widetilde{P}^{\widetilde{G}}$ determined by the elements of $\widetilde{G}$ constitute a group $\widetilde{\Phi}_{\widetilde{G}}$. The mapping just defined, $\widetilde{G} \to \widetilde{\Phi}_{\widetilde{G}}$, is an isomorphism.

Finally, we consider all pairs of the form (element of $\widetilde{P}^{\widetilde{G}}$, element of $\widetilde{\Phi}_{\widetilde{G}}$). Each such pair will be denoted by $(\{\alpha;a\},\Phi_\gamma)$ or, for brevity, by $(\{\alpha;a\},\gamma)$. The set of all these pairs, together with the product defined for any two of them as follows: $(\{\alpha;a\},\gamma)(\{\alpha;a'\},\delta) = (\{\alpha;a\}\{\alpha;a'\}^\gamma,\gamma\delta)$, thus constitutes the semidirect product of the group $\widetilde{P}^{\widetilde{G}}$ by the group $\widetilde{G}$, determined by the isomorphism $\widetilde{G} \leftrightarrow \widetilde{\Phi}_{\widetilde{G}}$. It is this semidirect product, $\widetilde{P}^{\widetilde{G}} \circledS_\Phi \widetilde{G}$, which is called the *wreath product*, $\widetilde{P} \circledw \widetilde{G}$, *of* $\widetilde{P}$ *by* $\widetilde{G}$: $\widetilde{P} \circledw \widetilde{G} \equiv \widetilde{P}^{\widetilde{G}} \circledS_\Phi \widetilde{G}$. The group $\widetilde{P}^{\widetilde{G}}$ regarded as a subgroup of $\widetilde{P} \circledw \widetilde{G}$ is normal in the latter, and the quotient group $(\widetilde{P} \circledw \widetilde{G})/\widetilde{P}^{\widetilde{G}}$ is isomorphic onto $\widetilde{G}$. If $\widetilde{P}$ and $\widetilde{G}$ are finite, $|\widetilde{P}| = m$, $|\widetilde{G}| = n$, then the order of $\widetilde{P} \circledw \widetilde{G}$ is nm^n.

We define the action of an element $([a_\alpha,a_\beta,\ldots,\ell_\lambda,\ldots],\tau) \equiv (\{\alpha;a\},\tau)$ of the wreath product $\widetilde{P} \circledw \widetilde{G}$ on a d-colour function $\phi = \{r_\alpha;c_s\}$ defined on $\widetilde{G}r_1$ (the group $\widetilde{G}$ will from now on be a discrete group of isometries, and $\widetilde{P}$ a transitive group of permutations) as follows:

$$\text{(A3)}\quad ([a_\alpha,b_\beta,\ldots,\ell_\lambda,\ldots],\tau)\{(r_\alpha,c_s),(r_\beta,c_t),\ldots,(r_\lambda,c_u),\ldots\}$$
$$= \{(G_\tau^{-1}r_\alpha,P_a c_s),(G_\tau^{-1}r_\beta,P_b c_t),\ldots,(G_\tau^{-1}r_\lambda,P_\ell c_u),\ldots\}.$$

According to this definition, the effect of the action of an element of $\widetilde{P} \circledw \widetilde{G}$ on the colour c_t of the point r_β depends on the permutation which, in the element $[a_\alpha,b_\beta,c_\gamma,\ldots]$ of the direct product $\widetilde{P}^{\widetilde{G}}$, is labelled by the element G_β of $\widetilde{G}$. This means that, although some other point r_λ may have the same colour $c_u = c_t$, the action of the same element of $\widetilde{P} \circledw \widetilde{G}$ may have a different effect on that colour than in the case of point r_β if $\ell_\lambda \neq b_\beta$.

We shall call any subgroup of $\tilde{P} \circledast \tilde{K}$ a W-*group*. A group $\tilde{U}$ will be called a *colour* W-*group* if it satisfies the conditions:

(W1) $\tilde{U}$ is a W-group such that each element of $\tilde{K}$ occurs exactly once as the K-part of its elements;

(W2) $\tilde{U}$ is an invariance group, in the sense of (A3), of some d-colour function ϕ.

As in the case of colour groups one can then easily show that $\tilde{K}$ is isomorphic onto $\tilde{U}$, and homomorphic onto $\widetilde{P^K}$. Each colour W-group $\tilde{U}$ is thus specified by some such homomorphism $\rho:\tilde{U}(\rho:\tilde{K}\to\widetilde{P^K})$, and each element of $\tilde{U}$ is of the form $(\{\alpha;a\}_\tau,\tau)$, where $\{\alpha;a\}_\tau$ is the element of $\widetilde{P^K}$ to which K_τ is mapped under ρ. Similarly, each element of a colour group $\tilde{B}(\rho:\tilde{K}\to\tilde{P})$ has the form (P_τ,τ).

Replacing (W1) by a weaker condition would mean considering W-groups each of which contains elements of the form $(\{\alpha;a\},\varepsilon)$, where $K_\varepsilon = E$ and $\{\alpha;a\}$ is not the unit element of $\widetilde{P^K}$, and are therefore not isomorphic onto $\tilde{K}$. A group consisting of only such elements may very well satisfy (W2). Such groups could be called *colour-only* W-*groups*. (Compare the definition of spin-only groups in Section 3.3.)

A colour W-group $\tilde{U}$ can be identified with a colour group $\tilde{B}$ if the P-part of each element of $\tilde{U}$ is of the form

$$(\{\alpha;a\}_\sigma,\sigma) = ([a_\alpha,a_\beta,\ldots,a_\lambda,\ldots]_\sigma,\sigma),$$

that is, if the function $\{\alpha;a\}_\sigma$ has a constant value a which depends only on K_σ. Such an identification is justified because the action (A3) of $\tilde{U}$ on a colour function ϕ is then identical with the action (A2) of the colour group $\tilde{B}_{\tilde{U}}$ whose elements are (a_σ,σ). We may then say that the colour W-group $\tilde{U}$ *is* the colour group $\tilde{B}_{\tilde{U}}$.

It turns out that the crystallographic colour W-point groups (that is, those with a point group $\tilde{K}$) are, in this sense, just colour groups, except for one cubic colour W-point group.

3.1. SPIN ARRANGEMENTS AND MAGNETIC GROUPS

Consider a crystal in a magnetically ordered state. Using a classical model, this means that the atoms of the crystal in such a state have, even in the absence of an external magnetic field, non-vanishing magnetic dipole moments which, in the simplest case, are of equal magnitudes but possibly of different orientations (the case of equal orientations is the ferromagnetic state). Although this model is, even from the classical point of view, too primitive for many purposes, it is sufficient for introducing *magnetic groups* and *spin groups*, which will be done presently. Instead of speaking of magnetic dipole moments, it is customary in this connection to call the latter *spins*; these are of course not the spins in the quantum mechanical sense (in some very special cases, a spin in the sense just introduced is the average value of the quantum mechanical magnetic dipole moment operator associated with the electron spin, in the sense of quantum statistical mechanics).

This very simple model of a magnetically ordered state of a crystal can be described by a function $\sigma = \{x;S\}$, where $x = (r,t)$ are points of a point set X in space-time, and S are spins, that is, vectors of a 3-dimensional Euclidean vector space Y, the spin space. The function σ will be called, as is customary, a *spin arrangement*. It will always be assumed in what follows that not all values of σ are null-vectors. However, both X and Y require a more precise specification. For, according to the standard electromagnetic theory, a magnetic dipole moment S has not only a well-defined behaviour under proper rotations of space (expressed by saying that magnetic dipole moments S are vectors), but also under space inversion (invariance of S) and under time inversion (multiplication of S by -1). In other words, the vector space Y is, not only a carrier space of a representation D_1^+ of the group O(3) of proper and improper rotations by a group of 3×3 orthogonal matrices R such that, to a rotation in O(3), there corresponds a matrix $\delta_R R$, where δ_R is the determinant of R, but Y is also a carrier space of the representation of the time-inversion group $\tilde{A}$ (consisting of the unit element E and time inversion E') such that E' is represented by a 3 × 3 unit matrix multiplied by -1.

To make full use of the fact that Y is thus the carrier of the representation $D_1^+ \times \Gamma^-$ of the group $O(3) \times \tilde{A}$, where Γ^- is the alternating representation of $\tilde{A}$, we treat a crystal as located in space-time, that is, we define the set X as the Cartesian product of $\tilde{F}r_1$, where $\tilde{F}$ is the symmetry space group of the crystal, and that one-dimensional Euclidean point space $E_t(1)$ which we call *time* in physics: $X = E_t(1) \times \tilde{F}r_1$. If $\tilde{\varepsilon}_t(1)$ is the group of all time isometries $(A|\tau)$, where τ is a time translation and A is an element of $\tilde{A}$, then the symmetry group of X is $\tilde{\varepsilon}_t(1) \times \tilde{F}$. Since $\tilde{F}$ is homomorphic onto its point group $\tilde{R}_F$, and $\tilde{\varepsilon}_t(1)$ is homomorphic onto $\tilde{A}$, the representation $\Gamma^- \times D_1^+$ of $\tilde{A} \times O(3)$ restricted to $\tilde{A} \times \tilde{R}_F$ can be regarded as a representation of $\tilde{\varepsilon}_t(1) \times \tilde{F}$, and Y the carrier space of the latter. We shall consider only time-independent spin arrangements σ; that is, σ will, by definition, not be affected by time translations and we may write: $\sigma = \{r;S\}$. Instead of $\tilde{\varepsilon}_t(1) \times \tilde{F}$ we need thus consider only the action of $\tilde{A} \times \tilde{F}$ on σ. Because Y is the carrier space of the representation $\Gamma^- \times D_1^+$ of $\tilde{A} \times \tilde{F}$, to $(A,(R|\tau))$ there corresponds a transformation $\eta_A \delta_R R$ of Y where $\eta_A = +1$ or -1 according as A = E or E'.

Taking all this into account means defining the action of $(A,F) \equiv (A,(R|v))$ on $\sigma = \{r;S\}$ as follows:

$$\text{(A4)} \qquad [\eta_A \delta_R R \| A,(R|v)]\{r;S\} = \{(R|v)^{-1} r; \eta_A \delta_R RS\}.$$

A group $\tilde{m}$ is then called a *magnetic group* if it satisfies the conditions:

(m1) $\tilde{m}$ is a subgroup of $\tilde{A} \times \tilde{K}$, where $\tilde{K}$ is a subgroup of a space group $\tilde{F}$, and $\tilde{m}$ is such that $\tilde{K}_m = \tilde{K}$ and $\tilde{A}_m = \tilde{A}$ (see (T1));

(m2) $\tilde{m}$ is an invariance group, in the sense of (A4), of some spin arrangement $\sigma = \{r;S\}$, where r belongs to $\tilde{F}r_1$.

According to whether $\tilde{K}$ is a space group, point group, etc., a magnetic group is called a *magnetic space group*, a *magnetic point group*, etc. (One can easily define also non-crystallographic magnetic groups. However, only crystallographic magnetic groups have extensively been used in physics, because the magnetically ordered state occurs only in crystals.)

Since no spin arrangement is invariant under the element (E',E) of $\tilde{A} \times \tilde{K}$, no magnetic group $\tilde{m}$ contains this element. Therefore, by an argument similar to that which immediately follows the definition of a colour group in Section 2.1, one concludes that $\tilde{m}$ must be isomorphic onto $\tilde{K}$, and homomorphic onto $\tilde{A}$, the kernel $\tilde{k}$ of the homomorphism being a subgroup $\tilde{D}[\tilde{K}]$ of index 2 in $\tilde{K}$. The elements of $\tilde{m}$ must thus be of the form (A_K,K), where $A_K = E$ or E', according to whether K is or is not in $\tilde{D}[\tilde{K}]$. Conversely, any group $\tilde{m}$ that satisfies (m1), with $\tilde{K}$ homomorphic onto $\tilde{A}$, also satisfies (m2). For example, let $\tilde{K} = \tilde{F}$; then such a group will be an invariance group of a spin arrangement σ defined on a crystal $\tilde{F}r_1$ (r_1 is supposed to be in a general position!) and such that the value of σ at the point $\tilde{F}r_1$ is $\eta_A\delta_R RS$ if $F = (R|v)$ is the F-part of the element (A_F,F) of that group, and S is the spin at the point r_1.

It is clear that, for each fixed crystallographic group $\tilde{K}$, one can establish a one-to-one correspondence between all magnetic groups $\tilde{m}$ and all two-colour (black-and-white) groups $\tilde{B}$ by pairing $\tilde{m}$ and $\tilde{B}$ for which $\tilde{D}[\tilde{K}] = \tilde{L}$. Therefore a list of all classes of magnetic groups can be interpreted as a list of all classes of black-and-white groups and vice versa, the classes in the two cases being defined in an analogous way. There are thus 1191 classes of magnetic space groups, 58 classes of magnetic point groups, etc.

Magnetic space groups (and also magnetic translation groups

and magnetic point groups) have extensively been used to interpret magnetic neutron diffraction data, and also nuclear magnetic resonance data, and to obtain from such data a description and classification of magnetically ordered crystals, in a way analogous to that leading to a description and classification of crystals by means of space groups starting out from X-ray diffraction data.

3.2. ELECTRIC DIPOLE MOMENT ARRANGEMENTS AND ELECTRIC GROUPS

If instead of a magnetically ordered state of a crystal, we consider an electrically ordered state, then the role played by spin arrangements $\sigma = \{r;S\}$ and magnetic groups $\tilde{m}$ will be taken over by electric dipole moment arrangements $\pi = \{r;P\}$ and groups $\tilde{p}$ which could be called *electric groups*. Here P is an electric dipole moment, that is, a vector which belongs to the carrier space of a representation $D_1^- \times \Gamma^+$ of the group $O(3) \times \tilde{A}$, Γ^+ being the identity representation of $\tilde{A}$. In other words, P is invariant under time inversion. Each electric group will then be a direct product $\tilde{K} \times \tilde{A}$, where $\tilde{K}$ is a crystallographic group.

Since the *vierergruppe* $\tilde{V}$ consisting of the unit element, space inversion, time inversion and the product of the last two, has 4 irreducible representations, and $O(3) \times \tilde{A} = SO(3) \times \tilde{V}$, one could, in addition to magnetic and electric groups, define another two kinds of groups, describing functions $\{r;W\}$, where W belongs to the carrier space of a representation $D_1^- \times \Gamma^-$ in one case, and of $D_1^+ \times \Gamma^+$ in the other case (which is the case where W is an electric current vector).

3.3. SPIN ARRANGEMENTS AND SPIN GROUPS

If we disregard the fact that, from the physical point of view, the vector space Y whose vectors are spins S is the carrier space of the representation $\Gamma^- \times D_1^+$ of $\tilde{A} \times \tilde{F}$, then we may consider, instead of the group $\tilde{A} \times \tilde{F}$, the group $\tilde{\Omega} \times \tilde{F}$, where $\tilde{\Omega}$ is the group of all orthogonal transformations of Y (that is, $\tilde{\Omega}$ is just another symbol for O(3)), and define the action of $(\Omega,F) \equiv (\Omega,(R|v))$ on a spin arrangement $\sigma = \{r;S\}$ as follows:

$$\text{(A5)} \qquad [\Omega\|(R|v)]\{r;S\} = \{(R|v)^{-1}r;\Omega S\}.$$

Since Y is no longer assumed to be a carrier space of a representation of a group of transformations of space-time, the definition (A5), contrary to the definition (A4), does not imply any relation between the orthogonal transformation $(R|o)$ of $\widetilde{Fr}_1$ and the orthogonal transformation Ω of Y. In other words, a spin arrangement σ is treated now as any function defined on $\widetilde{Fr}_1$, whose values are vectors in any Euclidean 3-dimensional vector space; for example, if vectors S were electric dipole moments P rather than magnetic dipole moments the definition (A5) would not have to be modified, while the definition (A4) would (see Section 3.2). Therefore the *spin groups* now to be defined would more appropriately be called *vector groups*.

A group $\tilde{s}$ is called a *spin group* if it satisfies the conditions:

(S1) $\tilde{s}$ is a subgroup of $\tilde{\Omega} \times \tilde{K}$, where $\tilde{K}$ is a crystallographic group, and $\tilde{s}$ is such that $\tilde{K}_s = \tilde{K}$; and $\tilde{\Omega}_s \neq \tilde{E}$;

(S2) $\tilde{s}$ is an invariance group, in the sense of (A5) of some spin arrangement $\sigma = \{r;S\}$.

Unlike a magnetic group, a spin group $\tilde{s}$ may very well contain elements of the form (Ω,E), where $\Omega \neq E$. All such elements of $\tilde{s}$ form a spin group $\tilde{s}_o$, which is a normal subgroup of $\tilde{s}$, and is called a *spin-only group* because it acts non-trivially only in spin space. There are, apart from a similarity transformation, three spin-only groups, because the spins of a spin arrangement may be confined to a one- or two-dimensional subspace of the spin space, or to neither. The Ω-parts of the elements of spin-only groups form correspondingly the groups, in "international notation", ∞m, m and 1 (trivial group). One can show that a spin group $\tilde{s}$ is necessarily a direct product group of a spin-only group $\tilde{s}_o$ and a group $\tilde{z}$ which does not contain any elements of the form (Ω,E), where $\Omega \neq E$, that is, $\tilde{s} = \tilde{s}_o \times \tilde{z}$. It follows then from (S1), by an argument already used in the case of colour groups and magnetic groups, that $\tilde{K}$ is isomorphic onto $\tilde{z}$, and that $\tilde{K}/\tilde{k}$ is isomorphic onto $\tilde{Q} = \tilde{Q}_z$; here $\tilde{k}$ is the normal subgroup of $\tilde{K}$ which consists of the K-parts of the elements (E,K) of $\tilde{z}$, and the subgroup $\tilde{Q}$ of $\tilde{\Omega}$ is the group constituted by all those orthogonal transformations of the spin space which are the Ω-parts of $\tilde{z}$. Therefore, for each natural homomorphism ρ_λ of $\tilde{K}$ onto $\tilde{Q}$ determined by the isomorphism λ of $\tilde{K}/\tilde{k}$ onto $\tilde{Q}$, with $\ker \rho_\lambda = \tilde{k}$, there is one spin group

$$\tilde{z} = \tilde{z}_k + (Q_2,K_2)\tilde{z}_k + \ldots + (Q_q,K_q)\tilde{z}_k,$$

where $\tilde{z}_k$ is the normal subgroup of $\tilde{z}$ constituted by the elements (E,K) whose K-parts form the group $\tilde{k}$, and K_α in (Q_α,K_α) is mapped to Q_α under ρ_λ, $\alpha = 2,3,\ldots,q$. The symbol $\tilde{z}(\rho_\lambda;\tilde{K}\to\tilde{Q})$ thus specifies $\tilde{z}$ completely.

One next shows (by an argument analogous to that used in Section 3.1, to show that $\tilde{m}$ satisfies (m2)) that each such group $\tilde{z}$ satisfies (S2). Since the groups $\tilde{z}$ thus satisfy both (S1) and (S2), they are spin groups, called *non-trivial spin groups*; according to whether $\tilde{K}$ is a space group, point group, etc., a non-trivial spin group is called *spin space group, spin point group*, etc.

How to establish a list of non-trivial spin groups will now be explained for the case of spin point groups. The method is analogous to Method II for obtaining all colour point groups. One first determines, for each fixed $\tilde{K}$ and each fixed normal subgroup $\tilde{k}$ of $\tilde{K}$, those crystallographic point groups, $\tilde{Q}_1,\tilde{Q}_2,\ldots,\tilde{Q}_i$, one from each class of the 32 classes of crystallographic point groups, which are isomorphic onto $\tilde{K}/\tilde{k}$ (there are 18 classes of non-isomorphic point groups, and $1 \le i \le 4$). For each $\tilde{Q}_j$, $j = 1,2,\ldots,i$, and each isomorphism λ of $\tilde{K}/\tilde{k}$ onto $\tilde{Q}_j$ there is then exactly one non-trivial spin point group $\tilde{z}(\rho_\lambda:\tilde{K}\to\tilde{Q}_j)$. However, one does not regard two spin groups $\tilde{z}(\rho_{\lambda_1}:\tilde{K}\to\tilde{Q}_j)$ and $\tilde{z}(\rho_{\lambda_2}:\tilde{K}\to\tilde{Q}_j)$ as different if the isomorphism λ_1 arises from the isomorphism λ_2 by an automorphism of $\tilde{Q}_j$ which can be brought about by a similarity transformation. More generally, two spin groups $\tilde{z}(\rho_1:\tilde{K}_1\to\tilde{Q}_1)$ and $\tilde{z}(\rho_2:\tilde{K}_2\to\tilde{Q}_2)$ are said to belong to the same *class of spin groups* if $\tilde{Q}_1$ and $\tilde{Q}_2$ are in the same class of crystallographic groups, if the same is true of $\tilde{K}_1$ and $\tilde{K}_2$, and if the similarity transformation of $\tilde{K}_1$ onto $\tilde{K}_2$ maps ker ρ_1 onto ker ρ_2. Taking $\tilde{K}$ from each of the 32 classes of point groups, one obtains in this way 566 classes of spin point groups.

The reason why the number of classes of spin point groups (566) is quite different from the number of classes of colour point groups (212) is essentially two-fold: (1) For a fixed $\tilde{K}$ and $\tilde{k}$, the number i of the point groups $\tilde{Q}_j$ isomorphic onto $\tilde{K}/\tilde{k}$ and the number m of the transitive groups $\tilde{P}_n$ of permutations isomorphic onto $\tilde{K}/\tilde{k}$ are, in general, different. (2) For fixed $\tilde{K}/\tilde{k}$, $\tilde{Q}_j$ and $\tilde{P}_n$, the spin groups are arranged into classes using a different criterion from that used to arrange the colour groups into classes.

Not only all spin groups but also all spin translation groups have been determined and appropriately classified, but not many spin space groups are known. It has been pointed out that the knowledge of spin translation groups may make the interpretation of incomplete magnetic diffraction data less ambiguous.

A magnetic group $\tilde{m}$ can always be identified with a non-trivial spin group $\tilde{z}_{\tilde{m}}$ determined as follows: if (A_k,K), where $K = (R|v)$, is an element of $\tilde{m}$, then the corresponding element of $\tilde{z}_{\tilde{m}}$ is (Ω_k,K), where $\Omega_k = \eta_A\delta_R R$. Then the action of $\tilde{m}$, in the sense of (A4), on a spin arrangement σ will be identical with the action of $\tilde{z}_{\tilde{m}}$, in the sense of (A5). One may thus say that $\tilde{z}_{\tilde{m}}$ *is* the magnetic group $\tilde{m}$. In this sense a magnetic group $\tilde{m}$ may be a proper subgroup of a non-trivial spin group that, in turn, may be a proper subgroup of a spin group: $\tilde{z}_{\tilde{m}} \subset \tilde{z} \subset \tilde{s}$; thus even a non-trivial spin symmetry group of a spin arrangement may be larger than the magnetic symmetry group of the latter. However, it is the magnetic symmetry group of a spin arrangement rather than its spin symmetry group, which determines the macroscopic symmetry of the magnetically ordered crystal.

A final remark: It is of course also possible to define the action of magnetic groups on spin arrangements, and of spin groups on vector functions, using the wreath product in a way similar to that introduced in Section 2.3 for colour groups.

4.1. MULTIPLE SYMMETRY GROUPS

If on the same crystal several functions are defined, which have different symmetry colour groups (in the sense of (A2)), then one may define invariance groups of such a system of colour functions in a straightforward way. Such groups have been called *multiple colour groups*.

A physically interesting case is that of two vector functions, defined on the same crystal, one being a spin arrangement σ and the other an electric dipole moment arrangement δ. Invariance groups (in the sense of (A4) for σ, and in the sense of an appropriate definition of action for δ) of systems of two such vector functions have been called *magneto-ferroelectric groups* or *magnetoelectric groups*. For lack of space, nothing more will here be said about these groups.

5.1. SOME SYNONYMS

Colour groups = colour symmetry groups = multicolour groups = Belov groups = Van der Waerden-Burckhardt groups; this last term is, occasionally, reserved for colour groups for which $\tilde{L} \neq \tilde{k}$.

2-colour groups = black-and-white groups = four-dimensional groups in the three-dimensional space = antisymmetry groups = Shubnikov groups = dichromatic symmetry groups.

Often the term "Shubnikov groups" is used for "magnetic space groups", and the term "Heesch groups" for "magnetic point groups".

Multiple two-colour groups = multiple antisymmetry groups = Zamorzaev groups.

Colour W-groups = colour W-symmetry groups.

5.2. SOME REFERENCES

It is impossible within the space prescribed for this paper to list even the most important references on the subject.

Fortunately, there is a book by Shubnikov and Koptsik (1974) which gives a detailed account of the work (mostly of Soviet crystallographers and physicists) on the topics of Sections 2.1 and 2.2 of this paper, and also to a minor extent on those of Section 3.1; the book also contains a very extensive bibliography. However, I would like to single out the following papers and books: Heesch (1930), which is historically the first important paper on metacrystallographic groups (here the term "four-dimensional groups in the three-dimensional space" was introduced), Shubnikov's book (1951), which reopened the subject; Belov, Neronova and Smirnova (1955), where a complete list of Shubnikov groups was published for the first time; Koptsik's book (1966) with extensive tables and diagrams of these groups; Belov and Tarkhova (1956), where for the first time d-colour groups ($d > 2$) were introduced; Van der Waerden and Burckhardt (1961), whose treatment of colour groups serves as a starting point for that given in Section 2.1; Harker (1976), who gives a very conveniently arranged list of all colour point groups; Koptsik and Kotzev (1974a), who were the first to point out the possibility of using the wreath product in the way described in Section 2.2, and investigated the colour

W-groups. The definition of the wreath product given in that section is based on that given by Kurosh (1967).

The term "magnetic space groups" seems to have been invented by Landau and Lifschitz (1951), who were the first to realize the importance of these groups for the theory of magnetic properties of solids. A survey of the properties of magnetic groups in connection with the problem of classification of all spin arrangements has been given by Opechowski and Guccione (1965), where also a list of all magnetic space groups can be found, derived and arranged in a way adapted to the needs of physics of solids; see also for Section 3.1: Opechowski and Dreyfus (1971); Section 3.2: Opechowski (1974); Section 3.3: Litvin and Opechowski (1974), Litvin (1973 and 1976), Koptsik and Kotzev (1974b). Zamorzaev and Sokolov (1957) were the first to consider multiple symmetry groups.

N.V. BELOV, N.N. NEROVA and T.S. SMIRNOVA, Krist. 2, 3 (1957); Sov. Phys. Cryst. 2, 311; see also Holser (1964).

N.V. BELOV and T.N. TARKHOVA, Krist. 1, 4 and 619 (1956); Sov. Phys. Cryst. 1, 5 and 487; see also Holser (1964).

D. HARKER, Acta Cryst. A32, 133 (1976).

H. HEESCH, Z. Krist. 73, 325 (1930).

W.T. HOLSER, editor: *"Colored Symmetry by A.V. Shubnikov, N.V. Belov and others"*, The MacMillan Co., New York, 1964.

V.A. KOPTSIK, *"Shubnikov's Groups"* (in Russian), Publ. of the Moscow University, 1966.

V.A. KOPTSIK and J.N. KOTZEV, Comm. Joint Inst. Nuclear Research, Dubna, USSR, P4-8068 (1974a).

V.A. KOPTSIK and J.N. KOTZEV, Comm. Joint Inst. Nuclear Research, Dubna, USSR, P4-8466 (1974b).

A.G. KUROSH, *"Theory of Groups"*, 2nd Edition (in Russian), 1967.

L.L. LANDAU and E.M. Lifschitz, *"Statisticheskaya Fizika"*, GITTL; Statistical Physics (Addison-Wesley, Reading, Mass.), 1958.

D.B. LITVIN, Acta Cryst. A29, 651 (1973).

D.B. LITVIN, Acta Cryst., in press.

D.B. LITVIN and W. OPECHOWSKI, Physica 76, 538 (1974).

W. OPECHOWSKI, Int. J. Magnetism 5, 317 (1974).

W. OPECHOWSKI and T. DREYFUS, Acta Cryst. A27, 470 (1971).

W. OPECHOWSKI and R. GUCCIONE, *"Magnetism"*, vol. IIA, G.T. Rado and H. Suhl, eds. (Academic Press, New York), Ch. 3, 1965.

A.V. SHUBNIKOV, *"Symmetry and Antisymmetry of Finite Figures"* (in Russian), Publ. of the Soviet Acad. of Science; English transl. in Holser (1964).

A.V. SHUBNIKOV and V.A. KOPTSIK, *"Symmetry in Science and Art"*, (Plenum Press, New York), 1974.

B.L. VAN DER WAERDEN and J.J. BURCKHARDT, Z. Krist. 115, 231 (1961).

A.M. ZAMORZAEV and E.I. SOKOLOV, Krist. 2, 9; Sov. Phys. Cryst. 2, 5 (1957).

THE SP(3,$\mathbb{R}$) MODEL OF NUCLEAR COLLECTIVE MOTION

G. Rosensteel and D.J. Rowe

1. INTRODUCTION

The Bohr-Mottelson model of nuclear rotations and quadrupole vibrations is firmly established as a successful collective model [1,2]. The model explains the observed properties of spatially collective states for a wide class of nuclei and, as Professor Moshinsky showed in his talk, an analytic solution can be given to the problem of constructing states with good angular momentum [3]. This is a very desirable state of affairs in the development of any theory. Nevertheless, this model is phenomenological and, in its standard formulation, cannot be naturally related to the microscopic theory of the nucleus. In particular, vectors in the Bohr-Mottelson model which define the states of a quantum mechanical liquid drop cannot be naturally identified with wave functions in $\mathcal{L}^2(\mathbb{R}^{3N})^a$, the Hilbert space of N nucleon anti-symmetric states (spin-isospin are temporarily suppressed in the explanation of spatial collective modes). Professor E. Wigner in his concluding remarks for the 1970 Solvay Conference emphasized the importance for nuclear structure physics of the problem of incorporating the Bohr-Mottelson model into the microscopic theory of the nucleus, i.e. the state space is $\mathcal{L}^2(\mathbb{R}^{3N})^a$ [4]. In this talk, the solution to this problem is given through an algebraic approach.

In this introduction, we briefly discuss the conditions necessary for a successful algebraic collective model and illustrate these general remarks by the rotational model $[\mathbb{R}^5]so(3)$. In 2, microscopic rotational states are characterized in terms of the Casimir invariants of $[\mathbb{R}^5]so(3)$; this criterion is then applied to the su(3) model with interesting results. The cm(3) algebraic model is shown in 3 to be an algebraic formulation of the Bohr-Mottelson theory. Finally, in 4, the $sp(3,\mathbb{R})$ model is derived as the adaptation to the shell model of the Bohr-Mottelson theory.

Algebraic models

Algebraic models afford a lucid and succinct formulation, both phenomenologically and microscopically, of collective phenomena. The assumption of the algebraic approach to collective problems is that a single irreducible representation of a Lie algebra of observables is adequate to describe collective effects qualitatively. This will be the case provided the Lie algebra contains the *relevant* collective observables and if it is a spectrum generating algebra (S.G.A.). The latter requirement is the dynamical condition that the commutator of any observable in the algebra with the Hamiltonian is a function only of the observables in the algebra [5,6,7]. The practical success of an algebraic model depends upon the choice of the relevant observables. If only the few degrees of freedom essential to the collective phenomena are contained in the algebra, then the model will be a considerable simplification of the full many-body problem with 3N degrees of freedom. In addition, to each irreducible representation defining a phenomenological model, the algebraic approach naturally yields a microscopic theory by reducing the representation of the group on many-particle state space $\mathcal{L}^2(\mathbb{R}^{3N})^a$ into its irreducible components.

A basic example in the context of collective rotational motion is due to Ui [8] and Weaver, Biedenharn and Cusson [9].

Consider a system of N nucleons with position and momentum coordinates $x_{j\alpha}$ and $p_{j\alpha} = -i \frac{\partial}{\partial x_{j\alpha}}$, $1 < j < N$, $\alpha = x,y,z$, respectively. Let $L_{\alpha\beta}$ denote the total angular momentum and $Q^{(2)}_{\alpha\beta}$ the traceless quadrupole tensor,

$$L_{\alpha\beta} = \sum_{j=1}^{N} (x_{j\alpha}p_{j\beta} - x_{j\beta}p_{j\alpha})$$
$$Q^{(2)}_{\alpha\beta} = \sum_{j=1}^{N} (x_{j\alpha}x_{j\beta} - \frac{1}{3}\delta_{\alpha\beta} \sum_{\gamma} x^2_{j\gamma}). \qquad (1)$$

The real Lie algebra spanned by the skew-adjoint operators $iL_{\alpha\beta}$ and $iQ^{(2)}_{\alpha\beta}$ is isomorphic to a semidirect sum $[\mathbb{R}^5]so(3)$. Here $\mathbb{R}^5$, the ideal of the sum, is the five-dimensional abelian Lie algebra spanned by the elements of the quadrupole tensor and so(3) is the angular momentum algebra.

Clearly, $[\mathbb{R}^5]so(3)$ is a Lie algebra of collective observables relevant to rotational motion. It is also a S.G.A. with respect to the rotor Hamiltonian,

$$H = \tfrac{1}{2} \sum_{\alpha\beta} L_{\alpha} \mathcal{I}^{-1}_{\alpha\beta} L_{\beta},$$

where $(\mathcal{I}_{\alpha\beta})$ is the inertia tensor which is a function of $(Q^{(2)}_{\alpha\beta})$. The exponentiation of $[\mathbb{R}^5]so(3)$ on $\mathcal{L}^2(\mathbb{R}^{3N})^a$ is a Lie group of unitary operators $[\mathbb{R}^5]SO(3) = \exp([\mathbb{R}^5]so(3))$. This is a semidirect product group with the normal subgroup $\mathbb{R}^5$ abelian; hence, its irreducible unitary representations are given by the inducing construction [10]. The irreps one obtains are indexed by two real numbers $\bar{Q}^{(2)}_{\Delta K=0}$ and $\bar{Q}^{(2)}_{\Delta K=2}$ and two discrete parameters ε_1 and ε_2 equal to either 0 or 1. The $\varepsilon_1 = \varepsilon_2 = 0$ irreps are most relevant to even-even nuclei. An orthonormal basis for the carrier space of each of the $\varepsilon_1 = \varepsilon_2 = 0$ irreps is given by $\{|KJM\rangle\}$, where the quantum numbers range over

$$K = 0, \quad J = 0,2,4,\ldots$$
$$K \text{ even}, \quad J = K, K+1, K+2, \ldots \quad . \tag{2}$$

Here, JM are the angular momentum quantum numbers. The reduced matrix elements of $Q^{(2)}$ are given by

$$\langle K'J' || Q^{(2)} || KJ \rangle = \sqrt{2J+1}\, (JK2\Delta K | J'K') Q^{(2)}_{\Delta K}, \quad \Delta K = K'-K. \tag{3}$$

This prediction for the spectrum of states and reduced matrix elements of the quadrupole moment is just that of the collective rotational model [2]. In addition to its clarity, the advantage to this algebraic formulation of the rotational model is its straightforward relation with the microscopic theory. Indeed, the reduction of $[\mathbb{R}^5]SO(3)$ into its irreducible components on $\mathcal{L}(\mathbb{R}^{3N})^a$ yields the transformation to the well-known collective and intrinsic coordinates of Villars [11].

However, the desired microscopic theory is one which exhibits both the collective and single-particle aspects of the nuclear wave function. For this purpose, Elliott developed the su(3) model [12], which is the adaptation of the $[\mathbb{R}^5]so(3)$ algebra to the considerations of the shell model. In addition to the angular momentum, su(3) is spanned by the second rank tensor

$$\mathcal{Q}^{(2)}_{\alpha\beta} = \tfrac{1}{2} \sum_{j=1}^{N} [(x_{j\alpha}x_{j\beta}+p_{j\alpha}p_{j\beta}) - \tfrac{1}{3}\delta_{\alpha\beta} \sum_{\gamma} (x^2_{j\gamma}+p^2_{j\gamma})]. \tag{4}$$

The operator $\mathcal{Q}^{(2)}$ has the same in-shell matrix elements as $Q^{(2)}$, a relation denoted here by $\mathcal{Q}^{(2)} = Q^{(2)}|\text{shell}$. But its inter-shell matrix elements are all zero. Thus the su(3) model is a truncation of $[\mathbb{R}^5]so(3)$ to a single shell. The su(3) model is very well known; we recall only that the irreps of su(3) exhibit a rotational band structure similar to the $[\mathbb{R}^5]so(3)$ model, cf. eq. (2), but truncated to some maximum J characteristic of the su(3) irrep.

2. SHAPE OBSERVABLES [13]

In addition to the requirements that the Lie algebra be spanned by relevant collective observables and be a spectrum generating algebra, the model restriction to irreducible representations needs to be justified. This restriction can be tested by determining if the Casimir invariants of the algebra take on constant values. In the case of su(3), the quadratic invariant must be constant in order that the su(3) Hamiltonian, with the quadrupole-quadrupole interaction $\underset{\sim}{Q}^{(2)} \cdot \underset{\sim}{Q}^{(2)}$, should be diagonalized. For $[\mathbb{R}^5]so(3)$, however, there is a more basic geometric reason.

Consider the three principal moments (eigenvalues) λ_α of the traceless quadrupole tensor $Q^{(2)}$. The λ_α are shape operators measuring the intrinsic deformation of the nucleus. Each of the λ_α is a solution to the cubic secular equation

$$\begin{aligned} 0 &= \det(\lambda I - Q^{(2)}) \\ &= \lambda^3 + a_2\lambda + a_3 \end{aligned} \tag{5}$$

where

$$\begin{aligned} a_2 &= -\frac{1}{2}\sum_{\alpha\beta} Q^{(2)}_{\alpha\beta} Q^{(2)}_{\beta\alpha} = -\frac{1}{12}[Q^{(2)} \times Q^{(2)}]^0 \\ a_3 &= -\det Q^{(2)} = \frac{1}{108}\sqrt{\frac{7}{2}}\,[Q^{(2)} \times Q^{(2)} \times Q^{(2)}]^0 . \end{aligned} \tag{6}$$

Note that the coefficients a_2 and a_3 are invariants of $[\mathbb{R}^5]so(3)$. In fact, a_2 and a_3 form an integrity basis.

A state Φ can be said to be rotational if it has a well-defined intrinsic structure with a sharp quadrupole shape. Hence, a rotational state is an approximate eigenfunction of the three shape operators λ_α. But, for a state to be an approximate eigenstate of all three λ_α, it is necessary and sufficient for that state to be an approximate simultaneous eigenstate of the coefficients of the secular equation, viz. a_2 and a_3. This gives then an effective test for a state to be rotational: the fluctuations in a_α must be small compared to the mean value.

Definition. A state Φ is rotational if the moments of a_2 and a_3 satisfy

$$\frac{\langle \Phi | a_\alpha^2 \Phi \rangle - \langle \Phi | a_\alpha \Phi \rangle^2}{\langle \Phi | a_\alpha \Phi \rangle^2} << 1. \tag{7}$$

A typical value for the left hand side of the above inequality is 1/10 for a rare earth nucleus.

In order that a set of rotational states form a rotational band, we further require that the states take on constant values for the invariants a_2 and a_3. Hence, the restriction to irreps of $[\mathbb{R}^5]so(3)$ is equivalent to the assumption that the states of the model form a rotational band.

Incidentally, other properties of the intrinsic shape can be characterized in terms of a_2 and a_3. For example, axially-symmetric states Φ satisfy

$$\frac{\langle \Phi | (-\frac{1}{27} a_2^3 - \frac{1}{4} a_3^2) \Phi \rangle}{\langle \Phi | \frac{1}{4} a_3^2 \Phi \rangle} << 1.$$

su(3) Model

The above criterion for rotational states may also be applied to the Elliott su(3) model. In this case, one requires that su(3) rotational states be eigenstates of $a_2|_{shell}$ and $a_3|_{shell}$. One can show that

$$\sum_{\alpha\beta} Q_{\alpha\beta}^{(2)} Q_{\beta\alpha}^{(2)} |_{shell} = \frac{5}{3} H_0 + C_{su(3)}^{(2)} - \frac{1}{2} L^2$$

$$\sum_{\alpha\beta\gamma} Q_{\alpha\beta}^{(2)} Q_{\beta\gamma}^{(2)} Q_{\gamma\alpha}^{(2)} |_{shell} = \frac{35}{18} H_0 + C_{su(3)}^{(2)} + \frac{7}{4} C_{su(3)}^{(3)} - \frac{7}{8} L^2 + \frac{3}{4} \sum_{\alpha\beta} L_\alpha Q_{\alpha\beta}^{(2)} L_\beta \tag{8}$$

where H_0 is the harmonic-oscillator Hamiltonian, $C^{(2)}_{su(3)}$ and $C^{(3)}_{su(3)}$ are the quadratic and cubic invariants of su(3) and

$$X = \sum_{\alpha\beta} L_\alpha Q^{(2)}_{\alpha\beta} L_\beta$$

is the Bargmann-Moshinsky operator [14].

Hence, the Elliott states $|(\lambda\mu)KLM\rangle$, although eigenstates of $a_2|_{shell}$, are not eigenstates of $a_3|_{shell}$ since X is not diagonal in the Elliott basis. We conclude that the states in the Elliott basis are in general not rotational even with the approximation of truncation to a single shell.

However, the closest possible approximation to rotational states that can be achieved in the truncated shell model basis of the su(3) model are the Bargmann-Moshinsky states $|(\lambda\mu)xLM\rangle$, which are eigenstates of X,

$$X|(\lambda\mu)xLM\rangle = x|(\lambda\mu)xLM\rangle .$$

These states are eigenstates of $a_2|_{shell}$ and $a_3|_{shell}$. The eigenvalues x have been determined for most cases of interest [15].

Although the criterion for rotational states resolves the "missing label" problem for su(3), it poses the interesting question as to what physical significance attaches to the x quantum number. In particular, what is the relationship between x and K, the projection of the angular momentum on the intrinsic z-axis.

In order to bring out the connection between

$$\sum_{\alpha\beta} L_\alpha Q^{(2)}_{\alpha\beta} L_\beta$$

and the K observable, it is necessary to compute

$$\sum_{\alpha\beta} L_\alpha Q^{(2)}_{\alpha\beta} L_\beta$$

in the collective and intrinsic coordinates of Villars [11]. In these coordinates we have

$$Q^{(2)}_{\alpha\beta} = \sum_A R_{\alpha A} R_{\beta A} \lambda_A(\xi), \qquad R \in SO(3) \tag{9a}$$

while

$$I_A = \sum_\alpha L_\alpha R_{\alpha A} \tag{9b}$$

are the angular momentum operators in the intrinsic frame,

$$\begin{aligned} I_3 \Psi^L_{MK} &= K \Psi^L_{MK} \\ I_\pm \Psi^L_{MK} &= (L(L+1)-K(K\pm 1))^{\frac{1}{2}} \Psi^L_{MK\pm 1}. \end{aligned} \tag{10}$$

Hence,

$$\begin{aligned} \sum_A I_A^2 &= L^2 \\ \sum_A \lambda_A I_A^2 &= \sum_{\alpha\beta} L_\alpha Q^{(2)}_{\alpha\beta} L_\beta \\ \sum_A \lambda_A^2 I_A^2 &= \sum_{\alpha\beta\gamma} L_\alpha Q^{(2)}_{\alpha\beta} Q^{(2)}_{\beta\gamma} L_\gamma. \end{aligned} \tag{11}$$

If these equations are solved for I_3^2, one obtains a very complicated expression. However, when acting on axially-symmetric rotational states, I_3^2 has the same action as

$$K^2 = \frac{1}{3} L^2 - \frac{1}{3} \left(\frac{1}{2} a_3\right)^{-1/3} \sum_{\alpha\beta} L_\alpha Q^{(2)}_{\alpha\beta} L_\beta. \tag{12}$$

Since it is only for axially-symmetric rotational states that K is a good quantum number, the relatively simple operator K^2 suffices for most purposes.

In order to test the operator K^2, we may compute K^2 in the $[\mathbb{R}^5]so(3)$ model. If the interband transitions are small (experimentally a factor 10-50 less than the intraband transitions, $\bar{Q}^{(2)}_{\Delta K=2} \ll \bar{Q}^{(2)}_{\Delta K=0}$, for rare earth nuclei), each state $|KJM\rangle$ is an approximate eigenstate of $\mathcal{K}^2$ belonging to the eigenvalue K^2,

$$\langle KJM|\mathcal{K}^2|KJM\rangle \doteq K^2. \tag{13}$$

In Table 1, the K quantum number has been determined for su(3) rotational (Bargmann-Moshinsky) states from the approximation,

$$\langle (\lambda\mu)xLM|\mathcal{K}^2|(\lambda\mu)xLM\rangle \doteq \frac{1}{3}L(L+1) - \frac{1}{3}(\frac{1}{2}\alpha_3)^{-1/3}x \tag{14}$$

where $\alpha_3 = a_3|_{shell}$. In view of the truncation to $0\hbar\omega$ shell model space, the emergence of bands with approximately constant values for K is quite remarkable. Note that in Mg^{24}, the states separate into bands with K close to the integral values 0, 2, 4, predicted by Elliott.

3. cm(3) MODEL

We have examined the algebraic model for collective rotational motion $[\mathbb{R}^5]so(3)$ and its shell model adaptation su(3). Now we want to include vibrations and obtain the Bohr-Mottelson model. To do this, it is convenient to construct explicitly the irreducible representations of $[\mathbb{R}^5]so(3)$ by the inducing method. When this construction is expressed in physical terms, the generalization to the Bohr-Mottelson theory becomes transparent.

The rotational and Bohr-Mottelson models suppose that the nucleus can be considered, for the explanation of certain collective effects, to be specified by its surface [1,2]. The model nucleus is defined by giving its radius $R(\theta,\psi)$ as a function of orientation angles (θ,ψ). In a multipole expansion, $R(\theta,\psi)$ is determined by a set of deformation parameters $\alpha_{\lambda\mu}$,

$$R(\theta,\psi) = R_0[1+ \sum_{\lambda\mu} \alpha^*_{\lambda\mu}Y_{\lambda\mu}(\theta,\psi)]. \tag{15}$$

For the explanation of quadrupole transitions and vibrations, only the $\alpha_{2\mu}$ parameters are involved. With this limitation, the

TABLE 1. Square root of the expectation value of K^2 with respect to the Bargmann-Moshinsky states.

Mg^{24}: $(\lambda,\mu) = 84$

L	0	2	4	6	6				
K	0.0	0.3	1.0i	2.4i	4.6i				
L	2	3	4	5	6	7	8	9	10
K	2.0	2.0	2.4	2.0	2.9	1.5	3.1	2.7i	2.3
L	4	5	6	7	8	9	10	11	12
K	3.8	3.9	4.1	4.2	4.4	4.4	4.9	4.2	5.2

TABLE 2. Ratio of B(E2:L→L-2) to B(E2:2→0) for the most deformed bands in each of the subspaces including states up to $0\hbar\omega$, $2\hbar\omega$ and $4\hbar\omega$.

	L	4	6	8
B(E2:L→L-2) / B(E2:2→0)	$0\hbar\omega$	1.27	1.07	0.64
	$2\hbar\omega$	1.34	1.29	1.07
	$4\hbar\omega$	1.37	1.39	1.28
	Rotational	1.43	1.57	1.65

rotational and Bohr-Mottelson models have supposed that the configuration space of the nucleus, vis-à-vis collective quadrupole phenomena, is parameterized by the five deformation parameters $\alpha_{2\mu}$.

We may regard these five parameters as the spherical coordinates on the manifold $\mathcal{Q}_0$ of traceless three by three symmetric real matrices

$$\mathcal{Q}_0 = \{q^{(2)} \in M_3(\mathbb{R}) \,|\, q^{(2)t} = q^{(2)}, \mathrm{tr}\, q^{(2)} = 0\}. \qquad (16)$$

A reformulation of the collective models is now suggested. Interpret the points of $\mathcal{Q}_0$ as the quadrupole moments $Q^{(2)}$.

SO(3) orbits and induced representations of $[\mathbb{R}^5]SO(3)$

The rotation group acts on $\mathcal{Q}_0$ by

$$q^{(2)} \xmapsto{R} Rq^{(2)}R^t, \quad q^{(2)} \in \mathcal{Q}_0, \quad R \in SO(3). \qquad (17)$$

For a rigid body, motion is constrained to an orbit of this action. Each such orbit contains a unique traceless diagonal matrix, $\mathrm{diag}(\lambda_1,\lambda_2,\lambda_3)$ with $\lambda_2 \le \lambda_1 \le \lambda_3$ (λ_α are the principal moments). In the generic case, $\lambda_2 \ne \lambda_1 \ne \lambda_3$ and the isotropy subgroup is

$$M = \{\mathrm{diag}(1,1,1), \mathrm{diag}(1,-1,-1), \mathrm{diag}(-1,1,-1), \mathrm{diag}(-1,-1,1)\}. \qquad (18)$$

An orbit is then identified with the coset space SO(3)/M via the diffeomorphism,

$$q^{(2)} = R^t \begin{pmatrix} \lambda_1 & & \\ & \lambda_2 & \\ & & \lambda_3 \end{pmatrix} R \longmapsto MR \in SO(3)/M. \qquad (19)$$

In addition, SO(3) now acts on SO(3)/M by

$$MR \xmapsto{x} MRx^t, \quad x \in SO(3). \qquad (20)$$

The quantized state space for this constrained motion is $\mathcal{L}^2(SO(3)/M)$. An orthonormal basis for $\mathcal{L}^2(SO(3)/M)$ is given by

$$\Psi^L_{KM}(R) = \begin{cases} D^L_{0M}(R), & K = 0, \quad L = 0,2,4,\ldots \\ \frac{1}{\sqrt{2}} (D^L_{KM}(R) + (-)^{L-K} D^L_{-KM}(R)), & K \text{ even}, \end{cases} \tag{21}$$

$$L = K, K+1, K+2, \ldots$$

since Ψ^L_{KM} is a complete set on SO(3) satisfying

$$\Psi^L_{KM}(mR) = \Psi^L_{KM}(R) \quad \text{all} \quad R \in SO(3), \quad m \in M. \tag{22}$$

Then, with respect to the action given in (20), the vector Ψ^L_{KM} is a good LM angular momentum state. The traceless quadrupole moment $Q^{(2)}_{\alpha\beta}$ acts on $\mathcal{L}^2(SO(3)/M)$ as a multiplication operator via the identification

$$(Q^{(2)}_{\alpha\beta}\Psi)(R) = \left[R^t \begin{pmatrix} \lambda_1 & & \\ & \lambda_2 & \\ & & \lambda_3 \end{pmatrix} R\right]_{\alpha\beta} \Psi(R). \tag{23}$$

This action of $\mathbb{R}^5$ and SO(3) on $\mathcal{L}^2(SO(3)/M)$ is an irreducible representation of $[\mathbb{R}^5]SO(3)$. In fact, the above derivation is formally the same as the inducing construction for the $\varepsilon_1 = \varepsilon_2 = 0$ series. The advantage to this derivation is that the connection with the collective rotational model is immediate. Moreover, the generalization to vibrations is straightforward.

SL(3) orbits and induced representations of $CM(3) = [\mathbb{R}^6]SL(3)$

Rotational motion was obtained in the collective model by constraining to SO(3) orbits in $(\mathcal{Q}_0)$. For vibrational motion, it is natural to let GL(3) act on $(\mathcal{Q}_0)$. However, since a traceless matrix is not mapped into a traceless matrix by the action of $(\mathcal{Q}_0)$ for GL(3), we generalize to the configuration space

$$\mathcal{Q} = \{q \in M_3(\mathbb{R}) \mid q^t = q\}, \tag{24}$$

the points of which are identified with

$$Q_{\alpha\beta} = \sum_{j=1}^{N} x_{j\alpha} x_{j\beta}.$$

GL(3) now acts on $\mathcal{Q}$ by

$$q \xmapsto{g} gqg^t, \quad q \in \mathcal{Q},\ g \in GL(3). \tag{25}$$

The Bohr-Mottelson model restricts consideration to incompressible (liquid drop) motion; correspondingly we constrain motion to orbits of the subgroup SL(3) of GL(3). Each orbit contains a unique multiple of the identity, diag (G_0, G_0, G_0). If $G_0 \neq 0$, then the isotropy subgroup is SO(3). An orbit is then identified with the coset space SL(3)/SO(3) by

$$q = g^{-1}\begin{pmatrix} G_0 & & \\ & G_0 & \\ & & G_0 \end{pmatrix}(g^t)^{-1} \longmapsto SO(3)g \in SL(3)/SO(3). \tag{26}$$

Moreover, SL(3) acts on SL(3)/SO(3) by

$$SO(3)g \xmapsto{x} SO(3)gx^{-1}, \quad x \in SL(3). \tag{27}$$

The corresponding quantized state space for this constrained motion is $\mathcal{L}^2(SL(3)/SO(3))$. The action of SL(3) on this state space is

$$(U(x)\Psi(g) = \Psi(gx), \quad x \in SL(3), \tag{28}$$

with $\Psi \in \mathcal{L}^2(SL(3)/SO(3))$, i.e., $\Psi(R_g) = \Psi(Rg)$, all $R \in SO(3)$, $g \in SL(3)$. The quadrupole moment acts on $\mathcal{L}^2(SL(3)/SO(3))$ as a multiplication operator,

$$(Q_{\alpha\beta}\Psi)(g) = G_0(g^t g)^{-1}_{\alpha\beta}\Psi(g). \tag{29}$$

A coordinate system for SL(3)/SO(3) and its relation to the familiar β,γ deformation parameters of the Bohr-Mottelson model can be given through the polar decomposition. One shows that the map

$$\begin{aligned} D \times SO(3)/M &\to SL(3)/SO(3) \\ (d,MR) &\mapsto SO(3)d\cdot R \end{aligned} \tag{30}$$

is a diffeomorphism where D is the subgroup of diagonal matrices

$$D = \{d=\text{diag}(d_1,d_2,d_3)\,|\,\det d=1, d_2>d_1>d_3>0\}. \tag{31}$$

Then SO(3)/M is the coset space of the rotational model and D is coordinatized by the β,γ parameters

$$d_k = \exp[-\frac{1}{2}\sqrt{\frac{5}{\pi}}\,\beta\,\cos(\gamma - k\,\frac{2\pi}{3})], \tag{32}$$

with $\beta > 0$ and $0 < \gamma < \frac{\pi}{3}$.

It is evident from the derivation that this model is the Bohr-Mottelson theory, albeit expressed in the terse algebraic style. Moreover, the action of $Q_{\alpha\beta}$ and sl(3) on $\mathcal{L}^2(SL(3)/SO(3))$ is an irreducible representation of cm(3) = $[\mathbb{R}^6]$sl(3), where $\mathbb{R}^6$ is the six dimensional abelian ideal spanned by the $Q_{\alpha\beta}$. Indeed, the argument given above is formally the same as the inducing construction for cm(3) irreps [16].

We also want to point out that the kinetic energy for these constrained motions can be naturally given. Each orbit of the action of the group G (= SO(3), SL(3) or GL(3)) on $\mathbb{R}^{3N}$ defines a submanifold of $\mathbb{R}^{3N}$ on which there is an induced metric. This metric defines the (free particle Hamiltonian) Laplace-Beltrami operator. In the case of SO(3), the rigid body Hamiltonian is derived; whereas, for SL(3), the Bohr-Mottelson kinetic energy is obtained. More details are given in a thesis [17].

The microscopic counterpart to the Bohr-Mottelson theory is

now given by reducing $CM(3) = \exp(cm(3)) = [\mathbb{R}^6]SL(3)$ into irreducible components on $\mathcal{L}^2(\mathbb{R}^{3N})^a$. If $\mathcal{L}^2(\mathbb{R}^{3N})^a$ is decomposed into

$$\mathcal{L}^2(\mathbb{R}^{3N})^a \cong \mathcal{L}^2(GL_+(3)) \otimes \mathcal{H}_{int}, \tag{33}$$

given by the splitting into orbits of $GL_+(3)$ and a smooth transversal (orbit representatives), called the intrinsic submanifold, then CM(3) acts only on the orbits, i.e., $\mathcal{L}^2(GL_+(3))$. The direct integral decomposition of CM(3) on $\mathcal{L}^2(GL_+(3))$ is straightforward [17,18,19].

Derived from the algebraic approach based on cm(3), the microscopic analogue to the Bohr-Mottelson theory is not completely satisfactory. What is most desirable is a microscopic theory whereby both the collective and single particle aspects of the nuclear wave function can be simultaneously investigated, in the context of the shell model. For the Bohr-Mottelson theory to be realized in the shell model, one considers the symplectic Lie algebra $sp(3,\mathbb{R})$.

4. $sp(3,\mathbb{R})$ MODEL

The smallest Lie algebra containing both cm(3) and su(3) is the Lie algebra of the noncompact real symplectic group $sp(3,\mathbb{R})$ of dimension 21. This real Lie algebra is spanned by the skew-adjoint one-body bilinear products in the position and momentum observables [20]. In addition to the subalgebra, isomorphic to $gl(3,\mathbb{R})$, spanned by the angular momentum $iL_{\alpha\beta}$ and the stretching momentum

$$iS_{\alpha\beta} = i \sum_{j=1}^{N} (X_{j\alpha}P_{j\beta}+X_{j\beta}P_{j\alpha}-i\delta_{\alpha\beta}I), \tag{34}$$

$sp(3,\mathbb{R})$ contains the quadrupole moments $iQ_{\alpha\beta}$ and the quadratics in the momenta

$$iK_{\alpha\beta} = i \sum_{j=1}^{N} P_{j\alpha} P_{j\beta}. \tag{35}$$

Relevant subalgebras of sp(3,ℝ) are shown in Figure 1.

The action of this Lie algebra can be most easily exhibited in terms of bilinear products of oscillator boson operators. The complexification of sp(3,ℝ) is spanned by

$$A_{\alpha\beta} = \sum_{j=1}^{N} a^{+}_{j\alpha} a^{+}_{j\beta}, \quad B_{\alpha\beta} = \sum_{j=1}^{N} a_{j\alpha} a_{j\beta}, \quad C_{\alpha\beta} = \sum_{j=1}^{N} a^{+}_{j\alpha} a_{j\beta}. \tag{36}$$

Note. The $C_{\alpha\beta}$ span just u(3).

The irreducible representations involved in the direct sum decomposition of sp(3,ℝ) on $\mathcal{L}^2(\mathbb{R}^{3N})^a$ are those in the discrete series [21,22]. If one starts with the su(3) states $|(\lambda\mu)xLM\rangle$ and augments this $0\hbar\omega$ subspace with the $2\hbar\omega$ states $A_{\alpha\beta}|(\lambda\mu)xLM\rangle$, $4\hbar\omega$ states $A_{\alpha\beta}A_{\gamma\delta}|(\lambda\mu)xLM\rangle$, etc., then the resulting states together span an irreducible discrete series representation of sp(3,ℝ). In Figure 2, the su(3) irreps occurring in the decomposition of the sp(3,ℝ) representations whose $0\hbar\omega$ levels transform according to (0,0) and (8,0) under su(3) are given for the first few oscillator levels.

5. SUMMARY AND APPLICATIONS

In this talk, a number of algebraic models have been discussed which together represent the various aspects of collective motion associated with nuclear quadrupole dynamics. These models are united in the sp(3) model and are adapted to the nuclear shell model in a way that permits practical microscopic calculations.

One application that immediately comes to mind is to attempt to generate pure rotational states in a shell-model basis. For this we should diagonalize the invariants of $[\mathbb{R}^5]$so(3) within an irreducible sp(3,ℝ) space, which generates states of sharp

FIGURE 1. Subalgebras of the symplectic Lie algebra sp(3, $\mathbb{R}$).

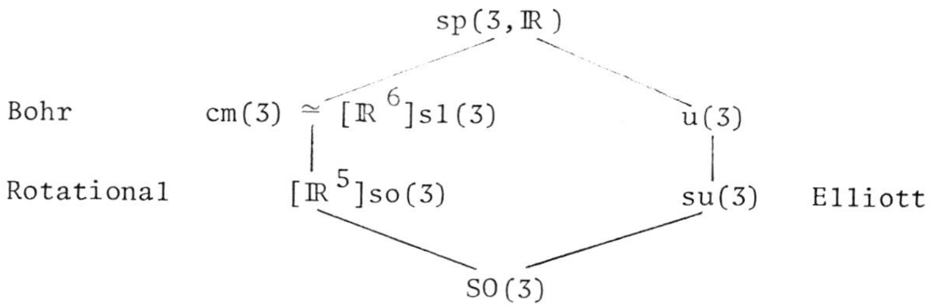

FIGURE 2. su(3) spectrum for the sp(3, $\mathbb{R}$) irreps whose $0\hbar\omega$ subspace transforms according to (0,0) and (8,0) under su(3).

	⋮		⋮	
(6,0), (2,2), (0,0)	$6\hbar\omega$	$0^3, 2^3, 3, 4^2, 6$	$6\hbar\omega$	
(4,0), (0,2)	$4\hbar\omega$	$0^2, 2^2, 4$	$4\hbar\omega$	(12,0), (10,1), (6,3), (4,4), $(8,2)^2$, (7,1), (6,0)
(2,0)	$2\hbar\omega$	0, 2	$2\hbar\omega$	(10,0), (8,1), (6,2)
(0,0)	$0\hbar\omega$	0	$0\hbar\omega$	(8,0)
(λ,μ)		L		(λ,μ)

quadrupole shape. Since this would require an infinite number of configurations, it is relevant to investigate the extent to which pure rotational states can be formed within a truncated subspace. We therefore diagonalized the quadratic invariant a_2 of $[\mathbb{R}^5]so(3)$ in the 3 truncated shell-model subspaces of vectors up to an including those of the $0\hbar\omega$, $2\hbar\omega$ and $4\hbar\omega$ shells, respectively. The ratios of E2 transitions are given, for the states of the most deformed band in each of the 3 subspaces, in Table 2. It is evident that these ratios must approach the rotational model prediction as the subspace is enlarged. From the table, it can be seen that even by $4\hbar\omega$, a substantial change from the $0\hbar\omega$ su(3) prediction toward the rotational model limit is achieved.

Another application is to generate rotational-vibrational states. To this end, one would seek eigenstates of a Hamiltonian

$$H = T + V(a_2, a_3),$$

where T is the nuclear kinetic energy and V, a polynomial in the quadrupole shape operators, is a potential energy which can be designed to have a minimum at some suitable equilibrium shape. This Hamiltonian is in the enveloping algebra of $sp(3,\mathbb{R})$ and can therefore be diagonalized within an irreducible $sp(3,\mathbb{R})$ subspace.

A third application is to investigate the influence of shell effects on collective motion by seeking eigenstates of the Hamiltonian

$$H = H_{HO} + V(a_2, a_3)$$

where H_{HO} is the harmonic oscillator shell-model Hamiltonian. This Hamiltonian is again in the enveloping algebra of $sp(3,\mathbb{R})$.

Finally the $sp(3,\mathbb{R})$ model can be used simply to provide basis states for conventional shell-model calculations with the most realistic microscopic Hamiltonians available. Apart from labelling basis states, the primary purpose served by the $sp(3,\mathbb{R})$

model would then be to prescribe what states should be included in the shell-model space to admit the possibility of collective correlations.

REFERENCES

1. A. BOHR, Mat. Phys. Medd. Dan. Vid. Selsk. 26, no. 14 (1952).
2. K. ADLER, A. BOHR, T. HUUS, B. MOTTELSON and A. WINTHER, Rev. Mod. Phys. 28, 432 (1956).
3. E. CHACON, M. MOSHINSKY and R.T. SHARP, J. Math. Phys. 17, 668 (1976).
4. E. WIGNER, *in* Symmetry Properties of Nuclei (Gordon and Breach, 1970).
5. Y. DOTHAN, Phys. Rev. D2, 2944 (1970).
6. R. HERMANN, J. Math. Phys. 13, 833, 838 (1972).
7. A. JOSEPH, Commun. Math. Phys. 36, 325 (1974).
8. H. UI, Prog. Theoret. Phys. 44, 153 (1970).
9. O.L. WEAVER, L.C. BIEDENHARN and R.Y. CUSSON, Ann. Phys. 77, 250 (1973).
10. G.W. MACKEY, Bull. Amer. Math. Soc. 69, 628 (1963).
11. F. VILLARS and G. COOPER, Ann. Phys. 56, 224 (1970).
12. J.P. ELLIOTT, Proc. Roy. Soc. A245, 128, 562 (1958).
13. G. ROSENSTEEL and D.J. ROWE, "On the Shape of Deformed Nuclei", Ann. Phys. (to be published).
14. V. BARGMANN and M. MOSHINSKY, Nucl. Phys. 23, 177 (1961).
15. B.R. JUDD, W. MILLER, Jr., J. PATERA and P. WINTERNITZ, J. Math. Phys. 15, 1787 (1974).
16. G. ROSENSTEEL and D.J. ROWE, Ann. Phys. 96, 1 (1976).
17. G. ROSENSTEEL, "On the Algebraic Formulation of Collective Models", 1975, Ph.D. Thesis, Univ. of Toronto.
18. W. ZICKENDRAHT, J. Math. Phys. 12, 1663 (1971).
19. P. GULSHANI and D.J. ROWE, Can. J. Phys. 54, 970 (1976).
20. S. GOSHEN and H.J. LIPKIN, Ann. Phys. 6, 301 (1959).
21. R. GODEMENT, Séminaire Cartan, 1958.
22. G. ROSENSTEEL and D.J. ROWE, "The Discrete Series of Sp(n,$\mathbb{R}$)", Int. J. Theor. Phys. (to be published).
23. G. ROSENSTEEL and D.J. ROWE, "The Nuclear Sp(3,R) Model", Phys. Rev. Letters (to be published).

A GROUP THEORETIC DESCRIPTION OF THE MAGNETIC PHASE TRANSITIONS IN THE AB_2O_4-TYPE SPINELS*†

Alfred K. Agyei and Joseph L. Birman

I. INTRODUCTION

The Landau theory of second-order phase transitions [1] and its various modifications [2-4] are unable to deal adequately with phase transitions that occur in structurally complex crystals like the spinels because they are macroscopic theories. We have recently formulated the Landau theory to take the microscopic crystal structure into account [5]. In the present paper we apply the new theory to the magnetic transitions observed in the AB_2O_4-type cubic spinels. (Here A is a divalent metal A^{2+}, B is a trivalent metal B^{3+}, and O is oxygen.) Here we give sufficient details (Sec. II) to allow the indicated application to be made to these spinels (Sec. III).

II. SUMMARY OF THE NEW THEORY

In the present theory, the microscopic structure of a crystal is taken into account by considering the transition from the

* Supported in part by NSF, AROD, and FRAP-CUNY.

† Based in part on a thesis presented to the Graduate School of Arts and Science, New York University, in partial fulfillment of the requirements for the Ph.D. degree by A.K. Agyei. (1976)

standpoint of the phase transformations of the individual sublattices, a procedure which is justifiable [5] on the basis of a Landau-type free-energy expansion in which the contributions of all the sublattices are taken into account, and the inter-sublattice interaction is assumed much smaller than intrasublattice contribution to the free energy.

Thus, we consider the phase transformation of the ν-th sublattice (made up of atoms of the kind A_ν) of a crystal of Federov space group G_0, point group P_0, and translation group T_0. Let $H_{0\nu}$ be the site point group of the sublattice and let q_ν be the index of $H_{0\nu}$ in P_0. The conventional unit cell of the crystal will contain at least q_ν of A_ν atoms. Locate the coordinate origin at one of the A_ν-atom sites. Then for the transitions of interest here the factor group $F_{0\nu} = G_0/T_0$ can always be written in the form

$$F_{0\nu} = \sum_{m=0}^{q_\nu - 1} \{R^m | \underset{\sim}{w}_m\} H_{0\nu}, \tag{2.1}$$

where $R \in P_0 - H_{0\nu}$ and the $\underset{\sim}{w}_m$ ($\underset{\sim}{w}_0 = 0$) are the position vectors of the q_ν atoms A_ν in the unit cell. This coset decomposition, which is a transitive permutation group defined on the q_ν atoms A_ν forms the basis of our theory, and will henceforth be referred to as the *transformation* group of the ν-th sublattices.

Now let $\rho_\nu(\underset{\sim}{r})$ be the density function of the ν-th sublattice. Following Landau, we can expand it in terms of any set of normalized functions $\psi_i^{(\nu)}(\underset{\sim}{r})$ complete under the group G_0

$$\rho_\nu(\underset{\sim}{r}) = \rho_{0\nu}(\underset{\sim}{r}) + \sum_i c_i^{(\nu)} \psi_i^{(\nu)}(\underset{\sim}{r}), \tag{2.2}$$

where $\rho_{0\nu}(\underset{\sim}{r}) = \psi^{(\nu)}(\underset{\sim}{r})$ is the function invariant under G_0. For the sublattice transition to occur at the temperature T_c, at least one of the quantities $c_i^{(\nu)}$ must be nonzero below T_c. Further, for the daughter (lower symmetry) crystal phase to have a perfect lattice of nontrivial symmetry, the $c_i^{(\nu)}$ must be localized on the A_ν atoms. Let $\rho_\nu(\underset{\sim}{r})$ be referred to the crystal unit cell, and let d_0 be the number of $c_i^{(\nu)}$ localized on each A_ν atom. Then the number of $c_i^{(\nu)}$ (and, hence, of $\psi_i^{(\nu)}(\underset{\sim}{r})$) entering into the expansion (2.2) is equal to $d_0 q_\nu$.

Thus, each of the sets $\{c_i^{(\nu)}\}$ and $\{\psi_i^{(\nu)}(\underset{\sim}{r})\}$ break up into q_ν subsets $\{c_{m\ell}^{(\nu)}\}$ and $\{\psi_{m\ell}^{(\nu)}(\underset{\sim}{r})\}$ $(m=0,1,\ldots,q_\nu-1;\ \ell=1,\ldots,d_0)$. Consideration of the transformation properties of the $\psi_{m\ell}^{(\nu)}(\underset{\sim}{r})$ shows that the functions of the m-th subset $\{\psi_{m\ell}^{(\nu)}(\underset{\sim}{r})\}$ should be centered about the m-th A_ν atom and should be such that $\psi_{m\ell}^{(\nu)}(r) = \{R^m|\underset{\sim}{w}_m\}\psi_{0\ell}^{(\nu)}(\underset{\sim}{r})$, where the $\psi_{0\ell}^{(\nu)}(\underset{\sim}{r})$ are the functions centered about the origin.

The functions $\psi_{0\ell}^{(\nu)}(\underset{\sim}{r})$ form a basis for a representation, $D_{0\nu}$, of the *site* point group $H_{0\nu}$, while the functions $\psi_{m\ell}^{(\nu)}(\underset{\sim}{r})$ form a basis for a representation, $\Gamma_{0\nu}$, of $F_{0\nu}$. Let us choose the $\psi_{0\ell}^{(\nu)}(\underset{\sim}{r})$ such that $D_{0\nu}$ is a physically irreducible representation. Then the representation is of a definite tensorial nature, and will give rise to only one type of phase transition (magnetic, electric, or structural). Further, the matrices of $\Gamma_{0\nu}$ consist of blocks that are matrices of $D_{0\nu}$, and are such that, if the $D_{0\nu}$-matrix blocks are each replaced by unity, then the resulting matrices form a transitive imprimitive representation of $F_{0\nu}$. For this reason, we call the representation $\Gamma_{0\nu}$ a *generalized transitive imprimitive representation* (GTIR) of $F_{0\nu}$.

Thus, in our theory, cell-preserving transitions that give rise to localized physical quantities are associated with certain GTIR of the transforming-sublattice transformation groups $F_{0\nu}$. In the usual approach ordinary crystal point group representations are used.

The characters of $\Gamma_{0\nu}$ are expressible in terms of the characters, $\chi^D(h \in H_{0\nu})$, of $D_{0\nu}$, and it can easily be shown [5], that, for a transition accompanied by a slight distortion of the crystal lattice, the group $G_{\nu\gamma} = T_0 \wedge S_{\nu\gamma}$, where

$$S_{\nu\gamma} = \sum_{m=0}^{q_\nu - 1} \{R^m | \underset{\sim}{V}_m\}_{\nu\gamma} \tag{2.3}$$

is always an admissible Fedorov space group of the daughter crystal provided

$$n_s = \frac{1}{N_s} \sum_{h \in H_{\nu\gamma}} \chi^D(h) > 0. \tag{2.4}$$

In (2.3) the $\underset{\sim}{V}_m = \underset{\sim}{w}_m + \Delta\underset{\sim}{w}_m$ ($\Delta\underset{\sim}{w}_m << \underset{\sim}{w}_m$) are the A_ν-atom position vectors in the daughter crystal, and are expressible [5] in terms of $\underset{\sim}{w}_m$ and $\Delta\underset{\sim}{w}_1$; in (2.4) N_s is the order of $H_{\nu\gamma}$.

The condition (2.4) is the admissibility condition for the "site" transition $H_{0\nu} \to H_{\nu\gamma}$. We therefore call it the *site-subduction* criterion. Using it and the Koster Tables [6], we can easily find the admissible $H_{\nu\gamma}$, of which there may be more than one (the index γ then numbers them). Then taking the semi-direct products $T_0 \wedge S_{\nu\gamma}$, we obtain the admissible daughter-sublattice (Fedorov) space groups $G_{\nu\gamma}$.

Now let us make three observations. First, notice that (2.1) may not be the only admissible coset decomposition of $F_{0\nu}$ relative to $H_{0\nu}$. If this is so, then to find all the admissible $G_{\nu\gamma}$, we must find and consider all the essentially nonequivalent coset decompositions, i.e., all the decompositions that yield different $G_{\nu\gamma}$.

Second, notice that above we have considered only the Fedorov space group, G_0, of the crystal and that the $G_{\nu\gamma}$ are also Fedorov space groups. For electric and structural transitions this procedure is evidently adequate. It is also adequate for magnetic transitions. Thus, let M_0 be the magnetic (Shubnikov) space group of the parent crystal. Then, we can always write M_0 in the form

$$M_0 = G_0 + G_0 R_\theta^{(0)} \theta, \tag{2.5}$$

where G_0 is a Fedorov space group, $R_\theta^{(0)}$ is a point-group operator, and θ is the time-reversal operator. As a result, if G_0 is the space group considered above, then the transition $G_0 \to G_{\nu\gamma}$ implies the transition $M_0 \to M_{\nu\gamma} = G_{\nu\gamma} + G_{\nu\gamma} R_\theta \theta$, where $R_\theta \in (H_{0\nu} - H_{\nu\gamma})$, does not necessarily coincide with $R_\theta^{(0)}$, and is such that it reverses the direction of the magnetic moment $\underset{\sim}{m}_0^{(\nu)}$ arising at the origin. The adequacy of the procedure stems from the fact that the site-subduction criterion gives the direction of $\underset{\sim}{m}_0^{(\nu)}$; therefore, we can always determine R_θ and hence, $M_{\nu\gamma}$. If R_θ exists, then $M_{\nu\gamma}$ is a type-III Shubnikov space group; if it does not, then $M_{\nu\gamma}$ is a type-I Shubnikov space group.

Third, if the transforming crystal contains more than one transforming sublattice, we should find all the admissible sublattice transitions and then determine which of them are compatible and can therefore occur simultaneously in the crystal. Clearly, transitions undergone by different sublattices will be compatible if they give rise to daughter sublattices with isomorphic crystallographic space groups $G_{\nu\gamma}^{c} = M_{\nu\gamma}\ (\theta{=}E)$, where $\theta = E$ denotes the replacement of θ by the identity operator E. We call this condition the *sublattice compatibility criterion*. It permits determination of the various admissible crystal composite transitions.

III. CELL-PRESERVING MAGNETIC TRANSITIONS IN THE CUBIC SPINELS AB_2O_4

The space group of the cubic spinels is O_h^7. The unit cell contains eight molecules with the A and B atoms occupying the sites:

$$\text{A:(8a)}\quad T_d\!:\!\{\underset{\sim}{w}_0, \underset{\sim}{w}_1\} = 000;\ \tfrac{1}{4}\tfrac{1}{4}\tfrac{1}{4},\ \text{F.C.} \tag{3.1}$$

$$\text{B:(16d)}\quad D_{3d}\!:\!\{\underset{\sim}{w}'_\nu\} = \tfrac{5}{8}\tfrac{5}{8}\tfrac{5}{8};\ \tfrac{5}{8}\tfrac{7}{8}\tfrac{7}{8};\ \tfrac{3}{8}\tfrac{7}{8}\tfrac{1}{8};\ \tfrac{3}{8}\tfrac{5}{8}\tfrac{3}{8};\ \text{F.C.} \tag{3.2}$$

This arrangement is the normal spinel arrangement. It is also possible for α% of the T_d sites to be occupied by B atoms, the displaced A atoms then going over to occupy the empty B-atom sites. The spinel is then said to be α% "inverted".

Assuming the spinel AB_2O_4 to be normal, we can distinguish three types of spinels: We call the normal spinel AB_2O_4 a *type-A* spinel if the A atom is paramagnetic, while the B atom is diamagnetic; a *type-B* spinel if B is paramagnetic and A is diamagnetic; and a *type-C* spinel if both A and B are paramagnetic. In the present section we determine the magnetic transitions admissible in these types of spinels, restricting ourselves to those transitions that require a tetragonal distortion of the crystals, since this is the distortion most often observed in these cubic crystals.

2.1. Cell-preserving magnetic transitions admissible in the type-A cubic spinels

The essentially nonequivalent coset decompositions of $F_{0A} = O_h^7/T_0$ relative to the sublattice site point group T_d are

$$F_{0A}(F_{0A2}) = T_d + \{C_{2b}(I)|\tfrac{1}{4}\ \tfrac{1}{4}\ \tfrac{1}{4}\}T_d. \tag{3.3}$$

From these coset decompositions we can, using the above-described procedure, deduce the (tetragonal) daughter-sublattice space groups:

$$G_{1(2)} = S_4^{(z)2} + \{C_{2b}(I)|\tfrac{1}{4}\ \tfrac{1}{4}\ \tfrac{1}{4}\}S_4^{(z)2} = D_{2d}^{12}(C_{4h}^6) \tag{3.4}$$

and the corresponding Shubnikov space groups

$$M_1 = D_{2d}^{12} + D_{2d}^{12}C_{2x}\theta = I4_1'/a'm'd, \tag{3.5}$$

$$M_2 = C_{4h}^6 + C_{4h}^6C_{2x}\theta = I4_1/am'd'. \tag{3.6}$$

In (3.4) the superscript z indicates that the magnetic moment arising at the origin is directed along the (fourfold) z axis. It follows, therefore, that a type-A spinel can undergo a cell-preserving simple antiferromagnetic (CPSA) transition to a phase of Shubnikov space group $I4_1'/am'd'$, or a cell-preserving simple ferromagnetic (CPSF) transition to a phase of Shubnikov space group $I4_1/am'd'$.

Experimentally*, the following type-A spinels are found to undergo CPSA transitions at temperatures ranging from 4 to 27°K: $CoA\ell_2O_4$, $MnA\ell_2O_4$, $MnRh_2O_4$, $NiRh_2O_4$, and $CoRh_2O_4$. No type-A AB_2O_4 spinel has thus far been observed to undergo a ferromagnetic transition.

2.2. Cell-preserving magnetic transitions admissible in the type-B cubic spinels

The paramagnetic atoms in these spinels are located at the D_{3d} sites (3.2). The only admissible coset decomposition of the factor group $F_{OB} = O_h^7/T_0$ relative to D_{3d} is

$$F_{OB} = \sum_{m=0}^{3} \{C_{4z}^{\nu}|\underset{\sim}{w}_{\nu}\}D_{3d}, \tag{3.7}$$

where the $\underset{\sim}{w}_m$ are the vectors (3.2) expressed relative to the site $(\frac{5}{8}\frac{5}{8}\frac{5}{8})$.

From (3.7) we obtain for the daughter crystal arising in the transition to the tetragonal magnetic state the type-I Shubnikov space group

$$G = M = \sum_{\nu=0}^{3} \{C_{4z}^{\nu}|\underset{\sim}{V}_{\nu}\}C_{2h}^{(b)3} = I4_1/amd. \tag{3.8}$$

Here the superscript b denotes the direction $[\bar{1}10]$, along which the magnetic moment $\underset{\sim}{M}_0$ arising at the origin should, in accordance with the site subduction criterion, be directed. It follows from

---
* See [5] for references.

this and the coset decomposition (3.8) that the magnetic moments in the crystal unit cell must lie along the directions [$\bar{1}$10], [$\bar{1}\bar{1}$0], [1$\bar{1}$0], and [110], so that the daughter crystal has a helical antiferromagnetic structure. We call [5] a transition that gives rise to such a structure a cell-preserving helical antiferromagnetic (CPHA) transition.

Experimentally*, the following type-B cubic spinels have been found to undergo CPHA transitions at temperatures between 10 and 16°K: $ZnFe_2O_4$, $MgCr_2O_4$, and $ZnCr_2O_4$. The actual space groups of the daughter crystals have not been determined experimentally, but each of these crystals is found to undergo a tetragonal distortion prior to the magnetic transition. Further, the magnetic moments are found to lie in the (001) plane, which is in accord with the theoretical prediction.

2.3. The ferrimagnetic transitions observed in the type-C AB_2O_4 cubic spinels

The magnetic transitions observed in the type-A and type-B AB_2O_4 spinels have close critical temperatures T_c (between 4 and 27°K). Furthermore, the daughter sublattices arising in the two sublattice transitions have the same crystallographic space group $G^c = D_{4h}^{19}$. Consequently, a composite magnetic transition in which the A-atom sublattice undergoes a CPSA transition to a sublattice of Shubnikov space group $I4_1'/a'm'd$, (3.5), while the B-atom sublattice undergoes a CPHA transition to a sublattice of Shubnikov space group $I4_1/amd$, (3.8), is admissible in a type-C AB_2O_4 cubic spinel and we consider it as quite probable.

This composite transition will be purely antiferromagnetic if the spinel is normal. If the spinel is α% (α > 0) inverted and the A^{2+} and B^{3+} ions (which are respectively the ionic states of the A and B atoms in these compounds) have different magnetic

---
* See [5] for references.

moments m_{2+} and m_{3+}, then the daughter crystal will possess a non-zero magnetization, and the composite transition will be a ferrimagnetic transition.

Such a ferrimagnetic transition is admissible because: (1) the T_d sites in the normal spinel admit of ferromagnetic and antiferromagnetic transitions requiring the same (tetragonal) distortion of the crystal lattice (see (3.5) and (3.6)); and (2) if the degree of inversion $\alpha = 50\%$, then the A and B ions at the D_{3d} sites (which sites admit of no ferromagnetic transition to a tetragonal state) can be arranged such that the daughter sublattice has the allowed helical antiferromagnetic structure.

2.3a. Saturation magnetization

Thus, we find that a 50% inverted type-C AB_2O_4 cubic spinel can undergo a ferrimagnetic transition. The net molecular saturation magnetization $\underset{\sim}{m}_s^{th}$ arising in the transition will be directed along the fourfold axis of the tetragonal daughter crystal, and will have a magnitude given by

$$m_s^{th} = \tfrac{1}{2}|m_{2+}-m_{3+}|. \tag{3.9}$$

To find m_s^{th} then we need to know only the intrinsic magnetic moments, m_{2+}, m_{3+}, of the A^{2+}, B^{3+} ions. In Table I we give the magnetic moments of the ions frequently occurring in compounds with the spinel structure. As far as possible we have chosen the results of measurements performed on strongly ionic crystals with simple crystallographic structures, so that we can regard the obtained magnetic moments as the intrinsic moments of the ions. Using these values, we have computed the m_s^{th} values for the type-C AB_2O_4 spinels that have been observed to undergo ferrimagnetic transitions. These theoretical values are given in the eighth column of Table II.

Also given in Table II for comparison are the experimentally

TABLE I. Magnetic moments of ions frequently occurring in compounds with the spinel structure. (For references see the Thesis [5].)

Ion	(in μ_B)	Crystal on which m_s measurements were made	T_c (°K)
Co^{2+}	3.0	CoF_2	50
Fe^{2+}	4.6	FeF_2	90
Mn^{2+}	5.0	MnF_2	75
Ni^{2+}	2.1	$NiSO_4$	37
V^{2+}	3.8	$V(bipy)_3(ClO_4)$	-
Co^{3+}	4.4	CoF_3	460
Fe^{3+}	5.0	FeF_3	394
Mn^{3+}	4.0	MnF_3	43
Cr^{3+}	2.8	Cr_2O_3	318
V^{3+}	0.8	ZnV_2O_4	

TABLE II. Experimentally observed ferrimagnetic and structural transitions occurring in type-C cubic spinels and comparison of the observed molecular saturation magnetization values m_s^{exp} with the theoretical values m_s^{th} computed from (3.9) with the m_{2+} and m_{3+} values given in Table I. (For references, see the Thesis [5].)

Spinel	Experimental data					
	Structural transition			Ferrimagnetic transition		m_s^{th}
	Distortion	T_m (°K)	Ref.	T_c (°K)	m_s^{exp} (in μ_B)	(μ_B)
$FeCr_2O_4$	tetragonal	~111	25	88	0.84	0.90
FeV_2O_4	"	127	26	109	1.95	1.90
$NiCr_2O_4$	"	~308	27	78	0.33	0.35
$NiCo_2O_4$				~500	1.25	1.15
$MnCo_2O_4$				203	0.04	0.30
$MnCr_2O_4$				43	1.20	1.10
MnV_2O_4				56	2.10	2.10
CoV_2O_4				145	1.33	1.10
$CoCr_2O_4$				98	0.18	0.10
VCo_2O_4				158	0.21	0.30
VF_2O_4				440	0.72	0.60

obtained molecular saturation magnetization values m_s^{exp}. As can be seen from the Table, the agreement between the theoretical and experimental values is excellent.

Thus, our theory is able to account for these previously unexplained transitions under the assumption that the spinels are 50% inverted. Of the spinels listed in Table II, two, $NiCo_2O_4$ and CoV_2O_4 are known to be inverted, but their degrees of inversion are not known. Perhaps the present theory will arouse the interest of experimenters in the exact determination of this quantity for these spinels.

REFERENCES

1. L.D. LANDAU and E.M. LIFSCHITZ, *Statistical Physics*, Pergamon, London, 1969.
2. O.V. KOVALEV, Sov. Phys. - Solid State 5, 2309 (1964).
3. E. ASCHER, Phys. Lett. 20, 352 (1966).
4. F.E. GOLDRICH and J.L. BIRMAN, Phys. Rev. 167, 528 (1968).
5. A.K. AGYEI, Ph.D. Thesis, New York University (1976). Available from University Microfilms, Ann Arbor, Michigan.
6. G.F. KOSTER et al., *Properties of the Thirty-Two Point Groups*, M.I.T. Press, Cambridge, Mass., 1963.

APPLICATIONS OF CRYSTAL CLEBSCH-GORDAN COEFFICIENTS*

Rhoda Berenson and Joseph L. Birman

In this paper we will illustrate several uses of crystal Clebsch-Gordan coefficients. In particular we will demonstrate that these coefficients provide an efficient and straightforward method for calculating the independent elements of tensors such as those which arise in effective Hamiltonian theory and in Raman scattering (including the influence of "morphic effects").

We will first briefly review those aspects of the theory of Clebsch-Gordan coefficients which will be used in later discussions. If D^{ℓ} and $D^{\ell'}$ are two irreducible representations of a group G, then the direct product representation $D^{\ell} \otimes D^{\ell'}$ can be reduced into a direct sum of irreducible representations $D^{\ell''}$:

$$D^{\ell}(R) \otimes D^{\ell'}(R) = \sum_{\ell''} C_{\ell\ell'\ell''} D^{\ell''}(R) \tag{1}$$

where $C_{\ell\ell'\ell''}$ indicates the number of times that $D^{\ell''}$ appears in the reduction. R is an arbitrary group element.

The Clebsch-Gordan coefficients are elements of the unitary matrix $U^{\ell\times\ell'}$ which brings the direct product $D^{\ell} \otimes D^{\ell'}$ into a

* Supported in part by grants from NSF-DMR74-21991-A01, AROD-DAHC04-75-G-0052, and FRAP (CUNY) 10753N.

fully reduced matrix Δ.

$$D^{\ell}(R) \otimes D^{\ell'}(R) = U^{\ell\times\ell'}\Delta(R)U^{\ell\times\ell'\,-1} \tag{2}$$

and

$$\Delta_{\ell''\gamma\mu'',\bar{\ell}''\bar{\gamma}\bar{\mu}''} = \delta_{\ell''\bar{\ell}''}\,\delta_{\gamma\bar{\gamma}}\,D^{\ell''}_{\mu''\bar{\mu}''} \tag{3}$$

where $\gamma = 1,\ldots,C_{\ell\ell'\ell''}$ indicates the multiplicity of $D^{\ell''}$ in Δ. Eq. (2) can therefore be written as

$$\sum_{\substack{\ell''\gamma\mu''\\ \bar{\ell}''\bar{\gamma}\bar{\mu}''}} U^{\ell\times\ell'}_{\mu\mu',\ell''\gamma\mu''}\Delta(R)_{\ell''\gamma\mu'',\bar{\ell}''\bar{\gamma}\bar{\mu}''}\,U^{\ell\times\ell'\,-1}_{\bar{\ell}''\bar{\gamma}\bar{\mu}'',\bar{\mu}\bar{\mu}'} \tag{4}$$
$$= (D^{\ell}(R)\otimes D^{\ell}(R))_{\mu\mu',\bar{\mu}\bar{\mu}'}\,.$$

Using eq. (3) in eq. (4), multiplying both sides by $D^{\bar{\ell}}(R)^*_{\nu''\bar{\nu}''}$, summing on R and using the orthogonality theorem we have

$$\sum_{\gamma} U^{\ell\times\ell'}_{\mu\mu',\ell''\gamma\mu''}U^{\ell\times\ell'*}_{\bar{\mu}\bar{\mu}',\ell''\gamma\bar{\mu}''} = \frac{|\ell''|}{g}\sum_{R} D^{\ell}(R)_{\mu\bar{\mu}}\,D^{\ell'}(R)_{\mu'\bar{\mu}'}\,D^{\ell''}(R)^*_{\mu''\bar{\mu}''} \tag{5}$$

where $|\ell''|$ is the dimension of $D^{\ell''}$ and g is the order of G.

Clebsch-Gordan coefficients have been calculated by Koster et al. [1] for the thirty-two point groups and by the authors for several space group representations. In addition a straightforward method for calculating these coefficients for space groups has been described [2]. Of most interest to physicists, however, is the application of Clebsch-Gordan coefficients to physical problems. Clebsch-Gordan coefficients were first used for calculating the symmetry of matrix elements of rotation group operators [3]. Koster then extended this to matrix elements of other group operators [4].

If we consider a matrix element of an operator O^{j}_{α} between an initial state ψ^{i}_{β} and a final state ψ^{ℓ}_{λ} for a system of symmetry

G, then O^j, ψ^i and ψ^ℓ transform as D^j, D^i and D^ℓ, respectively, and

$$\langle \psi^\ell_\lambda | O^j_\alpha | \psi^i_\beta \rangle = \sum_\gamma a_{ji\ell\gamma} U^{j\times i*}_{\alpha\beta,\ell\gamma\lambda} \tag{6}$$

where $a_{ji\ell\gamma}$ is a constant independent of α, β and λ. Hence, the matrix element $\langle \psi^\ell_\lambda | O^j_\alpha | \psi^i_\beta \rangle$ is equal to a constant times a Clebsch-Gordan coefficient if there is no multiplicity and is equal to a linear combination of Clebsch-Gordan coefficients when there is multiplicity.

This procedure has been used recently [5] to simplify the calculation of effective Hamiltonians which arise in effective mass $\underset{\sim}{k}\cdot\underset{\sim}{p}$ theory [6,7]. In this case one is interested in the matrix elements of the effective Hamiltonian near a degenerate eigenvalue. This Hamiltonian may depend on wave vector, strain tensor, external field, etc. The set of any one of these quantities is denoted by K and we can write

$$K = \{K^{(1)}\} + \ldots + \{K^{(i)}\} + \ldots + \{K^{(k)}\} \tag{7}$$

where each $K^{(i)}$ spans an irreducible representation of the appropriate space group G. It can be shown [7] that the Hamiltonian, $\mathcal{H}(K)$, can be written as

$$\mathcal{H}(K) = \sum_i a_i \delta_{i,jj*} \sum_\ell X^{(i)}_\ell K^{(i)*}_\ell \tag{8}$$

where a_i is a constant and

$$\delta_{i,jj*} \begin{cases} = 1 \text{ if } D^i \text{ is contained in } D^j \otimes D^{j*} \\ = 0 \text{ otherwise.} \end{cases} \tag{9}$$

The matrix elements $\mathcal{H}_{\alpha\beta}(K)$ between degenerate states ψ^j_α and ψ^j_β are given by

$$\mathcal{H}_{\alpha\beta}(K) = \langle \psi^j_\alpha | \mathcal{H}(K) | \psi^j_\beta \rangle \tag{10}$$

or

$$\mathcal{H}_{\alpha\beta}(K) = \sum_i a_i \delta_{i,jj*} \sum_\ell \langle \psi^j_\alpha | X^i_\ell | \psi^j_\beta \rangle K_\ell^{(i)*}. \tag{11}$$

Hence the $|\ell_j|^2$ elements of $\mathcal{H}_{\alpha\beta}(K)$ are given in terms of the independent elements $X^i_{\ell,\alpha\beta}$. But from eq. (6) we see that

$$\begin{aligned} X^i_{\ell,\alpha\beta} &= \sum_\gamma a_{ijj\gamma} U^{ij}_{\ell\beta,j\gamma\alpha}{}^* \\ &= \sum_\gamma a_{jj*i\gamma} U^{j*j}_{\alpha\beta,i\gamma\ell} \end{aligned} \tag{12}$$

so that

$$\mathcal{H}_{\alpha\beta}(K) = \sum_{i\ell\gamma} a_{jj*i\gamma} U^{j*j}_{\alpha\beta,i\gamma\ell} K_\ell^{(i)*}. \tag{13}$$

Since the crystal Clebsch-Gordan coefficients are readily available, eq. (13) provides a relatively simple method for constructing the effective Hamiltonian even in cases of high degeneracy. This has been applied to finding the form of the Hamiltonian at the points X and R in O_h^3 structures.

Clebsch-Gordan coefficients can also be used to simplify the calculation of scattering tensors. As an example we will first consider Raman scattering [8]. In this case an incident photon of (wave vector, polarization, frequency) = $(\underset{\sim}{k}_i, \underset{\sim}{\varepsilon}_i, \omega_i)$ is scattered producing a scattered photon $(\underset{\sim}{k}_s, \underset{\sim}{\varepsilon}_s, \omega_s)$ and creating or destroying one or more optic phonons. The total scattering intensity can be written

$$I \propto \sum_{\alpha\beta} |\varepsilon_{s\alpha} P_{\alpha\beta} \varepsilon_{i\beta}|^2 \tag{14}$$

where α and β are cartesian components and $P_{\alpha\beta}$ is the Raman scattering tensor. $P_{\alpha\beta}$ is dependent on ionic positions and can be expanded in a Taylor series in ionic displacements, or, equivalently, in phonons $Q(^{\ell}_{\sigma})$ where $Q(^{\ell}_{\sigma})$ belongs to row σ of representation D^{ℓ}. Hence

$$P_{\alpha\beta} = P^{0}_{\alpha\beta} + \sum_{\ell\sigma} P^{(1)}_{\alpha\beta}(^{\ell}_{\sigma})Q(^{\ell}_{\sigma}) + \sum_{\substack{\ell\sigma \\ \ell'\sigma'}} P^{(2)}_{\alpha\beta}(^{\ell\ell'}_{\sigma\sigma'})Q(^{\ell}_{\sigma})Q(^{\ell'}_{\sigma'}) + \cdots \quad (15)$$

where

$$P^{(1)}_{\alpha\beta}(^{\ell}_{\sigma}) = \left.\frac{\partial P_{\alpha\beta}}{\partial Q(^{\ell}_{\sigma})}\right|_0 ; \quad P^{(2)}_{\alpha\beta}(^{\ell\ell'}_{\sigma\sigma'}) = \left.\frac{\partial^2 P_{\alpha\beta}}{\partial Q(^{\ell}_{\sigma})\,\partial Q(^{\ell'}_{\sigma'})}\right|_0 . \quad (16)$$

$P^{(1)}_{\alpha\beta}$ gives rise to one phonon scattering, $P^{(2)}_{\alpha\beta}$ gives rise to two phonon scattering, etc.

Since $P^{(1)}_{\alpha\beta}$ is evaluated at equilibrium it is invariant under symmetry operations of the crystal. Therefore $P^{(1)}_{\alpha\beta}(^{\ell}_{\sigma})Q(^{\ell}_{\sigma})$ transforms as D^{ℓ}. But $P_{\alpha\beta}$ is a symmetric second rank tensor that transforms as the symmetrized vector product $[D^{V}\otimes D^{V}]_2$. Hence, for any group operation P_R

$$\begin{aligned} P_R \sum_{\ell\sigma} P^{(1)}_{\alpha\beta}(^{\ell}_{\sigma})Q(^{\ell}_{\sigma}) &= \sum_{\ell\sigma\sigma'} P^{(1)}_{\alpha\beta}(^{\ell}_{\sigma})D^{\ell}(R)_{\sigma'\sigma}Q(^{\ell}_{\sigma'}) \\ &= \sum_{\bar{\alpha}\bar{\beta}} \sum_{\ell\sigma} [D^{V}_{\bar{\alpha}\alpha} \otimes D^{V}_{\bar{\beta}\beta}]_2 P^{(1)}_{\bar{\alpha}\bar{\beta}}(^{\ell}_{\sigma})Q(^{\ell}_{\sigma}). \end{aligned} \quad (17)$$

Since the $Q(^{\ell}_{\sigma})$ are independent,

$$\sum_{\sigma} P^{(1)}_{\alpha\beta}(^{\ell}_{\sigma})D^{\ell}(R)_{\bar{\mu}\sigma} = \sum [D^{V}(R)_{\bar{\alpha}\alpha} \otimes D^{V}(R)_{\bar{\beta}\beta}]_2 P^{(1)}_{\bar{\alpha}\bar{\beta}}(^{\ell}_{\bar{\mu}}). \quad (18)$$

Multiplying both sides of eq. (18) by $D^{\ell}(R)^{*}_{\bar{\mu}\mu}$ and summing on $\bar{\mu}$

$$P^{(1)}_{\alpha\beta}(^{\ell}_{\mu}) = \sum_{\bar{\ell}\bar{\beta}\bar{\mu}} [D^{V}(R)_{\bar{\alpha}\alpha}\, D^{V}(R)_{\bar{\beta}\beta}]_2 D^{\ell}(R)^{*}_{\bar{\mu}\mu}\, P^{(1)}_{\bar{\alpha}\bar{\beta}}(^{\ell}_{\bar{\mu}}) \quad (19a)$$

$$= \frac{1}{|\ell|} \sum_{\bar{\alpha}\bar{\beta}\bar{\mu}\gamma} U^{[v\times v]_2}_{\bar{\alpha}\bar{\beta},\ell\gamma\bar{\mu}} U^{[v\times v]_2 *}_{\alpha\beta,\ell\gamma\mu} P^{(1)}_{\bar{\alpha}\bar{\beta}}\binom{\ell}{\bar{\mu}}. \tag{19b}$$

Eq. (19a) could be used directly to calculate $P^{(1)}_{\alpha\beta}\binom{\ell}{\sigma}$. Actually eq. (19a) is a set of equations, one for each symmetry operation R, and the conventional calculation then involves solving these equations to find relationships among the tensor elements. However, eq. (19b) is more useful since it is identically satisfied by writing

$$P^{(1)}_{\alpha\beta}\binom{\ell}{\mu} = \sum_{\gamma} A_{v\ell\gamma} U^{[v\times v]_2}_{\alpha\beta,\ell\gamma\mu}. \tag{20}$$

Hence, if there is no multiplicity $P^{(1)}_{\alpha\beta}$ is equal to a constant times a Clebsch-Gordan coefficient and in general is equal to a linear combination of Clebsch-Gordan coefficients.

Similarly, for higher order processes, for example "morphic" [9] induced scattering, Clebsch-Gordan coefficients greatly simplify calculations. In these processes the scattering occurs under the influence of an external force such as an electric field, field gradient, stress, etc. If this external force F^j_λ transforms as row λ of representation D^j then the first order Raman tensor can be written as

$$P^{(1)}_{\alpha\beta} = \sum_{\ell\sigma} \left[\frac{\partial P_{\alpha\beta}}{\partial Q\binom{\ell}{\sigma}}\bigg|_0 Q\binom{\ell}{\sigma} + \frac{\partial^2 P_{\alpha\beta}}{\partial Q\binom{\ell}{\sigma}\partial F^j_\lambda}\bigg|_0 Q\binom{\ell}{\sigma} F^j_\lambda \right]. \tag{21}$$

If we write

$$\frac{\partial^2 P_{\alpha\beta}}{\partial Q\binom{\ell}{\sigma}\partial F^j_\lambda}\bigg|_0 = P^j_{\alpha\beta,\lambda}\binom{\ell}{\sigma} \tag{22}$$

then for any symmetry operation R, the "morphic" induced tensor $P^j_{\alpha\beta,\lambda}\binom{\ell}{\sigma}$ transforms so that

$$P^{j}_{\alpha\beta,\lambda}\binom{\ell}{\sigma} = \sum_{\bar{\alpha}\bar{\beta}\bar{\lambda}\bar{\sigma}} [D^{v}(R)_{\bar{\alpha}\alpha}\, D^{v}(R)_{\bar{\beta}\beta}]_{2} D^{\ell}(R)^{*}_{\bar{\sigma}\sigma}\, D^{j}(R)^{*}_{\bar{\lambda}\lambda}\, P^{j}_{\bar{\alpha}\bar{\beta},\bar{\lambda}}\binom{\ell}{\bar{\sigma}}. \qquad (23)$$

Depending on the type of external force involved, D^{j} may be an irreducible representation or a sum or product of representations. For example, for electric field induced scattering F^{j}_{λ} transforms as the vector representation D^{v} and eq. (23) becomes

$$\begin{aligned} P^{E}_{\alpha\beta,\lambda}\binom{\ell}{\sigma} &= \sum_{\bar{\alpha}\bar{\beta}\bar{\lambda}\bar{\sigma}} [D^{v}(R)_{\bar{\alpha}\alpha}\, D^{v}(R)_{\bar{\beta}\beta}]_{2} D^{\ell}(R)^{*}_{\bar{\sigma}\sigma}\, D^{v}(R)^{*}_{\bar{\lambda}\lambda}\, P^{E}_{\bar{\alpha}\bar{\beta},\bar{\lambda}}\binom{\ell}{\bar{\sigma}} \\ &= \sum_{\bar{\alpha}\bar{\beta}\bar{\lambda}\bar{\sigma}} \sum_{mp\bar{p}} \frac{1}{|m|} U^{[v\times v]_{2}}_{\bar{\alpha}\bar{\beta},m\bar{p}}\, U^{[v\times v]_{2}*}_{\alpha\beta,mp}\, U^{v\times\ell *}_{\bar{\lambda}\bar{\sigma},m\bar{p}}\, U^{v\times\ell}_{\lambda\sigma,mp}\, P^{E}_{\bar{\alpha}\bar{\beta},\bar{\lambda}}\binom{\ell}{\bar{\sigma}} \end{aligned} \qquad (24)$$

This equation is identically satisfied by writing

$$P^{E}_{\alpha\beta,\lambda}\binom{\ell}{\sigma} = \sum_{mp} A_{\ell m v}\, U^{v\times\ell}_{\lambda\sigma,mp}\, P^{(1)}_{\alpha\beta}\binom{m}{p}. \qquad (25)$$

Thus the elements of the electric field induced Raman tensor are given as linear combinations of the elements for the intrinsic first order Raman tensor and the coefficients of these linear combination are again Clebsch-Gordan coefficients.

Similarly for electric field gradient induced scattering where F^{j}_{λ} transforms as $D^{v}\otimes D^{v}$ we find that

$$P^{\nabla E}_{\alpha\beta,\lambda\mu}\binom{\ell}{\sigma} = \sum_{kp} A_{\ell k v}\, U^{v\times\ell}_{\mu\sigma,kp}\, P^{E}_{\alpha\beta,\lambda}\binom{k}{p}. \qquad (26)$$

Other "morphic" tensors can be obtained in an analogous fashion. In particular, since the strain transforms as $[D^{v}\otimes D^{v}]_{2}$, the strain induced tensor is obtained from $P^{\nabla E}_{\alpha\beta,\lambda\mu}$ by symmetrizing on the indices $\lambda\mu$. The strain gradient induced tensors would then be calculated as a linear combination of strain induced tensors. In a similar manner one could obtain tensors for Raman scattering in the presence of a magnetic field or a magnetic field gradient. Brillouin tensors have been calculated using the same method.

REFERENCES

1. G. KOSTER, J.O. DIMMOCK, R.G. WHEELER and H. STATZ, *Properties of the Thirty-Two Point Groups* (MIT, Cambridge, Mass. 1963).

2. R. BERENSON and J.L. BIRMAN, J. Math. Phys. 16, 227 (1975); R. BERENSON, I. ITZKAN and J.L. BIRMAN, ibid. 16, 236 (1975).

3. E.P. WIGNER, *Group Theory and Applications to the Quantum Mechanics of Atomic Spectra* (Academic Press, New York 1959), p. 115.

4. G.F. KOSTER, Phys. Rev. 109, 227 (1958).

5. J.L. BIRMAN, T.K. LEE and R. BERENSON, Phys. Rev. B14 (1976).

6. J.M. LUTTINGER, Phys. Rev. 102, 1030 (1956).

7. G.L. BIR and G.E. PICUS, *Symmetry and Strain-Induced Effects in Semiconductors* (Wiley, New York, 1974).

8. J.L. BIRMAN and R. BERENSON, Phys. Rev. B9, 4512 (1974); J.L. BIRMAN, Phys. Rev. B9, 4518 (1974).

9. L.B. HUMPHREYS and A.A. MARADUDIN, Solid State Comm. 11, 1003 (1972).

ON THE USE OF THE SO(4,2) DYNAMICAL GROUP FOR THE STUDY OF THE GROUND STATE OF A HYDROGEN ATOM IN A HOMOGENEOUS MAGNETIC FIELD

J. Čížek and E.R. Vrscay

I. INTRODUCTION

In the last decade, considerable attention has been given to the study of an elementary yet rather complicated problem: the hydrogen atom in a strong magnetic field[1]. A great variety of variational and perturbation approaches has been applied to this case. When the eigenfunctions of the free (unperturbed) hydrogen atom are used as a basis set, difficulties arise in both approaches due to the continuum part of the spectrum.

However there is an alternative approach based on the use of the SO(4,2) dynamical group[2] which includes as its most important feature the non-unitary "tilting" transformation. By means of this operation, the original hamiltonian with mixed spectrum may be replaced by a non-hermitian hamiltonian having only a discrete spectrum. An additional advantage of this approach lies in the complete algebraization of the whole procedure. The nonhermiticity of the transformed hamiltonian presents no serious difficulties.

II. PROCEDURE

The Schrödinger equation for the problem considered has the

form

$$\hat{H}\psi = E\psi, \tag{1}$$

where

$$\hat{H} = \frac{\hat{p}^2}{2} - \frac{1}{r} + \frac{\gamma^2}{8}(x^2+y^2)$$

and

$$E = E_0 + \Delta\varepsilon .$$

In these formulae E_0 is the energy of the hydrogen atom in the absence of an external magnetic field and $\gamma = H/c$ where H represents the magnetic field intensity.

In order to transform this hamiltonian, we consider the generators of the so(4,2) Lie algebra[2], written in vector form: $\vec{J}$ (angular momentum), $\vec{A}$ (related to Runge-Lenz vector), $\vec{T}$ (generators of so(2,1) Lie algebra) and $\vec{B}$ and $\vec{\Gamma}$ which are defined by appropriate commutation relations involving the former operators.

The tilting transformation modifies an arbitrary hermitian operator $\hat{L}$ and wave function ψ according to the formulae:

$$\hat{L}_M = e^{i\theta\hat{T}_2}\hat{L}e^{-i\theta\hat{T}_2}$$

and

$$\psi_M = e^{i\theta\hat{T}_2}\psi$$

where the subscript denotes the modified operator and wave function and θ is a parameter. Multiplication of the Schrödinger equation (1) by r followed by a tilting operation yields a modified eigenvalue equation for the ground state (where θ is trivially zero) in the form

$$(\hat{K} + \frac{\gamma^2}{8}\hat{V} - \Delta\varepsilon\hat{S})\psi_M = 0 \tag{2}$$

where

$$\hat{K} = \hat{T}_3 - 1,$$

$$\hat{V} = (\hat{T}_3-\hat{T}_1)[(\hat{B}_1-\hat{A}_1)^2 + (\hat{B}_2-\hat{A}_2)^2]$$

and

$$\hat{S} = \hat{T}_3 - \hat{T}_1.$$

It can be shown that the non-hermitian operator $\hat{K}$ has the eigenvalue spectrum i=0,1,2,3,... . The matrix elements of $\hat{V}$ and $\hat{S}$ in the basis of eigenfunctions of $\hat{K}$ may be evaluated by group theoretical methods[2]. These matrix elements are nonzero only if they are near the diagonal, namely: $v_{ij} = 0$ if $|i-j| > 3$ and $s_{ij} = 0$ if $|i-j| > 1$.

A look at equation (2) shows that the problem is formally similar to a secular problem on a non-orthogonal basis. Both perturbation and variational methods may be employed for the determination of $\Delta\varepsilon$.

The perturbation theory in this case is only slightly more complicated than the usual method for an orthogonal basis. Introducing the following standard power series expansions

$$\psi_M = \sum_{n=0}^{\infty} \chi_n \lambda^n,$$

$$\Delta\varepsilon = \sum_{n=1}^{\infty} \Delta\varepsilon_n \lambda^n,$$

where $\lambda = \frac{\gamma^2}{8}$, and substituting them into (2), we obtain the following equations for the perturbation energies and wave functions:

$$\hat{K}\chi_n + \hat{V}\chi_{n-1} - \sum_{j=1}^{n} \Delta\varepsilon_j \hat{S}\chi_{n-j} = 0$$

$$\Delta\varepsilon_n \langle \chi_0|\hat{S}|\chi_0\rangle = \langle \chi_0|\hat{V}|\chi_{n-1}\rangle - \sum_{j=1}^{n-1} \langle \chi_0|\hat{S}|\chi_{n-j}\rangle \Delta\varepsilon_j .$$

Clearly, in the case $\hat{S} = 1$, the equations reduce to the usual formulae of perturbation theory.

It is worthwhile to mention that only finite summations are involved in the expansion of the χ_n in terms of the eigenfunctions of $\hat{K}$. This is due to the fact that in both the $\hat{V}$ and $\hat{S}$ matrices the elements are packed around the diagonals, as indicated above.

In the variational procedure, where the infinite matrices $\hat{V}$ and $\hat{S}$ are truncated to finite ones, orthogonalization of the basis is first performed. This is a rather simple procedure but the packing of matrix elements of $\hat{V}$ around the diagonal is lost. The subsequent diagonalization procedure is straightforward.

III. RESULTS

In this study, the perturbation procedure described above has been applied to the hydrogen atom system and the terms $\Delta\varepsilon_n$ have been calculated to 40th order. The difficulty in using this approach is that the series has zero radius of convergence. Nevertheless, the Padé approximant technique may be used for the summation of the series.

Relatively poor results are obtained when [N,N] Padé approximants are used for direct evaluation of the energy. The difficulty is that the energy is proportional to γ as $\gamma \to \infty$. However, more satisfactory results are obtained if we evaluate the so-called ionization energy of the atom, which represents the difference between the energy of the lowest state of the free electron in the external homogeneous magnetic field and the energy of the lowest state of the hydrogen atom placed in the same field. As $\gamma \to \infty$, the ionization energy is proportional only to $\ell n^2\gamma$. For a comparison, we shall use the results of Cabib et al[3]. For $\gamma = 1$,

(γ=5) they obtained an ionization energy E_I = 0.83121 ± 0.00005, (E_I=1.38040±0.00005). Using Padé approximants for E_I and considering $\Delta\varepsilon_n$ up to order 30, results virtually identical to those of ref. 3 were obtained for γ = 1, while for γ = 5 an error of 4% of the ionization energy was found.

In the variational approach with a matrix truncated to n = 28 and ℓ = 12 (producing a basis consisting of 154 functions) results virtually identical to those of ref. 3 were obtained both for γ = 1 and γ = 5.

IV. CONCLUSIONS

The accuracy of the Padé estimate of the ionization energy is surprising. This accuracy may even be improved provided that a quantity more appropriate than E_I be used for a Padé treatment. Several possibilities are being studied. The accuracy of the variational calculation is so satisfactory that we hope by proper choice of basis to match the asymptotic results for $\gamma \to \infty$.

Finally, it is interesting to compare our results with those of Brandi[4], where hydrogenlike functions were used for the variational approach in the interval $\gamma \in [0,1]$. For γ = 1 and the same basis as was used by Brandi (n=6, ℓ=2) our error is less than 0.1% as compared to his 7% error of the ionization energy. This is a clear indication of how the continuous part of the spectrum is of primary importance. This problem was discussed from a different point of view in a classical paper of Lowdin[5]. Detailed results of our calculations will be published elsewhere.

ACKNOWLEDGEMENTS

We are grateful to Professor M.L. Glasser and Mr. B.G. Adams for helpful discussions and to Mrs. V. Fris for help with the numerical computations. This work has been supported by National Research Council of Canada grants, which are hereby gratefully acknowledged.

REFERENCES

1. H.C. PRADDAUDE, Phys. Rev. A6, 1321 (1972) and references therein.

2. A.O. BARUT, Dynamical groups and generalized symmetries in quantum theory, University of Canterbury Publications, Christchurch, N.Z. 1972; B.G. WYBOURNE, Classical groups for physicists, John Wiley and Sons, N.Y. 1974; M. BEDNAR, Ann. Phys. N.Y. 75, 305 (1973).

3. D. CABIB, E. FABRI and G. FIORIO, Nuovo Cimento 10B, 185 (1972).

4. H.S. BRANDI, Phys. Rev. A11, 1835 (1975).

5. P.O. LOWDIN and H. SHULL, Phys. Rev. 101, 1730 (1956).

GROUP THEORY AROUND LIGAND FIELD THEORY

Maurice R. Kibler

Some basic considerations concerning the Wigner-Racah algebra of a chain $O_3 \supset G_1 \supset G_2 \ldots$, where G_1 is a (usually finite) point symmetry group, are reviewed, in Section 1, and applied, in Section 2, to ligand field theory (a branch of chemical and solid state physics). A brief discussion of the ligand field parameters leads to the introduction, in Section 3, of Lie groups both in ligand field theory and in the theory of atomic spectra.

SECTION 1

The interest for quantum chemistry and molecular and solid state physics of chains $O_3 \supset G_1 \supset G_2 \ldots$ ($O_3 \supset G_k$ for short) is now well established (1-4). One of the G_k's, say G_i, is a high symmetry group (e.g., icosahedral or cubical) leaving geometrically invariant a somewhat idealized molecular aggregate. $G_{i+1} \equiv G$ may be then a low symmetry group (e.g., tetragonal or trigonal) used for a more realistic description of our aggregate. The other G_k's are introduced for classifying the relevant state vectors and physical interactions as completely as possible. The total number n of G_k's does not generally exceed 3. Different chains $O_3 \supset G_k$ with the same G_1 often correspond to different quantization axes. In some cases, the parent group O_3 is a symmetry group for a molecular aggregate restricted to a free atom. If we are interested in the electronic properties of an aggregate with an odd number of electrons, it is necessary for labelling the relevant state vectors to introduce the chain $\bar{O}_3 \supset \bar{G}_1 \supset \bar{G}_2 \ldots$ ($\bar{O}_3 \supset \bar{G}_k$ for short). ($\bar{H}$ stands for the double group of H.)

Quantitative determination of properties for our aggregate is simplified by using:

(i) $\bar{O}_3 \supset \bar{G}_k$ symmetry adapted vectors (SAV's), on some (usually finite dimensional) Hilbert space $\mathcal{E}$, providing a convenient basis for the physical states,

(ii) $O_3 \supset G_k$ symmetry adapted operators (SAO's), on $\mathcal{E}$, in terms of which the physical interactions can be developed, and

(iii) $\bar{O}_3 \supset \bar{G}_k$ symmetry adapted Clebsch-Gordan coefficients (SACGc's) for coupling SAV's and SAO's.

(i) The $\bar{O}_3 \supset \bar{G}_k$ SAV's are obtained as linear combinations of $\bar{O}_3 \supset \bar{O}_2$ vectors $|\tau jm)$, where τ is an out-$\bar{O}_3 \supset \bar{O}_2$ chain-label (which may correspond to $U_N \supset O_3$). More precisely, from the $\bar{O}_3$ irreducible tensorial set (ITS) $\{|\tau jm);\ m = -j, -j+1, \ldots, j\}$ spanning the standard matrix representation $\mathcal{D}^{(j)}$ of $\bar{O}_3$, we can generate another orthonormal $\bar{O}_3$ ITS $\{|\tau ja\Gamma\gamma);\ a\Gamma\gamma \text{ ranging}\}$ spanning a nonstandard representation of $\bar{O}_3$, where

$$|\tau ja\Gamma\gamma) = \sum_m |\tau jm)(jm|ja\Gamma\gamma).$$

Γ stands for an irreducible representations class (IRC) of one of the $\bar{G}_k$'s, say $\bar{G}_{i+1} \equiv \bar{G}$. Further, a is a label of type $a(\bar{O}_3\bar{G}_1)\Gamma(\bar{G}_1)a(\bar{G}_1\bar{G}_2)\Gamma(\bar{G}_2) \ldots a(\bar{G}_i\bar{G})$, where a(AB) denotes a branching multiplicity label to be used when the IRC Γ(B) of the group B appears several times in Γ(A) of A. Finally, γ is a label of type $a(\bar{G}\bar{G}_{i+2})\Gamma(\bar{G}_{i+2}) \ldots a(\bar{G}_{n-1}\bar{G}_n)\Gamma(\bar{G}_n)\gamma(\bar{G}_n)$, where $\gamma(\bar{G}_n)$ is an index to be used when the dimension of $\Gamma(\bar{G}_n)$ is greater than 1. The coefficients $(jm|ja\Gamma\gamma)$ are chosen in such a way that $\{|\tau j a \Gamma \gamma); \gamma$ ranging$\}$ be a $\bar{G}$ ITS spanning a given unitary representation D^{Γ} associated to Γ. These coefficients are thus determined from projection operator techniques. In other words, the $\bar{O}_3 \supset \bar{G}_k$ SAV $|\tau j a \Gamma \gamma)$ is an eigenvector of J^2 and of the Wigner projection operator

$$\sum_{R \in \bar{G}} D^{\Gamma}(R)^*_{\gamma\gamma} P_R .$$

Alternatively, the $\bar{O}_3 \supset \bar{G}_k$ SAV's can be taken to be eigenvectors of J^2 and of a normal (generally Hermitian) operator Q defined in the universal enveloping algebra su_2 of SU_2 (and consequently in u_N) and invariant under one of the G_k's, say $G_i(4)$. This state labelling operator is to be chosen for its mathematical and/or physical interest. Integrity bases to obtain Q have been given for the different subgroups of $O_3(4)$. According to Wigner's theorem, the eigenvectors of J^2 and Q are also eigenvectors of

$$\sum_{R \in \bar{G}} D^{\Gamma}(R)^*_{\gamma\gamma} P_R .$$

In that case, the label a is to be replaced by an eigenvalue λ of Q. We are thus led to super SAV's (SSAV's) of type $|\tau j \lambda \Gamma \gamma)$. [We assume the $\lambda\Gamma\gamma$'s completely label the state vectors within the IRC (j) of SU_2.] Clearly, each SSAV is of the form

$$|\tau j \lambda \Gamma \gamma) = \sum_a |\tau j a \Gamma \gamma)(j\Gamma)_{a\lambda},$$

where the unitary matrix $(j\Gamma)$ diagonalizes the matrix of Q set up on $\{|\tau j a \Gamma \gamma); a$ ranging$\}$.

(ii) The transformation $(jm|ja\Gamma\gamma)$ defining our $\bar{O}_3 \supset \bar{G}_k$ SAV's may be used to obtain $O_3 \supset G_k$ SAO's from $O_3 \supset O_2$ operators T^k_q. More precisely, from the O_3 ITS $\{T^k_q; q = -k, -k+1, \ldots, k\}$ spanning $\mathcal{D}^{(k)}$ of O_3, we can generate the G ITS $\{T^k_{a\Gamma\gamma}; \gamma$ ranging$\}$ spanning D^{Γ} of G, where

$$T^k_{a\Gamma\gamma} = \sum_q T^k_q (kq|ka\Gamma\gamma).$$

As an example of importance for applications we have the following particular G harmonics

$$y^k_{a_o\Gamma_o} = \sum_q y^k_q (kq|ka_o\Gamma_o),$$

where Γ_o is the identity IRC of G and y^k_q the q-th spherical harmonic of order k. Remark that Q can be obtained by applying the operator equivalents method [i.e., within a subspace of constant angular momentum: $u \sim J_u$ with $u = x,y,z$ up to symmetrization (5)] to

$$\sum_{ka_o} c(ka_o\Gamma_o) y^k_{a_o\Gamma_o},$$

where $c(ka_o\Gamma_o) \in \mathbb{C}$.

(iii) The $\bar{O}_3 \supset \bar{G}_k$ SACGc's are defined in terms of the $\bar{O}_3 \supset \bar{O}_2$ Wigner coefficients $(j_1j_2m_1m_2|jm)$ by (3)

$$(j_1j_2a_1\Gamma_1\gamma_1a_2\Gamma_2\gamma_2|ja\Gamma\gamma) = \sum_{m_1m_2m} (j_1m_1|j_1a_1\Gamma_1\gamma_1)^* (j_2m_2|j_2a_2\Gamma_2\gamma_2)^*(j_1j_2m_1m_2|jm)(jm|ja\Gamma\gamma).$$

The f function defined through

$$f\begin{pmatrix} j_1 & j_2 & j \\ a_1\Gamma_1\gamma_1 & a_2\Gamma_2\gamma_2 & a\Gamma\gamma \end{pmatrix} = (-)^{2j}(j_2ja_2\Gamma_2\gamma_2a\Gamma\gamma|j_1a_1\Gamma_1\gamma_1)^*/(2j_1+1)^{\frac{1}{2}}$$

proved useful in numerical calculations (1,3). Indeed the interest for f clearly appears in the $\bar{O}_3$ Wigner-Eckart theorem in an $\bar{O}_3 \supset \bar{G}_k$ basis (3)

$$(\tau_1 j_1 a_1 \Gamma_1 \gamma_1 | T^k_{a\Gamma\gamma} | \tau_2 j_2 a_2 \Gamma_2 \gamma_2) = (\tau_1 j_1 \| T^k \| \tau_2 j_2) \, f\begin{pmatrix} j_1 & j_2 & k \\ a_1\Gamma_1\gamma_1 & a_2\Gamma_2\gamma_2 & a\Gamma\gamma \end{pmatrix}.$$

Considerable attention has been paid to the $\bar{O}_3 \supset \bar{G}_k$ SACGc's, f function, and symmetrized f function (3). Let us just mention it is possible in most cases to choose the SAV's so that the corresponding SACGc's be real numbers of the form $\pm(p/q)^{\frac{1}{2}}$ $(p,q \in \mathbb{N})$. Note that studying the $\bar{O}_3 \supset \bar{G}_k$ SACGc's may lead to useful information on the $\bar{O}_3 \supset \bar{O}_2$ 3-jm Wigner symbols. For instance the accidental (with respect to $SO_3 \supset SO_2$) zero value for $\begin{pmatrix} 3 & 2 & 3 \\ -2 & 0 & 2 \end{pmatrix}$ can be explained (from an inner Kronecker product existence condition for O) by considering an associated $SO_3 \supset O \supset D_4$ SACGc.

Let us mention other coefficients related to the $\bar{O}_3 \supset \bar{G}_k$ SACGc's. Following Racah's factorization lemma we have

$$(j_1 j_2 a_1 \Gamma_1 \gamma_1 a_2 \Gamma_2 \gamma_2 | j a \Gamma \gamma) = \sum_{\beta} (j_1 a_1 \Gamma_1 + j_2 a_2 \Gamma_2 | j a \beta \Gamma)(\Gamma_1 \Gamma_2 \gamma_1 \gamma_2 | \beta \Gamma \gamma),$$

where $(\Gamma_1 \Gamma_2 \gamma_1 \gamma_2 | \beta \Gamma \gamma)$ is a $\bar{G}$ CGc. (β is an inner Kronecker product multiplicity label.) In many instances this relation provides us with a simple way of computing CGc's and 3-$\Gamma\gamma$ symbols for $\bar{G}$. The isoscalar factors (+1) and the symmetrized isoscalar factors have been extensively studied (6). Finally when going from the SAV's to the SSAV's, the SACGc's are replaced by super SACGc's (SSACGc's) according to

$$(j_1 j_2 \lambda_1 \Gamma_1 \gamma_1 \lambda_2 \Gamma_2 \gamma_2 | j \lambda \Gamma \gamma) = \sum_{a_1 a_2 a} (j_1 \Gamma_1)^*_{a_1 \lambda_1} (j_2 \Gamma_2)^*_{a_2 \lambda_2} (j_1 j_2 a_1 \Gamma_1 \gamma_1 a_2 \Gamma_2 \gamma_2 | j a \Gamma \gamma)(j\Gamma)_{a\lambda}.$$

Particular SSACGc's have been studied by various authors (1,2,4). Note that the SSACGc's $(jk\lambda\Gamma\gamma a_o \Gamma_o | j\lambda\Gamma\gamma)$ are the eigenvalues (up to constants) of the operator Q deduced from $y^k_{a_o\Gamma_o}$. Note also that many SSACGc's are different from $\pm(p/q)^{\frac{1}{2}}$.

SECTION 2

We now apply these ideas to ligand field theory (LFT). LFT deals with optical and magnetic properties of an $n\ell^x$ ion (x valence electrons) embedded in a molecular, solid state or biological environment on which acts an external magnetic field. The nearest neighbours of the ion (the so-called ligands) are generally responsible for the observed properties. Nearly 50 years ago group theory (principally character theory) was applied (principally in a qualitative way) to LFT (7). Applications of group theory of a more quantitative nature have been achieved by several authors during the two last decades (1,3,6,8), principally by Tanabe, Sugano, and Kamimura [see (8)], Griffith [see (8)], and Schönfeld, Flato, and Rosengarten [see (1)].

The Hamiltonian $\mathcal{H}$ for our ion may be realistically approximated as

$$\mathcal{H} = \mathcal{H}_e + \mathcal{H}_{so} + \mathcal{H}_{\ell f} + \mathcal{H}_Z,$$

where $\mathcal{H}_e$ describes the electrostatical interaction within $n\ell^x$, $\mathcal{H}_{so}$ the spin-orbit interaction within $n\ell^x$, $\mathcal{H}_{\ell f}$ the LF interaction between $n\ell^x$ and the ligands, and $\mathcal{H}_Z$ the Zeeman interaction between $n\ell^x$ and the magnetic field. Let $G_{i+1} \equiv G$ be the group leaving invariant the aggregate constituted by the ion and its ligands. According to the Curie-Becquerel principle $\mathcal{H}_{\ell f}$ is G-invariant. We often have

$$\mathcal{H}_{\ell f} = \mathcal{H}_{G_i} + \mathcal{H}_G,$$

where $\mathcal{H}_{G_i}$ is invariant under the high symmetry group G_i and $\mathcal{H}_G$ under the low symmetry group G (although not being G_i-invariant). For example $G_{i=1} = O$ and $G_{i+1=2}$ ($\equiv$ G) = D_3 so that we can take

$G_3 = C_3$ yielding the chain $SO_3 \supset O \supset D_3 \supset C_3$ or $SU_2 \supset \bar{O} \supset \bar{D}_3 \supset \bar{C}_3$ according as x is even or odd.

There are at least 4! coupling schemes for calculating the matrix elements of $\mathcal{H}$ within the $n\ell^x$ subspace. The best known are the strong, medium, and weak field coupling schemes which correspond to the limiting cases $\mathcal{H}_{\ell f} > \mathcal{H}_e > \mathcal{H}_{so}$ (strong field), $\mathcal{H}_e > \mathcal{H}_{\ell f} > \mathcal{H}_{so}$ (medium field), and $\mathcal{H}_e > \mathcal{H}_{so} > \mathcal{H}_{\ell f}$ (weak field), respectively. The levels of $\mathcal{H}$ computed in any of the 4! schemes are the same once the diagonalization of $\mathcal{H}$ is performed within the (finite) $n\ell^x$ subspace [of dimension $(4\ell+2)!/(4\ell+2-x)!x!$]. Therefore we choose the weak field coupling scheme (WFCS) because it leads to the easiest calculations. The WFCS state vectors may be taken as $\bar{O}_3 \supset \bar{G}$ SAV's of the form $|n\ell^x\alpha SLJa\Gamma\gamma)$. The $\mathcal{H}_e + \mathcal{H}_{so}$ matrix in the WFCS easily follows from the one for the free ion

$$(n\ell^x\alpha'S'L'J'a'\Gamma'\gamma'|\mathcal{H}_e+\mathcal{H}_{so}|n\ell^x\alpha SLJa\Gamma\gamma) = \delta(J'J)\delta(a'a)\delta(\Gamma'\Gamma)\delta(\gamma'\gamma)$$
$$[\delta(S'S)\delta(L'L)\delta(SLJ)E_e(n\ell^x\alpha'\alpha SL)+E_{so}(n\ell^x\alpha'S'L'\alpha SLJ)],$$

where E_e and E_{so} are functions of the Slater-Condon-Shortley parameters $F^{(k)}$ and spin-orbit parameter $\zeta_{n\ell}$, respectively. The matrix elements of $\mathcal{H}_{\ell f}$ between our $\bar{O}_3 \supset \bar{G}$ SAV's are readily calculated by expressing $\mathcal{H}_{\ell f}$ in terms of $O_3 \supset G$ SAO's. We thus obtain

$$(n\ell^x\alpha'S'L'J'a'\Gamma'\gamma'|\mathcal{H}_{\ell f}|n\ell^x\alpha SLJa\Gamma\gamma) = \delta(S'S)\delta(\Gamma'\Gamma)\delta(\gamma'\gamma)$$
$$\sum_{\substack{k=0,2,\ldots,2\ell \\ a_o}} D(ka_o\Gamma_o)(\ell^x\alpha'SL'J'\|U^k\|\ell^x\alpha SLJ)\,f\begin{pmatrix} J' & J & k \\ a'\Gamma & a\Gamma & a_o\Gamma_o \end{pmatrix},$$

where U^k is a Racah unit tensor and $D(ka_o\Gamma_o)$ a LF parameter. Finally after developing $\mathcal{H}_Z$ in $O_3 \supset G$ SAO's we get

$$(n\ell^{x}\alpha'S'L'J'a'\Gamma'\gamma'|\mathcal{H}_Z|n\ell^{x}\alpha SLJa\Gamma\gamma) = \delta(\alpha'\alpha)\delta(S'S)\delta(L'L)(-)^{J-J'}\beta$$

$$(\ell^{x}\alpha SLJ'\|kL+g_S S\|\ell^{x}\alpha SLJ) \sum_{a_1\Gamma_1\gamma_1} H^{1}_{a_1\Gamma_1\gamma_1} f\begin{pmatrix} J & J' & 1 \\ a\Gamma\gamma & a'\Gamma'\gamma' & a_1\Gamma_1\gamma_1 \end{pmatrix}^{*},$$

where $H^{1}_{a_1\Gamma_1\gamma_1}$ is the $a_1\Gamma_1\gamma_1$-th component of the magnetic field and k the orbital reduction factor. The reduced matrix elements $(\| \ \|)$ as well as E_e and E_{so} are obtainable from Racah's method, compiled tables or existing programs. Consequently the only task in building the $\mathcal{H}$ matrix in the WFCS is to compute the relevant f coefficients. This has been done for numerous chains (1,3).

The energy levels for our ion are parametrically dependent on the $F^{(k)}$'s, $\zeta_{n\ell}$, $D(ka_o\Gamma_o)$'s, $k(\sim 1)$, $g_S(\sim 2)$, and $H^{1}_{a_1\Gamma_1\gamma_1}$'s. The $F^{(k)}$'s, $\zeta_{n\ell}$, $D(ka_o\Gamma_o)$'s, and k are phenomenological parameters to be obtained by a least squares fitting procedure from experimental data (optical spectra, e.s.r. spectra, susceptibilities,...). By way of illustration consider an nf^{x} ion in a cubical LF ($G_i = O$) with a tetragonal distortion ($G = D_4$). The various parameters are then $F^{(0)}$, $F^{(2)}$, $F^{(4)}$, $F^{(6)}$, ζ_{nf}, $D(0A_1A_1)$, $D(4A_1A_1)$, $D(6A_1A_1)$, $D(2EA_1)$, $D(4EA_1)$, $D(6EA_1)$, and k. [$D(4A_1A_1)$ and $D(6A_1A_1)$ are relative to $\mathcal{H}_O$ while $D(2EA_1)$, $D(4EA_1)$, and $D(6EA_1)$ to $\mathcal{H}_{D_4}$.] Note that in the general case we can assume without loss of generality $F^{(0)} = D(0\Gamma_o) = 0$ in so far as the properties of the aggregate are confined to arise from the configuration $n\ell^{x}$.

SECTION 3

The $D(ka_o\Gamma_o)$ parametrization of $\mathcal{H}_{\ell f}$ presents several advantages. The different linearly independent $ka_o\Gamma_o$-contributions to $\mathcal{H}_{\ell f}$ of multipolar order k are clearly emphasized. This can be used to obtain the relative importance of $\mathcal{H}_{G_i}$ and of (the distortion) $\mathcal{H}_G$. In addition, the center of gravity rule applies to each G_i level perturbed by (i) any $ka_o\Gamma_o$-contribution which is not G_i-invariant and thus (ii) by $\mathcal{H}_G$.

The LF parameters $D(ka_o\Gamma_o)$ may be interpreted in the frame of the point charge electrostatic model. In this respect the basic relation is

$$D(ka_o\Gamma_o) \sim \sum_J I_k(J) Y^k_{a_o\Gamma_o}(\Theta_\tau, \Phi_J)^*,$$

where the sum runs over the ligands and $I_k(J)$ is a (phenomelogical) radial parameter. (Clearly the I_k parametrization corresponds to a multipolar expansion of $\mathcal{H}_{\ell f}$ considered as arising from an electrostatic potential.) Given a set of parameters $D(ka_o\Gamma_o)$ we may get a (possibly nonunique) set of parameters $I_k(J)$. This latter set can be discussed in turn in the frame of the so-called angular overlap model by using (9)

$$e_\lambda(J) = (-)^\lambda (2\ell+1) \sum_k \begin{pmatrix} \ell & k & \ell \\ 0 & 0 & 0 \end{pmatrix} \begin{pmatrix} \ell & k & \ell \\ -\lambda & 0 & \lambda \end{pmatrix} I_k(J).$$

The e_λ (angular overlap model) parametrization corresponds to an expansion of $\mathcal{H}_{\ell f}$ of the form

$$\sum_{\lambda=0,1,\ldots,\ell} \sum_J V_\lambda(J) e_\lambda(J),$$

where V_λ ($\equiv V_\sigma, V_\pi, V_\delta, \ldots$) couples only the orbitals λ [$\equiv \sigma(\lambda=0)$, $\pi(\lambda=1)$, $\delta(\lambda=2), \ldots$]. In particular $V_\sigma(J)$ can be replaced by the effective operator

$$V_\sigma(J) = (4\pi/(2\ell+1)) \sum_{i=1,2,\ldots,x} \delta(\Omega_i - \Omega_J),$$

where $\delta(\Omega_i - \Omega_J)$ is a surface delta interaction. The operator $V_\lambda(J)$ describes the λ character of the (weak covalent) bond between the partly filled shell ion and the J-th ligand. Consequently, semiquantitative information about the chemical bonding in the aggregate under consideration may be gained from the knowledge of the parameters $e_\lambda(J)$. The spectrum of

$$\sum_J V_\lambda(J)e_\lambda(J)$$

often presents numerous accidental (with respect to G) degeneracies. As an example consider the aggregate ML (of symmetry $C_{\infty v}$), M being a nd^1 ion. Such an aggregate generally has 3 LF levels: 1 singlet $W(A_1)$ and 2 doublets $W(E_1)$ and $W(E_2)$. In the case where $\mathcal{H}_{\ell f} = V_\pi e_\pi$, we just get 2 levels: $W(E_1)$ and $W(E_2) = W(A_1)$. Let $\{\varphi_3,\varphi_4\}$ and $\{\varphi_5\}$ be the ITS's of real d orbitals spanning the representations associated to E_2 and A_1, respectively. It can be proved V_π varies as $(\varphi_3)^2 + (\varphi_4)^2 + (\varphi_5)^2$ so that the accidental (with respect to $C_{\infty v}$) degeneracy of φ_3, φ_4, and φ_5 is explained via the introduction of an O_3 group. Other accidental degeneracies introduced by

$$\sum_J V_\lambda(J)e_\lambda(J)$$

acting within the $n\ell^x$ manifold can be explained by considering O_3 and O_4 groups and more generally subgroups of $GL_{2\ell+1}$ (9).

The e_λ parametrization has been transcribed to the case of atomic spectroscopy by replacing $I_k(J)$ by the Slater-Condon-Shortley parameter $F^{(k)}$ and $e_\lambda(J)$ by an atomic parameter $\mathcal{E}^\lambda$ (10). The $\mathcal{E}^\lambda$ parametrization thus obtained presents several peculiarities. It corresponds to an expansion of $\mathcal{H}_e$ of the form

$$\sum_\lambda \mathcal{V}^\lambda \mathcal{E}^\lambda,$$

where $\mathcal{V}^\lambda$ may be written as

$$\mathcal{V}^\lambda = (-)^\lambda (2\ell+1) \sum_k \begin{pmatrix} \ell & k & \ell \\ 0 & 0 & 0 \end{pmatrix} \begin{pmatrix} \ell & k & \ell \\ -\lambda & 0 & \lambda \end{pmatrix} \sum_{i>j=1,2,\ldots,x} (v_i^{(k)} \cdot v_j^{(k)})$$

in terms of the Judd unit tensors $v^{(k)}$. In particular $\mathcal{V}^\sigma$ can be replaced by the effective operator

$$\mathcal{V}^\sigma = (4\pi/(2\ell+1)) \sum_{i>j=1,2,\ldots,x} \delta(\Omega_i - \Omega_j).$$

Further, the energy of the terms ^{2S+1}L of $n\ell^x$ often assume a simple form when expressed in terms of the parameters $\mathcal{E}^{\lambda}$; for instance $^{1}S(n\ell^2) = \mathcal{E}^{\sigma}$ and $^{3}P(n\ell^2) = \mathcal{E}^{\pi}$. Finally, the spectrum of V^{λ} often presents numerous accidental (with respect to O_3) degeneracies. As an example the degeneracies introduced by V^{σ} within a nd^x configuration coalesce with the Laporte-Platt degeneracies which have been explained by Judd by using the chain $SU_5 \supset O_5$ (10).

ACKNOWLEDGMENTS

The author is much indebted to Professors L. Michel, J. Patera, R.T. Sharp, and P. Winternitz for helpful discussions. He also acknowledges the financial support from the France-Quebec Scientific Exchange Program and the kind hospitality extended to him at the Centre de Recherches Mathématiques de l'Université de Montréal.

REFERENCES

1. W. LOW, *"Quantum Electronics I and II"*, Columbia Univ. Press, New York, 1960 and 1961; in *"Spectroscopic and Group Theoretical Methods in Physics"* (F. Bloch, S.G. Cohen, A. de-Shalit, S. Sambursky, and I. Talmi, Eds.), North-Holland, Amsterdam, 1968; W. LOW and G. ROSENGARTEN, J. Mol. Spectrosc. 12, 319 (1964); M. FLATO, J. Mol. Spectrosc. 17, 300 (1965).
2. J. MORET-BAILLY, Cah. Phys. 15, 237 (1961); J. MORET-BAILLY, L. GAUTIER, and J. MONTAGUTELLI, J. Mol. Spectrosc. 15, 355 (1965); F. MICHELOT and J. MORET-BAILLY, J. Phys. 36, 451 (1975); J.P. CHAMPION, G. PIERRE, F. MICHELOT, AND J. MORET-BAILLY, Can. J. Phys. XX, XXX (1976).
3. M. KIBLER, J. Mol. Spectrosc. 26, 111 (1968); Int. J. Quantum Chem. 3, 795 (1969); C.R. Acad. Sci. B 268, 1221 (1969); J. Math. Phys. 17, 855 (1976); J. Mol. Spectrosc. XX, XXX (1976); M.R. KIBLER and P.A.M. GUICHON, Int. J. Quantum Chem. 10, 87 (1976); M.R. KIBLER and G. GRENET, Int. J. Quantum Chem. (in press); M.R. KIBLER and P.A.M. GUICHON, to be published.

4. J. PATERA and P. WINTERNITZ, J. Math. Phys. 14, 1130 (1973); B.R. JUDD, W. MILLER, Jr., J. PATERA, and P. WINTERNITZ, J. Math. Phys. 15, 1787 (1974); J. PATERA, P. WINTERNITZ, and H. ZASSENHAUS, J. Math. Phys. 16, 1597 (1975); W. McKAY, J. PATERA, and R.T. SHARP, Comp. Phys. Com. 10, 1 (1975); J. PATERA and P. WINTERNITZ, J. Chem. Phys. XX, XXX (1976); R.P. BICKERSTAFF and B.G. WYBOURNE, J. Phys. A 9, 1051 (1976).

5. A. ABRAGAM and M.H.L. PRYCE, Proc. Roy. Soc. A 205, 135 (1951); K.W.H. STEVENS, Proc. Phys. Soc. A 65, 209, 311 (1952); R.J. ELLIOTT and K.W.H. STEVENS, Proc. Roy. Soc. A 215, 437 (1952); 218, 553 (1953); 219, 387 (1953); B. BLEANEY and K.W.H. STEVENS, Rep. Prog. Phys. 16, 108 (1953).

6. TANG AU-CHIN, SUN CHIA-CHUNG, KIANG YUAN-SUN, DENG ZUNG-HAU, LIU JO-CHUANG, CHANG CHIAN-ER, YAN GO-SEN, GOO ZIEN, and TAI SHU-SHAN, Sci. Sinica 15, 610 (1966); T. BUCH, J. Math. Phys. 13, 1892 (1972); T. LULEK, Acta Phys. Polon. A 43, 705 (1973); 48, 501 (1975); J.C. DONINI and B.R. HOLLEBONE, Theoret. Chim. Acta XX, XXX (1976).

7. H.A. BETHE, Ann. Physik 3, 133 (1929); R. SCHLAPP and W.G. PENNEY, Phys. Rev. 42, 666 (1932); J.H. VAN VLECK, J. Chem. Phys. 3, 807 (1935).

8. Y. TANABE and S. SUGANO, J. Phys. Soc. Japan 9, 753, 766 (1954); Y. TANABE and H. KAMIMURA, J. Phys. Soc. Japan 13, 394 (1958); S. SUGANO and Y. TANABE, J. Phys. Soc. Japan 13, 880 (1958); J.S. GRIFFITH, Trans. Faraday Soc. 54, 1109 (1958); 56, 193, 996 (1960); Mol. Phys. 3, 79, 285, 457, 477 (1960).

9. M.R. KIBLER, Chem. Phys. Lett. 7, 83 (1970); 8, 142 (1971); J. Chem. Phys. 55, 1989 (1971); 61, 3859 (1974); Int. J. Quantum Chem. 9, 403 (1975).

10. M.R. KIBLER, Int. J. Quantum Chem. 9, 421 (1975).

FINITE REPRESENTATIONS OF THE UNITARY GROUP AND THEIR APPLICATIONS IM MANY-BODY PHYSICS

P. Kramer

In part A of this note the definition and the algebraic properties of certain square symbols appearing in the finite irreducible representations of the group U(n) are given. In part B we link this concept to properties of the symmetric group S(N) and its representations. It is claimed that the recent developments of the technique of S(N) for many-body physics [1] are closely related to the properties of these square symbols.

A. A unitary irreducible representation (IR) D^h of U(n) is determined by a partition h of N; its reduction under restriction to subgroups $U(n) \supset U(n-1) \times U(1)$, $U(n-1) \supset U(n-2) \times U(1)$ etc. is described by a Gelfand pattern q of partitions. To given h and q there belongs a weight $w = (w_1 w_2 \ldots w_n)$ which yields the IR of the subgroup $A(n) = U(1) \times \ldots \times U(1)$ of U(n).

Given two weights $\bar{w}$ and w whose components have the same sum N we define a matrix $k = (k_{ij})$ of integers as a magic square subject to the row and column conditions

$$\sum_{j}^{n} k_{ij} = \bar{w}_i \qquad \sum_{i}^{n} k_{ij} = w_j .$$

We introduce for k and w the notation

$$k! = \prod_{ij}^{n} (k_{ij})! \qquad w! = \prod_{i}^{n} (w_i)!.$$

For complex n×n matrices z we denote the transposed by ${}^t z$, the adjoint by z^+, the complex conjugate by z^* and apply the same notation to matrix representations.

The double Gelfand states $P^h(z)$ are polynomial functions of n^2 complex variables $z = (z_{ij})$ such that under application of elements u_1, u_2 of U(n) one has

$$P^h(u_1 z u_2) = D^h(u_1) \cdot P^h(z) \cdot D^h(u_2).$$

Moreover these polynomials are orthonormal in the Bargmann space [2] $F(n^2)$, that is,

$$\int [P^{h'}_{\bar{q}'q'}(z)]^* P^h_{\bar{q}q}(z) d\mu(z) = \delta(\bar{q}'h'q', \bar{q}hq).$$

A theorem stated by Louck [3] for U(n) and extended later by Brunet and Seligman [4] to GL(n,C) and further to all complex n×n matrices assures that

$$D^h_{\bar{q}q}(z) = \lambda_h P^h_{\bar{q}q}(z).$$

This allows us to apply techniques of Bargmann space to the IR of these matrix groups. The Bargmann space possesses the complete orthonormal basis

$$[k!]^{-\frac{1}{2}} z^k$$

where $z^k = \prod (z_{ij})^{k_{ij}}$. We define a square symbol by expanding $P^h(z)$ in this basis

$$P^h_{\bar{q}q}(z) = \sum_k \begin{bmatrix} h & q \\ \bar{q} & k \end{bmatrix} [k!]^{-\frac{1}{2}} z^k.$$

Clearly the elements of k are restricted in the way indicated above. Since this is a unitary transformation one obtains for the square symbols

$$\sum_k \begin{bmatrix} h & q \\ \bar{q} & k \end{bmatrix} \begin{bmatrix} h' & q' \\ \bar{q}' & k \end{bmatrix}^* = \delta(\bar{q}hq, \bar{q}'h'q') \tag{1}$$

$$\sum_{\bar{q}hq} \begin{bmatrix} h & q \\ \bar{q} & k \end{bmatrix} \begin{bmatrix} h & q \\ \bar{q} & k' \end{bmatrix}^* = \delta(k,k'). \tag{2}$$

We write now for the IR $D^h(z)$

$$D^h(z) = \lambda_h \sum_k \begin{bmatrix} h & q \\ \bar{q} & k \end{bmatrix} [k!]^{-\frac{1}{2}} z^k \tag{3}$$

and derive properties of the square symbols from this expression. We shall make use of the fact that identical polynomials have identical coefficients.

To analyze the representation condition we first define algebraic coefficients $[k^a k^b k^c]$ by the generating function

$$[k^c!]^{-\frac{1}{2}} (a\cdot b)^{k^c} = \sum_{k^a k^b} [k^a!]^{-\frac{1}{2}} a^{k^a} [k^b!]^{-\frac{1}{2}} b^{k^b} [k^a k^{bt} k^c]. \tag{4}$$

The explicit expansion yields

$$[k^a k^b k^c] = [k^a! k^b! k^c!]^{\frac{1}{2}} \sum \prod_{ist} [c_{ist}!]^{-1} \tag{5}$$

where the sum runs over all integers satisfying the equations

$$\sum_t c_{ist} = k^a_{is} \qquad \sum_i c_{ist} = k^b_{st} \qquad \sum_s c_{ist} = k^c_{ti}$$

and where the weights are related by

$$w^a = w^{-b} \qquad w^b = w^{-c} \qquad w^c = w^{-a}.$$

A geometrical interpretation of these relations can be given in terms of a magic cube with a summation condition for the integers

c_{ist} belonging to a given site. The representation condition

$$D^h(a \cdot b) = D^h(a) \cdot D^h(b)$$

yields then the algebraic relation

$$\sum_{k^c} \lambda_h \begin{bmatrix} h & q \\ \bar{q} & t_k{}^c \end{bmatrix} [k^a{}_k{}^b{}_k{}^c] = \sum_{q'} \lambda_h \begin{bmatrix} q & q' \\ \bar{q} & k^a \end{bmatrix} \lambda_h \begin{bmatrix} h & q \\ q' & k^b \end{bmatrix} .$$

Next consider the Kronecker product of two representations with ρ being a multiplicity label. Using

$$z^{k'} z^{k''} = z^{k'+k''}$$

one finds for the square symbols

$$\sum_{k'k''} \left[\frac{k!}{k'!k''!}\right]^{\frac{1}{2}} \lambda_{h'} \begin{bmatrix} h' & q' \\ \bar{q}' & k' \end{bmatrix} \lambda_{h''} \begin{bmatrix} h'' & q'' \\ \bar{q}'' & k'' \end{bmatrix} \delta(k'+k'',k)$$
$$= \sum_{\bar{q}hq\rho} \langle h'\bar{q}'h''\bar{q}'' | h\bar{q}\rho \rangle \lambda_h \begin{bmatrix} h & q \\ \bar{q} & k \end{bmatrix} \langle hq\rho | h'q'h''q'' \rangle . \tag{7}$$

Finally we consider the adjoint and the complex conjugate representation of GL(n,C).

By assuming

$$(D^{-1})^+(z) = D((z^{-1})^+)$$

we find

$$\lambda_h \begin{bmatrix} h & q \\ \bar{q} & k \end{bmatrix} = \lambda_h^* \begin{bmatrix} h & \bar{q} \\ q & t_k \end{bmatrix}^* \tag{8}$$

and similarly from the condition

$$D^*(z) = D(z^*)$$

we get

$$\lambda_h \begin{bmatrix} h & q \\ \bar{q} & k \end{bmatrix} = \lambda_h^* \begin{bmatrix} h & q \\ \bar{q} & k \end{bmatrix}^* . \tag{9}$$

We turn now to subgroups of U(n). Under the subduction

$U(n) \downarrow U(n-1) \times U(1)$ we find

$$\lambda_h \begin{bmatrix} h & q \\ \bar{q} & k \end{bmatrix}_{U(n)\downarrow U(n-1)\times U(1)} = \lambda_h \begin{bmatrix} h & q \\ \bar{q} & k \end{bmatrix}_{U(n-1)} .$$

Next we consider the semidirect product K(n) formed from the group A(n) of diagonal unitary matrices and the group S(n) [5]. Subducing D^h to this subgroup one gets contributions for a given permutation matrix p provided that $k = k(p)$ has entry zero in the same position as p, hence

$$D^h(ap) = \lambda_h \begin{bmatrix} h & q \\ \bar{q} & k(p) \end{bmatrix} [k(p)!]^{-\frac{1}{2}} (ap)^{k(p)} .$$

In this case the magic cube yields

$$[k^a k^b k^c] = [w!]^{\frac{1}{2}}$$

and the multiplication law of the corresponding square symbols becomes

$$\lambda_h [w!]^{-\frac{1}{2}} \begin{bmatrix} h & q \\ \bar{q} & k(p_c) \end{bmatrix} = \sum_{q'} \lambda_h [w!]^{-\frac{1}{2}} \begin{bmatrix} h & q' \\ \bar{q} & k(p_a) \end{bmatrix} \begin{bmatrix} h & q \\ q' & k(p_b) \end{bmatrix} \lambda_h [w!]^{-\frac{1}{2}} \tag{10}$$

where $p_c = p_a p_b$ is the multiplication of the permutations. Hence the square symbols with "permutational" structure themselves form a reducible representation. In [5] it was shown how IR of K(n) can be obtained as induced representation. The IR are characterized by an ordered weight and IR of subgroups of S(n) which belong to multiple weight components. The reduction to IR of S(n) follows by reducing these induced representation via a generalized reciprocity theorem.

B. The square symbols discussed in part A were obtained from IR of U(n) or GL(n,C). We shall state here certain reinterpretations of concepts and quantities with respect to the IR of S(N). With the weights $\bar{w}$ and w we can associate the subgroups

$$\bar{H} = S(\bar{w}_1) \times S(\bar{w}_2) \times \ldots \times S(\bar{w}_n)$$

$$H = S(w_1) \times S(w_2) \times \ldots \times S(w_n)$$

of S(N). The double cosets of S(N) with respect to these subgroups are described in terms of symbols which correspond precisely to a matrix k [1]. Considering now the IR d^h of S(N) reduced to the subgroups $\bar{H}$ and H respectively and assuming the IR of the subgroups to be the symmetric ones, it can be shown that the square symbols of part A can be expressed as

$$\begin{bmatrix} h & q \\ \bar{q} & k \end{bmatrix} = \left[\frac{|h|_{S(N)}}{N!} \frac{\bar{w}!w!}{k!}\right]^{\frac{1}{2}} d^h_{\bar{q}q}(Z_k). \tag{11}$$

Here Z_k is the generator of the double coset described by the matrix k. The representation d^h is labeled by IR of a sequence of subgroups which are arranged in a Gelfand pattern q. Taking the matrix k in diagonal form one concludes that

$$\lambda_h = \left[\frac{N!}{|h|_{S(N)}}\right]^{\frac{1}{2}}. \tag{12}$$

The equations (8) and (9) become the unitarity and reality relations for the IR of S(N).

These relations are of great interest for the physics of many-particle systems. Here one has to deal with high-dimensional IR of S(N) for which new techniques are required. It turns out that the concept of double cosets is very important as it is directly linked to the exchange properties. As an example consider a system of N particles occupying n different single-particle states φ_i, i = 1,...,n which have scalar products

$$\int \varphi_i^*(x)\varphi_j(x)d^3x = \varepsilon_{ij}.$$

Adapting this system to the IR h of S(N) and allowing for different configurations, one finds for the scalar product of the many-body states

$$\int \bar{\psi}^*(\bar{h}\bar{q})\psi(hq)\ \Pi\ d^3x$$
$$= D^{h}_{\bar{q}q}(\varepsilon)\delta(\bar{h},h) \tag{13}$$
$$= \delta(\bar{h},h)\lambda_h \sum_k \begin{vmatrix} h & q \\ \bar{q} & k \end{vmatrix} [k!]^{-\frac{1}{2}}\varepsilon^k.$$

Here the numbers k_{ij} determine precisely how many particles have been shifted under the permutation Z_k from state j to state i.

The systematic evaluation of matrix elements of operators which has been derived in [1] from the point of view of S(N) can now be given in terms of algebraic coefficients of U(n) and leads to important simplifications. On the other hand it is clear that the properties known for S(N) can serve to bring out new features of the IR of U(n) like Regge symmetry [6].

REFERENCES

1. P. KRAMER and T.H. SELIGMAN, Nuclear Physics A123, 161 (1969), A136, 545 (1969), A186, 49 (1972).
2. V. BARGMANN, Commun. Pure Appl. Math. 14, 187 (1961).
3. J. LOUCK, Amer. J. Phys. 38, 3 (1970).
4. M. BRUNET and T.H. SELIGMAN, Rep. Math. Phys. 8, 233 (1975).
5. P. KRAMER, J. Math. Phys. 9, 639 (1968).
6. P. KRAMER, Z. Physik 216, 68 (1968). P. KRAMER and T.H. SELIGMAN, Z. Physik 219, 105 (1969).

ALGEBRAIC AND GEOMETRIC METHODS OF QUANTISATION OF THE ISOTROPIC HARMONIC OSCILLATOR

Maria Lasocka and Jan Olszewski

It is well-known that the n-dimensional isotropic harmonic oscillator possesses a symmetry represented by the unitary group SU(n). In fact its Hamiltonian is (we put m=1, ω=1, $\hbar$=1):

$$H = \sum_{i=1}^{n} H_i = \tfrac{1}{2} \sum_{i=1}^{n} (P_i^2+Q_i^2) = \sum_{k=1}^{n} (a_k^+ a^k + \tfrac{1}{2}) \qquad (1)$$

where the operators a^k, a_k^+ are defined by:

$$a^k = \frac{1}{\sqrt{2}} (Q_k + iP_k)$$
$$a_k^+ = \frac{1}{\sqrt{2}} (Q_k - iP_k) \qquad (2)$$

and obey the commutation relations:

$$[a^k, a_\ell^+] = \delta_\ell^k, \quad [a_k^+, a_\ell^+] = [a^k, a^\ell] = 0 \qquad (3)$$

The operators:

$$A_k^\ell = \tfrac{1}{2}(a_k^+ a^\ell + a^\ell a_k^+) \qquad (4)$$

obey the commutation relations:

$$[A_k^\ell, A_m^n] = \delta_m^\ell A_k^n - \delta_k^n A_m^\ell \tag{5}$$

and form the Lie algebra of U(n). Since the Hamiltonian

$$H = \sum_{k=1}^{n} A_k^k = \sum_{k=1}^{n} (a_k^+ a^k + \tfrac{1}{2}) \tag{6}$$

commutes with all the operators A_k^ℓ we can form traceless operators:

$$A'^{\ell}_k = A_k^\ell - \frac{1}{n}\delta_k^\ell H \tag{7}$$

which also satisfy the commutation relations (5).

The set of traceless operators (7) forms a basis for the Lie algebra of the group SU(n) of dimension n^2-1. The eigenvectors belonging to the eigenvalues of H will then form bases for irreducible representations of SU(n). Not all the irreducible representations of SU(n) appear in the investigated system. The only representations present are the symmetric representations of the (N,0,0,...) type. We get the normalized basic vector of the irreducible representation space for $N \geq 1$ (N is an integer) in the following way:

$$\frac{1}{\sqrt{\nu_1!\nu_2! \dots \nu_n!}} (a_1^+)^{\nu_1} \dots (a_n^+)^{\nu_n} |0\rangle = |\nu_1, \nu_2 \dots \nu_n\rangle \tag{8}$$

where

$$\nu_1 + \nu_2 + \dots + \nu_n = N, \quad \nu_i = 0,1,2,\dots .$$

The dimension of an irreducible representation is given by the formula:

$$\dim(N,0) = \frac{(N+n-1)!}{N!(n-1)!} . \tag{9}$$

The corresponding eigenvalues of the Hamiltonian are:

$$E_N = (N + \frac{n}{2}). \tag{10}$$

It is well-known that the SU(n) group is a symmetry group of the harmonic oscillator for one fixed energy level and thus it does not reproduce the full energy spectrum within one irreducible representation. To get the full energy spectrum we have to extend the discussed SU(n) algebra [1].

In the case of the harmonic oscillator we have now to find an algebra containing SU(n) as a subalgebra and having a representation of SU(n) based on the space of all vectors (8). At least two such algebras exist. One of them is the non-compact algebra Sp(n,R) defined by the generators (4) and:

$$\begin{aligned} A^0_{ij} &= \tfrac{1}{2}\{a^+_i, a^+_j\} \\ A^0_0 &= H \qquad\qquad \{\ \}\text{ is an anticommutator} \\ A^{ij}_0 &= -\tfrac{1}{2}\{a^i, a^j\} \end{aligned} \tag{11}$$

with commutators:

$$\begin{aligned} [A^i_j, A^{\ell k}_0] &= -\delta^\ell_j A^{ki}_0 - \delta^k_j A^{\ell i}_0 \\ [A^i_j, A^0_{\ell k}] &= \delta^i_\ell A^0_{jk} + \delta^i_k A^0_{j\ell} \\ [A^0_{\ell k}, A^{mn}_0] &= \delta^m_k A^n_\ell + \delta^n_k A^m_\ell + \delta^m_\ell A^n_k + \delta^n_\ell A^m_k \qquad n \neq k \neq \ell \neq m \\ [A^0_0, A^{ij}_0] &= -2A^{ij}_0; \quad [A^0_0, A^0_{ij}] = 2A^0_{ij} \\ [A^0_{k\ell}, A^0_{mn}] &= [A^{k\ell}_0, A^{mn}_0] = 0. \end{aligned} \tag{12}$$

It is easy to show that the bilinear Casimir operator of Sp(n,R) can be expressed in the following form:

$$Q_2 = B_2 + \frac{H^2}{n} + \tfrac{1}{2}\{A_0^{ij}, A_{ij}^0\}. \tag{13}$$

The value of Q_2 is fixed once we fix the representation of Sp(n,R) as a direct sum of the symmetric representations of SU(n). It reads:

$$Q_2 = -\frac{n}{2}\,(n+\tfrac{1}{2}). \tag{14}$$

B_2 is the Casimir operator of SU(n) which for the symmetric representations has the following eigenvalues:

$$B_2 = \frac{n-1}{n}\,N(n+N). \tag{15}$$

The representations belonging to a fixed energy eigenstate can be obtained by inserting N = 0,1,2,... ($N = \nu_1+\nu_2+\ldots+\nu_n$).

The anticommutator from the formula can be expressed in the form:

$$\tfrac{1}{2}\{A_0^{ij}, A_{ij}^0\} = -H^2 - \frac{n}{2}\,(\frac{n}{2}+1) \tag{16}$$

from which we get:

$$H^2 = \frac{n}{n-1}\,[B_2 + \frac{n}{2}\,(\frac{n}{2} - \tfrac{1}{2})] = (N + \frac{n}{2})^2. \tag{17}$$

To obtain the representations of the harmonic oscillator we have to take two infinite irreducible representations of the Sp(n,R) group. The first of them is a direct sum of all the symmetric representations of the SU(n) algebra with N = 0,2,4,... the second one is a similar sum for N = 1,3,5,... . Using the method described above one obtains a full algebraic model for the simple dynamical harmonic oscillator system. One can use these results to construct an operator (built from the creation and anihilation operators) which can produce a state corresponding to an arbitrary energy level of the 3-dimensional oscillator from the vacuum state - the basic state of the oscillator. To do this we introduce operators of the spherical basis:

$$a_{\pm 1} = \mp \frac{1}{\sqrt{2}} (a_x \pm i a_y); \quad a_0 = a_z$$

$$a^*_\mu = (-1)^\mu (a_{-\mu})^+; \quad \mu = \pm 1, 0 \tag{18}$$

which obey the following commutations relations:

$$[a_\mu, a^+_{\mu'}] = (-1)^\mu \delta_{\mu,-\mu'} \qquad \delta_{\mu,-\mu'} = \begin{cases} 1 & \mu+\mu' = 0 \\ 0 & \mu+\mu' \neq 0 \end{cases}$$
$$[a^+_\mu, a^+_{\mu'}] = [a_\mu, a_{\mu'}] = 0 \tag{19}$$

The Hamiltonian:

$$H = \sum_\mu (-1)^\mu a^+_\mu a_\mu + \frac{3}{2} \tag{20}$$

fulfills the following commutation relations:

$$[H, a^+_\mu] = a^+_\mu \qquad [H, a_\mu] = -a_\mu . \tag{21}$$

If we denote by L_0 the "z" component of the angular momentum operator we have also:

$$[L_0, a_\mu] = \mu a_\mu; \qquad [L_0, a^+_\mu] = \mu a^+_\mu . \tag{22}$$

It follows from the commutation relations (21) (22) that the a^+_μ operators create single energy quanta, and μ-quanta of angular momentum projection. The a_μ operators annihilate one energy quantum and create μ-quanta of the angular momentum projection. Using the Eckart-Wigner theorem and properties of the Clebsch-Gordan coefficients we can show that:

$$|n+1, \ell', m'\rangle = N^+_{\ell'\ell} \sum_\mu C^{\ell 1, \ell'}_{m\mu} a^+_\mu |n, \ell, m\rangle \tag{23a}$$

$$|n-1, \ell', m'\rangle = N^-_{\ell'\ell} \sum_\mu C^{\ell 1, \ell'}_{m\mu} a_\mu |n, \ell, m\rangle \tag{23b}$$

where $m' = m+\mu$, $\ell' = \ell\pm1$, and $N^{+}_{\ell\ell'}$, $N^{-}_{\ell\ell'}$ are normalization constants. Action of the operators (23a) (23b) on the state may be visualized as follows:

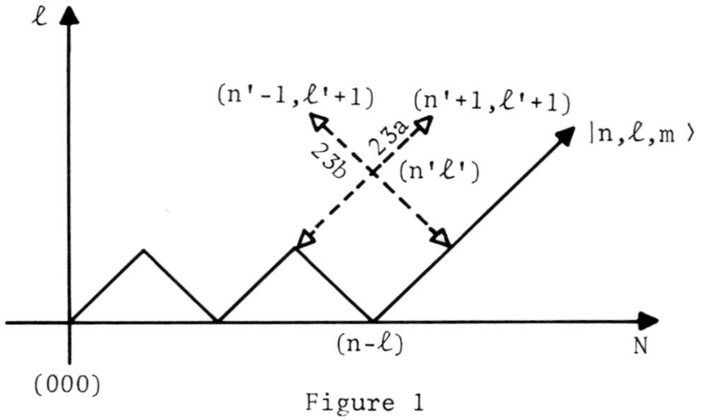

Figure 1

In figure 1, we show the harmonic oscillator states. Each point represents $(2\ell+1)$ angular momentum states of the harmonic oscillator. To construct an arbitrary state we form the operator:

$$\chi^{m}_{n,\ell} = K_{n\ell}\, a^{+(n-\ell)}\, y^{m}_{\ell}(\hat{a}^{+}) \tag{24}$$

where $K_{n\ell}$ is a normalization constant and: $(a^{+})^{n-\ell} = (a^{+}\cdot a^{+})^{\nu}$ $2\nu = n-\ell$ is a spherical scalar operator. The fact that the form of $\chi^{m}_{n\ell}$ operator is correct becomes obvious if we repeat the operation (23a) for the state $n=0$, $\ell=0$, $m=0$ until we get the state defined by the quantum numbers $|n,\ell,m\rangle$, as is visualized in figure 1. To express matrix elements of any operator as vacuum averages we use the following relation:

$$\langle n_1,\ell_1,m_1|\mathcal{O}|n_2,\ell_2,m_2\rangle = \langle 0,0,0|\chi^{m_1+}_{n_1\ell_1}\,\mathcal{O}\,\chi^{m_2}_{n_2\ell_2}|0,0,0\rangle . \tag{25}$$

This method has been applied by authors to compute the matrix elements for the dipole transitions of the harmonic oscillator. The results obtained by this method are identical to those obtained using standard methods of quantum mechanics confirming the usefulness of the algebraic method.

The purpose of this paper is to compare the algebraic method with the method of geometrical quantization [2]. By geometrical quantization we mean the procedure which allows us to obtain the quantum description of the investigated dynamical system starting from the classical dynamical system described in phase space. The phase space of the finite dimensional classical Hamiltonian system is a C^{∞}-differentiable manifold X with given 2-form ω fulfilling the condition $d\omega = 0$. To apply the method of geometrical quantization we have to choose the polarization F of the symplectic form ω. By polarization we mean the complex involutive distribution on the complex extension of phase space X (dim X = 2n) which is maximal isotropic. The quantization allows us to obtain Hilbert space H^F which we may call F-representation of wave function. Classical variables $\varphi \in C^{\infty}(X)$ are called quantizable if the vector fields ζ_{φ} defined by:

$$\omega(\eta,\zeta_{\varphi}) = \eta\varphi \tag{26}$$

for all the vector fields η on X are complete. Polarization F which contains vector fields ζ_{φ} defines the space H^F in which the operator φ acts as an operator of multiplication by function φ. Then the Hilbert space H^F defines a φ-representation which may be regarded as a generalization of position or momentum representations, known from quantum mechanics.

Phase space of the harmonic oscillator is a set:

$$X := \{(p,q) \in R^{2n};\ (p,q) \neq 0;\ p,q \in R^n\} \tag{27}$$

with the given symplectic form:

$$\omega = h^{-1} \sum_{j=1}^{n} dp_j \wedge dq_j \tag{28}$$

where h is a constant chosen to get dimensionless ω with p,q having the dimensions of momentum and position respectively.

The Hamiltonian being a classical function of energy is written:

$$H = \frac{1}{2m} p^2 + \frac{1}{2} m\nu^2 q^2 \tag{29}$$

and the corresponding vector field generated by the classical motions is given by:

$$\zeta_H = h \sum_{j=1}^{n} \left(\frac{p_j}{m} \frac{\partial}{\partial q_j} - m\nu^2 q_j \frac{\partial}{\partial p_j}\right). \tag{30}$$

To get the energy representation H^F we have to choose a polarization F which contains ζ_H. Then for every vector field $\eta \in F$ we have:

$$\eta H = \omega(\eta, \zeta_H) = 0; \tag{31}$$

hence the polarization is tangent to the surface of constant energy. For the harmonic oscillator the surface of constant energy is a (2n-1)-dimensional sphere:

$$p^2 + m\nu^2 q^2 = 2mE \tag{32}$$

which we can write in a form more suitable for the following discussion:

$$|z|^2 = 2mE \tag{33}$$

by introducing complex coordinates:

$$z_j = p_j - im\nu q_j \tag{34}$$

and defining:

$$|z|^2 = \sum_{j=1}^{n} |z_j|^2 \tag{35}$$

Classical motions can be expressed by:

$$z(t) = e^{i\nu t} z(0). \tag{36}$$

Now we can map the energy surface onto a complex projective plane

$P^{n-1}(C)$ by a mapping:

$$\pi : (z_1,\ldots,z_n) \to [z_1,\ldots,z_n] \tag{37}$$

where [z] denotes a set of all the non-zero complex scalar multiples of a variable "z". Each classical orbit is mapped in this way on a point of the projective space $P^{n-1}(C)$. The symplectic form ω is invariant along classical orbits and its confinement to the surface of constant energy induces a symplectic form ω_E on $P^{n-1}(C)$. In the subset $\{[z] \in P^{n-1}(C); z_n \neq 0\}$ we adopt new coordinates:

$$Z_1 = \frac{z_1}{z_n},\ldots,Z_{n-1} = \frac{z_{n-1}}{z_n}. \tag{38}$$

In the new coordinates the form ω_E reads:

$$\omega_E = -\frac{iE}{\nu h}\,\frac{\sum_{j=1}^{n-1} dZ_j \wedge d\bar{Z}_j}{(1+|Z_1|^2+\ldots+|Z_{n-1}|^2)^2}\ . \tag{39}$$

The complex distribution F can be projected as complex distribution F_E on $P^{n-1}(C)$ which fulfills the condition $\omega_E(F_E,F_E) = 0$. Polarization F_E is a polarization of (2n-2)-dimensional symplectic manifold $P^{n-1}(C)$ with a symplectic form ω_E.

An obvious choice for polarization F_E is:

$$F_E = \langle \frac{\partial}{\partial\bar{Z}_1},\ldots,\frac{\partial}{\partial\bar{Z}_{n-1}} \rangle \tag{40}$$

which corresponds to the use of complex structure on $P^{n-1}(C)$. Polarization F can be generated by the following vector fields:

$$\zeta = \sum_{j=1}^{n} (z_j \frac{\partial}{\partial z_j} - z_j \frac{\partial}{\partial\bar{z}_j}) = \frac{1}{i\nu h}\,\zeta_H;\ \eta_{jk} = z_j \frac{\partial}{\partial\bar{z}_k} - z_k \frac{\partial}{\partial\bar{z}_j}. \tag{41}$$

The mappings:

$$X \to S^{2n-1} \times R^1; \quad z \to (\frac{z}{|z|}; |z|)$$
$$\pi : S^{2n-1} \to P^{n-1}(C); \quad z \to [z_1,\ldots,z_n] \tag{42}$$

define two foliations, the leaf of the first one being $E \equiv S^{2n-1}$, the leaf of the second one $D \equiv S^1$ corresponding to the classical orbit. Complex manifold foliated with respect to D (being the space of all classical orbits) can be written as: $X/D = X/E + E/D$, where E/D is identified with definite projective space $P^{n-1}(C)$ with the usual complex structure, and X/E is identified with R^+ as was shown in [3]. The wave function in F-representation may be represented in a general form $T(\nu)^{\frac{1}{2}}(\mu)^{\frac{1}{2}}$, where T is a solution of the set of partial differential equations of first order of the general form:

$$\zeta T + 2\pi i\langle \theta,\zeta \rangle T = 0 \qquad \zeta \in F \tag{43}$$

and

ν - is a measure defined by some holomorphic (n-1)-form on each complex manifold $E/D \equiv P^{n-1}(C)$

μ - induces 1-form on 1-dimensional manifold $X/E \equiv R^+$.

One can show that, in the case of polarization F defined by (41), the wave function in the F-representation confined to one coordinate neighbourhood on the complex manifold X has the form:

$$\psi = e^{i\nu Mt} f(z)|Z|^M \delta(r - \sqrt{-2m\nu\hbar M})(dZ_1 \wedge\ldots\wedge dZ_{n-1})^{\frac{1}{2}}(dr)^{\frac{1}{2}} \tag{44}$$

where $M \leq 0$ and f is a holomorphic function. As is seen from (44) the wave function ψ extends to the function defined on the whole manifold X iff:

$$M = -(N + \frac{n}{2})$$

where $N > 0$ and integer and f is some polynomial of order not exceeding N. Wave functions in the F-representation then take the form:

$$\psi_N = e^{-i\nu(N+n/2)t} f(z)\,|Z|^{-(N+n/2)}\,\delta(r - \sqrt{2m\nu\hbar(N+n/2)}) \times (dZ_1 \wedge \ldots \wedge dZ_{n-1})^{\frac{1}{2}}(dr)^{\frac{1}{2}}. \tag{45}$$

To quantize the function $\varphi \in C^{\infty}(x)$ we have to multiply by $1/2\pi i$ the infinitesimal operator of the action of vector field ζ_φ on the wave function. In particular, for the energy function H and time t we have the relation:

$$\zeta_H = -h\,\frac{\partial}{\partial t}\,. \tag{46}$$

To quantise the function H we should take according to the method described above:

$$-\frac{1}{2\pi i}\,h\,\frac{\partial}{\partial t} = i\hbar\,\frac{\partial}{\partial t}\,. \tag{47}$$

Applying the described procedure for the harmonic oscillator for which the Hamiltonian has the form:

$$H = \frac{p^2}{2m} + \tfrac{1}{2}m\nu^2 q^2 = \frac{r^2}{2m} \tag{48}$$

and using the wave function given by formula (45) we get:

$$i\hbar\,\frac{\partial\psi_N}{\partial t} = \nu\hbar(N + \tfrac{n}{2})\psi_N \tag{49}$$

and

$$\frac{r^2}{2m}\,\psi_N = \frac{1}{2m}\,(\sqrt{2m\nu\hbar(N+n/2)})^2\psi_N = \nu\hbar(N + \tfrac{n}{2})\psi_N. \tag{50}$$

From the formulae (49) (50) it follows that:

1. ψ_N obeys the general Schrödinger equation and
2. ψ_N is an eigenstate of the Hamiltonian for the eigen-

value:

$$E_N = (N + \frac{n}{2})\hbar\nu. \tag{51}$$

It is not easy to compare directly the algebraic method with a geometric one, as the former deals mainly with the symmetry group of the oscillator Hamiltonian on the quantum level, by explicit construction of its Lie algebra representation, while in the latter one starts with the classical phase space of the oscillator, performs its double foliation $X/D = X/E + E/D$, introduces a suitable polarization F by means of which the F-representation for the wave function ψ_N is obtained. In this latter method explicit use of symmetry group of the oscillator Hamiltonian is not made. To exhibit a deeper link between the two methods it is most convenient to make use of the hermitian symmetry space formalism (see e.g. [4,5]). In particular, we shall use the well-known interpretation of the complex space $P^{n-1}(C)$ as the hermitian symmetric space of the type A III (according to Helgason's classification [4]), which can be identified with the homogeneous space $SU(n)/S(U(1) \times U(n-1))$ of the oscillator symmetry group SU(n) in the following way. Let us consider the complex vector space C^n with its canonical basis $(e_0, e_1, \ldots, e_{n-1})$ and hermitian (and kaehlerian) inner product on C^n defined by the bilinear form:

$$h(z,w) = \sum_{k=0}^{n-1} z^k \bar{w}^k = g'(z,w) + ig'(z,iw) \tag{52}$$

where

$$g'(z,w) = \mathrm{Re}\,h(z,w)$$

can be interpreted as a natural inner product in R^{2n} identified with C^n. Observe that $\omega(z,w) := g'(z,iw)$ may be considered as a 2-form defining a symplectic structure on R^{2n}. The unit sphere S^{2n-1} may be identified with the subset $\{z \in C^n;\ h(z,z) \equiv g'(z,z) = 1\}$. The tangent space $T_z(S^{2n-1})$ may be identified with a subset $\{w \in C^n;\ g'(z,w) = 0\}$ (by parallel displacement in C^n).

Let T'_z denote the subspace in T_z of codimension 1 defined by the second condition $g'(iz,w) = 0$ so that $\dim T'_z = 2n-2$. Now if $\pi : z \to [z]$ is the canonical projection of C^n onto $P^{n-1}(C)$ with $[z] := \{\lambda z;\ \lambda \in C\}$ then it is easy to see that the projection π defines on S^{2n-1} the principal bundle structure over $P^{n-1}(C)$, with the structure group S^1, and that there exists a connection on S^{2n-1} with respect to which the subspace T'_z is the horizontal subspace of $T_z(S^{2n-1})$, for each z. This means that the projection π induces a linear isomorphism of T'_z onto $T_p(P^{n-1}(C))$, $p = \pi(z)$ for each z. By means of these isomorphisms the riemannian, the almost complex and symplectic structures defined on S^{2n-1}, may be transferred on $P^{n-1}(C)$, being S^1-invariant. Now, there is a transitive (but not effective!) action of the SU(n) group on the projective space $P^{n-1}(C)$, with respect to which the riemannian metric and the corresponding measure are invariant. The isotropy subgroup of a point $p_0 = \pi(e_0)$ is a subset:

$$\left\{\begin{pmatrix} e^{i\theta}, & 0 \\ 0\ , & b \end{pmatrix} : b \in U(n-1),\ \det b = e^{-i\theta}\right\} \tag{53}$$

of matrices of SU(n) and constitutes its subgroup $S(U(1) \times U(n-1))$. By denoting this subgroup by H, and SU(n) by G we obtain the well-known diffeomorphism f of G/H onto $P^{n-1}(C)$. It is easy to show that this diffeomorphism is in fact the symmetric space isomorphism (see e.g. [5]) and defines on $P^{n-1}(C)$ the structure of hermitian symmetric space. The Lie algebra $\underline{G}$ of the group $G = SU(n)$ has a canonical decomposition $\underline{G} = \underline{H}+\underline{M}$ where:

$$\underline{H} = \underline{U}(1) + \underline{U}(n-1) = \left\{\begin{pmatrix} \lambda, & 0 \\ 0, & B \end{pmatrix};\ \lambda+\bar{\lambda} = 0;\ B \in U(n-1)\right\}$$
$$\underline{M} = \left\{\begin{pmatrix} 0 & -\bar{\zeta}^T \\ \zeta & 0 \end{pmatrix} : \zeta \in C^{n-1}\right\} \tag{54}$$

and the action of adH on $\underline{M}$ is completly described by the natural action of the U(n-1) subgroup on C^{n-1}. The above described interpretation of the projective space $P^{n-1}(C)$ as a hermitian symmetric space $SU(n)/S(U(1)\times U(n-1))$ (for the details of construction see [5]) allows us to compare in the most natural way the algebraic and geometric methods of treating (and quantizing) the n-dimensional isotropic oscillator. As was shown in the preceding paragraph, the surface of constant energy is the unit sphere S^{2n-1} (in some convenient units) in the phase space $X = R^{2n}$, the set of classical orbits with this value of energy possessing the structure of the space $P^{n-1}(C)$ (the classical motion $z(t) = e^{i\nu t}z(0)$ may be obtained as a result of action of the structural group S^1, $\theta = \nu t$, on each fiber of the projection π). We thus obtain a precise mode of action of the symmetry group $SU(n)$ on $P^{n-1}(C)$, identified with the set of classical motions, by transferring the natural action of $G = SU(n)$ on the homogeneous space G/H, $g_0 : gH \to g_0 gH$ and the full variety of geometric structures (riemannian metric, Killing fields, and so on) induced by the Lie algebra $\underline{G}$ and its canonical decomposition $\underline{G} = \underline{H}+\underline{M}$ can be defined in a natural way.

The process of quantization is performed by considering induced representations, the construction of which is based on various homogeneous vector bundles associated with the principal bundle G/H and defined by various 1-dimensional representations ρ_r of the subgroup H (homomorphisms of the group H into the set of complex numbers of modulus one):

$$\rho_r(h) = e^{ir\theta}; \quad h = \begin{pmatrix} e^{i\theta}, & 0 \\ 0 , & b \end{pmatrix}, \quad b \in U(n-1), \ \det b = e^{i\theta} \tag{55}$$

where r can be any integer number. As can be shown (see e.g. [6]) the sections of these bundles can be identified with the complex-valued functions on the group G with the "homogeneity condition":

$$\psi(gh) = \rho(h^{-1})\psi(g) \qquad g \in G, \quad h \in H \tag{56}$$

which, in turn, can be identified with functions of n complex arguments $z_0, z_1, \ldots, z_{n-1}$ obeying the following "homogeneity condition":

$$\psi(\lambda z) = \left(\frac{\lambda}{|\lambda|}\right)^{-r}\psi(z) \qquad \lambda \in C, \qquad r\text{-integer}. \tag{57}$$

We can obtain, in particular, the "oscillator representations" by using for $\psi(z)$ the "homogeneous polynomials" of the type:

$$\psi(z) = A^{i_1,\ldots,i_N} z_{i_1},\ldots,z_{i_N} |z|^{-N}$$
$$A^{i_1,\ldots,i_N} \text{ - symmetric tensor} \tag{58}$$

on which the regular representation $(T(g)\psi)(z) = \psi(g^{-1}z)$ acts irreducibly (acting effectively on symmetric tensors

$$A^{i_1,\ldots,i_N}$$

according the tensor representation of the type $(N,0,\ldots,0)$). In this manner we can reproduce all the results of the algebraic method and most of the results of the geometric method, as well. Moreover, in this frame better understanding of the role of the "non-compact extension" of the symmetry group SU(n) of the oscillator, described at the beginning of this paper can be reached. As can be shown, the fact that the symmetric space $P^{n-1}(C) = SU(n)/S[U(1)\times U(n-1)]$ is of rank 1 makes it possible to construct new homogeneous vector bundles, over $P^{n-1}(C)$, whose respective induced representation has the property of the "ladder representation" reproducing the whole set of quantum states of the n-dimensional isotropic oscillator, in the sense explained in the first part of this work. Details of these constructions as well as other examples with some general ideas of their application to the description of internal symmetries of hadrons will be described in a later publication.

REFERENCES

1. P. BUDINI, "Algebraic formulation of dynamical problem" Lecture given at the International Universitatswochen fur Kernphysik 1967.

2. B. KOSTANT, "Quantization and unitary representations" Lecture Notes in Mathematics, Vol. 170, Springer Berlin 1970.

3. D.J. SIMMS, "Geometric quantization of the harmonic oscillator", presented to the 2nd International Colloquium on Group Theoretical Method in Physics, Nijmegen 1973.

4. S. HELGASON, "Differential Geometry and Symmetric Spaces", Academic Press, N.Y. 1962.

5. S. KOBAYASHI and K. NOMIZU, "Foundations of differential geometry, Vol. II, Interscience Publ. N.Y. 1968.

6. R. HERMANN, "Lie group for physicists", A. Benjamin Inc. N.Y. 1966.

THE ISING ALGEBRA

A.I. Solomon

INTRODUCTION

This note introduces a Lie algebra connected with the solution of a classical problem in statistical mechanics, the Ising model. The model consists of a two-dimensional array of spins, each of which interacts with its nearest neighbours only. It is not the purpose to give an exposition of the Ising model here; such a description may be found in most texts on statistical mechanics. The standard treatments show that such two-dimensional lattice problems may be reduced to one-dimensional form by the introduction of an elegant device called the transfer matrix T. The solution then involves diagonalization of this transfer matrix, the thermodynamic free energy being determined by the maximum eigenvalue of T. Generally T has the form of a product of exponentials. In the particular case of interest here, the Ising lattice, the transfer matrix has the form

$$T = e^A e^B$$

where

$$A = a \sum_{i=1}^{N} X_i X_{i+1}$$

and

$$B = b \sum_{i=1}^{N} Z_i .$$

The notation is that X_i is an operator representing a Pauli matrix σ^x on the ith site of a cyclic linear lattice of N sites;

$$X_i = I \otimes I \otimes \ldots \otimes \sigma^x \otimes \ldots \otimes I, \qquad \sigma^x = \begin{bmatrix} & 1 \\ 1 & \end{bmatrix}$$

and the Z_i are similarly defined $2^N \times 2^N$ matrices. The cyclic invariance is expressed by $X_N X_{N+1} \equiv X_N X_1$. The real coefficients a and b are functions of the two coupling constants and temperature.

The value of the Lie algebraic approach is now manifest. If one can determine the Lie algebra generated by A and B, then by the Campbell-Baker-Hausdorff theorem T is simply the exponential of some element C of this algebra. And now C may be obtained by evaluating the product in any convenient faithful representation. In the original formulation of the problem above, the matrices A and B are $2^N \times 2^N$ dimensional. It is not too difficult to obtain a rotation algebra, essentially sO(2N), which contains the required matrices A and B. (See, for example, the article by the present writer in the Proceedings of the third conference of this series.) As this has a representation whose dimension is of order N^2 a considerable reduction has been effected. However the rotation algebra is still far too large to be a candidate for the Lie algebra generated by A and B as it contains non-translation invariant terms. Imposing the requirement of translation invariance reduces the algebra to one of dimension of order N. Further, this can be represented by 2×2 matrices whose entries are polynomials, and so the solution of the Ising problem can essentially be reduced to the diagonalization of a 2×2 matrix.

In this note we shall give an abstract construction for a certain Lie algebra, which we hope is of interest in its own right. Finally we indicate briefly the relevance of this algebra to the solution of the Ising problem.

THE G_N ALGEBRA

Let M be a ring of real matrices. (We shall only consider real 2 × 2 matrices in the present application, so for definiteness the reader may confine himself to these.) Consider the polynomial extension $M[\lambda]$; that is, the ring of polynomials of finite degree in λ whose coefficients are elements of M. The polynomial $\lambda^n - 1$ (more precisely, $I(\lambda^n-1)$ where I is the unit-matrix of M) generates an ideal I_n in $M[\lambda]$, where n is a fixed, positive integer. We may therefore construct the quotient ring

$$Q_n = M[\lambda]/I_n.$$

Less formally, Q_n consists of polynomials in λ of degree less than n with coefficients in M; on multiplication of two such polynomials their degrees are added modulo n.

We now equip a certain subset of Q_n with a Lie algebra structure. Define $G_n(M)$ to be the Lie algebra over the real numbers generated by the subset

$$\{m\lambda^r - \tilde{m}\lambda^{n-r} : m \in M, r=0,1,2,\ldots,n-1\}$$

(the tilde refers to ordinary matrix transposition). For elements a and b in $G_n(M)$, the Lie bracket [a,b] is defined by ab-ba, with the multiplication ab being that inherited from Q_n. We shall show that this algebra is isomorphic to a translation-invariant subalgebra of sO(2N), containing the required elements A and B.

RELEVANCE TO THE ISING MODEL

By means of the Jordan-Wigner transformation, discussed in the previously-mentioned note of the author, one can construct an sO(2N) algebra which contains the operators $\Sigma X_i X_{i+1}$ and ΣZ_i. (The algebra is, more precisely, sO(2N) ⊕ sO(2N); however, we omit fine points in this brief exposition.) Thus any element may be written

$$\Sigma \omega_{rs} L_{rs}$$

where the L_{rs} are the N(2N-1) generators. If we write the coefficient matrix ω in blocked form

$$\omega = \left[\begin{array}{c|c} \omega_1 & \omega_2 \\ \hline \omega_3 & \omega_4 \end{array}\right]$$

then the ω_i are N × N matrices whose components are the coefficients of the physical operators on the N-site lattice. Translation invariance is then expressed by demanding that

$$(I \otimes \Delta)\omega(I \otimes \Delta^{-1}) = \omega$$

where Δ is the N × N cyclic matrix,

$$\Delta = \begin{bmatrix} 0 & 1 & & & \\ & 0 & 1 & & \\ & & \cdot & \cdot & \\ & & & \cdot & 1 \\ 1 & & & & 0 \end{bmatrix}$$

which obeys $\Delta^N = I$, $\Delta\tilde{\Delta} = I$. The most general antisymmetric ω obeying this constraint is

$$\omega = \sum_{r=0}^{N-1} (m_r \otimes \Delta^r - \tilde{m}_r \otimes \Delta^{-r});$$

the m_r are arbitrary 2 × 2 real matrices. If we choose the standard representation for the L_{rs},

$$\hat{L}_{rs} = e_{rs} - e_{sr}$$

where the 2N × 2N matrices e_{rs} have components $(e_{rs})_{ij} = \delta_{ri}\delta_{sj}$, the general translation invariant element of sO(2N) is in fact represented (up to a factor of 2) by ω itself. The algebraic monomorphism (one-to-one homomorphism) ϕ between our G algebra and sO(2N) is now immediate;

$$\phi : G_N \to sO(2N)$$

$$m_r\lambda^r - \tilde{m}_r\lambda^{N-r} \to m_r \otimes \Delta^r - \tilde{m}_r \otimes \Delta^{N-r}.$$

This correspondence tells us that we may calculate eigenvalues of sO(2N) elements from the corresponding G_N elements, taking λ to be each Nth root of unity in turn. Further, the sO(2N) theory tells us that the λ^r term corresponds to (r+1)-body interactions, and in fact we may represent

$$\frac{1}{2i}\,\Sigma Z_i = \begin{bmatrix} & 1 \\ -1 & \end{bmatrix}$$

$$\frac{1}{2i}\,\Sigma X_i X_{i+1} = \begin{bmatrix} & -\lambda^{-1} \\ \lambda & \end{bmatrix}$$

by passing from the original $(2^N \times 2^N)$ representation to the G_n representation via the $2N \times 2N$ sO(2N) representation (using the ϕ map).

We conclude by sketching the diagonalization of the transfer matrix using this approach. In the G_N representation

$$\exp 2ia\begin{bmatrix} & -\lambda^{-1} \\ \lambda & \end{bmatrix}\exp 2ib\begin{bmatrix} & 1 \\ -1 & \end{bmatrix} \to \exp 2i\begin{bmatrix} & \mu \\ -\mu & \end{bmatrix}$$

gives the diagonalization, by some rotation. We readily calculate, from the invariance of traces,

$$\cosh 2\mu = \cosh 2a \cosh 2b - \cos q \sinh 2a \sinh 2b,$$

where

$$\lambda + \lambda^{-1} = 2 \cos q.$$

This is reflected in the sO(2N) representation by

$$T \to \exp \left[\begin{array}{c|c} & \begin{matrix}\mu_1 & & \\ & \ddots & \\ & & \mu_N\end{matrix} \\ \hline \begin{matrix}-\mu_1 & & \\ & \ddots & \\ & & -\mu_N\end{matrix} & \end{array}\right]$$

where the μ_i are obtained by letting λ run through the N N^{th} roots of unity. In the original representation we have

$$\exp a\Sigma X_i X_{i+1} \exp b\Sigma Z_i \to \exp \Sigma\mu_i Z_i$$

and the maximum eigenvalue (the one of thermodynamic interest) is given by $\pi_i e^{\mu i}$.

ACKNOWLEDGEMENTS

I should like to thank Professor D.D. Betts and Professor Y. Takahashi for the hospitality of the Theoretical Physics Institute of University of Alberta, where these notes were written, and acknowledge with pleasure the stimulating atmosphere of the 15th Séminaire de Mathématiques Supérieures of the University of Montreal.

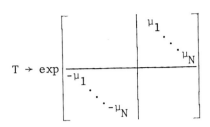

THE GROUP $O(3)_\Lambda(T_2 \times \bar{T}_2)$ AND THE HYDROGEN ATOM

J. Yadegar

INTRODUCTION

In this talk we consider the group $O(3)_\Lambda(T_2 \times \bar{T}_2)$ in which the generators additional to those of O(3) form a reducible tensor representation of O(3), $\{T(\frac{1}{2},\pm\frac{1}{2}),\bar{T}(\frac{1}{2},\pm\frac{1}{2})\}$.

$O(3)_\Lambda(T_2 \times \bar{T}_2)$ is an interesting group since its enveloping algebra contains, in addition to an invariant, X, three O(3) scalar operators, $\hat{Y}_o$ and $\hat{Y}_\pm$, which themselves generate a group isomorphic to O(3) whose Casimir is identical to that of the original O(3). Hence for any fixed IUR, $O(3)_\Lambda(T_2 \times \bar{T}_2)$ contains an O(4) subgroup and, moreover, this IUR contains, precisely once, each IUR of O(4) of type $\ell_1 = \ell_2 = \ell$. These are just the IUR of O(4) realized by the hydrogen atom (see for instance Hughes (1967)). $O(3)_\Lambda(T_2 \times \bar{T}_2)$ is therefore an alternative to O(4,1) as the spectral group of the hydrogen atom.

One may go even further and show that $O(3)_\Lambda(T_2 \times \bar{T}_2)$ is the dynamical group of the hydrogen atom, in the sense of Barut and Kleinert (1966). If we identify our ℓ with $\frac{1}{2}(n-1)$, n being the principal quantum number, we find that the n-shifting operators of the above authors are precisely the ℓ-shifting operators which may be constructed from $T_2 \times \bar{T}_2$. Eight of these, together with

$\hat{Y}_o$, $\hat{Y}_\pm$, L^2 and the generators ℓ_o, $\ell_\pm$ of O(3) satisfy, for a fixed IUR of $O(3)_\Lambda(T_2\times\bar{T}_2)$, the commutation relations of O(4,2). Hence the IUR of O(4,2) used by Barut and Kleinert may be replaced by a fixed, but arbitrary, IUR of the far more economical group $O(3)_\Lambda(T_2\times\bar{T}_2)$. All the matrix elements of the O(4,2) operators are in fact contained in those of the four operators $T(\frac{1}{2},\pm\frac{1}{2})$ and their hermitian conjugates $\bar{T}(\frac{1}{2},\mp\frac{1}{2})$.

2. THE GROUP $O(3)_\Lambda(T_2\times\bar{T}_2)$

The basis for the Lie algebra of $O(3)_\Lambda(T_2\times\bar{T}_2)$ is chosen to consist of the O(3) generators ℓ_o, $\ell_\pm$ together with two pairs of operators both of which transform under commutation with the ℓ_o, $\ell_\pm$ as irreducible two dimensional tensor representations of O(3). We denote these by $q_{\pm\frac{1}{2}} \equiv T(\frac{1}{2},\pm\frac{1}{2})$ and $\bar{q}_{\pm\frac{1}{2}} \equiv \bar{T}(\frac{1}{2},\pm\frac{1}{2})$. Their commutation relations are

$$[\ell_o,\ell_\pm] = \pm\ell_\pm, \quad [\ell_+,\ell_-] = 2\ell_o$$

$$[\ell_o,\overset{(-)}{q}_{\pm\frac{1}{2}}] = \pm\tfrac{1}{2}\overset{(-)}{q}_{\pm\frac{1}{2}}, \quad [\ell_\pm,\overset{(-)}{q}_{\mp\frac{1}{2}}] = \overset{(-)}{q}_{\pm\frac{1}{2}} \tag{1}$$

$\overset{(-)}{q}_{\pm\frac{1}{2}}$ mutually commute.

The hermiticity properties of the operators are given by

$$\ell_o^\dagger = \ell_o, \quad \ell_\pm^\dagger = \ell_\mp, \quad q_{\pm\frac{1}{2}}^\dagger = \pm\bar{q}_{\mp\frac{1}{2}}. \tag{2}$$

The group has one Casimir given by

$$X = q_{\frac{1}{2}}\bar{q}_{-\frac{1}{2}} - q_{-\frac{1}{2}}\bar{q}_{\frac{1}{2}} \tag{3}$$

and three O(3)-scalar operators:

$$Y_+ = -(2q_{\frac{1}{2}}q_{-\frac{1}{2}}\ell_o+q_{-\frac{1}{2}}q_{-\frac{1}{2}}\ell_+-q_{\frac{1}{2}}q_{\frac{1}{2}}\ell_-) \tag{4}$$

$$Y_- = (2\bar{q}_{\frac{1}{2}}\bar{q}_{-\frac{1}{2}}\ell_o+\bar{q}_{-\frac{1}{2}}\bar{q}_{-\frac{1}{2}}\ell_+-\bar{q}_{\frac{1}{2}}\bar{q}_{\frac{1}{2}}\ell_-) \tag{5}$$

$$Y_o = (q_{\frac{1}{2}}\bar{q}_{-\frac{1}{2}}\ell_o + q_{-\frac{1}{2}}\bar{q}_{\frac{1}{2}}\ell_o + q_{-\frac{1}{2}}\bar{q}_{-\frac{1}{2}}\ell_+ - q_{\frac{1}{2}}\bar{q}_{\frac{1}{2}}\ell_-). \tag{6}$$

From (2) one may check that X is a positive definite hermitian operator and

$$Y_o^\dagger = Y_o, \quad Y_\pm^\dagger = Y_\mp. \tag{7}$$

Also Y's satisfy the commutation relations

$$[Y_o, Y_\pm] = \pm XY_\pm, \quad [Y_+, Y_-] = 2XY_o \tag{8}$$

and

$$Y_+Y_- + Y_o^2 - XY_o = X^2L^2. \tag{9}$$

The IUR of $O(3)_\Lambda(T_2\times\bar{T}_2)$ are labelled by the eigenvalue x of X and states denoted by $|x;\ell,k,m\rangle$; k is the only additional parameter needed to completely specify states of the same ℓ and m values in case of ℓ degeneracy.

From ℓ_o, $\ell_\pm$ and $\overset{(-)}{q}_{\pm\frac{1}{2}}$ one can construct shift operators which change ℓ and m by $\pm\frac{1}{2}$. They are

$$\overset{(-)}{O}{}^{\pm\frac{1}{2},\frac{1}{2}}_{\ell,m} = (m\pm\ell+\tfrac{1}{2}(1\pm1))\overset{(-)}{q}_{\frac{1}{2}} + \overset{(-)}{q}_{-\frac{1}{2}}\ell_+ \tag{10}$$

$$\overset{(-)}{O}{}^{\pm\frac{1}{2},-\frac{1}{2}}_{\ell,m} = -(m\mp\ell-\tfrac{1}{2}(1\pm1))\overset{(-)}{q}_{-\frac{1}{2}} + \overset{(-)}{q}_{\frac{1}{2}}\ell_-. \tag{11}$$

3. <u>THE IUR OF $O(3)_\Lambda(T_2\times\bar{T}_2)$</u>

By application of

$$\bar{O}^{-\frac{1}{2},-\frac{1}{2}}_{\underline{\ell},m}, \quad O^{\frac{1}{2},\frac{1}{2}}_{\underline{\ell}-\frac{1}{2},m-\frac{1}{2}} \quad \bar{O}^{-\frac{1}{2},-\frac{1}{2}}_{\underline{\ell},m}$$

and

$$O^{-\frac{1}{2},\frac{1}{2}}_{\underline{\ell}-\frac{1}{2},m-\frac{1}{2}} \quad \bar{O}^{-\frac{1}{2},-\frac{1}{2}}_{\underline{\ell},m}(\underline{\ell} = \ell_{min})$$

on $|x;\ell,k,m\rangle$ one obtains the relation $\underline{\ell}x = 0$ from which two distinct classes of IUR arise.

A. $x = 0;\ \ell = 0, \frac{1}{2}, 1, \ldots$

One is able to show that for this class the IUR contain no more than the IUR of the O(3) subgroup.

B. $x > 0$ and $\ell = 0$

Define

$$\hat{Y}_o = \frac{1}{x} Y_o, \quad \hat{Y}_\pm = \frac{1}{x} Y_\pm. \tag{12}$$

These normalized operators satisfy (7) and the commutation relations of an $O^Y(3)$ group, in contradistinction to $O^\ell(3)$. From (9)

$$\hat{Y}_+\hat{Y}_- + \hat{Y}_o^2 - \hat{Y}_o = L^2 = \ell_+\ell_- + \ell_o^2 - \ell_o. \tag{13}$$

Thus k is chosen to be the eigenvalue of $\hat{Y}_o$. Clearly $k = -\ell, \ldots, \ell$.

One can derive the following equations

$$O_{\ell,m}^{\pm\frac{1}{2},\pm\frac{1}{2}}|x;\ell,k,m\rangle = \left[\frac{x(2\ell+1)(\ell\pm k+\frac{1}{2}\pm\frac{1}{2})(\ell+m+\frac{1}{2}\pm\frac{1}{2})}{2(\ell+\frac{1}{2}\pm\frac{1}{2})}\right]^{\frac{1}{2}}|x;\ell\pm\tfrac{1}{2},k+\tfrac{1}{2},m\pm\tfrac{1}{2}\rangle \tag{14}$$

$$\bar{O}_{\ell,m}^{\pm\frac{1}{2},\pm\frac{1}{2}}|x;\ell,k,m\rangle = \pm\left[\frac{x(2\ell+1)(\ell\mp k+\frac{1}{2}\pm\frac{1}{2})(\ell+m+\frac{1}{2}\pm\frac{1}{2})}{2(\ell+\frac{1}{2}\pm\frac{1}{2})}\right]^{\frac{1}{2}}|x;\ell\pm\tfrac{1}{2},k-\tfrac{1}{2},m\pm\tfrac{1}{2}\rangle \tag{15}$$

$$O_{\ell,m}^{\pm\frac{1}{2},\mp\frac{1}{2}}|x;\ell,k,m\rangle = \pm\left[\frac{x(2\ell+1)(\ell\pm k+\frac{1}{2}\pm\frac{1}{2})(\ell-m+\frac{1}{2}\pm\frac{1}{2})}{2(\ell+\frac{1}{2}\pm\frac{1}{2})}\right]^{\frac{1}{2}}|x;\ell\pm\tfrac{1}{2},k+\tfrac{1}{2},m\mp\tfrac{1}{2}\rangle \tag{16}$$

$$\bar{O}_{\ell,m}^{\pm\frac{1}{2},\mp\frac{1}{2}}|x;\ell,k,m\rangle = \left[\frac{x(2\ell+1)(\ell\mp k+\frac{1}{2}\pm\frac{1}{2})(\ell-m+\frac{1}{2}\pm\frac{1}{2})}{2(\ell+\frac{1}{2}\pm\frac{1}{2})}\right]^{\frac{1}{2}}|x;\ell\pm\tfrac{1}{2},k-\tfrac{1}{2},m\mp\tfrac{1}{2}\rangle \tag{17}$$

$$\begin{aligned} q_{\pm\frac{1}{2}}|x;\ell,k,m\rangle &= \left[\frac{x(\ell+k+1)(\ell\pm m+1)}{2(2\ell+1)(\ell+1)}\right]^{\frac{1}{2}}|x;\ell+\tfrac{1}{2},k+\tfrac{1}{2},m\pm\tfrac{1}{2}\rangle \\ &\pm\left[\frac{x(\ell-k)(\ell\mp m)}{2\ell(2\ell+1)}\right]^{\frac{1}{2}}|x;\ell-\tfrac{1}{2},k+\tfrac{1}{2},m\pm\tfrac{1}{2}\rangle \end{aligned} \tag{18}$$

$$\bar{q}_{\pm}|x;\ell,k,m\rangle = \left[\frac{x(\ell-k+1)(\ell\pm m+1)}{2\ell(2\ell+1)(\ell+1)}\right]^{\frac{1}{2}}|x;\ell+\tfrac{1}{2},k-\tfrac{1}{2},m\pm\tfrac{1}{2}\rangle$$
$$\mp\left[\frac{x(\ell+k)(\ell\mp m)}{2\ell(2\ell+1)}\right]^{\frac{1}{2}}|x;\ell-\tfrac{1}{2},k-\tfrac{1}{2},m\pm\tfrac{1}{2}\rangle \tag{19}$$

4. $O(3)_{\Lambda}(T_2\times\bar{T}_2)$ AND THE HYDROGEN ATOM

Our states $|x;\ell,k,m\rangle$ correspond to Stark or parabolic states of the hydrogen atom, $|n_1,n_2,M\rangle$. In fact following relations guarantee this statement,

$$\ell = \tfrac{1}{2}(n_1+n_2+M),\quad k = \tfrac{1}{2}(M+n_2-n_1),\quad m = \tfrac{1}{2}(M-n_2+n_1) \tag{20}$$

whose inverse is

$$M = m+k,\quad n_1 = \ell-\tfrac{1}{2}k+\tfrac{1}{2}m-\tfrac{1}{2}|m+k|,\quad n_2 = \ell+\tfrac{1}{2}k-\tfrac{1}{2}m-\tfrac{1}{2}|m+k|. \tag{21}$$

Using (14) to (17) one is able to show that

$$O^{\frac{1}{2},-\frac{1}{2}}|n_1,n_2,M\rangle = \left[\frac{xn(n_2+1)(n_2+|M|+1)}{(n+1)}\right]^{\frac{1}{2}}|n_1,n_2+1,M\rangle \tag{22}$$

$$O^{-\frac{1}{2},-\frac{1}{2}}|n_1,n_2,M\rangle = \left[\frac{xn n_1(n_1+|M|)}{(n-1)}\right]^{\frac{1}{2}}|n_1-1,n_2,M\rangle \tag{23}$$

$$O^{\frac{1}{2},\frac{1}{2}}|n_1,n_2,M\rangle = \left[\frac{xn(n_1+M+1)(n_2+M+1)}{(n+1)}\right]^{\frac{1}{2}}|n_1,n_2,M+1\rangle,\quad M\geq 0$$
$$= \left[\frac{xn(n_1+1)(n_2+1)}{(n+1)}\right]^{\frac{1}{2}}|n_1+1,n_2+1,M+1\rangle,\quad M<0 \tag{24}$$

$$O^{-\frac{1}{2},\frac{1}{2}}|n_1,n_2,M\rangle = -\left[\frac{xn n_1 n_2}{(n-1)}\right]^{\frac{1}{2}}|n_1-1,n_2-1,M+1\rangle,\quad M\geq 0$$
$$= -\left[\frac{xn(n_1-M)(n_2-M)}{(n-1)}\right]^{\frac{1}{2}}|n_1,n_2,M+1\rangle,\quad M<0 \tag{25}$$

with similar expressions for the actions of $\bar{O}$'s which may be obtained from the above by taking hermitian conjugates.

Define operators $N^{\pm}_{1,2}$, $A^{\pm}_{\pm}$, $A^{\mp}_{\pm}$ and N by

$$\begin{aligned}
N_1^+ &= -\left(\frac{n+1}{xn}\right)^{\frac{1}{2}} \bar{O}^{\frac{1}{2},\frac{1}{2}} & N_2^+ &= \left(\frac{n+1}{xn}\right)^{\frac{1}{2}} O^{\frac{1}{2},-\frac{1}{2}} \\
N_1^- &= -\left(\frac{n-1}{xn}\right)^{\frac{1}{2}} O^{-\frac{1}{2},-\frac{1}{2}} & N_2^- &= \left(\frac{n-1}{xn}\right)^{\frac{1}{2}} \bar{O}^{-\frac{1}{2},\frac{1}{2}} \\
A_+^+ &= 2\left(\frac{n+1}{xn}\right)^{\frac{1}{2}} O^{\frac{1}{2},\frac{1}{2}} & A_+^- &= 2\left(\frac{n-1}{xn}\right)^{\frac{1}{2}} O^{-\frac{1}{2},\frac{1}{2}} \\
A_-^- &= 2\left(\frac{n-1}{xn}\right)^{\frac{1}{2}} \bar{O}^{-\frac{1}{2},-\frac{1}{2}} & A_-^+ &= -2\left(\frac{n+1}{xn}\right)^{\frac{1}{2}} \bar{O}^{\frac{1}{2},-\frac{1}{2}}
\end{aligned} \tag{26}$$

$$N = 2R+1$$

where R is the formal operator with eigenvalue ℓ.

Using (22) to (25) and the actions of $\bar{O}$'s one may verify that operators given in (26) have precisely the same actions on $|n_1,n_2,M\rangle$ as do those of the same name introduced by Barut and Kleinert (1966). In fact their expressions are only valid for $M \geq 0$ whereas here M takes all possible values. Barut and Kleinert show that the matrix elements of the dipole operator can be given in terms of those of the O(4,2) operators, and therefore call O(4,2) the dynamical group of the hydrogen atom. We see from the above that we may also call $O(3)_\Lambda(T_2\times\bar{T}_2)$ the dynamical group of the hydrogen atom. The action of generators of $O(3)_\Lambda(T_2\times\bar{T}_2)$ on $|n_1,n_2,M\rangle$ are given by

$$\ell_o|n_1,n_2,M\rangle = \tfrac{1}{2}(M-n_2+n_1)|n_1,n_2,M\rangle \tag{27}$$

$$\begin{aligned}
\ell_+|n_1,n_2,M\rangle &= [n_2(n_1+M+1)]^{\frac{1}{2}}|n_1,n_2-1,M+1\rangle, \quad M \geq 0 \\
&= [(n_1+1)(n_2-M)]^{\frac{1}{2}}|n_1+1,n_2,M+1\rangle, \quad M < 0
\end{aligned} \tag{28}$$

$$\begin{aligned}
\ell_-|n_1,n_2,M\rangle &= [(n_2+1)(n_1+M)]^{\frac{1}{2}}|n_1,n_2+1,M-1\rangle, \quad M > 0 \\
&= [n_1(n_2-M+1)]^{\frac{1}{2}}|n_1-1,n_2,M-1\rangle, \quad M \leq 0
\end{aligned} \tag{29}$$

$$q_{-\frac{1}{2}}|n_1,n_2,M\rangle = \left[\frac{x(n_2+1)(n_2+|M|+1)}{n(n+1)}\right]^{\frac{1}{2}}|n_1,n_2+1,M\rangle - \left[\frac{xn_1(n_1+|M|)}{n(n-1)}\right]^{\frac{1}{2}}|n_1-1,n_2,M\rangle \tag{30}$$

$$\bar{q}_{\frac{1}{2}}|n_1,n_2,M\rangle = \left[\frac{x(n_1+1)(n_1+|M|+1)}{n(n+1)}\right]^{\frac{1}{2}}|n_1+1,n_2,M\rangle - \left[\frac{xn_2(n_2+|M|)}{n(n-1)}\right]^{\frac{1}{2}}|n_1,n_2-1,M\rangle \tag{31}$$

$$\begin{aligned} q_{\frac{1}{2}}|n_1,n_2,M\rangle &= \left[\frac{x(n_1+M+1)(n_2+M+1)}{n(n+1)}\right]^{\frac{1}{2}}|n_1,n_2,M+1\rangle \\ &+ \left[\frac{xn_1n_2}{n(n-1)}\right]^{\frac{1}{2}}|n_1-1,n_2-1,M+1\rangle, \quad M \geq 0 \\ &= \left[\frac{x(n_1+1)(n_2+1)}{n(n+1)}\right]^{\frac{1}{2}}|n_1+1,n_2+1,M+1\rangle \\ &+ \left[\frac{x(n_1-M)(n_2-M)}{n(n-1)}\right]^{\frac{1}{2}}|n_1,n_2,M+1\rangle, \quad M < 0 \end{aligned} \tag{32}$$

$$\begin{aligned} \bar{q}_{-\frac{1}{2}}|n_1,n_2,M\rangle &= \left[\frac{x(n_1+M)(n_2+M)}{n(n+1)}\right]^{\frac{1}{2}}|n_1,n_2,M-1\rangle \\ &+ \left[\frac{x(n_1+1)(n_2+1)}{n(n+1)}\right]^{\frac{1}{2}}|n_1+1,n_2+1,M-1\rangle, \quad M > 0 \\ &= \left[\frac{xn_1n_2}{n(n-1)}\right]^{\frac{1}{2}}|n_1-1,n_2-1,M-1\rangle \\ &+ \left[\frac{x(n_1-M+1)(n_2-M+1)}{n(n+1)}\right]^{\frac{1}{2}}|n_1,n_2,M-1\rangle, \quad M \leq 0. \end{aligned} \tag{33}$$

A more detailed account of this work is given by Hughes and

Yadegar (1976(a)).

We conclude this article by mentioning that the Lie algebra of SU(3) also contains ℓ_o, $\ell_\pm$, $q_{\pm\frac{1}{2}}$ and $\bar{q}_{\pm\frac{1}{2}}$, together with an O(3) scalar operator $p_o \equiv T(0,0)$, ℓ_o, $\ell_\pm$ being generators of SU(2) subgroup. By changing the commutation relations one is able to give an analysis of SU(3) in an SU(2) basis in which Y_o, $Y_\pm$ are still O(3) scalar operators, although they no longer generate an O(3) group. So $Y_\pm$ can be used instead of p_o (the hypercharge operator of the Octet model for Hadrons) as state labelling operators (Hughes and Yadegar, 1976(b)).

The author of this paper would like to state that this work has been executed in collaboration with Dr. J.W.B. Hughes of the same address as that of the author.

REFERENCES

1. A.O. BARUT and H. KLEINERT, Phys. Rev., 156, 1541-5.
2. J.W.B. HUGHES, 1973c, J. Phys. A. Math. Nucl. Gen., 6, 445-52.
3. J.W.B. HUGHES and J. YADEGAR, (a) - J. Phys. A: Math. Gen., 9, 1569-1580 (1976).
4. J.W.B. HUGHES and J. YADEGAR, (b) - J. Phys. A: Math. Gen., 9, 1581-1593 (1976).

Part II

Coherent States, Supersymmetry, Gauge Fields, Relativity

SPONTANEOUS BREAKING OF EXCEPTIONAL GROUPS*

Feza Gürsey

I. INTRODUCTION

Following the possible identification of the "charge space" of internal quantum numbers of lepton and quark fields with the family of exceptional (octonionic) quantum mechanical spaces[1] initiated by Jordan, Wigner and von Neumann[2], one is led to consider spontaneously broken gauge field theories - patterned after the Weinberg-Salam model[3] or the more general models[4] unifying strong, weak and electromagnetic interactions - based on exceptional groups like E_6 [5] or E_7 [6]. For instance, a local two component spinor field belonging to the fundamental 56-dimensional representation of E_7 unifies 6 colored Dirac quarks and 10 Dirac leptons into a single multiplet while the photon, the eight color gluons, the 3 intermediate vector bosons $(Z,W^{\pm})$, 31 additional color singlet heavier vector bosons and 90 superheavy leptoquark colored vector bosons are unified into the 133-dimensional representation of the same gauge group.

The group E_7 being pseudoreal[7] is free of Adler-Bell-Jackiw

* Research (Yale Report COO-3075-165) supported in part by the U.S. Energy Research and Development Administration under Contract No. EY-76-C-02-3075.

anomalies[8] and requires the left handed parts of Dirac spinors and their charge conjugates to be put in the same two-component spinor representation. Thus the vector bosons of the adjoint representation are coupled to some left handed currents and some right handed currents. When the right and left handed currents coupled to the same boson are related by parity conjugation, that particular interaction is parity invariant. Otherwise, it violates parity. The physical picture is obtained by introducing scalar and pseudoscalar Higgs fields belonging to one or more multiplets of the gauge group and minimizing the Higgs potential in the standard way[9] to obtain a physical vacuum. The vacuum expectation values of the Higgs fields now give masses to the physical quark and lepton fields that are linearly related to the original fields through Higgs dependent coefficients. The physical fields transform linearly under a subgroup H of the gauge group G and nonlinearly under the complete group G, resulting in the spontaneous breaking of G into H. The vector bosons associated with H remain massless while those corresponding to the coset G/H acquire mass. At this point we can make contact with phenomenology through the spectrum of leptons, quarks and vector bosons and effective Lagrangians resulting from the interactions through a set of vector bosons. If we take the group E_7 as a model, then many predictions become possible for e^+e^- processes at high energies as well as lepton-nucleon, lepton-lepton scattering cross sections including neutrino and antineutrino high energy scattering and neutral current effects in atomic physics. All these predictions seem so far in reasonable agreement with experiment[6]. We conclude that gauge theories based on exceptional groups remain viable candidates for the unification of strong, electromagnetic and weak interactions. Hence there is enough motivation to study further their mathematical structure and the possible modes of their spontaneous symmetry breaking.

In this talk I shall first show how the groups F_4, E_6 and E_7

arise as the octonionic generalization of the more familiar groups SU(3), SU(3)×SU(3) and SU(6), with the octonions providing the additional color degrees of freedom. I then review the algebraic properties of the special directions obtained through spontaneous symmetry breaking of the unitary groups mentioned above. These are expressed by means of the Michel-Radicati algebra[10], the Freudenthal algebra[11] or a ternary algebra[12]. The algebras and the special directions are further generalized to exceptional groups. Finally the result is compared with the hierarchy of interactions suggested by experiment.

II. THE COLOR-FLAVOR CONSTRUCTION OF THE EXCEPTIONAL GROUPS

The exceptional groups F_4, E_6, E_7 and E_8 and some of their subgroups can be treated in a unified way by a method developed by Freudenthal[13], Rozenfeld[14] and Tits[15]. They are regarded as Lie algebras that can be constructed out of Jordan algebras J_3^i of 3×3 hermitian matrices over the four Hurwitz algebras (i=1 for real numbers, i=2 for complex numbers, i=4 for quaternions and i=8 for octonions) and a commuting set H^j of Hurwitz algebras and the automorphism algebras of these algebras. Thus, they fit in a square called the magic square with entries J_3^i and H^j.

TABLE I. The magic square.

		J_3^1	J_3^2	J_3^4	J_3^8
H^1	0	SO(3)	SU(3)	Sp(6)	F_4
H^2	0	SU(3)	SU(3)×SU(3)	SU(6)	E_6
H^4	SU(2)	Sp(6)	SU(6)	SO(12)	E_7
H^8	G_2	F_4	E_6	E_7	E_8

In Table I, the row below J_3^i corresponds to the automorphism groups of the Jordan algebras J_3^i. The column to the right of H^j represents the automorphism groups of Hurwitz algebras H^j. The group at the intersection (i j) is associated with the Lie algebra L_{ij} given by:

$$L_{ij} = \text{Aut } J_3^i + J_3^{i(0)} \otimes H^{j(0)} + \text{Aut } H^j \qquad (2.1)$$

where $J_3^{i(0)}$ is the traceless 3×3 hermitian matrix J_3^i and $H^{i(0)}$ is the purely imaginary part of the Hurwitz number H^j. Thus, for the first row we have $L_{i1} = \text{Aut } J_3^i$. The dimension of the magic square group is

$$d(L_{ij}) = d(L_{i1}) = [d(J_3^i)-1][d(H^j)-1] + d(\text{Aut } H^j) \qquad (2.2)$$

where

$$d(L_{11}) = 3, \quad d(L_{12}) = 8, \quad d(L_{14}) = 21, \quad d(L_{18}) = 52, \qquad (2.3a)$$

$$d(\text{Aut } H^1) = d(\text{Aut } H^2) = 0, \ d(\text{Aut } H^4) = 3, \ d(\text{Aut } H^8) = 14, \ (2.3b)$$

$$d(J_3^1) = 6, \quad d(J_3^2) = 9, \quad d(J_3^4) = 15, \quad d(J_3^8) = 27, \qquad (2.3c)$$

$$d(H^j) = j. \qquad (2.3d)$$

This is Tits' construction corresponding to the reduction of the Lie algebras L_{ij} with respect to the subalgebras Aut J_3^i × Aut H^j, the cosets being represented by $J_3^{i(0)} \otimes H^{j(0)}$. Thus, the exceptional groups F_4, E_6, E_7 and E_8 are reduced respectively with respect to the subgroups $SO(3)\times G_2$, $SU(3)\times G_2$, $Sp(6)\times G_2$ and $F_4\times G_2$. Hence, this construction displays the exceptional groups as the octonionic generalizations of the groups in the first column when we adjoin the seven octonionic imaginary units $e_1,\ldots,e_7$ to the real numbers.

Now, the exceptional groups of the last column can be equally regarded as the extensions of the Lie algebras of the second column by an antisymmetrical algebra over six of the octonionic units $e_1,\ldots,e_6$. This algebra M_6 (called a Malcev algebra) is defined as[16]

$$e_a \wedge e_b = [e_a\ e_b\ e_7] \tag{2.4}$$

where

$$[e_i\ e_j\ e_k] = (e_i\ e_j)e_k - e_i(e_j\ e_k) \tag{2.5}$$

is the associator for the octonion algebra given by

$$e_i e_j = -\delta_{ij} + \phi_{ijk} e_k. \tag{2.6}$$

The coefficients ϕ_{ijk} are antisymmetrical and equal to unity for (ijk) = (123), (246), (435), (367), (651), (572), and (714). The automorphism group of (2.6) is G_2 while that of (2.4) is the SU(3) subgroup of G_2. We identify this SU(3) group with the color SU(3) group $SU^c(3)$ of elementary particle physics[17]. Hence the exceptional groups of the last column can be characterized by:

$$L_{i8} = L_{i2} + J_3^i \otimes M_6 + \text{Aut } M_6, \tag{2.7}$$

so that

$$d(L_{i8}) = d(L_{i2}) + 6\ d(J_8^i) + 8. \tag{2.8}$$

Indeed, by using $d(J_3^i)$ from (2.3c) we obtain 52, 78, 133 and 248 for the dimensions of the groups of the last column. The groups L_{i2} may be called the flavor groups. The color group being Aut M this alternative construction of exceptional groups may be called the color-flavor construction.

The color-flavor construction may be cast in a form similar to the Tits form (2.2) by introducing the Freudenthal algebras for complex 3×3 Jordan matrices that are hermitian with respect to the conjugation for Hurwitz algebras. If A and B are elements of complex Jordan matrices we define the Freudenthal product by

$$A \times B = A \cdot B - \tfrac{1}{2} A \,\mathrm{Tr}\, B - \tfrac{1}{2} B \,\mathrm{Tr}\, A - \tfrac{1}{2} I [\mathrm{Tr}(A \cdot B) - \mathrm{Tr}\, A \,\mathrm{Tr}\, B], \tag{2.9}$$

where

$$A \cdot B = \tfrac{1}{2}(AB+BA) \tag{2.10}$$

is the Jordan product. Now the groups L_{i2} of the second column are the automorphism groups for the Freudenthal algebra of complex Jordan matrices which we may call $\mathcal{F}_3^i$. Thus, instead of (2.7) we can write

$$L_{i8} = \mathrm{Aut}\, \mathcal{F}_3^i + \mathcal{F}_3^i \otimes M_6 + \mathrm{Aut}\, M_6. \tag{2.11}$$

This has a structure similar to the Tits formula, with the Malcev algebra and the Freudenthal algebra replacing respectively the octonion algebra and the Jordan algebra. It follows that the groups F_4, E_6 and E_7 can be regarded as the Malcev generalizations of the corresponding flavor groups SU(3), SU(3)×SU(3) and SU(6). Now, the Malcev algebra M_6 has a 2-dimensional sub Malcev algebra M_2 with elements e_3 and e_6 (alternatively, e_1 and e_4 or e_2 and e_5) such that

$$e_3 \wedge e_3 = e_6 \wedge e_6 = e_3 \wedge e_6 = 0. \tag{2.12}$$

Then, the Lie algebras of the third column of the magic square can be written as

$$L_{i4} = \mathrm{Aut}\, \mathcal{F}_3^i + \mathcal{F}_3^i \otimes M_2 + \mathrm{Aut}\, M_2. \tag{2.13}$$

Since Aut M_2 is SO(2) or U(1) and M_2 has dimension 2 we obtain

$$d(L_{i4}) = d(L_{i2}) + 2d(J_3^i) + 1, \tag{2.14}$$

giving 21, 35, 66 and 133 as the dimensions of the Lie algebras of the third column for which the color group is just U(1).

In the Tits case the Lie product between the elements of the Lie algebra is expressed by defining a product of the coset elements $J_3^{i(0)} \otimes H^{j(0)}$ through the traceless Jordan product. In the color-flavor case a similar antisymmetrical product can be defined for the elements $F_3^i \otimes M$ through the Freudenthal product.

III. SPECIAL ALGEBRAS ASSOCIATED WITH REPRESENTATIONS OF THE MAGIC SQUARE GROUPS

Let us start with the Jordan algebra of the 3×3 hermitian matrices J_3^i over H^i, i.e. real, complex, quaternionic and octonionic numbers. If A is such a matrix we have

$$A^\dagger = \bar{A}^T, \qquad C = A \cdot B = \tfrac{1}{2}(AB+BA), \tag{3.1}$$

where T is transposition and the bar denotes conjugation with respect to H^2, H^4 and H^8 under which imaginary units change sign. Then, if we denote the associator with respect to the Jordan product by

$$[ABC] = (A \cdot B) \cdot C - A \cdot (B \cdot C) \tag{3.2}$$

we have

$$[ABC] = -[CBA], \qquad [ABA^2] = 0, \tag{3.3}$$

so that the Jordan algebra is commutative, non-associative but power associative.

We now define the wedge product (proportional to the Michel-Radicati product) as the traceless part of A•B

$$A \vee B = A \cdot B - \frac{1}{3} I \; \mathrm{Tr}(A \cdot B) \tag{3.4}$$

where I is the 3×3 unit matrix. This product is defined for traceless Jordan matrices $J_3^{i(0)}$ and is commutative, but not power associative. Since Tr A and Tr(A•B) are left invariant by the groups of the first column the automorphism groups of the wedge algebra for $J_3^{i(0)}$ are just the groups of the first column of the magic square with L_{i1} as their Lie algebras. The elements of the wedge algebra $J_3^{i(0)}$ are representations of SO(3), SU(3), Sp(6) and F_4 of respective dimensions 5, 8, 14 and 26. The wedge product is a generalization of Gell-Mann's d-product for SU(3).

In connection with the groups of the second column we consider the Freudenthal algebra of 3×3 matrices $\mathcal{F}_3^i$ over complex numbers, bicomplex numbers (with two commuting imaginary units) biquaternions (complex quaternions) and complex octonions. The $\mathcal{F}_3^1$ is complex and symmetrical, $\mathcal{F}_3^2$ is hermitian with respect to one of the imaginary units only, $\mathcal{F}_3^4$ and $\mathcal{F}_3^8$ are hermitian with respect to quaternionic and octonionic conjugation only. Thus we have

$$d(\mathcal{F}_3^i) = d(J_3^i), \tag{3.5}$$

so that $\mathcal{F}_3^i$ has $d(J_3^i)$ complex elements. The Freudenthal product for two such matrices is defined by (2.9). It is therefore defined for the 6-dimensional representations of SU(3), for the $(3,\bar{3})$ representation of SU(3)×SU(3), for the 15 dimensional representation of SU(6), and the 27 dimensional representation of E_7. We note that all these representations are complex. The Freudenthal product gives the complex conjugate representation so that

$$\mathcal{F}_3^i \times \mathcal{G}_3^i \approx \mathcal{H}_3^{i\dagger} = \mathcal{H}_3^{i*}, \tag{3.6}$$

and we have

$$\text{SU(3)}: \quad 6\times 6 = \bar{6}, \tag{3.7a}$$

$$\text{SU(3)}\times\text{SU(3)}: \quad (3,\bar{3})\times(3,\bar{3}) = (\bar{3},3), \tag{3.7b}$$

$$\text{SU(6)}: \quad 15\times 15 = \overline{15} \tag{3.7c}$$

$$E_6: \quad 27\times 27 = \overline{27}. \tag{3.7d}$$

Since such products are invariant under the groups of the second column, the corresponding Lie algebras L_{i2} are the automorphism algebras of the Freudenthal product. We can also define a scalar product between an element $\mathcal{F}$ of the Freudenthal algebra and a complex conjugate element G^* by writing

$$(G^*,F) = \tfrac{1}{2}\mathrm{Tr}(G^*\cdot F). \qquad (3.8)$$

The scalar product corresponds to the compositions

$$\mathrm{SU}(3):\quad \bar{6}\times 6 = 1, \qquad (3.9a)$$

$$\mathrm{SU}(3)\times\mathrm{SU}(3):\quad (\bar{3},3)\times(3,\bar{3}) = 1, \qquad (3.9b)$$

$$\mathrm{SU}(6):\quad \overline{15}\times 15 = 1, \qquad (3.9c)$$

$$E_6:\quad \overline{27}\times 27 = 1. \qquad (3.9d)$$

The Lie algebras L_{i2} also leave the scalar product invariant.

Finally we consider the small representations of respective dimensions 14, 20, 32, and 56 of the Lie algebras L_{i3} of the groups Sp(6), SU(6), SO(12) and E_7 of the third column. Let S denote such a representation. We cannot define a binary product for the elements S since S×S does not contain S. But we note that the symmetrical part of the direct product S×S contains the adjoint representation of L_{i3} and $L_{i3}\times S$ contains S. It follows that we can define a ternary product of the elements S which will be symmetrical and transform like S.

The representations S are self conjugate. They can be expressed by means of a singlet and a Jordan matrix representation of the groups of their second column together with complex conjugate representations. We obtain

$$S^i = f + F^i + G^{i*} + g^*, \qquad (3.10)$$

giving for the dimension of S^i

$$d(S^i) = 2 + 2d(J_3^i), \qquad (3.11)$$

and reproducing the numbers 14, 20, 32 and 56. The symmetrical ternary product[18]

$$(S_1 S_2 S_3) = S_4, \tag{3.12}$$

can be obtained from the symmetrical cubing operation

$$(S\ S\ S) = T, \tag{3.13}$$

by polarization of S. The groups of the third column of the magic square now emerge as the automorphism groups of this ternary algebra.

The ternary algebra can be simply expressed by means of spin 3/2 matrices. Indeed the groups of the third column admit as subgroups the groups of the first column and SU(2) which is the automorphism group of H^4, the algebra of quaternions. Under $G_{i1} \otimes SU(2)$ the elements S^i of the ternary algebra decompose as follows:

$$S^i = (J_3^{i(0)},2) + (1,4). \tag{3.14}$$

Thus under SU(2) S^i is a doublet of traceless Jordan matrices plus a spin 3/2 representation of a singlet under the groups of the first column. Now, if we write S^T as the row

$$S^T(\mathfrak{f},\mathcal{F},\mathcal{G}^*,g^*), \tag{3.15}$$

then, it can be shown that the cubing operation (3.13) takes the form

$$(S\ S\ S) = \Sigma_i S\tilde{S}^{\dagger}\Sigma_i S, \qquad (i=1,2,3), \tag{3.16}$$

where Σ_i are the three 4×4 spin 3/2 rotation matrices and the row $\tilde{S}^T$ defined by

$$\tilde{S}^T = (-g,\mathcal{G},-\mathcal{F}^*,\mathfrak{f}^*) \tag{3.17}$$

transforms like S^T under the groups of the third column. A summation over i is implied and the products of the elements of S in the expression of (3.16) are either the Freudenthal product or the scalar product defined respectively by (2.9) and (3.8).

IV. ALGEBRAIC PROPERTIES OF SPECIAL DIRECTIONS GIVEN BY SPONTANEOUS SYMMETRY BREAKING

Let the spontaneous symmetry breaking arise from the minimization of the Higgs potential $V(\Phi)$ of one scalar field Φ that belongs to one of the representations for which one of the algebras discussed in the preceding section is defined. In a renormalizable theory $V(\Phi)$ is at most of the 4th degree in Φ and depends on the invariants constructed out of Φ. Then the equation

$$\partial V/\partial\Phi = 0, \tag{4.1}$$

is an invariant equation of at most third degree in Φ. If an invariant bilinear product can be defined, by separating the solution $\Phi=0$ we can rewrite (4.1) as a bilinear equation in Φ. The symmetry of G is then reduced to the symmetry of the subgroup H that leaves the solution of the invariant algebraic equation obtained from (4.1) invariant.

Let us apply this procedure to the groups of the first column. For Φ we take the traceless Jordan matrix $J_3^{i(0)}$. Then Φ is the 5, 8, 14 or 26 dimensional representation of the respective groups SO(3), SU(3), Sp(6) or F_4. The invariants we can construct are:

$$I_2 = \tfrac{1}{2}\mathrm{Tr}(\Phi\cdot\Phi), \tag{4.2}$$

and

$$I_3 = \mathrm{Det}\ \Phi = \frac{1}{3}\ \mathrm{Tr}(\phi\cdot\phi\vee\phi), \tag{4.3}$$

so that

$$V(\phi) = -\mu^2 I_2 + \kappa I_3 + \lambda I_2^2. \tag{4.4}$$

Using

$$\delta I_2 = \mathrm{Tr}(\phi\cdot\delta\phi), \quad \delta I_3 = \mathrm{Tr}(\phi\vee\phi\cdot\delta\phi), \tag{4.5}$$

we obtain

$$\delta V = \mathrm{Tr}\{(-\mu^2\phi+2\lambda I_2\phi+\kappa\phi\vee\phi)\cdot\delta\phi\} = 0. \tag{4.6}$$

Since $\delta\phi$ is arbitrary we must have

$$\phi\vee\phi = \kappa^{-1}(\mu^2-2\lambda I_2)\phi. \quad (4.7)$$

Now we note that ϕ satisfies the secular equation

$$\phi^2\cdot\phi - I_2\phi - I_3 I = 0, \quad (4.8)$$

which leads to

$$\mathrm{Tr}(\phi^2\cdot\phi^2) = 2I_2^2. \quad (4.9)$$

Hence taking the trace of the square of both sides of (4.7) and using (4.9) we find

$$\frac{2}{3} I_2^2 = \kappa^{-2}(\mu^2-2\lambda I_2)^2 2I_2, \quad (4.10)$$

which gives I_2 as a function of κ, μ and λ

$$I_2 = I_2(\phi,\lambda,\mu) = \frac{\mu^2}{2} + (\kappa^2/24\lambda^2)[1\pm(1+24\lambda\mu^2/\kappa^2)^{\frac{1}{2}}]. \quad (4.11)$$

Then (4.7) takes the form of the generalized Michel-Radicati equation

$$\phi\vee\phi = \pm\frac{1}{\sqrt{3}} I_2^{\frac{1}{2}}(\kappa,\lambda,\mu)\phi. \quad (4.12)$$

By a transformation of the groups of the first column we can always bring the solution ϕ_0 of (4.12) to the standard form

$$\Phi_0 = \mp I_2^{\frac{1}{2}}\lambda_8 \quad (4.13)$$

where λ_8 is the Gell-Mann matrix associated with the hypercharge direction. The subgroups of the groups of the first column that leave Φ_0 invariant are found to be SO(2), SU(2)×U(1), SO(5) ≈ Sp(4) and SO(9) which are maximal symmetry breaking groups for

SO(3), SU(3), Sp(6) and F_4 respectively. Thus a Higgs field that belongs to the 26-dimensional representation of F_4 breaks it spontaneously to SO(9) (or rather its covering group spin 9) which has as subgroup $SU(2)\times U(1)\times SU^c(3)$. A more complicated Higgs sector is needed for breaking F_4 down to the phenomenological group.

The breaking we have considered has a well known geometrical interpretation[19]. The coset space $F_4/SO(9)$ which is 16 dimensional can be represented by a pair of octonions that can be interpreted as the octonionic coordinates of a point in a generalized (compactified) Moufang projective plane. The projective geometry in this coset space is known to be non-Desarguesian.

We can generalize this procedure to the groups of the second column by choosing a Higgs field F that is a complex Jordan matrix. Then, we have 4 invariants:

$$I_2 = \tfrac{1}{2}\,\mathrm{Tr}(F^\dagger\cdot F), \qquad I_4 = \tfrac{1}{2}\,\mathrm{Tr}\{(F^\dagger\times F)\cdot(F^\dagger\times F)\}, \tag{4.14}$$

$$I_3 + iJ_3 = \mathrm{Det}\;F = \frac{1}{3}\,\mathrm{Tr}\{(F\times F)\cdot F\}. \tag{4.15}$$

For the Higgs potential we take

$$V(F) = -\mu^2 I_2 + \kappa_1 I_3 + \kappa_2 J_3 + \lambda_1 I_2^2 + \lambda_2 I_4. \tag{4.16}$$

In the case of the group SU(3), I_4 is not independent of I_2. Minimization of V(F) leads to a general relation of the type

$$F^\dagger \times (F\times F) - \tfrac{1}{2}F\;\mathrm{Tr}(F^\dagger\cdot F) = -c^2 F, \tag{4.17}$$

where c is a certain function of μ, κ_1, κ_2, λ_1 and λ_2. Through a transformation of the groups of the second column it is possible to bring F into a diagonal form with all the diagonal elements having the same phase δ. Then δ and the absolute values ρ_0, σ_0, τ_0 of the diagonal elements can be expressed as functions of the

4 invariants. For Det F $\neq 0$, this is solved by means of

$$F\times F = kF^{\dagger}. \tag{4.18}$$

Using Springer's identity

$$(F\times F) \times (F\times F) = F \text{ Det } F, \tag{4.19}$$

from Eq. (4.17) we get the constraint

$$k^{-1} \text{ Det } F - I_2 = -c^2, \tag{4.20}$$

while from (4.18), using (4.15) we obtain

$$\text{Det } F = \frac{2}{3} kI_2, \tag{4.21}$$

so that

$$c^2 = \frac{1}{3} I_2. \tag{4.22}$$

If we seek a solution with Det F = 0, then k=0 and we find

$$F\times F = 0, \tag{4.23}$$

which is solved by $\sigma_0 = \tau_0 = 0$. Thus, the solution F_0 is obtained from an idempotent matrix by application of the group E_6 in the octonionic case. These idempotents are left invariant by respectively the subgroups SU(2), SU(2)×SU(2)×U(1), SU(4)×SU(2)×U(1), SO(10)×SO(2) of the groups of the second column SU(3), SU(3)×SU(3), SU(6) and E_6.

Solutions with Det F $\neq 0$ give

$$\rho_0^2 = \sigma_0^2 = \tau_0^2. \tag{4.24}$$

The solutions with positive ρ_0, σ_0 and τ_0 give directions left invariant by the groups of the first column, so that E_6 is broken down to F_4 spontaneously. Which solution occurs for a given

Higgs potential depends on the relative values of the constants in (4.16).

Turning to the groups of the third column we choose the Higgs field Φ to belong to the elements of the ternary algebra corresponding to the 14, 20, 32 and 56 dimensional representations of Sp(6), SU(6), SO(12) and E_7. The quadratic invariant associated with the element S in (3.10) reads

$$I_2 = (\tilde{S},S) = f^*f + \mathrm{Tr}\, \mathcal{F}^{*}\cdot\mathcal{F} + \mathrm{Tr}\, \mathcal{G}^{*}\mathcal{G} + g^*g. \qquad (4.25)$$

An independent quartic element is

$$I_4 = (T,S), \qquad (4.26)$$

where T is given by (3.13) and (3.16). There is no cubic invariant. The Higgs potential now reads

$$V(S) = -\mu^2 I_2 + \lambda_1 I_2^2 + \lambda_2 I_4. \qquad (4.27)$$

Minimization can only lead to the algebraically invariant relation

$$(S\ S\ S) = k\tilde{S}, \qquad (4.28)$$

where k is a function of the invariants and the constants μ, λ_1, λ_2. For special solutions that give $I_4 = 0$ we have

$$(S\ S\ S) = 0. \qquad (4.29)$$

This equation is solved by

$$S^T = \lambda(1,J,J\times J,\mathrm{Det}\ J). \qquad (4.30)$$

By a transformation which belongs to the coset $E_7/(E_6\times U(1))$ λ can be made real and J can be made equal to zero, so that only the real part of the f component survives in the expression of S. Such a special S_0 is seen to be left invariant by E_6. Thus E_7

can be spontaneously broken down to E_6 by a 56 dimensional Higgs field. For solutions corresponding to non-vanishing I_4 the surviving subgroup is smaller, the smallest one can obtain being SO(8).

V. CONCLUDING REMARKS

We have shown that spontaneous symmetry breaking of the exceptional groups leads in general to algebraic relations for the values of the Higgs field that minimize the Higgs potential. Those relations generalize the Michel-Radicati relations and are expressed in terms of a binary product for the groups occurring in the first and second columns of the magic square. The wedge product and the Freudenthal product are associated respectively with the first column which includes F_4 and the second column which includes E_6. The third column which includes E_7 is associated with a ternary product. Choosing the Higgs field to bellong to the small representations of F_4, E_6 and E_7 we find algebraic relations that define special directions in the spaces of these exceptional groups.

Those directions are left invariant by subgroups of F_4, E_6, E_7 which all have the color group $SU^c(3)$ as a subgroup. What remains to be done is to study more complicated Higgs structures that can reduce the F_4, E_6 or E_7 symmetries down to the phenomenologically suggested group $SU(2)\times U(1)\times SU^c(3)$.

It is a pleasure to acknowledge the hospitality of Montreal University where part of this work was done and to thank Dr. P. Sikivie, Dr. P. Ramond, Professor L. Michel and my colleagues at Yale for the many ideas that have found their way to the subject of this talk.

REFERENCES

1. F. GÜRSEY, in Kyoto International Symposium on Mathematical Physics, ed. H. Araki, p. 189 (Springer 1975); F. GÜRSEY, in Group Theoretical Methods in Physics, Ed. A. Janner et al., p. 225 (Springer, 1976); F. GÜRSEY, in New Pathways in High Energy Physics I, Ed. A. Perlmutter, p. 231 (Plenum Press 1976).

2. P. JORDAN, J. VON NEUMANN and E.P. WIGNER, Ann. Math. 35, 29 (1934).

3. S. WEINBERG, Phys. Rev. Lett. 19, 1264 (1967); A. SALAM, in "Elementary Particle Theory", Ed. N. Svartholm, (Wiley, New York 1969).

4. H. GEORGI and S.L. GLASHOW, Phys. Rev. Lett. 32, 433 (1974); H. GEORGI, H. QUINN and S. WEINBERG, Phys. Rev. Lett. 33, 451 (1974); H. FRITZSCH and P. MINKOWSKI, Ann. Phys., N.Y. 93, 193 (1974).

5. F. GÜRSEY, P. RAMOND and P. SIKIVIE, Phys. Rev. D12, 2166 (1975), and Phys. Lett. 60B, 177 (1976).

6. F. GÜRSEY and P. SIKIVIE, Phys. Rev. Lett. 36, 775 (1976); P. RAMOND, Nucl. Phys. B110, 214 (1976); P. SIKIVIE and F. GÜRSEY, "Quark and Lepton Assignments in the E_7 Model", to be published.

7. See M.L. MEHTA, J. Math. Phys. 7, 1824 (1966); M.L. MEHTA and P.K. SRIVASTAVA, ibid, 1833 (1966).

8. S.L. ADLER, Phys. Rev. 177, 2426 (1969); J.S. BELL and R. JACKIW, Nuovo Cimento 60, 47 (1969); H. GEORGI and S.L. GLASHOW, Phys. Rev. D6, 429 (1972). See also Ref. 5 and Ref. 6.

9. See for instance J. BERNSTEIN, Rev. Mod. Phys. 46, 7 (1974); E.S. ABERS and B.W. LEE, Phys. Reports 9C, Nb 1 (1974).

10. L. MICHEL and L.A. RADICATI, Ann. of Phys. 66, 758 (1971) and in "Evolution of Particle Physics", p. 191 (Academic Press, New York 1970).

11. H. FREUDENTHAL, Oktaven, Ausnahmengruppen und Oktavengeometrie (Utrecht, 1951).

12. J.R. FAULKNER, Trans. Amer. Math. Soc. 155, 397 (1971) and ibid 167, 49 (1972). See also R.B. BROWN, J. Reine Angew. Math. 236, 79 (1969).

13. H. FREUDENTHAL, Advances in Mathematics 1, 145 (1965).

14. B.A. ROZENFELD, Proc. Colloq. Utrecht, p. 135 (1962).

15. J. TITS, Proc. Colloq. Utrecht, p. 175 (1962).

16. A. MALCEV, Mat. Sbornik 78, 569 (1955); A.A. SAGLE, Trans. Amer. Math. Soc. 101, 426 (1961). For the Malcev algebra of color, see M. GÜNAYDIN, J. Math. Phys. 17, 1875 (1976).

17. M. GÜNAYDIN and F. GÜRSEY, Phys. Rev. D9, 3387 (1974); F. GÜRSEY, in "Proc. of the Johns Hopkins Workshop", ed. by G. Domokos and S. Kövesi-Domokos, p. 15 (Baltimore, MD. 1974). See also R. CASALBUONI, G. DOMOKOS and S. KÖVESI-DOMOKOS, Nuovo Cimento 33A, 432 (1976).

18. See Ref. 12, and for a more general treatment see G.B. SELIGMAN, in Rational Methods in Lie Algebras, p. 140 and p. 258 (M. Dekker, New York, 1976).

19. P. JORDAN, Abh. Math. Sem. Univ. Hamburg, 16, 74 (1949); A. BOREL, Comptes Rendus Ac. Sci. 230, 1378 (1950).

COMPLEX SPACE-TIMES WITH NULL STRINGS

Jerzy F. Plebanski and Ivor Robinson

The Goldberg-Sachs theorem [1] has played a distinguished role in the modern history of relativity, providing, when skillfully used, a convenient starting point in the search for algebraically degenerate solutions. However from the point of view of the Einstein equations, understood as complex equations on analytic functions, this theorem which emphasized the role of the shear-free congruences of null geodesics was in a sense misleading.

What complex Einstein equations like more than geodesics are 2-surfaces, and particularly totally null 2-surfaces which can be for brevity called null strings*.

The work of Alfred Schild and one of the present authors [2], concerned with conformal extensions of the Goldberg-Sachs theorem, already recognizes the technical significance of the self-dual

---

* One of the reasons for this is perhaps the fact that while a complex riemannian V_4 consists of the pair: the differential (analytic) variety M_4 and Hodge's involutory star $(* : \Lambda^p \rightarrow \Lambda^{4-p})$, permitting the definition of the inner product, $\alpha \rfloor \beta : = *(\alpha \wedge *\beta)$, α and $\beta \in \Lambda^1$, the fundamental star as applied to the 2-forms is a *conformally invariant* operation.

bi-vectors*. It was however much later that, by abandoning the confinement to real space-times, S. Hacyan and the other present author [3] recognized that the Goldberg-Sachs theorem is in fact an intersection of the two more basic "complex" theorems concerning null strings in complex V_4's.

What are null strings? One defines them as 2-surfaces in the complex riemannian space V_4 which possess a tangent space spanned by a pair of null vectors which are mutually orthogonal. One shows then that these tangent spaces are automatically parallelly propagated along the surface, so that the surface can be justly called totally geodesic.

Now, the theorem established in [3] asserts that the Einstein-flat complex V_4 is one-sidedly degenerate in the sense amply explained in [4] and used in [5] (see also [6]) if and only if the complex space-time admits the existence of a null string. Technically (using the notation and the formalism of [7], extended to the complex case as in [4]) we orient the tetrad members e^1 and e^3 along the tangents to null strings, and denote by $C^{(a)}$'s the components of the left (heavenly) part of the conformal curvature. Then we have

$$\Gamma_{422} = \Gamma_{424} = 0 \Longleftrightarrow C^{(5)} = C^{(4)} = 0, \qquad (1)$$

provided that Einstein vacuum equations $G_{ab} = 0$ are fulfilled. The left algebraic degeneracy assumed, we then find from the structure equations for the null tetrad $e^a, de^a = e^b \wedge \Gamma^a_{\ b}$, $[ds^2 := 2e^1 \otimes e^2 + 2e^3 \otimes e^4]$, that:

$$de^1 = \alpha \wedge e^1 + \beta \wedge e^3, \quad de^3 = \gamma \wedge e^1 + \delta \wedge e^3, \qquad (2)$$

$$\alpha,\beta,\gamma,\delta \in \Lambda^1.$$

---

* The existence of a null self-dual solution of Maxwell equations was shown - modulo field equations - to be equivalent to the degeneracy of the self-dual part of the Weyl tensor.

But this is precisely the thesis of the Frobenius theorem (see eg. [8], or for more elementary treatment [9]) and thus (in a singly connected region of V_4) there exist scalars such that:

$$e^1 = f^1_u du + f^1_v dv, \quad e^3 = f^3_u du + f^3_v dv \qquad (3)$$
$$\det(f^i_j) \neq 0.$$

The surfaces defined as intersections (in arbitrary coordinates) of $u(x^\mu) = \text{const}$, $v(x^\mu) = \text{const}$, should be thought of as our congruence of null strings.

Now, taking this as our starting point, i.e., assuming only the (complex) Einstein vacuum equations in V_4 and the conformal curvature at least minimally degenerate from one side, we undertake the task of integrating the Einstein equations in all remaining generality. Surprisingly enough, we can report that we have been successful in this task. Let $C_{ab} := R_{ab} - \frac{1}{4}g_{ab}R$. According to what was said above, assuming one-sided algebraic degeneration $[C^{(5)} = C^{(4)} = 0]$ we can show, by using the field equations $C_{44} = C_{42} = C_{22} = 0$, that a canonical tetrad can always be selected so that:

$$\begin{aligned} e^1 &= \phi^{-2}du, \quad & e^3 &= dx + \mathcal{P}du + \mathcal{R}dv, \\ e^2 &= \phi^{-2}dv, \quad & e^4 &= dy + \mathcal{R}du + \mathcal{Q}dv, \end{aligned} \qquad (4)$$

where $\{x^\mu\} = \{uvxy\}$ are local coordinates and $\mathcal{P}, \mathcal{Q}, \mathcal{R}$ and ϕ are structural functions. Then, from the first three field equations mentioned above, one easily shows, using the remaining freedom of choice for coordinates, that we can restrict ourselves to the consideration of the two distinct cases only where:

$$\begin{aligned} &\text{I)} \qquad && \phi := 1, \\ &\text{II)} \qquad && \phi := x+y. \end{aligned} \qquad (5)$$

The invariant distinction between these two cases is this: with the surface element of our null string, $\Sigma := du \wedge dv = \frac{1}{2}\Sigma_{ab}e^{a} \wedge e^{b}$, one easily shows that the expansion form defined by:

$$\theta := \theta_{a}e^{a} = \tfrac{1}{2}(u^{;a}_{;a}dv - v^{;a}_{;a}du) \tag{6}$$

vanishes in Case I and is different from zero in Case II. Observe that $\Sigma_{ab;c}\Sigma^{cd} + \Sigma_{ab}\theta^{d} = 0$ and that with $\theta=0$ equations $x^{a}_{;r}\Sigma^{rb} = 0$ have a tetrad of independent solutions.

Considering now the field equation R=0 and integrating the second triplet of field equations $C_{41} = C_{12} = C_{32} = 0$ we find correspondingly that in Cases I and II there exist functions such that:

$$\text{I)} \qquad \mathcal{P} = -\Theta_{yy}+xf, \quad \mathcal{Q} = -\Theta_{xx}+yg, \quad \mathcal{R} = \Theta_{xy} \tag{7}$$

and

$$\text{II)} \qquad \mathcal{P} = \partial_{y}\phi^{4}\partial_{y}\Pi + \tfrac{1}{4}\mu\phi^{3} \tag{8}$$

$$\mathcal{Q} = \partial_{x}\phi^{4}\partial_{x}\Pi + \tfrac{1}{4}\mu\phi^{3}$$

$$\mathcal{R} = -\tfrac{1}{2}(\partial_{x}\phi^{4}\partial_{y}+\partial_{y}\phi^{4}\partial_{x})\Pi + \tfrac{1}{4}\mu\phi^{3}$$

where Θ and Π are general functions of four coordinates and f, g and μ are functions of u and v only. In Case II one then shows that using the remaining freedom of choice for coordinates, we can assume without losing generality that μ is constant, which considerably simplifies the further considerations.

The last triplet of field equations $C_{33} = C_{31} = C_{11} = 0$, can be integrated up to the very end in both cases. Indeed, like the previous two triplets, it can be shown to reduce to a system of equations of the type

$$\left.\begin{matrix}\partial_{x}\partial_{x}\\ \partial_{x}\partial_{y}\\ \partial_{y}\partial_{y}\end{matrix}\right\} \text{acting on the same function} = 0 \tag{9}$$

and is therefore easily integrated, yielding in Cases I and II conditions which can be described as follows:

$$\text{I)}\quad \Theta_{xx}\Theta_{yy} - \Theta_{xy}\Theta_{xy} + \Theta_{xu} + \Theta_{yv} + f(\Theta_x - x\Theta_{xx}) + g(\Theta_y - y\Theta_{yy}) = \tfrac{1}{2}xy(f_v+g_u) + \alpha x + \beta y, \tag{10}$$

where α and β can depend on u and v only and

$$\text{II)}\quad \phi^4(F_xG_y - F_yG_x) + \phi(G_u+F_v) + \tfrac{1}{4}\mu(\partial_x+\partial_y)\phi^5(\partial_x+\partial_y)\Pi = \tfrac{1}{2}\nu(x-y) + \lambda \tag{11a}$$

where ν and λ are arbitrary functions of u and v only and:

$$F := \phi^{-2}\partial_y\phi^3\Pi \qquad G := \phi^{-2}\partial_x\phi^3\Pi. \tag{11b}$$

The left (heavenly) components of the conformal curvature are then given respectively by:

$$\text{I)}\quad C^{(5)} = C^{(4)} = C^{(3)} = 0; \quad C^{(2)} = f_v - g_u, \tag{12}$$
$$C^{(1)} = (\partial_u+f)\{\alpha - \tfrac{1}{2}yC^{(2)}\} + (\partial_v+g)\{\beta - \tfrac{1}{2}xC^{(2)}\},$$

and

$$\text{II)}\quad C^{(5)} = C^{(4)} = 0, \quad C^{(3)} = -2\mu\phi^3, \quad C^{(2)} = -2\nu\phi^5, \tag{13}$$
$$C^{(1)} = 2\phi^7[(x\nu_v - y\nu_u) + 2\nu(\partial_x - \partial_y)\phi^3\Pi + (\partial_u+\partial_v)\{\lambda - \mu\phi^{-\frac{1}{2}}(\partial_x+\partial_y)\phi^{3/2}\Pi\}].$$ [10]

The right (hellish) components of the conformal curvature are then given correspondingly in the two cases by:

$$\text{I)}\quad \bar{C}^{(5)} = 2\Theta_{yyyy} \qquad \bar{C}^{(1)} = 2\Theta_{xxxx} \tag{14}$$
$$\bar{C}^{(3)} = 2\Theta_{yyxx}$$
$$\bar{C}^{(4)} = 2\Theta_{yyyx} \qquad \bar{C}^{(2)} = 2\Theta_{xxxy}$$

and

$$\text{II)} \quad \bar{C}^{(5)} = -2\phi^2 \mathcal{P}_{yy} \qquad \bar{C}^{(1)} = -2\phi^2 \mathcal{Q}_{xx} \tag{15}$$

$$3\bar{C}^{(3)} = \phi^2 [4\mathcal{R}_{xy} - (\mathcal{Q}_{yy} + \mathcal{P}_{xx})]$$

$$\bar{C}^{(4)} = \phi^2 [\mathcal{R}_{yy} - \mathcal{P}_{xy}] \qquad \bar{C}^{(2)} = \phi^2 [\mathcal{R}_{xx} - \mathcal{Q}_{yx}].$$

[Of course in (15), $\mathcal{P}$, $\mathcal{Q}$ and $\mathcal{R}$ are just abbreviations for the right hand members of (8).]

The results outlined above exhibit a remarkable fact: the Einstein empty space equations and (at least) one sided minimal algebraic degeneration reduce the problem of determining the (degenerate) "complex gravity" to the study of *one* scalar equation only.

One can show that in both cases I and II, if one wants to determine the characteristic surfaces related to the equations (10) and (11) on $\Theta(xyuv)$ and $\Pi(xyuv)$ respectively, by applying standard techniques (see, e.g. Courant-Hilbert [11]) one finds that the jumps of the second derivatives of the functions Θ and Π are precisely permitted only on the eikonal surfaces of complex space-time, i.e.,

$$g^{\mu\nu} S_{,\mu} S_{,\nu} = 0 \tag{16}$$

where $g^{\mu\nu}$ is just the riemannian metric defined by our ds^2.

We have grounds to believe that the present results are only a part of a major truth concerning the structure of solutions of Einstein equations. We believe that there are functions, say Θ and $\bar{\Theta}$, which describe completely all vacuum metrics of the type $G \otimes G$. These functions would be gravitational counterparts of the electromagnetic Hertz potentials.

As far as the electromagnetic field is concerned, already on the level of our present results, there is no difficulty in adjoining it, as described by a Hertz potential, to metrics which carry a congruence of null strings. The situation is here somewhat similar to that known from real algebraically degenerate V_4's where in the presence of an electromagnetic field one cannot apply (claiming completeness) the Goldberg-Sachs theorem, but where one postulates the congruence of non-shearing null geodesics as an "Ansatz" which facilitates the derivation of the physically interesting solutions.

In the real case, one correlates eigenvectors of the electromagnetic field with the principal degenerate null direction. Quite similarly, postulating our congruence of null strings it is convenient to restrict ourselves to a field such that $\Sigma_{ab}F^{ab} = 0$. Then, the left and right electromagnetic fields,

$$\omega^{(\pm)} := \tfrac{1}{2}\{f_{\mu\nu} \pm f_{\mu\nu}\}dx^{\mu} \wedge dx^{\nu}, \tag{17}$$

which together with our tetrad (4) fulfill Einstein-Maxwell equations are:

$$\omega^{+} = \varepsilon(du\wedge dx + dv\wedge dy) + (\delta + x\varepsilon_{v} - y\varepsilon_{u})du\wedge dv \tag{18}$$

$$\omega^{-} = \phi^{2}[H_{xx}e^{2}\wedge e^{3} + H_{yy}e^{1}\wedge e^{4} + H_{xy}(e^{1}\wedge e^{2} + e^{4}\wedge e^{3})]$$

where H is any solution of

$$[\partial_u\partial_x + \partial_v\partial_y - \mathcal{P}\partial_x\partial_x - 2\mathcal{R}\partial_x\partial_y - \mathcal{Q}\partial_y\partial_y]H = 0, \tag{19}$$

and ε and δ are disposable functions of u and v only. One can show that then the Einstein equations integrate as before causing the presence of electromagnetic terms in the equations which generalize (10) and (11) for both types.

The details of this will be given in a paper on this subject by the present authors together with Dr. A. Garcia where also the inclusion of the cosmological constant will be described.

It will be perhaps of some interest to enumerate different types of solutions of Einstein vacuum equations reduced to our equations (10) and (11).

In obvious symbolism, we have basically seven distinct spaces:

Diagram I

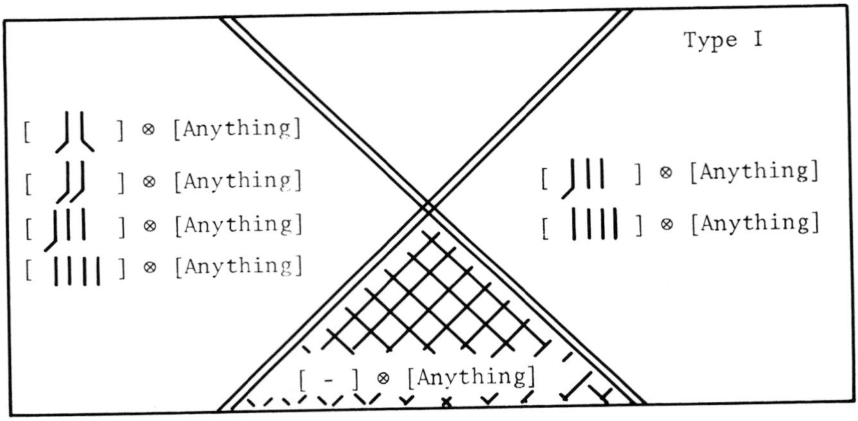

This diagram is to be interpreted as follows: within the type I we have the obvious three cases [⅃||] ⊗ [A] (if $C^{(2)} \neq 0$), [|||] ⊗ [A] (if $C^{(2)} = 0$, $C^{(1)} \neq 0$) and [-] ⊗ [A] if all $C^{(a)} = 0$. The type II provides for types [⅃L]⊗[A] (if $C^{(3)} \neq 0$) and:

$$\delta := 2[C^{(2)}]^2 - 3C^{(1)}C^{(3)} \neq 0 \quad [12], \tag{20}$$

then [⅃⅃] ⊗ [A] (if $C^{(3)} \neq 0$ and $\delta = 0$), [||L] ⊗ [A] (if $C^{(3)} = 0$, $C^{(2)} \neq 0$), [||||] ⊗ [A] (if $C^{(3)} = C^{(2)} = 0$, $C^{(1)} \neq 0$); finally, within the type I, the case [-] ⊗ [A] because of results [4] must overlap with the case [-] ⊗ [A] of the type I. [One should notice, however, that the intersection of type I and II solutions also includes solutions which are degenerate on both sides with the property that the expansion vector vanishes from one side only.]

These particular spaces [-] ⊗ [A], Newman called "heavens" at GR7 Conference, Tel Aviv, 1974, and one of us described in the terms of the first and second heavenly equations for the key functions Ω and Θ, [4]. The basic results of the present work contain, of course, the equation for the key function Θ as the most special case. For that reason, if one were tempted to follow the previous terminology one could perhaps to call the group of complex space-times from the diagram I, HH metrics with the H metric being the special case, the second H referring to "higher" or "hyper" heavens. Egs. (10-11), correspondingly, could be called HH equations.

It is to be pointed out that although [-] ⊗ [A], the H spaces, are interesting, by their nature they certainly do not permit any "earthly" real cross-section. The complex minimally degenerate metrics described in this paper, however, by their very nature, do contain all results of the theory of real algebraically degenerate solutions as corresponding real cross-sections.

Helpful discussions with J.D. Finley III and I. Ozsvath are warmly appreciated by both authors.

REFERENCES

1. J. GOLDBERG and R. SACHS, Acta Phys. Polon. Suppl. 22, 13 (1962).
2. I. ROBINSON and A. SCHILD, J. Math. Phys. 4, 484 (1963).
3. J.F. PLEBANSKI and S. HACYAN, J. Math. Phys. 16, 2403 (1975).
4. J.F. PLEBANSKI, J. Math. Phys. 16, 2395 (1975).
5. J.D. FINLEY and J.F. PLEBANSKI, J. Math. Phys. 17, 585 (1976).
6. C.W. FETTE, ALLEN I. JANIS and EZRA T. NEWMAN, J. Math. Phys. 17, 660 (1976).
7. G.C. DEBNEY, R.P. KERR and A. SCHILD, J. Math. Phys. 10, 1842 (1969).
8. W. SLEBODZINSKI, Exterior Forms and Their Applications, Monografie Matematyczne, Warszawa, 1970, PWN.

9. H. FLANDERS, Differential Forms with Applications to the Physical Sciences, Ac. Press, New York and London, 1963.

10. This expression for $C^{(1)}$ which involves the iterated use of eq. (11), was derived by Drs. J.D. FINLEY III, and A. GARCIA, whose effort and computational skills are greatly appreciated.

11. R. COURANT and D. HILBERT, Methods of Mathematical Physics, Inter-Science Publishers, New York (1962).

12. Within the type II, [⌋⌋] ⊗ [A] metrics are obtained by integrating $\delta = 0$, with $C^{(1)}$, $C^{(2)}$ and $C^{(3)}$ from (13); equation obtained is a linear equation of the second order on $\phi^3\Pi$ and can be easily integrated; a solution of the inhomogeneous equation can be simply found, linear in x and y with coefficients depending on u and v only. An obvious coordinate transformation then changes the homogeneous equation into the diffusion equation in two dimensions, $F_{xy}+F_t = 0$.

INTELLIGENT SPIN STATES AND HEISENBERG EQUALITY

C. Aragone, E. Chalbaud and S. Salamó

We discuss both the difference between three important sets of states related to the SU(2) algebra and the properties of the Heisenberg uncertainty relation associated with the elements of this algebra. These states are: a) Minimum uncertainty states, b) Bloch states (atomic or coherent spin states) and c) intelligent spin states.

Some dynamical properties associated with the intelligent spin states are shown.

1. BLOCH STATES, MINIMUM UNCERTAINTY STATES AND INTELLIGENT SPIN STATES

The SU(2) algebra, defined by the commutation relations

$$[J_{\varepsilon},J_{-\varepsilon}] = 2\varepsilon J_3, \qquad [J_3,J_{\varepsilon}] = \varepsilon J_{\varepsilon} \tag{1}$$

where $J_{\varepsilon} = J_1 + i\varepsilon J_2$ and $\varepsilon = \pm 1$, provide us an important set of vectors, the Bloch states[1,2]. These states are defined as:

$$|\tau\rangle = \frac{1}{(1+|\tau|^2)^j} e^{\tau J_+}|-j\rangle = e^{\tau J_+} e^{\ell n(1+|\tau|^2)J_3} e^{-\tau^* J_-}|-j\rangle . \tag{2a}$$

or equivalently by:

$$|\tau\rangle = R(\tau)|-j\rangle ; \quad \tau = \tan\frac{\theta}{2}\, e^{-i\phi} \tag{2b}$$

in which $R(\tau)$ is a rotation through an angle θ about an axis $\hat{n} = \sin\phi\hat{e}_1 - \cos\phi\hat{e}_2$. The inner product of two different Bloch states is given by:

$$\langle\tau_1|\tau_2\rangle = \frac{(1+\tau_1^*\tau_2)^{2j}}{(1+|\tau_1|^2)^j(1+|\tau_2|^2)^j} . \tag{3}$$

The Heisenberg equality for the SU(2) operators is

$$\Delta J_1^2 \Delta J_2^2 = \tfrac{1}{4}\, |\langle J_3\rangle|^2. \tag{4}$$

In order to study this equality, we define two functionals, the uncertainty of any state ψ, $I(\psi)$:

$$I(\psi) \equiv \frac{\langle\psi|\Delta J_1^2|\psi\rangle\langle\psi|\Delta J_2^2|\psi\rangle}{\langle\psi|\psi\rangle^2} \tag{5a}$$

and the half-commutator squared $C(\psi)$:

$$C(\psi) \equiv \tfrac{1}{4}\, \frac{|\langle\psi|J_3|\psi\rangle|^2}{\langle\psi|\psi\rangle^2}. \tag{5b}$$

Now we introduce two kind of states:

a) States that minimize $I(\psi)$ are called Minimum Uncertainty States (M.U.S.).

b) States $|W\rangle \ni I(W) = C(W)$ are called Intelligent Spin States (I.S.S.).

It has been proven[3] that the unitary I.S.S. for the SU(2) algebra are those states that satisfy the following equation:

$$J_\alpha|W\rangle \equiv (J_1 - i\alpha J_2)|W\rangle = (\langle W|J_1|W\rangle - i\alpha\langle W|J_2|W\rangle)|W\rangle \equiv W|W\rangle$$

$$\langle W|W\rangle = 1 \tag{6}$$

where α is an arbitrary real number. After solving this equation we obtain[4] for $|W\rangle$:

$$|W_N(\tau_\alpha)\rangle = \hat{a}_N \sum_{\ell=0}^{N} \binom{N}{\ell}(2j-\ell)!(-2\tau_\alpha J_+)^\ell|\tau_\alpha\rangle, \quad 0 \le N \le 2j \tag{7}$$

where

$$\tau_\alpha^2 \equiv \frac{\gamma_+}{\gamma_-}; \quad \gamma_\varepsilon \equiv \tfrac{1}{2}(1-\varepsilon\alpha)$$

and

$$\varepsilon = \pm 1.$$

For the eigenvalues W we have: $W_N = 2\gamma_+\tau_\alpha^{-1}(j-N)$. Notice that τ_α has to be real or pure imaginary.

We can enlarge the definition of our I.S.S. by allowing τ_α to cover all the complex plane τ. We shall call these new states the generalized I.S.S. Of course the G.I.S.S. do not necessarily satisfy the Heisenberg equality. The reason for introducing them is evident later on. A straightforward calculation shows that the G.I.S.S. can be written as

$$|W_n(\tau)\rangle = a_n Y_1 \partial_y^n e^{\tau y} e^{(\ell n y^{-2})J_3}|-j\rangle$$

$$|W_n(\tau)\rangle = a_n Y_1 \partial_y^n K(y,\tau)|-j\rangle = a_n Y_1 \partial_y^n p_j(y,y,\tau)|\tau_y\rangle \tag{8}$$

where

$$n = 2j - N; \quad p_j(y,z,\tau) = |yz + |\tau|^2(y-2)(z-2)|^j$$

$$\tau_y = \tau(1-2y^{-1}); \quad Y_1 f(z) \equiv f(1),$$

and

$$a_n = [Z_1 Y_1 \partial_y^n \partial_z^n p_{zj}(y,z,\tau)]^{-\frac{1}{2}} \equiv [p_{zj}^{nn}]^{-\frac{1}{2}}.$$

Notice the resemblance of this expression with the Bloch states defined earlier. An important result follows from the last equation: The set of all G.I.S.S. contains the Bloch states, therefore {G.I.S.S.} can be thought as a refinement of them. In particular it is worth mentioning that $|W_n(\tau)\rangle = |\mu\rangle$ ($|\mu\rangle$ being an arbitrary Bloch state if and only if $n = 0$, $\mu = -\tau$ or $n = 2j$, $\mu = \tau$.

In order to see difference between the M.U.S. and the I.S.S., let us denote by H_j the space spanned by the I.S.S. Now let $B \subset H_j$, defined through:

$$B \equiv \{|\tau\rangle ; \tau = \tan\frac{\theta}{2}\, e^{-i\rho}\}.$$

Clearly an I.S.S. $\in B$ is going to be an I.S.S. in H_j; on the contrary $|\psi\rangle$ a M.U.S. in B does not necessarily imply that $|\psi\rangle$ is a M.U.S. in H_j.

Using the results of Lieb[5] we get for the functionals $I(\tau)$ and $C(\tau)$:

$$I(\tau) = \frac{j}{4}\,(1-\sin^2\frac{\theta}{2}\,\sin^2\phi)(1-\sin^2\theta\,\cos^2\phi)$$

$$C(\tau) = \frac{j^2}{4}\cos\theta. \tag{9}$$

It is completely straightforward to see that in the τ plane the points where $I(\tau)$ achieves stationary values are:

$$\tau_n = e^{-\frac{in\pi}{4}}, \qquad n = 0,1,\ldots,7,$$

while the functional $C(\tau)$ has minimum values on the whole circle of radius 1. The results are shown in the figure below:

$\tau = \tan \frac{1}{2}\, \theta e^{-i\phi}$.

0 ≡ Points where $I(\tau)$ has a minimum.

x ≡ Points where $I(\tau)$ has saddle points.

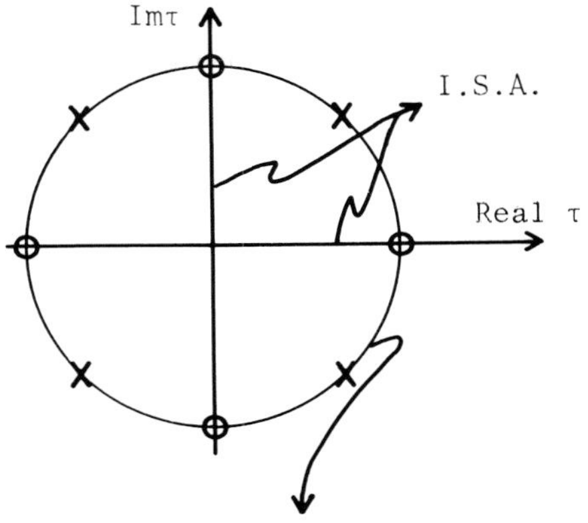

2. SOME DYNAMICAL PROPERTIES OF THE I.S.S.

Let us discuss the time evolution of a non-relativistic system of spin j having a magnetic moment γ, interacting with a magnetic field of the form[6]:

$$\underset{\sim}{B}(t) = 2B_{\perp} |\cos(2\omega, t)\hat{x} + \sin(2\omega, t)\hat{y}| + 2B_{\parallel}\hat{z}. \tag{10}$$

The Hamiltonian is then given by:

$$H(t) = -\hbar\gamma \underset{\sim}{J}\cdot\underset{\sim}{B}(t). \tag{11}$$

In the two-dimensional representation of SU(2) one can evaluate the time evolution operator (which satisfies the equation $i\hbar\dot{U} = HU$). That gives:

$$U(t) = \begin{pmatrix} \cos^2\psi e^{i\omega_- t} + \sin^2\psi e^{-i\omega_+ t} & i \sin^2\psi \sin(\omega_2 t) e^{-i\omega_1 t} \\ i \sin^2\psi \sin(\omega_2 t) e^{i\omega_1 t} & \cos^2\psi e^{-i\omega_- t} + \sin^2\psi e^{i\omega_+ t} \end{pmatrix} \tag{12}$$

where:

$$\omega_{\pm} = \omega_2 \pm \omega_1, \quad \omega_2 = [\gamma^2 B_{\perp}^2 + (\gamma B_{\parallel} + \omega_1)^2]^{\frac{1}{2}}$$

and

$$\sin 2\psi = \gamma B_{\perp}\omega_2^{-1}.$$

If we assume that the initial state has been prepared in an I.S.S., $|W_n(\tau)\rangle$, at a later time t, the system will be described by the state $|W_n(t,\tau)\rangle \equiv U(t)|W_n(\tau)\rangle$.

Then the following question arises: Is $|W_n(t,\tau)\rangle$ an I.S.S.? We know that the evolution of a Bloch state under the Hamiltonian (11) keeps being a Bloch state[2]. To answer this question we have to use some results concerning the group SL(2,C)[7]. If $\ell \in SL(2,C)$, ℓ can be uniquely decomposed as $\ell = kz$ (for $\ell_4 \neq 0$), where

$$\ell = \begin{pmatrix} \ell_1 & \ell_2 \\ \ell_3 & \ell_4 \end{pmatrix}, \qquad k = \begin{pmatrix} p^{-1} & q \\ 0 & p \end{pmatrix} \quad \text{and} \quad z = \begin{pmatrix} 1 & 0 \\ Z & 1 \end{pmatrix}. \tag{13}$$

Using now the fact that $k(p,q) = e^{qp^{-1}J_+} e^{-2\ell n p J_3}$, we see that the element $k(y,\tau)$ which occurs in equation (8) has the same structures as $k(p\cdot q)$. Also it is easy to prove that $k(y,t)$ for $|y| \neq 1$ contains a Lorentz boost, excluding the possibility of its being a rotation.

Now we can write the state $|W_n(t,\tau)\rangle$ in a way that resembles an I.S.S. This can be done if we define $\hat{\ell}$ by

$$\hat{\ell} = \begin{pmatrix} \hat{\ell}_1 & \hat{\ell}_2 \\ \hat{\ell}_3 & \hat{\ell}_4 \end{pmatrix} = U(t)k(y,\tau), \quad \text{with} \quad K(Y,\tau) = \begin{pmatrix} y^{-1} & \tau(y-2) \\ 0 & y \end{pmatrix}. \tag{14}$$

We find for $\hat{\ell}_1$, $\hat{\ell}_2$, $\hat{\ell}_3$ and $\hat{\ell}_4$:

$$\hat{\ell}_1 = y^{-1}[\cos^2\psi e^{i\omega_- t} + \sin^2\psi e^{-i\omega_+ t}]$$

$$\hat{\ell}_2 = [\tau(y-2)\cos^2\psi + y \sin\psi \cos\psi]^{i\omega_- t}$$

$$+ [\tau(y-2)\cos^2\psi - y \sin\psi \cos\psi]e^{i\omega_+ t}$$

$$\hat{\ell}_3 = y^{-1}\sin\psi \cos\psi(e^{i\omega_+ t} - e^{i\omega_- t})$$

$$\hat{\ell}_4 = [y \sin^2\psi + \tau(y-2)\sin\psi \cos\psi]e^{i\omega_+ t}$$

$$+[y \cos^2\psi - \tau(y-2)\sin\psi \cos\psi]e^{-i\omega_- t} . \qquad (15)$$

With these results we are able to write the state $|W_n(t,\tau)\rangle$ as:

$$|W_n(t,\tau)\rangle = a_n(\tau)Y_1\partial_y^n\hat{\ell}(y,t,\tau)|-j\rangle$$

$$|W_n(t,\tau)\rangle = a_n Y_1 \partial_y^n \hat{\ell}_4^{2j} e^{\hat{\ell}_2\hat{\ell}_4^{-1}} |-j\rangle \qquad (16)$$

or in terms of the Bloch states $|\hat{\tau}\rangle \equiv |\hat{\ell}_2\hat{\ell}_4^{-1}\rangle$ as:

$$|W_n(t,\tau)\rangle = a_n(\tau)Y_1\partial_y^n e^{2ij\arg\hat{\ell}_4}(|\hat{\ell}_2|^2+|\hat{\ell}_4|^2)|\hat{\tau}\rangle . \qquad (17)$$

From the above equation we can prove that for $B_\perp = 0$, the resulting state is an I.S.S. for t satisfying the following relation:

$$2B_\parallel t_m = \frac{m\pi}{2}, \quad m=0,1,2,\ldots .$$

For n = 0 we recover the results given by Arecchi et al[2].

The relevance of these new states in the description of a point like laser is discussed by Aragone, Chalbaud and Salamó[8].

REFERENCES AND FOOTNOTES

1. F. BLOCH, Phys. Rev. 70, 460 (1946). J.M. RADCLIFFE, J. Phys. A4, Gen. Phys. 313 (1971).

2. F. ARECCHI, E. COURTENS, R. GILMORE, and H. THOMAS, Phys. Rev. A6, 2211 (1972).

3. See for instance K. GOTTFRIED, "Quantum Mechanics", Vol. I: Fundamentals (W.A. Benjamin, New York, 1966).
4. C. ARAGONE, G. GUERRI, S. SALAMÓ, and J.L. TANI, J. Phys. A: Math. Nucl. Gen. 7 L149 (1974).
5. E. LIEB, Commun. Math. Phys. 31, 327 (1973).
6. R. GILMORE, "Lie Groups, Lie Algebras, and Some of their Applications" (Wiley, New York, 1974).
7. N. SCIARRINO and M. TOLLER, J. Math. Phys. 8, 1252 (1967).
8. C. ARAGONE, E. CHALBAUD, and S. SALAMÓ, Jour. Math. Phys. 17, 1963 (1976).

SOME ASPECTS OF GRADED LIE ALGEBRAS

Nigel Backhouse

The theory of graded (or pseudo) Lie algebras has recently been given a boost by the realization that it is at the heart of supersymmetry. In this paper we show that a number of the concepts and results which arise in Lie algebra theory carry over to the wider context of graded Lie algebras.

THE UNIVERSAL ENVELOPING ALGEBRA

The basic definitions relating to graded Lie algebras (GLA) are given in the review [1], and a fairly exhaustive list of references is to be found in the newsletters of Kaplansky [2].

Let $L = L_0 \oplus L_1$ be a GLA (we only consider even/odd gradings) with sign function σ and grading automorphism π. We can form the universal enveloping algebra U(L) in the obvious way. That is, we first form the tensor algebra T(L) = {linear combinations of $x_1 \otimes x_2 \otimes \ldots \otimes x_n$, for all $x_i \in L$ and for all n}, and then factor out the two-sided ideal J(L) generated by the linear span of elements of the form $x \otimes y - (-1)^{\sigma(x)\sigma(y)} y \otimes x - [x,y]$, for all x, $y \in L_0 \cup L_1$. Now U(L) contains a faithful image of L, which we do not distinguish from L, and inherits a multiplication from T(L), denoted by juxtaposition, for which the bracket operation in L is given by the linear extension of $[x,y] = xy - (-1)^{\sigma(x)\sigma(y)}$

yx, for all x, $y \in L_0 \cup L_1$.

U(L) is a filtered algebra, but also has a natural even/odd grading. To see this we observe that the grading automorphism π of L extends to an involutive automorphism of T(L) which preserves the ideal J(L) and hence gives rise to an involutive automorphism π' of U(L). Now put $U(L)_0 = \{u \in U(L):\pi'u = u\}$ and $U(L)_1 = \{u \in U(L):\pi'u = -u\}$, then $U(L) = U(L)_0 \oplus U(L)_1$ and we have $U(L)_0U(L)_0 \subseteq U(L)_0$, $U(L)_0U(L)_1 \subseteq U(L)_1$, $U(L)_1U(L)_0 \subseteq U(L)_1$ and $U(L)_1U(L)_1 \subseteq U(L)_0$. To pin down $U(L)_0$ and $U(L)_1$ we use the graded version of the Poincaré-Birkhoff-Witt (PBW) theorem, first proved by Ross [3]. This says that a basis for U(L) consists of a commutative identity together with monomials of the form $(y_1)^{i_1}(y_2)^{i_2},\ldots,(y_t)^{i_t}$, where $y_1,y_2,\ldots,y_s$ forms a basis for L_0, $y_{s+1},\ldots,y_t$ forms a basis for L_1, $i_1,\ldots,i_s$ are arbitrary integral exponents but $i_{s+1},\ldots,i_t$ can only take the values 0,1. Thus U(L) has dimension 2^{t-s} over $U(L_0)$, regarding the latter as a subalgebra of U(L) in the obvious way. Now a basis for $U(L)_1$ consists of those PBW basis elements where

$$\sum_{r=s+1}^{t} i_r$$

is odd. The remaining PBW basis elements form a basis for $U(L)_0$. This means that $U(L)_0$ is strictly larger than $U(L_0)$ if L_1 is non-trivial. We also remark that U(L) may possess zero divisors - they are absent in the case of ordinary Lie algebras.

In U(L), an associative algebra with unit, not only does ordinary multiplication make sense, but so does any combination of commutation and anticommutation. However, we find that the bracket operation in L extends in a very specific way to U(L). A simple way to achieve this is by inductively writing $(\mathrm{ad}x)yu = ((\mathrm{ad}x)y)u + (-1)^{\sigma(x)\sigma(y)}y(\mathrm{ad}x)u$ for $x,y \in L_0 \cup L_1$ and $u \in U(L)$ - the induction is on the degree of elements in U(L). It turns out, on examination, that we could have defined adx in one go as follows:

(i) If $x \in L_0$ then adx maps $U(L)_0$ into $U(L)_0$ and $U(L)_1$ into $U(L)_1$ according to $(\mathrm{ad}x)u = xu - ux$;

(ii) If $x \in L_1$ then adx maps $U(L)_0$ into $U(L)_1$ and $U(L)_1$ into $U(L)_0$ according to $(\mathrm{ad}x)u = xu \mp ux$, depending on whether $u \in U(L)_0$ or $U(L)_1$. We can of course further extend ad in a manner consistent with the grading to make sense of ad u, $u \in U(L)$.

We say that $u \in U(L)$ is invariant under L if $(\mathrm{ad}x)u = 0$ for all $x \in L$ - equivalently, if $(\mathrm{ad}v)u = 0$ for all $v \in U(L)$. The set of all such invariants, the centre ZU(L), forms a graded subalgebra of U(L) with respect to ordinary multiplication. We should remark that if we write $ZU(L) = (ZU(L))_0 \oplus (ZU(L))_1$, then the even part is not the centre of $U(L_0)$ and indeed it is only a subalgebra of the centre of $U(L)_0$. Thus we should distinguish between $(ZU(L))_0$, $ZU(L_0)$ and $ZU(L)_0$ - there are some further related algebras which should be considered in the full story.

In a recent paper, [4], we showed, for non-degenerate GLA, how to construct homogeneous, even invariants C_n for all degrees of homogeneity n - these are the natural generalizations of the Casimir invariants of the semi-simple Lie algebra theory. For an example consider the so-called di-spin algebra L, where $L_0 = \{e,h,f\}$, $L_1 = \{x,y\}$, and the relations are $[h,e] = 2e$, $[h,x] = x$, $[x,x] = e$, $[h,f] = -2f$, $[h,y] = -y$, $[y,y] = -f$, $[e,f] = h$, $[f,x] = y$, $[x,y] = -\frac{1}{2}h$ and $[e,y] = x$. All other brackets between generators are zero. L is a non-degenerate GLA in which L_0 is A_1 and L_1 carries the spin $\frac{1}{2}$ representation of L_0. Put $\alpha = fe + ef + \frac{1}{2}h^2$ and $\beta = xy - yx$. Then we find $C_2 = \frac{1}{3}(\alpha+\beta)$; C_3 can be reduced to $\frac{1}{4}C_2$; $C_4 = \frac{20}{3}C_2^2 - \frac{1}{18}C_2$. Now we show that ZU(L) consists of all the polynomials in $\alpha + \beta$.

First consider α and β, which mutually commute. α is the Casimir operator of degree 2 for the spin 1 representation of L_0 and β is the quadratic invariant for the spin $\frac{1}{2}$ representation of L_0. It follows from representation theory that all the invariants of L_0 within U(L) are the polynomials in α and β. We now have to show that if we further demand invariance under L_1, then we can only allow polynomials in the unique combination $\lambda = \alpha + \beta$. Since α and β commute, any invariant polynomial in α and β can be written in the form $P(\alpha,\beta) = f_0(\lambda) + f_1(\lambda)\beta + f_2(\lambda)\beta^2 + f_n(\lambda)\beta^n$, where n is an integer and each $f_i(\lambda)$, $i=0,1,\ldots,n$, is a polynomial in λ - actually we can choose n=1 because $\beta^2 = \frac{1}{2}(\gamma+\beta)$ but we prefer to allow n to be arbitrary for future generalizations. Now for $z = x$ or $y \in L_1$, repeatedly apply adz to $P(\alpha,\beta)$. On the left-hand side we get zero by invariance. Noting that (adz)(adz) = adu for some $u \in L_0$, which implies $(\mathrm{adz})(\mathrm{adz})\beta = 0$, we have after n applications of adz the equation $0 = b_n f_n(\lambda)((\mathrm{adz})\beta)^n$, where b_n is some non-zero constant. We know from the representation theory of the di-spin algebra, [1], that, for every integer N, L has an irreducible representation of dimension 2N + 1 in which λ is represented by $\frac{1}{2}N(N+1)$ times the identity and $(\mathrm{adz})\beta$, for $z = x$ or y, is nilpotent of degree not less than 2N + 1. So in this representation, $f_n(\lambda)$ is $f_n(\frac{1}{2}N(N+1))$ times the identity, $\mathrm{ad}((\mathrm{adz})\beta)^n$ is zero if $n \geq 2N + 1$, but non-zero if $n < 2N + 1$. Since $b_n \neq 0$ we have that $f_n(\frac{1}{2}N(N+1)) = 0$ for all N such that $2N + 1 > n$. This implies, since f_n is a finite degree polynomial, that it vanishes identically. Repeating this argument we find $f_i \equiv 0$ for all $i > 0$ and hence $P(\alpha,\beta) = f_0(\lambda)$, as required.

We can say some further things about the structure of the algebra of invariants. First, since $[L_1,L_1] = L_0$, anything invariant under L_1 is also invariant under L_0. Secondly, because $xy \equiv \beta(\mathrm{mod}\ L_0)$, we have that xy commutes with β, and then, because $\alpha \equiv \lambda(\mathrm{mod}\ \beta)$, we have that xy also commutes with α. Now $U(L)_0 = U(L_0) + U(L_0)xy$, by the PBW basis theorem. So we find $ZU(L)_0$ = {polynomials in α and β} which contains $ZU(L_0)$ = {polynomials in α} and $ZU(L) = (ZU(L))_0$ = {polynomials in $\lambda = \alpha + \beta$} as proper subalgebras.

We would like to have more general results concerning ZU(L) for arbitrary or special L. Our arguments in the special case of the di-spin algebra depend too heavily on explicitly known representations to be useful generally, though I suspect that the question of zero divisors will be of some importance. Finally observe that because L_1 is a representation space for the spin $\frac{1}{2}$ representation of L_0, so $L_1L_1 \subseteq U(L)_0$ is a representation space for the direct sum of the spin 1 and spin 0 representations. We can be more explicit than this, for a basis for the spin 1 representation is the symmetrized square of $L_1 = \{xx, xy+yx, yy\} = \{e,h,f\}$, putting us straight back into L_0. A basis for the spin 0 representation is the anti-symmetrized square of $L_1 = \{xy-yx\} = \{\beta\}$. Results of this nature are true for arbitrary GLA.

SCHUR'S LEMMA

As in [1] let End V = $(\text{End } V)_0 \oplus (\text{End } V)_1$ be the graded algebra of linear transformations on the graded complex vector space $V = V_0 \oplus V_1$. End V can be given the structure of a GLA in the obvious way. A homomorphism ϕ of $L = L_0 \oplus L_1$ into End V, for some V, is called a representation of L if $\phi(L_0) \subseteq (\text{End } V)_0$, $\phi(L_1) \subseteq (\text{End } V)_1$, and ϕ preserves the graded bracket structure. Given such a ϕ, we define its commutant $C(\phi) = \{T \in \text{End } V: [T,\phi(x)] = 0 \text{ for all } x \in L\}$. It is not hard to show that $C(\phi)$ is a graded subalgebra (both in the associative and Lie sense) of End V and we can write $C(\phi) = C(\phi)_0 \oplus C(\phi)_1$ where $C(\phi)_0 \subseteq (\text{End } V)_0$ and $C(\phi)_1 \subseteq (\text{End } V)_1$. Also it is clear that $\phi((ZU(L))_0) \subseteq C(\phi)_0$ and $\phi((ZU(L))_1) \subseteq C(\phi)_1$, and furthermore that $\phi(ZU(L))$ is a graded Abelian subalgebra of End V.

We say that ϕ is reducible if there exists a proper graded subspace U of V such that the restriction of ϕ to U is a representation of L. Otherwise ϕ is irreducible. The graded Schur's Lemma which we now state characterizes irreducible representations by their commutants.

If ϕ is irreducible then $C(\phi)_0$ consists of scalar multiples of the identity, just as in the ordinary case. However, it turns out, only for very special representations is $C(\phi)_1$ non-trivial. First note that both V_0 and V_1 carry ordinary representations of L_0, say ϕ_0 and ϕ_1. If ϕ_0 and ϕ_1 are inequivalent then certainly $C(\phi)_1$ is trivial. If ϕ_0 is equivalent to ϕ_1, then it is convenient to choose bases in V_0 and V_1 such that in matrix form ϕ_0, ϕ_1 are identical. Now, for $x \in L_1$, denote by $\phi_a(x)$, $\phi_b(x)$ the off-diagonal blocks of the matrix $\phi(x)$ which map respectively V_0 to V_1 and V_1 to V_0. We find that $C(\phi)_1$ is trivial unless $\phi_a(x) = \phi_b(x)$ for all $x \in L_1$ and that in this case $C(\phi)_1$ consists of all scalar multiples of the matrix

$$J = \left(\begin{array}{c|c} 0 & I \\ \hline -I & 0 \end{array}\right).$$

We have a situation which is somewhat analogous to the reality classification of group representations. The conditions which we have imposed on ϕ in order that $C(\phi)_1$ be non-trivial at first sight seem very stringent, but we do note that they are in fact satisfied by the self-representation of the (f,d) algebra of Gell-mann, Michel and Radicati, mentioned in [1].

REFERENCES

1. L. CORWIN, Y. NE'EMAN and S. STERNBERG, Rev. Mod. Phys. 47, 573 (1975).
2. I. KAPLANSKY, "Newsletter on graded Lie algebras, nos. 1-4", (Dept. Math., University of Chicago).
3. L. ROSS, Trans. Amer. Math. Soc. 120, 17 (1965).
4. N. BACKHOUSE, J. Math. Phys. (to be published).

U(n,n) AND GENERALIZED TWISTORS

H. Bacry*

INTRODUCTION

Twistors have been introduced by Penrose [1] in relation to General Relativity, Conformal Group and Massless Particles. The sympletic properties of twistor space have been used [2] to obtain by Kostant-Souriau quantization [3] some irreducible representations of the Poincaré group. Such representations have been given in a quite different context in ref. [4]. Here we intend to give a pedagogical introduction to twistor theory in a slightly more general version and without use of dotted and undotted indices.

1. THE U(n,n) GROUP

The U(n,n) group is isomorphic to the group of 2n × 2n complex matrices U which satisfy the following property

$$U\beta U^{+} = \beta$$

with

$$\beta = \begin{vmatrix} 0 & -i \\ i & 0 \end{vmatrix}$$

* Université d'Aix-Marseille II and Centre de Physique Théorique, CNRS, Marseille.

where i stands for a block n × n matrix equal to i times the unit matrix. By writing U in block form, we get for

$$U = \begin{vmatrix} A & B \\ C & D \end{vmatrix}$$

$$\begin{cases} AB^+ = BA^+ \\ CD^+ = DC^+ \\ AD^+ - BC^+ = 1. \end{cases}$$

(Note the existence of an outer automorphism:
$\{A,B,C,D\} \to \{A,-B,-C,D\}$.)

One of the interesting subgroups of U(n,n) is the group* inh SL(n,c) which is obtained by imposing $C = 0$ and $\det A = 1$. It is a simple exercise to show that it is composed of all matrices of the type

$$\begin{vmatrix} \Lambda & H\Lambda^{+-1} \\ 0 & \Lambda^{+-1} \end{vmatrix} \quad \text{with} \quad \det \Lambda = 1, \quad H = H^+.$$

Any element for which $\Lambda = 1$ will be called a translation. This comes from the fact that for n = 2, inh SL(2,C) is the covering group of the Poincaré group and $\Lambda = 1$ corresponds to space time translations.

The generalized twistor space T *is the complex* 2n-*dimensional vector space on which* U(n,n) *acts*. Any twistor θ reads, in block form,

$$\theta = \begin{pmatrix} \omega \\ \pi \end{pmatrix}.$$

Under a translation, only the π part of the twistor is invariant:

$$\omega \to \omega' = \omega + H\pi$$

$$\pi \to \pi' = \pi.$$

---

* About this class of groups, see ref. [7].

From U(n,n) definition, it readily follows that the quantity

$$\bar{\theta}\theta = \theta^{+}\beta\theta = 2\mathrm{Im}(\omega^{+}\pi)$$

is invariant.

Property. The twistor space has a complex symplectic structure which is invariant under U(n,n). In other words, U(n,n) acts as a group of canonical transformations.

We define the Poisson bracket of two functions on T as follows

$$\{f,g\} = \frac{\partial f}{\partial \omega^{a}} \frac{\partial g}{\partial \bar{\pi}_{a}} - \frac{\partial g}{\partial \omega^{a}} \frac{\partial f}{\partial \bar{\pi}_{a}} + \text{compl. conj.}$$

or. equivalently:

$$\{\omega^{a}, \bar{\pi}_{b}\} = \{\bar{\omega}^{a}, \pi_{b}\} = \delta^{a}_{b}$$

and all other brackets are zero (not only U(n,n) but also complex conjugation preserve the Poisson brackets). We will not give the proof of U(n,n) invariance here.

Let us denote by

$$\begin{vmatrix} S & R \\ -X & -S^{+} \end{vmatrix} \qquad (R=R^{+},\ X=X^{+})$$

the generators of U(n,n); they induce on the set of functions on T the generators

$$Q = R^{ab}\bar{\pi}_{a}\pi_{b} + X_{ab}\bar{\omega}^{a}\omega^{b} + S^{a}_{b}\bar{\pi}_{a}\omega^{b} + \bar{S}^{a}_{b}\pi_{a}\bar{\omega}^{b}.$$

It is a *real* quadratic form (due to the choice of β). The generators of inh SL(n,c) correspond to X = 0 and Tr S = 0. The generators of translations are the $\bar{\pi}_{a}\pi_{b}$'s.

2. QUANTIZATION

If we replace ω^a by

$$\frac{\partial}{\partial\bar{\pi}_a},$$

we obtain a "ladder" representation of U(n,n) on the Hilbert space of functions $f(\pi)$ with scalar product

$$\int \overline{f(\pi)}\, g(\pi) d^n\bar{\pi} d^n\pi .$$

This generalizes the ladder representation of U(2,2) given for instance in ref. [5]. In fact, to make this representation irreducible we must require the functions $f(\pi)$ to be eigenfunctions of the operator $\mathrm{Im}(\omega^+\pi)$, i.e.

$$i\bar{\pi}_a \frac{\partial f}{\partial\bar{\pi}_a} - i\pi_a \frac{\partial f}{\partial\pi_a} = \lambda f.$$

λ can be called the *generalized helicity*.

3. ACTION OF THE POINCARE GROUP ON TWISTOR SPACE

We could define three kinds of action:

a) action on the twistor space itself,

b) action on the space of "twistors up to a phase",

c) action on the twistor projective space (twistors up to a factor).

It is case b) which will retain our attention. Since $n = 2$, the corresponding space is of *real* dimension $4n-1 = 7$. It is a simple matter to decompose it into orbits. If we except the trivial orbit, we get:

α) For $\pi = 0$ ($\omega \neq 0$), there exists an inh SL(2,C) transformation which maps such an element on

$$\theta = \begin{vmatrix} 1 \\ 0 \\ 0 \\ 0 \end{vmatrix} .$$

β) For $\pi \neq 0$, there exists a transformation which maps such an element on

$$\theta = \begin{vmatrix} -i\lambda \\ 0 \\ 1 \\ 0 \end{vmatrix}$$

with λ *real*.

In this last case, the corresponding stabilizer is made of transformations such that

$$\Lambda = \begin{vmatrix} e^{i\varphi} & 0 \\ c & e^{-i\varphi} \end{vmatrix}, \qquad H = \begin{vmatrix} 0 & -i\bar{c}\lambda e^{i\varphi} \\ ic\lambda e^{-i\varphi} & \gamma \end{vmatrix}$$

$$c \in C, \qquad\qquad \gamma \in \mathbb{R} .$$

They form a 4-dimensional subgroup. Therefore the corresponding orbits are 6-dimensional. They are labelled by the *real* number $\lambda \geq 0$. We note the remarkable fact that the little group depends on λ (they are not conjugate). It follows that each orbit forms a single stratum.

These six-dimensional orbits are canonical symplectic homogeneous spaces in the sense of Kostant-Souriau theory when applied to the Poincaré group*. Each of them can be given the interpretation of phase space for a massless particle of helicity λ.

---

* All canonical symplectic homogeneous spaces of the Poincaré group have been classified by Arens [6].

REFERENCES

1. R. PENROSE and M.A.H. MacCALLUM, Phys. Reports C6, 241 (1972), and references therein.

2. N.M.J. WOODHOUSE, in Proceedings of the 4th International Colloquium on Group Theoretical Methods in Physics, Nijmegen, 1975, A. Janner, Editor (Springer, 1976).

3. B. KOSTANT, Lectures in Modern Analysis and Quantization (Springer Verlag, 1970). J.M. SOURIAU, Structure des Systèmes Dynamiques (Dunod, 1970).

4. H. BACRY and A. KIHLBERG, J. Math. Phys. 10, 2132 (1969), p. 2140.

5. G. MACK and I.T. TODOROV, J. Math. Phys. 10, 2078 (1969).

6. R. ARENS, J. Math. Phys. 12, 2415 (1971).

7. H. BACRY and A. KIHLBERG, Commun. Math. Phys. 1, 150 (1965).

SOME COMMENTS ON SYMPLECTIC STRUCTURES IN COMPLEX RIEMANNIAN GEOMETRY

Charles P. Boyer

In this talk I would like to discuss briefly the appearance of symplectic structures in complex Riemannian geometry. Before entering explicitly into this discussion, it is convenient to recall briefly Cartan's[(1)] algebrazation of systems of partial differential equations. Cartan shows how one can set any system of partial differential equations in terms of a closed ideal I of differential forms in the Grassmann algebra $\Lambda(M)$ of differential forms on a differential manifold M. Actually the ideal of forms is somewhat more general since to be equivalent to a system of partial differential equations one must also specify that the independent variables, say x^i, in the equations be independent in the ideal I, i.e. $dx^1 \wedge dx^2 \wedge \ldots \wedge dx^n \neq 0$. When this happens Cartan calls the ideal **in involution with respect to the variables** x^i. Now given such a system of equations or an ideal of differential forms, the integral submanifolds are given as follows: let I be generated by $\omega_i \in \Lambda(M)$ and let N be a manifold and i an immersion of N into M, then an integral submanifold is a pair (N,i) such that $i^*\omega_i = 0$ for all $\omega_i \in I$, where i* denotes the pullback with respect to i, i.e. $(i^*\omega)(p) = \omega(i(p))$, $p \in M$. The integral submanifolds are the solutions of the partial differential equations.

Now one is also often interested in symmetries of I. These are given by all local diffeomorphisms ϕ (class C^r): $M \to M$ such that $\phi^*\omega_i \in I$ for all ω_i in I. Such mappings patched together globally do not necessarily form a Lie group, but a pseudogroup. This is in fact essentially the definition of Lie pseudogroup. There does exist to some extent a classification[2] of infinite pseudogroup structures developed in fact by Cartan[3]. In modern language one has a manifold M and constructs the bundle of linear frames L(M) over M. The structure group of L(M) is GL(n). If one then considers the automorphisms of the bundle one obtains the pseudogroup of diffeomorphisms on M. Now one can classify certain types of pseudogroup structures by classifying the subgroups $H \subset GL(n)$. For each H we consider the problem of finding a subbundle $B \subset L(M)$ which has structure group H. In general this depends strongly on M and involves fibre bundle obstruction theory. We will now consider some simple examples: let M be R^n or C^n, i) take $H = SL(n)$ then the automorphisms of the corresponding subbundle B give just the volume preserving diffeomorphisms on M, ii) similarly if $H = Sp(n)$ then one gets the pseudogroup of canonical transformations.

Now we have studied[4] in collaboration with J.F. Plebanski the integral submanifolds of the complex Einstein equations in vacuum characterized by a non-vanishing self-dual conformal curvature. Plebanski[5] has shown that all such complex universes (heavens) are determined by one partial differential equation which in spinor notation takes the form[4]

$$\frac{1}{2}\frac{\partial^2 \Theta}{\partial p^A \partial p_B}\frac{\partial^2 \Theta}{\partial p_A \partial p^B} + \frac{\partial^2 \Theta}{\partial p^A \partial q_A} = 0 \qquad (1)$$

where p_A and q_A are complex spinors. We have found through prolongation techniques that there is a hierarchy of closed 1-forms (presumably infinite) associated with the integral submanifolds of (1). However, only two are needed to specify the system.

Alternatively, heavens can be written as the integral submanifolds of two 2-forms - a kind of intersecting symplectic structure. Now in many ways the variables q_A are special. Indeed the metric, connection, and curvature are respectively given by the second, third, and fourth derivatives with respect to only the spinor p_A. For example, the conformal curvature is simply

$$C_{ABCD} = \frac{\partial^4 \circledast}{\partial p^A \partial p^B \partial p^C \partial p^D} . \quad (2)$$

The fact that two variables split off so naturally coupled with the presumably infinite hierarchy of conservation laws (Pfaffian 1-forms) suggests the intriguing possibility that there may be soliton type solutions.

Another interesting point about (1) is the Monge-Ampère like structure of the first term. Strangely enough this term is related to the appearance of a symplectic structure. To see this we split off the linear term containing the q_A spinor by looking for a solution of (1) which reduces (1) to

$$\circledast_{xx} \circledast_{yy} - \circledast_{xy}^2 = 1. \quad (3)$$

Consider the case where all variables in (3) are real, then we can write the solutions of (3) as integral submanifolds of the ideal generated by

$$\begin{gathered} du \wedge dv - dx \wedge dy \\ du \wedge dx + dv \wedge dy. \end{gathered} \quad (4)$$

If we now define the complex variables $z_2 = v+ix$, $z_1 = u-iy$, (4) becomes the complex symplectic structure

$$dz_1 \wedge dz_2 . \quad (5)$$

The integral submanifolds of (5) which are obtained immediately give the solutions of (3) as long as $dx \wedge dy \neq 0$. Now consider the case when all variables in (3) are complex. Then introduce quaternions $q_1 = ue_0 - ye_2$, $q_2 = ve_0 + xe_2$, where $e_0 e_i = e_i e_0 = e_i$, $-e_i^2 = e_0^2 = 1$, (no sum on i) and $e_i e_j = e_k$ (cyclicly). Now all variables in (4) are complex and (4) can be written as a quaternionic symplectic structure

$$dq_1 \wedge dq_2. \tag{6}$$

The integral submanifolds are now easily obtained in terms of two complex functions each of one complex variable. We also mention that in these simple two dimensional cases (5) and (6) are volume forms and the integral submanifolds are the singular elements. Now the symmetries of (5) and (6) are just the complex and quaternionic canonical transformations, respectively. Now another interesting point is seen by making a linear canonical transformation[4] $y \to v$, $v \to -y$, on (4). Upon integration we get $\phi_{xx} + \phi_{vv} = 0$. Thus we see that the real Laplace equation is equivalent to a complex symplectic structure while the complex Laplace equation is equivalent to a quaternionic symplectic structure. Hence, the symmetries on T*(M) of the Laplace equation are just the complex or quaternionic canonical transformations. Thus we have been able to understand in a very simple way the nonlinear term in (1). The situation, however, appears much more complicated when the linear term containing the additional spinor q_A is added, and we have been able to give the general integral submanifolds in implicit form only, thus far[4]. Finally, we mention that the important Monge-Ampère structure is retained in the more general (minimal algebraic degeneracy) situation of Plebanski and Robinson[6]. Thus it is of quite definite interest to be able to understand the appearance of symplectic structures (quaternionic) in complex Riemannian geometry in complete detail.

REFERENCES

1. E. CARTAN, *Les systèmes différentiels extérieurs et leurs applications géométriques*, (Hermann, Paris, 1945).

2. S. KOBAYASHI, *Transformation Groups in Differential Geometry*, (Springer-Verlag, New York, 1972).

3. E. CARTAN, *Oeuvres Complètes*, Partie 2, Vol. 2.

4. C.P. BOYER and J.F. PLEBANSKI, Heavens and their Integral Varieties, preprint Comun. Tecn., Vol. 7, No. 127, Univ. de México, to appear in J. Math. Phys.

5. J.F. PLEBANSKI, J. Math. Phys. 16, 2395 (1975).

6. J.F. PLEBANSKI and I. ROBINSON, presented at this conference by J.F. Plebanski; Phys. Rev. Lett. 37, 493 (1976).

COHERENT STATES ON THE CONFORMAL GROUP AND POSITION OPERATOR

Zbigniew Haba

We shall assume that the symmetry group of physical space-time coincides with the conformal group [1]. In explicitly covariant theories (if the existence of an operator for the observable position is assumed) this means that the four-position transforms covariantly under a unitary representation of the conformal group U(C), i.e.

$$U(\Lambda,a)\hat{w}_\mu U^{-1}(\Lambda,a) = \Lambda_\mu^{\nu}\hat{w}_\nu + a_\mu \quad \text{under the Poincaré subgroup}$$

$$U(\ell)\hat{w}_\mu U^{-1}(\ell) = \ell^{-1}\hat{w}_\mu \quad \text{under scale transformations}$$

$$U(b)\hat{w}_\mu U^{-1}(b) = \frac{\hat{w}_\mu + b_\mu \hat{w}^2}{1+2b\hat{w}+b^2\hat{w}^2} \quad \text{under special conformal transformations.} \tag{1}$$

Generators of the subgroups will be denoted by $M_{\mu\nu}$, P_μ, D, and K_μ correspondingly.

It can be checked that the operator [2]

$$\hat{w}_\mu = -[M_{\nu\mu}P^\nu + (D+iN)P_\mu]P^{-2} \tag{2}$$

where

$$N = 2 + \sqrt{4+C_2} \qquad C_2 = \tfrac{1}{2}M_{\mu\nu}M^{\mu\nu} - 4iD - D^2 + P_\mu K^\mu \tag{3}$$

fulfils eqs. (1) in the infinitesimal form, if P_μ, $M_{\nu\mu}$, D, K_μ form the algebra of the conformal group C (isomorphic to SO(4,2))

$$
\begin{aligned}
&i[P_\mu,D] = P_\mu \\
&i[M_{\alpha\beta},P_\gamma] = g_{\alpha\gamma}P_\beta - g_{\beta\gamma}P_\alpha \\
&i[M_{\alpha\beta},M_{\mu\nu}] = g_{\alpha\mu}M_{\beta\nu} - g_{\beta\mu}M_{\alpha\nu} + g_{\alpha\nu}M_{\mu\beta} - g_{\beta\nu}M_{\mu\alpha} \\
&i[D,K_\alpha] = K_\alpha \\
&i[M_{\alpha\beta},K_\gamma] = g_{\alpha\gamma}K_\beta - g_{\beta\gamma}K_\alpha \\
&i[P_\alpha,K_\beta] = 2(-g_{\alpha\beta}D+M_{\alpha\beta}) .
\end{aligned}
\tag{4}
$$

Physical requirements imposed on the spectrum of momentum P_μ determine representation of the conformal algebra, which should be chosen in the definition of $\hat{w}_\mu$ eq. (1); the physical spectrum should fulfil the conditions $P^2 = m^2 \geq 0$, $P_0 \geq 0$. These conditions are fulfilled for the discrete series representations [3] with continuous mass spectrum (zero mass representation $P^2 = 0$ is excluded because of P^2 in the denominator in eq. (1)).

The position operator $\hat{w}_\mu$ is non-hermitian. It can be written in terms of its hermitian part X_μ and momenta

$$\hat{w}_\mu = X_\mu + i(N-2)P_\mu P^{-2}. \tag{5}$$

The hermitian part X_μ has no eigenstates and no self-adjoint extension [4]. We can show however that the non-hermitian operator $\hat{w}^\dagger_\mu$ has a complete set of eigenstates $|w\rangle$ (localized states) with eigenvalues w_μ in the physical forward tube π. This forward tube is the analyticity domain of Wightman functions [5]. On this domain the conformal group acts transitively. In fact the forward tube is a homogeneous hermitian symmetric space according to the terminology of Helgason [6]. It can be obtained as a coset space $SO(4,2)/SO(4) \otimes SO(2)$, where $SO(4) \sim SU(2) \otimes SU(2)$ is a maximal compact subgroup in SO(4,2). The eigenstates

$$\hat{w}_\mu^\dagger |w\rangle = w_\mu |w\rangle \tag{6}$$

can be written in the form [7] (here we restrict ourselves to the most degenerate discrete series with $C_2 = n(n-4)$ [3])

$$|w\rangle = 2^{2n-2}\det(E-iW)^{-n} \sum_{jmq_1q_2} \det[(E-iW)^{-1}(E+iW)]^m$$

$$D^j_{q_1q_2}[(E-iW)^{-1}(E+iW)]\,|jmq_1q_2\rangle \tag{7}$$

where

$$W = \sum_{\mu=0}^{3} w_\mu \sigma^\mu,$$

σ^μ - Pauli matrices, E - unit matrix and $D^j_{q_1q_2}$ are matrix elements of SU(2) extended on non-unitary matrices; finally $|jmq_1q_2\rangle$ are eigenstates of the generators of the maximal compact subgroup $SU(2) \otimes SU(2) \otimes O(2)$, which label the basis of SO(4,2) (cf. ref. [3]).

The states $|w\rangle$ are covariant under the conformal group, i.e.

$$U(\Lambda,a)|w\rangle = |\Lambda w+a\rangle \qquad \text{(spin equals zero)}$$

$$U(\lambda)|w\rangle = \lambda^{-n}|\lambda^{-1}w\rangle$$

$$U(b)|w_\mu\rangle = (1+2bw+b^2w^2)^{-n}\left|\frac{w_\mu+b_\mu w^2}{1+2bw+b^2w^2}\right\rangle \tag{8}$$

They form a complete set in the representation space H of the conformal algebra. Therefore each vector $|f\rangle \in H$ can be represented by a function $\langle w|f\rangle$ defined on the forward tube π. Moreover we can easily represent action of an arbitrary operator by its action on the states $|w\rangle$ by means of the resolution of the identity operator

$$1 = \int_{\pi} d^8w[(\mathrm{Im}w)^2]^{n-4}|w\rangle\langle w|. \tag{9}$$

In order to prove formula (9) we use the orthogonality of D^j functions and the resolution

$$1 = \sum_{jmq_1q_2} |jmq_1q_2\rangle\langle jmq_1q_2|. \tag{10}$$

From eq. (9) the scalar product in the space of analytic functions $\langle w|f\rangle$ (depending on $\bar{w}_\mu = x_\mu - iy_\mu$) can be obtained

$$\langle f|g\rangle = \int_{\pi} d^8w[(\mathrm{Im}w)^2]^{n-4}\langle f|w\rangle\langle w|g\rangle. \tag{11}$$

Together with eq. (8) this defines the realization of the discrete series representation of the conformal group in the Hilbert space $\mathcal{H}_n(\pi)$ of analytic functions defined on the forward tube π [8]. The states $|w'\rangle$ have now representation in the space $\mathcal{H}_n(\pi)$

$$\langle w|w'\rangle = 2^{2n-4}i^{-2n}[(w'-\bar{w})^2]^{-n}. \tag{12}$$

The scalar product (12) can be computed using the relation

$$\langle jmq_1q_2|j'm'q_1'q_2'\rangle = \delta_{jj'}\delta_{mm'}\delta_{q_1q_1'}\delta_{q_2q_2'}. \tag{13}$$

The function $\langle w|w'\rangle$ is the Bergman kernel function in $\mathcal{H}_n(\pi)$, which follows from the formula (9)

$$\langle w'|f\rangle = \int_{\pi} d^8w[(\mathrm{Im}w)^2]^{n-4}\langle w'|w\rangle\langle w|f\rangle. \tag{14}$$

The inverse theorem is also true. If we have a discrete series representation T of a group G in the Hilbert space of functions f(C) defined on one of the hermitian symmetric spaces M [6]

$$T(g)f(C) = \mu(C,g)f(Cg), \qquad C \in M = G/H, \quad g \in G, \tag{15}$$

then the kernel function $K_C(C')$ (which exists in this case [6]) is an eigenstate of $\hat{C}^\dagger$, where $\hat{C}$ is an operator of multiplication by the variable C. So, we have

$$\hat{C}^\dagger K_{C'}(C) = \bar{C}' K_{C'}(C) \tag{16}$$

and

$$T(g)\hat{C}^\dagger T^{-1}(g) = (\hat{C}g)^\dagger . \tag{17}$$

Eq. (16) can be easily obtained from the definition of the kernel function, which provides us with the formula

$$\langle K_{C'}(C) | Cf(C) \rangle = C'f(C') = C' \langle K_{C'} | f \rangle \tag{18}$$

at the same time

$$\langle K_{C'} | Cf \rangle = \langle C^\dagger K_{C'} | f \rangle \text{ for arbitrary } | f \rangle . \tag{19}$$

Eqs. (18) and (19) together give eq. (16), and formula (17) is now obvious.

We should further explain the physical interpretation of the operator $\hat{w}_\mu$. It is non-hermitian and its hermitian part has no self-adjoint extension. It is already a consequence of the von Neumann theorem [4] that *no* self-adjoint position operator, covariant under the Poincaré group, exists. Invariance under rotations determines uniquely (for irreducible representations of the Poincaré group) the Newton-Wigner position operator, which is not relativistically covariant. However, recent investigations on the foundations of quantum mechanics (see Srinivas [10] for a review) show that there is no reason to insist on self-adjoint operators. Only probability measure describing the physical experiments is essential for physical interpretation. In our case this probability measure can be obtained from the resolution of identity (9)

$$P(\Delta) = \int_\Delta d^4x \int_{\substack{y^2 \geq 0 \\ y_0 \geq 0}} d^4y (y^2)^{n-4} | x+iy \rangle \langle x+iy | . \tag{20}$$

$P(\Delta)$ has properties of the probability operator measure [11] (semi-spectral measure)

1. $P(M^4) = 1$, M^4 - the Minkowski space

2. $P(\bigcup_i \Delta_i) = \sum_i P(\Delta_i)$ if $\Delta_i \cap \Delta_j = 0$

3. $P(\Delta) = P(\Delta)^\dagger$. (21)

It is not a spectral measure, because $P(\Delta)P(\Delta') \neq P(\Delta \cap \Delta')$. Nevertheless it determines uniquely a decomposition of the real part X_μ of $\hat{w}_\mu$

$$X_\mu = \int x_\mu dP(x) \tag{22}$$

here x_μ is the mean value of X_μ in the state $|w\rangle$

$$x_\mu = \langle w|w\rangle^{-1} \langle w|X_\mu|w\rangle . \tag{23}$$

By means of the probability measure (20) we can compute the dispersion of position X_μ in the state $|w\rangle$ [11]

$$\sigma_w^2(X_\mu) = \langle w|w\rangle^{-1}[\int x_\mu^2 \langle w|dP(x)|w\rangle - (\int x_\mu \langle w|dP(x)|w\rangle)^2]. \tag{24}$$

The computations show that it is proportional to $\mathrm{Im}w_\rho \mathrm{Im}w^\rho$, with a coefficient depending on μ and n [7]. This shows that the imaginary part of w_μ gives dispersion of position in the state $|w\rangle$. It can be computed that $\sigma_w^2(P_\mu)$ is proportional to $(\mathrm{Im}w_\rho \mathrm{Im}w^\rho)^{-1}$. In this way we obtain the result

$$\sigma_w^2(X_0)\sigma_w^2(P_0) = \frac{n^2}{(n-1)(n-3)} \hbar^2 \quad \text{if} \quad \mathrm{Im}w_0 \neq 0,\ \mathrm{Im}w_i = 0,\ n > 3$$

$$\sigma_w^2(X_i)\sigma_w^2(P_i) = \frac{3n(3n-4)}{(n-1)(n-3)} \hbar^2 . \tag{25}$$

Here, we have written explicitly the dependence of the r.h.s. on $\hbar$ (Planck constant), which in previous formulas was assumed to be equal to 1.

The expression (25) shows that the states $|w\rangle$ are indeed coherent. We did not get exactly $\frac{1}{4}h^2$ as in the usual Heisenberg relations [12], but it can easily be seen that in the four-dimensional case there do not exist *covariant* coherent states, which minimize the product of dispersions to the value $\frac{1}{4}h^2$. On the formal level similarity of the states $|w\rangle$ with the usual coherent states [12] defined by the equation

$$\frac{1}{\sqrt{2}}(X+iP)|z\rangle = z|z\rangle \tag{26}$$

follows by comparison with eq. (4), where the mass $\frac{1}{n-2}P^2$ is not a constant but changes in a continuous way from zero to infinity.

Finally, we will compare the group theoretical structure of the coherent states $|w\rangle$ with the usual ones, which are defined on the nilpotent group N of canonical commutation relations

$$[X,P] = -i. \tag{27}$$

The group multiplication law for N can be written in the form

$$\begin{aligned}&\exp[it]\exp[i\alpha X]\exp[i\beta P]\exp[it']\exp[i\alpha' X]\\&\exp[i\beta' P] = \exp[i(t+t'-\alpha'\beta)]\exp[i(\alpha+\alpha')X]\exp[i(\beta+\beta')P]\\&t,\ \alpha,\ \beta\ \text{- real numbers.}\end{aligned} \tag{28}$$

Then the definition (16) of a coherent state is equivalent to (26) with

$$M = N/T \quad \text{where} \quad T = \exp[it]. \tag{29}$$

This definition of a coherent state can be further compared with one given by Perelomov [13]. Let $|\psi_0\rangle$ be a vector in the representation space $\mathcal{H}$ of a unitary representation U(G) of a group G. Let H be a stationary subgroup of the vector $|\psi_0\rangle$, i.e.

$$U(h)|\psi_0\rangle = e^{ir(h)}|\psi_0\rangle \qquad h \in H;$$

then the coherent state as defined by Perelomov is given by

$$|z\rangle = U(g)|\psi_0\rangle,$$

where

$$z \in G/H. \tag{30}$$

It can be seen that definitions (16) and (30) coincide for stationary subgroup H equal to maximal compact subgroup and G/H being a hermitian symmetric space.

We did not consider here the connection between the position operator $\hat{w}_\mu$ and localization of particles. In order to discuss this problem the conformal symmetry should be broken. We have considered the conformal symmetry breaking for free field theories [14]. We hope that still more interesting examples of broken conformal symmetry can be obtained giving e.g. the solution of the localization problem for extended and unstable particles.

REFERENCES AND FOOTNOTES

1. Conformal symmetry can be considered only as an approximate symmetry, which probably is fulfilled at high energies or at small distances, c.f. K. WILSON, Phys. Rev. 179, 1599 (1969).
2. This operator is a four-dimensional analogue of the two-dimensional operator introduced by F. GÜRSEY and S. ORFANIDIS, Phys. Rev. D7, 2414 (1973).
3. T. YAO, Journ. Math. Phys. 9, 1615 (1968); 12, 315 (1971).
4. It can be proven that Poincaré covariant position operator cannot be self-adjoint, c.f. J. VON NEUMANN, Ann. Math. 104, 570 (1931); H.J. BORCHERS, Comm. Math. Phys. 4, 315 (1967); G.C. HEGERFELDT, Phys. Rev. D10, 3320 (1974).
5. R.F. STREATER and A.S. WIGHTMAN, PCT, Spin and Statistics and All That, Benjamin, 1964, New York.
6. S. HELGASON, Differential Geometry and Symmetric Spaces, Academic Press, 1962, New York.
7. Z. HABA, Nuovo Cimento, 30A, 567 (1975).

8. This realization was obtained first by M.L. GRAEV, Dokl. Akad. Nauk USSR, 98, 517 (1954); Amer. Math. Soc. Transl., 66, 1 (1968), application to physics was studied by W. RÜHL, Comm. Math. Phys. 27, 53 (1972); 30, 287 (1973).

9. T.D. NEWTON and E.P. WIGNER, Rev. Mod. Phys. 21, 400 (1949). A.S. WIGHTMAN, Rev. Mod. Phys. 34, 845 (1962).

10. M.D. SRINIVAS, Journ. Math. Phys. 16, 1672 (1975).

11. Z. HABA and A.A. NOWICKI, Phys. Rev. D13, 523 (1976).

12. J.R. KLAUDER and E.C.G. SUDARSHAN, Fundamentals of Quantum Optics, Benjamin, 1968, New York.

13. A.M. PERELOMOV, Comm. Math. Phys. 26, 222 (1972); 44, 197 (1975).

14. Z. HABA, Nuovo Cimento, 32A, 174 (1976).

GAUGE THEORY OF THE CONFORMAL GROUP

J.P. Harnad and R.B. Pettitt

1. INTRODUCTION

The "gauge principle" has been the basis for generating unified field theories for elementary particle interactions[1] and may also be used to arrive at a formulation of General Relativity (G.R.)[2,3,4] (and variations thereof) in terms of elementary matter and gauge fields. The formal structure of all such theories may be elegantly described in the language of fibre bundles[5,6], such an approach serving both to clarify the essential differences between the "intrinsic" and the "space-time" based gauge field theories and also to provide a framework for extending the underlying geometrical structure of G.R.

The approach of elementary particle theorists has been to start with a Lagrangian, generating a set of field equations which are invariant under a group of internal-symmetry transformations of the rigid variety (i.e., no dependence upon the space-time point). Then, through a minimal replacement of ordinary by covariant derivatives (defined by the introduction of suitably transforming gauge fields) the new equations are made invariant under the non-rigid actions of the group (gauge transformations of the second type). When the same approach is applied to a non-intrinsic symmetry group (e.g. Lorentz transformations), which act upon the space-time points themselves, the notion of non-rigid actions becomes somewhat more subtle, and one is obliged to introduce independent (though differentiably related) reference frames attached to the various space-time points. The gauge fields are interpreted as connection components relating frames at different points and the underlying geometry, if these gauge fields are non-integrable, is non-Euclidean. Restricting to orthonormal frames only, upon which the Lorentz transformations act transitively, and which suffice to describe a Riemannian structure, one is led naturally to the tetrad formulation of G.R.

The notion of "independent frames" at different points is most naturally described in terms of fibre bundles and the construction of alternatives or extensions of G.R. based upon larger gauge groups becomes very tempting. In particular, the Einstein-Cartan theory with spin and torsion is very elegantly described as a gauge theory of the inhomogeneous Lorentz Group(3) and the Weyl unified theory of electromagnetism, and gravitation is the gauge theory of the Weyl group(7). If one goes to the conformal group, which involves non-linear transformations of space-time, the notion of frames must be generalized so the group can act freely (without fixed points) on the set of frames at each point, and one is led quite naturally to the "bundle of second-order frames"(8). Besides the usual arguments for studying conformal invariant field theories(9) one can motivate such a study from a gauge viewpoint by the fact that the conformal group is the smallest simple group containing the Weyl group - the (semi-) simplicity of the gauge group being an essential requirement for having a truly unified structure.

In the following sections we first outline the geometric structure underlying gauge theories in the language of fibre bundles. The particular case of the bundle $P^2(M)$ of second-order frames is then discussed, together with the sub-bundle of conformal frames. The transformation properties for the connection (gauge fields) and curvature forms are given. Next, the identities and conservation laws following from invariance under changes of local section in the bundle of conformal frames are derived. Since the structure group is not the entire conformal group, but the 11-parameter subgroup consisting of Lorentz, scaling and special conformal transformations, a separate discussion is given concerning the gauge theory related to the full conformal group.

2. FIBRE BUNDLES

2.1. Principal G-Bundles and Connections

Detailed definitions of the following structures may be found in standard textbooks[(10)]; therefore, only those properties relevant to the subsequent development will be mentioned here. The diagram below illustrates the structure of a *principal fibre bundle*, which generalizes the notion of independent frames at different points in a manifold:

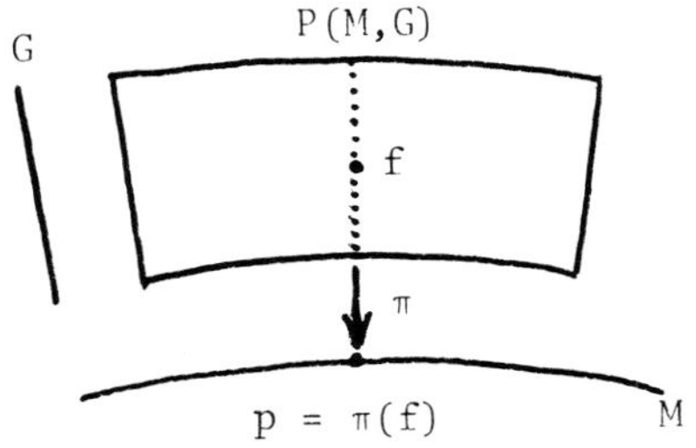

The bundle P(M,G) with base manifold M and structure group G is a differential manifold with the properties:

(i) G acts to the right on M ($f \to fg$, $f \in P$, $g \in G$) differentiably with no fixed points.

(ii) $M = P/\sim$ where $\sim$ is the equivalence relation defined by right translation R_g, $g \in G$.

(iii) G is isomorphic to the fibre $\pi^{-1}(p)$ over any $p \in M$ (i.e., $G \overset{\phi}{\Longrightarrow} \pi^{-1}(p)$, $\ni \phi^{-1}(fg) = \phi^{-1}(f)g$).

(This isomorphism varies differentiably from one point in M to another, so that locally, P is a direct product of G with open regions in M.)

The right action of G allows one to make a natural isomorphism between elements A of the Lie algebra $\mathcal{G}$ (left invariant vector fields over G) and the "fundamental vector fields" A* tangent to the fibres of P. Such vectors are called "vertical" and a connection for P(M,G) is a definition of a complimentary "horizontal" subspace Q_f of the tangent space at each point $f \in P$ such that

a) $$\tilde{X} = X_v + X_h \tag{2.1}$$

gives a unique decomposition for any vector $\tilde{X} \in T_f$ into a vertical part X_v tangent to the fibre and a horizontal part $X_h \in Q_f$.

b) The horizontal subspaces at different points are related by right translation:

$$Q_{fg} = R_{g*}Q_f. \tag{2.2}$$

The connection form ω is a $\mathcal{G}$-valued 1-form on P defined uniquely by:

$$\omega(A^*) = A \tag{2.3b}$$

$$\omega(X_h) = 0. \tag{2.3a}$$

Under right translation, it transforms under the adjoint representation:

$$R_g^*\omega = \mathrm{ad}(g^{-1})\omega. \tag{2.4}$$

The curvature Ω is a 2-form defined as the exterior covariant derivative of ω:

$$\Omega \equiv D\omega \tag{2.5}$$

(where $D\varphi(\tilde{X}_1,\ldots,\tilde{X}_{p+1}) = d\varphi(X_h,\ldots,X_{(p+1)h})$ for any p-form, $d\varphi$ being the ordinary exterior derivative). It follows that the curvature acts as follows

$$\Omega(\tilde{X},\tilde{Y}) = d\omega(\tilde{X},\tilde{Y}) + \tfrac{1}{2}[\omega(\tilde{X}),\omega(\tilde{Y})] \tag{2.6}$$

(Cartan Structure Equation)

2.2. Associated Vector Bundles

A vector space V upon which G acts through a linear representation $\rho(G)$ (to the left) serves to define the vector bundle E(M,G,V) associated with P(M,G). A point in E is defined by the equivalence class $\{[p,v] = [pg,\rho(g^{-1})v],\ p \in P,\ v \in V,\ g \in G\}$ of points in $P\times V$. Together with the projection $\bar{\pi}(p,v) \equiv \pi(p)$ this defines a vector bundle over M with fibre-type V. A ρ-field is a cross-section $\psi \in \Gamma(E)$ of the bundle E(M,G,V).

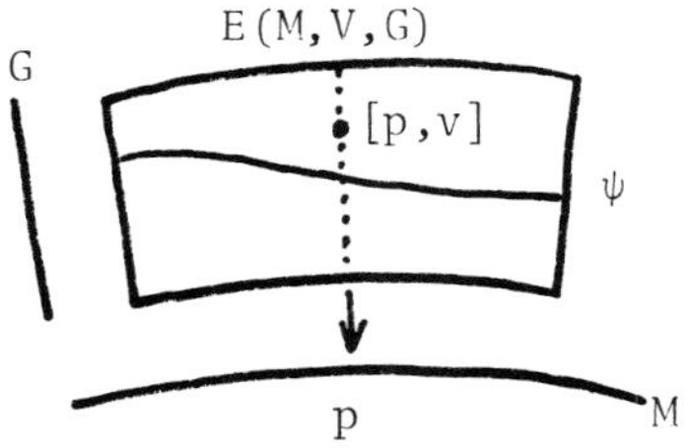

An equivalent way of defining ψ is as a ρ-invariant V-valued distribution on P.

$$\tilde{\psi}(pg) = \rho(g^{-1})\tilde{\psi}(p) \tag{2.7}$$

the two definitions being related by:

$$\psi(\pi(p)) = [p,\tilde{\psi}(p)]. \tag{2.8}$$

The covariant derivative of ψ is defined as:

$$\widetilde{\nabla_{\tilde{X}}\psi} = X_h\tilde{\psi}, \tag{2.9}$$

this having the properties:

$$\widetilde{\nabla_{\tilde{X}}\psi}(pg) = \rho(g^{-1})\widetilde{\nabla_{\tilde{X}}\psi}(p) \tag{2.10}$$

together with linearity and the satisfaction of Leibnitz' rule.

2.3. Local Sections

Finally, making a choice of local sections σ_A in P(M,G) allows us to define local fields, curvature and connection forms on the base manifold M by:

$$\psi_A \equiv \tilde{\psi} \circ \sigma_A \tag{2.11}$$

$$\nabla_X\psi_A \equiv \nabla_{\tilde{X}}\tilde{\psi} \circ \sigma_A = X\psi_A + \rho(\omega_A(X))\psi_A \tag{2.12}$$

$$(X = \pi_*(\tilde{X}) \in T_{\pi(p)})$$

$$\omega_A \equiv \omega \circ \sigma_{A*} \tag{2.13}$$

$$\Omega_A \equiv \Omega \circ \sigma_{A*} = d\omega_A + \tfrac{1}{2}[\omega_A,\omega_A]. \tag{2.14}$$

Here ψ_A is understood as a V-valued local distribution on $U_A \subset M$, where U_A is the open region where the section $\sigma_A : U_A \to p$ is defined while ω_A and Ω_A are G-valued local 1 and 2-forms on U_A.

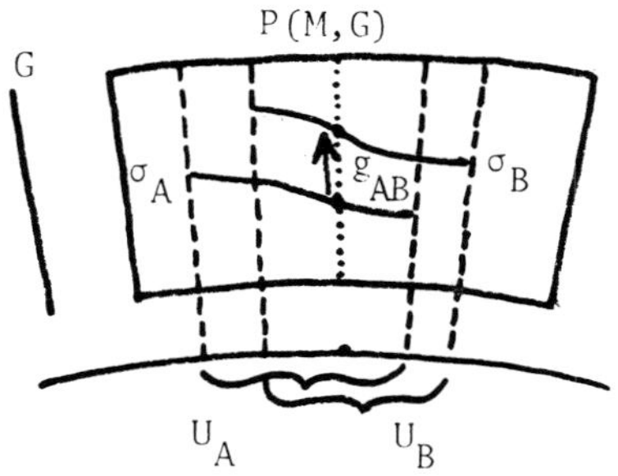

A change in local section $\sigma_A \to \sigma_B$ with transition function $g_{AB} : U_A \cap U_B \to G$ has the following effect on ψ_A, ω_A, Ω_A

$$\psi_B = \rho(g_{AB}^{-1})\psi_A \tag{2.15}$$

$$\nabla_X\psi_B = \rho(g_{AB}^{-1})\nabla_X\psi_A \tag{2.16}$$

$$\omega_B = \mathrm{ad}g_{AB}^{-1} + g_{AB}^{-1}dg_{AB} \tag{2.17}$$

$$\Omega_B = \mathrm{ad}g_{AB}^{-1}\Omega_A . \tag{2.18}$$

Choosing a local coordinate basis $\{\partial/\partial x^\mu\}$, $\{dx^\mu\}$ for vectors and 1-forms on M and $\{t_a\}$ for G, these quantities may be expressed as:

$$\omega_A = \omega_\mu^a t_a dx^\mu \tag{2.19}$$

$$\Omega_A = \tfrac{1}{2}R_{\mu\nu}^a t_a dx^\mu \wedge dx^\nu \tag{2.20}$$

$$\nabla_\mu\psi_A = \partial_\mu\psi_A + \omega_\mu^a T_a\psi_A \tag{2.21}$$

$$(T_a \equiv \rho(t_a))$$

$$R_{\mu\nu}^a = \omega_{\mu,\nu}^a - \omega_{\nu,\mu}^a + \tfrac{1}{2}f_{bc}^a\omega_\mu^b\omega_\nu^c \tag{2.22}$$

$$\psi_{;i} \equiv h^{\mu}_{i} \nabla_{\mu} \psi_A \tag{2.23}$$

$$(\text{where } h_i \equiv h^{\mu}_{i} \frac{\partial}{\partial x^{\mu}} = \text{non-holonomic basis} \tag{2.24}$$

$$b^i = b^i_{\mu} dx^{\mu} = \text{dual basis}). \tag{2.24'}$$

3. GAUGE INVARIANT FIELD THEORIES

3.1. Identities and Conservation Laws

Now, by a standard method of minimal replacement[6], one can go from a Lagrangian density $L(\psi, \partial_{\mu}\psi)$ (henceforth the subscript on ψ_A will be suppressed, a particular choice of local section σ_A being implicit) to $\mathcal{L}(\psi, \psi_{;i})$ where the second Lagrangian is an invariant for arbitrary (non-rigid) group transformations provided L is invariant under rigid ones. The consequences of such invariance may be expressed by the following set of identities and conservation laws.

(i) Relationship between dynamically and kinematically defined energy-momentum tensors:

$$\mathcal{T}^{\mu}_{\nu} = t^{\mu}_{\nu} - S^{\mu}_{a} \omega^{a}_{\nu} \tag{3.1}$$

where

$$t^{\mu}_{\nu} \equiv h^{\mu}_{i} \frac{\partial h}{\partial h^{\mu}_{i}} \tag{3.2}$$

is a generalization of the dynamical energy-momentum density.

$$\mathcal{T}^{\mu}_{\nu} \equiv -\delta^{\mu}_{\nu}\mathcal{L} + \frac{\partial \mathcal{L}}{\partial(\partial_{\mu}\psi)} \partial_{\nu}\psi \tag{3.3}$$

is the canonical (kinematically defined) energy-momentum tensor density and

$$S^{\mu}_{a} = \frac{\partial \mathcal{L}}{\partial \omega^{a}_{\mu}} \tag{3.4}$$

is the intrinsic current density coupling to the gauge field ω^a_μ.

(ii) Relationship between dynamically and kinematically defined current densities:

$$S^\mu_a = \frac{\partial\mathcal{L}}{\partial\partial_\mu\psi} T_a\psi \tag{3.5}$$

(intrinsic current associated with the one-parameter subgroup generated by t_a).

(iii) Covariant conservation laws for currents:

$$\nabla_\mu S^\mu_a = 0 \qquad \text{(non-linear part of } G) \tag{3.6}$$

$$\nabla_\mu S^{i\mu}_{\cdot j} = t^\mu_\nu b^i_\mu h^\nu_{\cdot j} \equiv t^i_j \qquad \text{(linear part of } G) \tag{3.7}$$

$$\begin{aligned}(\text{where } \nabla_\mu S^\mu_a &= \partial_\mu S^\mu_a - f^c_{ab}\omega^b_\mu S^\mu_c \\ &= \text{covariant divergence of } S^\mu_a).\end{aligned} \tag{3.8}$$

(iv) Covariant Divergence of Energy-Momentum Tensor:

$$\partial_\mu t^\mu_\nu + \Gamma^\alpha_{\mu\nu} t^\mu_\alpha = R^a_{\mu\nu} S^\mu_a \tag{3.9}$$

where

$$\begin{aligned}\Gamma^\alpha_{\mu\nu} &= (b^i_{\mu,\nu} + \omega^i_{\cdot j\nu} b^j_\mu) h^\alpha_{\cdot i} \\ &= -(h^\alpha_{\cdot i,\nu} - \omega^j_{\cdot i\nu} h^\alpha_{\cdot j}) b^i_\mu .\end{aligned} \tag{3.10}$$

Equation (3.9) has the interpretation that a coupling between the curvature and the intrinsic currents gives rise to a force density causing deviations from geodesic motion.

3.2. A Lagrangian Model

For groups G with conformal linear parts[(6)] the following invariant density may be used to describe the pure gauge field contribution to the Lagrangian.

$$\mathcal{L}_G = -\frac{N}{4}\, bR^a_{\mu\nu}R^b_{\sigma\tau}g^{\mu\sigma}g^{\nu\tau}g_{ab} \qquad (3.11)$$

where

$$g^{\mu\sigma} \equiv h^\mu_i h^\sigma_j \eta^{ij} \qquad (\eta = \text{Minkowski metric}) \qquad (3.12)$$

and

$$g_{ab} = \text{group metric} \quad (\text{Killing-Cartan}).$$

Adding this to the matter field Lagrangian $\mathcal{L}(\psi,\nabla_\mu\psi)$ and applying the variational procedure (treating ω^a_μ, h^μ_i and ψ as independent) gives rise to the following field equations:

$$N\nabla_\mu[bR^{\mu\nu}_a] = S^\nu_a \qquad (3.13)$$

$$Nb[R^a_{\nu\sigma}R^{\mu\sigma}_a - \tfrac{1}{4}S^\mu_\nu R^a_{\sigma\tau}R^{\sigma\tau}_a] = t^\mu_\nu \qquad (3.14)$$

$$\mathcal{L}[\psi] = 0. \qquad (3.15)$$

(where $b = \det\{b^i_\mu\}$).

Although (3.13) (together with the Bianchi-Identities) form a natural analog of Maxwell's equations and (3.14) is readily interpretable as relating matter energy-momentum density to that of the gauge fields, these field equations are quite dissimilar to those of Einstein, making the correspondence with G.R. difficult to see. Nevertheless (3.11) does present a concrete example of how gauge invariance might be implemented for groups of this type.

4. SECOND ORDER FRAME BUNDLE

To give an explicit realization of a principal bundle with the conformal group (minus translations) as structure group, we turn to the bundle of second order frames[8]. In analogy with first order frames, which consist of a basis of first order differential operators, second order frames may be considered as a complete set of first and second order differential operators. In a coordinate basis this may be denoted $\{\ell_\mu, \ell_{\mu\nu}\}$ with:

$$\ell_\mu = \frac{\partial}{\partial x^\mu} \qquad \ell_{\mu\nu} = \frac{1}{2}\frac{\partial^2}{\partial x^\mu \partial x^\nu} \tag{4.1}$$

($\{x^\mu\}$ here form a local coordinate system on an open set U of the manifold).

The changes in such a basis induced by a change of local coordinates $x^\mu \to x'^\mu$ depend upon the (inverse) Jacobean $\bar{J}^\sigma_\mu$ and its first derivative $\bar{J}^\sigma_{\mu\nu} = \bar{J}^\sigma_{\nu\mu}$ in a way that leads to identification of a group composition law following from successive transformations:

$$\{\bar{J}^\sigma_\mu, \bar{J}^\sigma_{\mu\nu}\}\cdot\{\bar{J}'^\sigma_\mu, \bar{J}'^\sigma_{\mu\nu}\} = \{\bar{J}^\sigma_\lambda \bar{J}'^\lambda_\mu, \bar{J}^\sigma_\lambda \bar{J}'^\lambda_{\mu\nu} + \bar{J}^\sigma_{\lambda\kappa}\bar{J}'^\lambda_\mu \bar{J}'^\kappa_\nu\}. \tag{4.2}$$

The group G^2 consisting of elements of the form $\{a^i_j, a^i_{jk}\}$ (such that a^i_j is non-singular and $a^i_{jk} = a^i_{kj}$) with (4.2) as composition law is the structure group of the bundle $P^2(M)$ of second-order frames. An arbitrary element of this bundle, over a point with coordinates $\{x^\mu\}$ is defined by the quantities $\{h_i, h_{ij}\}$ where

$$\left.\begin{aligned} h_i &= h^\mu_i \ell_\mu \\ h_{ij} = h_{ji} &= h^\mu_{ij}\ell_\mu + h^\mu_i h^\nu_j \ell_{\mu\nu} \end{aligned}\right\}. \tag{4.3}$$

This may be considered as defining a local section, suitable transformation properties being given to $\{h^\mu_i, h^\mu_{ij}\}$ to insure the invariance of (4.3). The action of an arbitrary transition function g_{HK} upon $\{h_i, h_{ij}\}$ gives rise to a new local section $\{k_i, k_{ij}\}$ given by:

$$\{k_i, k_{ij}\} = \{h_i, h_{ij}\}\cdot g_{HK} = \{h_\ell a^\ell_i,\ h_\ell a^\ell_{ij} + h_{\ell m} a^\ell_i a^m_j\} \tag{4.4}$$

where

$$g_{HK} = \{a^i_j, a^i_{jk}\}. \tag{4.5}$$

Eq. (4.3) can be understood as a particular case of (4.4) when the transition is from a coordinate (holonomic) frame to an arbitrary one through the transition function $\{h^\mu_i, h^\mu_{ij}\}$.

A basis for the algebra G^2 corresponding to G^2 is given by the elements:

$$\sigma^i_j \equiv \left.\frac{\partial g}{\partial a^j_i}\right|_{g=I} \qquad \tau^{ij}_k \equiv \left.\frac{\partial g}{\partial a^k_{ij}}\right|_{g=I} \tag{4.6}$$

in terms of which the commutators are:

$$\left.\begin{aligned} [\sigma^i_j, \sigma^k_\ell] &= \delta^i_k \sigma^k_j - \delta^k_j \sigma^i_\ell \\ [\sigma^i_j, \tau^{k\ell}_m] &= \delta^i_m \tau^{k\ell}_j - \delta^k_j \tau^{i\ell}_m - \delta^\ell_j \tau^{ki}_m \\ [\tau^{ij}_k, \tau^{\ell m}_m] &= 0 \end{aligned}\right\}. \tag{4.7}$$

Relative to this basis, the local connection form and curvature form may be expressed:

$$\omega_H = \omega^i_j \sigma^j_i + \omega^i_{jk} \tau^{jk}_i \tag{4.8}$$

$$\Omega_H = \tfrac{1}{2}[R^i_j \sigma^j_i + R^i_{jk} \tau^{jk}_i] \tag{4.9}$$

where

$$\begin{aligned} R^i_j &= 2[d\omega^i_j + \omega^i_r \wedge \omega^r_j] \\ R^i_{jk} &= 2[d\omega^i_{jk} + \omega^i_r \wedge \omega^r_{jk} - \omega^r_j \wedge \omega^i_{jk} - \omega^r_k \wedge \omega^i_{jr}]. \end{aligned} \tag{4.10}$$

These have the following transformation properties under a change of section induced by g_{HK}.

(Denoting by $\bar{a}^i_j$ the inverse of a^i_j)

$$\begin{aligned} \omega'^{\,i}_{j} &= \bar{a}^{i}_{\ell}\omega^{\ell}_{k}a^{k}_{j} + \bar{a}^{i}_{k}da^{k}_{j} \\ \omega'^{\,i}_{jk} &= \bar{a}^{i}_{\ell}[-a^{\ell}_{mk}\omega'^{\,m}_{j}-a^{\ell}_{jm}\omega'^{\,m}_{k}+\omega^{\ell}_{mn}a^{m}_{j}a^{n}_{k}+\omega^{\ell}_{m}a^{m}_{jk}+da^{\ell}_{jk}] \end{aligned} \tag{4.11}$$

$$\begin{aligned} R'^{\,i}_{j} &= \bar{a}^{i}_{\ell}R^{\ell}_{k}a^{k}_{j} \\ R'^{\,i}_{jk} &= \bar{a}^{i}_{\ell}[-a^{\ell}_{mk}R'^{\,m}_{j}-a^{\ell}_{jm}R'^{\,m}_{k}+\omega^{\ell}_{mn}R^{n}_{j}R^{n}_{k}+R^{\ell}_{m}a^{m}_{jk}]. \end{aligned} \tag{4.12}$$

Eqs. (4.11) and (4.12) may be used to obtain an interpretation of parallel transport of second order frames, when written for the case (4.3) of transition from a coordinate frame to an arbitrary one. Denoting by $\{\Gamma^{\mu}_{\sigma},\Lambda^{\mu}_{\sigma\tau}\}$ and $\{R^{\mu}_{\sigma},R^{\mu}_{\sigma\tau}\}$ the components of the local connection and curvature forms, respectively, relative to the coordinate basis, we have:

$$\begin{aligned} dh^{\mu}_{i} + \Gamma^{\mu}_{\sigma}h^{\sigma}_{i} &= \omega^{j}_{i}h^{\mu}_{j} \\ dh^{\mu}_{ij} + \Gamma^{\mu}_{\sigma}h^{\sigma}_{ij} + \Lambda^{\mu}_{\sigma\tau}h^{\sigma}_{i}h^{\tau}_{j} &= \omega^{k}_{ij}h^{\mu}_{k} + \omega^{k}_{i}h^{\mu}_{kj} + \omega^{k}_{j}h^{\mu}_{ik} \end{aligned} \tag{4.13}$$

$$\begin{aligned} R^{\mu}_{\sigma}h^{\sigma}_{i} &= R^{j}_{i}h^{\mu}_{j} \\ R^{\mu}_{\sigma}h^{\sigma}_{ij} + R^{\mu}_{\sigma\tau}h^{\sigma}_{i}h^{\tau}_{j} &= R^{k}_{ij}h^{\mu}_{k} + R^{k}_{i}h^{\mu}_{ij} + R^{k}_{i}h^{\mu}_{ik}. \end{aligned} \tag{4.14}$$

Applying both sides of (4.13) to the infinitesimal vector relating the point P with coordinates x^{μ} to that with coordinates $x^{\mu}+\delta x^{\mu}$, we obtain:

$$\begin{aligned} h^{\mu}_{i}(x) - \Gamma^{\mu}_{\sigma\nu}\delta x^{\nu}h^{\sigma}_{i} &= h^{\mu}_{j}(x+\delta x)\,[\delta^{j}_{i}-\omega^{j}_{i\nu}\delta x^{\nu}] \\ h^{\mu}_{ij}(x) - \Gamma^{\mu}_{\sigma\nu}\delta x^{\nu}h^{\sigma}_{ij} - \Lambda^{\mu}_{\sigma\tau\nu}\delta x^{\nu}h^{\sigma}_{i}h^{\tau}_{j} & \\ = h^{\mu}_{\ell m}(\delta^{\ell}_{i}-\omega^{\ell}_{i\mu}\delta x^{\mu})(\delta^{m}_{j}-\omega^{m}_{j\nu}\delta x^{\nu}) &- h^{\mu}_{k}\omega^{k}_{ij\mu}\delta x^{\mu}. \end{aligned} \tag{4.15}$$

Here the left-hand side must clearly be interpreted as the components of the parallel transferred frame $\{h_{i},h_{ij}\}$ at P and the right-hand side as the transform of the frame at the new point by the infinitesimal group element $\{\delta^i_j - \omega^i_{j\mu}\delta x^\mu,\ -\omega^k_{ij\mu}\delta x^\mu\}$ in accord with the usual notion of the connection as defining infinitesimal parallel transport. Similarly, applying (4.14) to the infinitesimal surface tensor with components $\delta\sigma^{\alpha\beta}$, we have:

$$h'^{\mu}_i \equiv h^\mu_i - R^\mu_{\sigma\alpha\beta}\delta\sigma^{\alpha\beta}h^\sigma_i = (\delta^j_i - R^j_{i\alpha\beta}\delta\sigma^{\alpha\beta})h^\mu_j \tag{4.16a}$$

$$\begin{aligned} h'^{\mu}_{ij} &\equiv h^\mu_{ij} - R^\mu_{\sigma\alpha\beta}\delta\sigma^{\alpha\beta}h^\sigma_{ij} - R^\mu_{\sigma\tau\alpha\beta}\delta\sigma^{\alpha\beta}h^\sigma_i h^\tau_j \\ &= (\delta^m_i - R^m_{i\alpha\beta}\delta\sigma^{\alpha\beta})(\delta^n_j - R^n_{j\delta\gamma}\delta\sigma^{\delta\gamma})h^\mu_{mn} - R^m_{ij\alpha\beta}\delta\sigma^{\alpha\beta}h^\mu_m. \end{aligned} \tag{4.16b}$$

Here $\{h'^{\mu}_i, h'^{\mu}_{ij}\}$ are the components of the second order frame parallel transferred around the perimeter of $\delta\sigma^{\alpha\beta}$, and the RHS of (4.16) relates this to the original frame through the infinitesimal group element $\{\delta^i_j - R^i_{j\alpha\beta}\delta\sigma^{\alpha\beta},\ -R^i_{jk\alpha\beta}\delta\sigma^{\alpha\beta}\}$. This is in accord with the usual interpretation of curvature as generator of the infinitesimal holonomy group[10].

5. CONFORMAL FRAMES AND CONNECTIONS

We now consider an eleven parameter subgroup $G_0 \subset G^2$ defined by elements of the form:

$$G_0 = \{a^i_j, a^i_j c_k + a^i_k c_j - \eta_{jk} a^i_\ell \eta^{\ell m} c_m\} \equiv (a^i_\ell, c_m) \tag{5.1}$$

where

$$a^i_j = a\Lambda^i_j \qquad \Lambda\eta\Lambda^\top = \eta. \tag{5.2}$$

These satisfy the composition law:

$$(a^i_j, c_k)\cdot(a'^i_j, c'_k) = (a^i_\ell a'^\ell_j, c'_k + c_m a'^m_k). \tag{5.3}$$

A basis of infinitesimal generators is given by:

$$d \equiv \frac{\partial g}{\partial a}\bigg|_{g=I}$$
$$m^{ij} \equiv \frac{\partial g}{\partial \Lambda_{ij}}\bigg|_{g=I} \tag{5.4}$$
$$k^{\ell} \equiv \frac{\partial g}{\partial c_{\ell}}\bigg|_{g=I}$$

which gives rise to the commutator algebra

$$[m^{ij},m^{k\ell}] = \eta^{ik}m^{j\ell} + \eta^{j\ell}m^{ik} - \eta^{jk}m^{i\ell} - \eta^{i\ell}m^{jk}$$
$$[k^{i},m^{j\ell}] = \eta^{i\ell}k^{j} - \eta^{ij}k^{\ell}$$
$$[k^{i},d] = k^{i} \tag{5.5}$$
$$[k^{i},k^{j}] = 0.$$

Eqs. (5.3) and (5.5) may be recognized, respectively, as the composition law and Lie algebra for the subgroup of the conformal group consisting of Lorentz transformations, dilations and special conformal transformations, the generators for the subgroups being m^{ij}, d and k^{i} respectively. This identification can be seen in another way by considering the bundle of "conformal frames" $P(M,G_0)$ with G_0 as structure group, which is obtained as a sub-bundle of $P^2(M)$ consisting of frames related to each other by transformations in G_0 only. (The existence of such a sub-bundle is equivalent to that of a "pseudo-metric" tensor field g with the identification of conformal frames as those for which the linear parts satisfy:

$$g(h_i,h_j) = \eta_{ij}h \tag{5.6}$$

(h being any scalar quantity).

Now, if $\rho(G_0) : V \to V$ is a representation of G_0 on the vector space V (and $\rho(\mathcal{G}_0)$ the corresponding representation of the Lie algebra $\mathcal{G}_0$), we may define a field ψ as a mapping $U_A \to V$ of the open region U_A in the space-time manifold into V determined by a cross-section of the associated vector bundle $E(M,G_0,V)$ and the local section $\sigma_A : U_A \to P$ of the principal bundle. A change of local section with transition element (a^i_j, c_k) has the following effect on the local field:

$$\psi' = \rho(a^i_j, c_\ell)^{-1}\psi \tag{5.7}$$

or, in infinitesimal form:

$$\psi' = \psi + i(\chi^{ij}\Sigma_{ij} + \chi\Delta + \chi_i\kappa^i)\psi \tag{5.8}$$

where

$$i\Sigma_{ij} = \rho(m_{ij}) \qquad i\Delta = \rho(d) \qquad i\kappa^i = \rho(k^i) \tag{5.9}$$

(raising and lowering of Roman indices defined by the Minkowski metric η^{ij}). In particular, for a coordinate change defined by an infinitesimal conformal transformation:

$$x^\mu \to x^\mu + \alpha^\mu + \lambda x^\mu + \varepsilon^\mu_\nu x^\nu + c^\mu x^2 - 2(x\cdot c)x^\mu \tag{5.10}$$

the corresponding change of coordinate section in $P^2(M)$ is accomplished by an infinitesimal transition function determined by the Jacobian and its first derivatives in (5.10) to give:

$$\begin{aligned} \chi^i_j &= \varepsilon^i_j + \lambda\delta^i_j - 2(c^i x_j - c_j x^i) + 2(c\cdot x)\delta^i_j \\ \chi_i &= c_i . \end{aligned} \tag{5.11}$$

Substitution of (5.11) into (5.8) gives exactly the transformation law for field representations of the conformal group arrived at more usually by the induced representation method[9]. The (local) connection and curvature forms for the bundle $P(M,G_0)$ are now referred to the basis (5.4) in the Lie algebra G_0:

$$\omega = \bar{\omega}^{i}_{j} m^{j}_{i} + \bar{\omega} d + \bar{\omega}_{i} k^{i} \tag{5.12}$$

$$2\Omega = \bar{R}^{i}_{j} m^{j}_{i} + \bar{R} d + \bar{R}_{i} k^{i} \tag{5.13}$$

with

$$\left.\begin{aligned} \bar{R}^{i}_{j} &= 2[d\bar{\omega}^{i}_{j} + \bar{\omega}^{i}_{r} \wedge \bar{\omega}^{r}_{j}] \\ \bar{R} &= 2d\bar{\omega} \\ \bar{R}_{i} &= 2[d\bar{\omega}_{i} + \bar{\omega}_{r} \wedge \bar{\omega}^{r}_{i} + \bar{\omega}_{i} \wedge \bar{\omega}] \end{aligned}\right\}. \tag{5.14}$$

The connection components, under a change of local section with transition function $(a\Lambda^{i}_{j} = a^{i}_{j}, c_{k})$ transform to:

$$\left.\begin{aligned} \bar{\omega}'^{i}_{j} &= \bar{\Lambda}^{i}_{m} \bar{\omega}^{m}_{\ell} \Lambda^{\ell}_{j} + \bar{\Lambda}^{i}_{m} d\Lambda^{m}{}_{j} \\ \bar{\omega}' &= \bar{\omega} + da \\ \bar{\omega}'_{i} &= \bar{\omega}_{m} a^{m}_{i} - \bar{\omega}_{i} c_{m} - \bar{\omega}' c_{i} + d(a^{m}_{i} c_{\mu}) \end{aligned}\right\} \tag{5.15}$$

with a similar transformation law for $(\bar{R}^{i}_{j}, \bar{R}, \bar{R}_{i})$, omitting the inhomogeneous terms. In infinitesimal form this becomes:

$$\left.\begin{aligned} \delta\bar{\omega}^{i}_{j} &= \chi^{i}_{k} \bar{\omega}^{k}_{j} - \chi^{k}_{j} \bar{\omega}^{i}_{k} - d\chi^{i}_{j} \\ \delta\bar{\omega} &= -d\chi \\ \delta\bar{\omega}_{i} &= -\bar{\omega}_{m} \chi^{m}_{i} - \bar{\omega}_{i} \chi + \bar{\omega}^{\ell}_{i} \chi_{\ell} + \bar{\omega} \chi_{i} - d\chi_{i} \end{aligned}\right\}. \tag{5.15'}$$

The covariant derivative of the field ψ is given explicity by:

$$\psi_{;i} \equiv h^{\mu}_{i} (\partial_{\mu} \psi + i\bar{\omega}_{jk\mu} \Sigma^{jk} \psi + i\bar{\omega}_{\mu} \Delta\psi + i\bar{\omega}_{j\mu} \kappa^{j} \psi). \tag{5.16}$$

We may then form invariant Lagrangians from the fields ψ and their covariant derivatives which define the coupling of matter to gauge fields. The particular form of the identities (3.1)-

(3.9) for this case is the following:

(i) Dynamical definition of (intrinsic) currents:

$$\mathcal{S}^{ij\mu} \equiv \frac{\delta\mathcal{L}}{\delta\bar{\omega}_{ij\mu}} = i\,\frac{\partial\mathcal{L}}{\partial(\partial_\mu\psi)}\,\Sigma^{ij}\psi \quad \text{(spin current)} \tag{5.17a}$$

$$\mathcal{D}^{\mu} \equiv \frac{\delta\mathcal{L}}{\delta\bar{\omega}_{\mu}} = i\,\frac{\partial\mathcal{L}}{\partial(\partial_\mu\psi)}\,\Delta\psi \quad \text{(dilation current)} \tag{5.17b}$$

$$K^{i\mu} \equiv i\,\frac{\delta\mathcal{L}}{\delta\bar{\omega}_{i\mu}} = i\,\frac{\partial\mathcal{L}}{\partial(\partial_\mu\psi)}\,\kappa^{i}\psi \quad \text{(special conformal current)} \tag{5.17c}$$

(ii) Relation between canonical and dynamical energy-momentum tensors:

$$T^{\mu}_{\nu} = t^{\mu}_{\nu} - \mathcal{S}^{ij\mu}\bar{\omega}_{ij\nu} - \mathcal{D}^{\mu}\bar{\omega}_{\nu} - K^{i\mu}\bar{\omega}_{i\nu}. \tag{5.18}$$

(iii) Covariant divergence of currents:

$$\nabla_\mu \mathcal{S}^{ij\mu} \equiv \partial_\mu \mathcal{S}^{ij\mu} - \omega^{[j}_{m\mu}\mathcal{S}^{i]m\mu} + \bar{\omega}^{[j}_{\mu}K^{i]\mu} = t^{[ij]} \tag{5.19a}$$

$$\nabla_\mu \mathcal{D}^{\mu} \equiv \partial_\mu \mathcal{D}^{\mu} - \bar{\omega}_{i\mu}K^{i\mu} = t^{i}_{i} \tag{5.19b}$$

$$\nabla_\mu K^{i\mu} \equiv \partial_\mu K^{i\mu} - \omega^{i}_{j\mu}K^{j\mu} - \omega_{\mu}K^{i\mu} = 0. \tag{5.19c}$$

(iv) Covariant divergence of energy momentum tensor:

$$\partial_\mu t^{\mu}_{\nu} + \Gamma^{\alpha}_{\mu}t^{\mu}_{\alpha} = R_{ij\mu\nu}\mathcal{S}^{ij\mu\nu} + F_{\mu\nu}\mathcal{D}^{\nu} + Q_{i\mu\nu}K^{i\mu} \tag{5.20}$$

where

$$\left.\begin{aligned} R_{ij\mu\nu} &\equiv \bar{R}_{ij}(\partial_\mu,\partial_\nu) \\ F_{\mu\nu} &\equiv \bar{R}(\partial_\mu,\partial_\nu) \\ Q_{i\mu\nu} &\equiv \bar{R}_{i}(\partial_\mu,\partial_\nu) \end{aligned}\right\}. \tag{5.21}$$

From eq. (5.20) we see that there exists a non-inertial force density coupling the part of the curvature in the Lorentz subalgebra (the Riemannian part) with the intrinsic spin current density, another term coupling the scaling current with the corresponding component of the curvature and a third coupling the intrinsic special conformal currents to the respective curvature components. The first coupling is well-known within the context of G.R. and may be derived from a classical model for a particle with spin(11). The second term is reminiscent of the Lorentz force density in electrodynamics and indeed, such an interpretation leads to the unified field theory of Weyl, based upon the non-Riemannian geometry obtained by allowing changes in scale under parallel transport. The last term rises only in a non-affine geometry, such as that based on the conformal group. However terms involving a coupling between the second moments of the energy-momentum tensor (such as the special conformal currents) and the first derivatives of the Riemannian curvature may be obtained by extending the particle model studied by Mathisson, Papapetrou and Taub to include such higher-order structure.

We may view eq. (5.20) as a generalization of such a coupling where the appropriate quantity $Q_{i\mu\nu}$ is an independant geometrical quantity, rather than one defined in terms of derivatives of the Riemannian curvature.

6. GAUGE THEORY FOR THE FULL CONFORMAL GROUP

In the preceding section, the structure group was really the 11-parameter subgroup $G_0 \subset G$. In a sense, this is the more natural structure to study, since the corresponding fields are the curved space generalizations of the field representations of the (full) conformal group obtained by the induced representation construction(9). However, G_0 is not semi-simple, and therefore a Lagrangian of the form (3.11) and the corresponding field equations do not lead to a geniunely unified field theory for the different gauge fields. (Indeed, certain currents would decouple completely and not appear as sources for the gauge fields.) We shall therefore procede to define a theory based upon fields ψ which are cross-sections of the bundle E(M,G,V) associated with the principal bundle P(M,G) with the full conformal group as structure group. This requires a representation $\bar{\rho}(G)$ on the vector space V. Such a representation cannot however be obtained from a representation $\rho(G_0)$ by merely mapping the elements of G/G_0 onto the identity (as may be done for the inhomogeneous Lorentz group given a representation of the homogeneous part) precisely because G is simple. This means that no natural correspondence between such fields and the ones treated in the previous section can be made. At the geometrical level, there exists a method, given a connection for $P(M,G_0)$ to construct a Cartan connection which takes its values in the full conformal algebra[8], and this in turn may be used to define a connection for the bundle P(M,G). Such an induced connection is however very non-unique, and the "natural" choice for inducing an affine connection starting from a linear one has no precise analogue for the conformal case.

We therefore treat the bundles P(M,G), E(M,G,V) and the corresponding fields independently of those accurring in the previous section. Furthermore, we discuss only the infinitesimal transformation laws, which are sufficient for obtaining the structure equations for the curvature as well as the conservation laws and identities.

Accordingly, we consider a vector space V upon which a representation $\bar{\rho}(G)$ of the conformal Lie algebra acts, with the basis elements $\{i\pi^\mu, i\Sigma^{\mu\nu}, i\Delta, i\kappa^\mu\}$ denoting, respectively (intrinsic) translations, Lorentz transformations, dilations and special conformal transformations. The commutator algebra is

$$\left.\begin{aligned}
&[\pi_\mu,\Delta] = i\pi_\mu \\
&[\Sigma^{\mu\nu},\Sigma^{\sigma\tau}] = -i(\eta^{\mu\sigma}\Sigma^{\nu\tau}+\eta^{\nu\tau}\Sigma^{\mu\sigma}-\eta^{\mu\tau}\Sigma^{\nu\tau}-\eta^{\nu\sigma}\Sigma^{\mu\tau}) \\
&[\Sigma^{\sigma\tau},\pi_\mu] = i(\delta^\sigma_\mu\pi^\tau-\delta^\tau_\mu\pi^\sigma) \\
&[\Sigma^{\sigma\tau},\kappa_\mu] = -i(\delta^\sigma_\mu\kappa^\mu-\delta^\tau_\mu\kappa^\sigma) \\
&[\kappa_\mu,\Delta] = -i\kappa_\mu \\
&[\kappa_\mu,\pi_\nu] = -2i(\eta_{\mu\nu}\Delta+\Sigma_{\mu\nu})
\end{aligned}\right\}. \tag{6.1}$$

A field ψ will be considered as a cross-section of the vector bundle $E(M,V,G)$, referred to a local cross-section σ of $P(M,G)$ as defined in (2.11). The change in ψ induced by the infinitesimal group element with parameters (relative to this basis for G) $\{\alpha^i,\varepsilon^{ij},\lambda,c^i\}$ is given by:

$$\delta\psi = i(\alpha^i\pi_i+\varepsilon^{ij}\Sigma_{ij}+\lambda\Delta+c_i\kappa^i)\psi. \tag{6.2}$$

(Again, Roman indices rather than Greek are used to emphasize that the basis for the algebra is not the holonomic one related to the coordinate system used, raising and lowering of indices again being defined by the Minkowski metric η^{ij}. Holonomic coordinates are still denoted by Greek indices.) The intrinsic currents carried by such a field will be denoted in the same way as in the previous section (eq. (5.17)), but these are now augmented by a fourth, *intrinsic* translational current defined as:

$$P^\mu_i \equiv i\,\frac{\partial\mathcal{L}}{\partial(\partial_\mu\psi)}\,\pi_i\psi. \tag{6.3}$$

The local connection form is now defined by its components relative to the basis in G; $\{\omega^i_\mu,\omega^{ij}_\mu,\omega_\mu,\bar{\omega}_{i\mu}\}$ (respectively, translation, Lorentz, dilation and special conformal components). The corresponding curvature components are defined by:

$$R^i \equiv P^i_{\mu\nu}dx^\mu \wedge dx^\nu \equiv 2[d\omega^i + \omega^k \wedge \omega_k{}^i - \omega^i \wedge \omega] \tag{6.4a}$$

$$R^i_j \equiv R^i_{j\mu\nu}dx^\mu \wedge dx^\nu \equiv 2[d\omega^i_j + \omega^i_k \wedge \omega^k_j + \omega^i \wedge \bar{\omega}_j] \tag{6.4b}$$

$$R \equiv F_{\mu\nu}dx^\mu \wedge dx^\nu \equiv 2[d\omega + \omega^i \wedge \bar{\omega}_i] \tag{6.4c}$$

$$\bar{R}_i \equiv Q_{i\mu\nu}dx^\mu \wedge dx^\nu \equiv 2[d\bar{\omega}_i + \bar{\omega}_i \wedge \omega + \bar{\omega}_j \wedge \omega^j{}_{-i}]. \tag{6.4d}$$

The infinitesimal change in the connection components (gauge fields) corresponding to (6.2) is given by:

$$\delta\omega^i = -d\alpha^i + \alpha^k\omega^i_k - \varepsilon^i_k\omega^k - \alpha^i\omega + \lambda\omega^i \tag{6.5a}$$

$$\delta\omega_{ij} = -d\varepsilon_{ij} + \varepsilon_{i\ell}\omega^\ell_j - \varepsilon_{j\ell}\omega^\ell_i - \alpha_j\bar{\omega}_i + c_i\omega_j \tag{6.5b}$$

$$\delta\omega = -d\lambda + c_i\omega^i - \alpha^i\bar{\omega}_i \tag{6.5c}$$

$$\delta\bar{\omega}_i = -dc_i + c_j\omega^j_i + \varepsilon^j_i\bar{\omega}_j + c_i\omega - \lambda\bar{\omega}_i \tag{6.5d}$$

(with the same expressions, omitting the inhomogeneous terms, giving the transformation of the curvature components).

An invariant Lagrangian $\mathcal{L}(\psi,\psi_{ji})$ may now be formed from the fields ψ and their covariant derivatives:

$$\psi_{ji} \equiv h^\mu_i(\partial_\mu\psi + i\omega_{jk\mu}\Sigma^{jk}\psi + i\omega_\mu\Delta\psi + i\omega^j_\mu\pi_j\psi + i\bar{\omega}_j\kappa^j\psi). \tag{6.6}$$

The identities and conservation laws following from the invariance are then:

(i) Dynamical definition of currents: For $S^{ij\mu}$, $\mathcal{D}^\mu$, $K^{i\mu}$ the same as in eq. (5.17), plus:

$$P^\mu_i \equiv \frac{\delta\mathcal{L}}{\delta\omega^i_\mu} = i\,\frac{\partial\mathcal{L}}{\partial(\partial_\mu\psi)}\,\pi_i\psi \quad \text{(Intrinsic translational current)} \tag{6.7}$$

(ii) Relation between canonical and dynamical energy-momentum tensors: (cf. eq. (5.18))

$$T^{\mu}_{\nu} = t^{\mu}_{\nu} - S^{ij\mu}\omega_{ij\nu} - \mathcal{D}^{\mu}\omega_{\nu} - K^{i\mu}\bar{\omega}_{i\nu} - P^{\mu}_{i}\omega^{i}_{\nu}. \qquad (6.8)$$

(iii) Covariant divergence of currents: (cf. eq. (5.19))

$$\nabla_{\mu}S^{ij\mu} \equiv \partial_{\mu}S^{ij\mu} - \omega^{[j}_{m\mu}S^{i]m\mu} + \bar{\omega}^{[j}_{\mu}K^{i]\mu} + \omega^{[j}_{\mu}P^{i]\mu} = t^{[ij]} \qquad (6.9a)$$

$$\nabla_{\mu}\mathcal{D}^{\mu} \equiv \partial_{\mu}\mathcal{D}^{\mu} - \bar{\omega}_{i\mu}K^{i\mu} + \omega^{i}_{\mu}P^{\mu}_{i} = t^{i}_{i} \qquad (6.9b)$$

$$\nabla_{\mu}K^{i\mu} \equiv \partial_{\mu}K^{i\mu} - \omega^{\,i}_{j\mu}K^{j\mu} - \omega_{\mu}K^{i\mu} - 2\omega^{i}_{\mu}\mathcal{D}^{\mu} - 2\omega^{j}_{\mu}S^{i}_{j\mu} = 0 \qquad (6.9c)$$

$$\nabla_{\mu}P^{\mu}_{i} \equiv \partial_{\mu}P^{\mu}_{i} + \omega_{\mu}P^{\mu}_{i} - \omega^{\,j}_{i\mu}P^{\mu}_{j} + 2\bar{\omega}_{i\mu}\mathcal{D}^{\mu} + 2\bar{\omega}_{j\mu}S^{j\mu}_{i} = 0. \qquad (6.9d)$$

(iv) Covariant divergence of energy momentum tensor:

$$\partial_{\mu}t^{\mu}_{\nu} + \Gamma^{\alpha}_{\mu\nu}t^{\mu}_{\nu} = R_{ij\mu\nu}S^{ij\mu} + F_{\mu\nu}\mathcal{D}^{\nu} + P^{i}_{\mu\nu}P^{\mu}_{i} + Q_{i\mu\nu}K^{i\mu} \qquad (6.10)$$

We note that these relations differ in form from eqs. (5.17)-(5.20) by the addition of terms in the "intrinsic" translational current P^{μ}_{i}, and the corresponding curvature $P^{i}_{\mu\nu}$. This latter quantity in the case of the bundle of affine frames may be identified uniquely with the torsion provided the connection is defined through a Cartan connection for the bundle of linear frames with the translational part of the connection form equal to the cannonical 1-form[10] θ. Such an identification is impossible here since the simplicity of the conformal algebra implies that the components of ω in the translational subalgebra of G may not satisfy the transformation law required of θ under right translation by the group.

Finally, if we examine the Lagrangian model of section 3.2 applied to this case, we obtain the following gauge field Lagrangian:

$$\mathcal{L}_G = \frac{b}{4\gamma} g^{\mu\sigma} g^{\nu\tau} \{3R_{ij\mu\nu} R^{ij}_{\sigma\tau} + 2P^i_{\mu\nu} Q_{i\sigma\tau} - 2F_{\mu\nu} F_{\sigma\tau}\} \tag{6.11}$$

leading to the following field equations:

$$\partial_\mu (bQ^{\mu\nu}_i) - J^\nu_i = \gamma P^\nu_i \tag{6.12a}$$

$$-3\partial_\mu (bR^{ij\mu\nu}) - J^{ij\nu} = \gamma S^{ij\nu} \tag{6.12b}$$

$$-\partial_\mu (bP^{\mu\nu}_i) - \bar{J}^\nu_i = \gamma K^\nu_i \tag{6.12c}$$

$$2\partial_\mu (bF^{\mu\nu}) - J^\nu = \gamma \mathcal{D}^\nu \tag{6.12d}$$

$$\delta^\mu_\nu \mathcal{L}_G - \frac{b}{\gamma}(Q^i_{\nu\sigma} P^{\mu\sigma}_i + P_{i\nu} Q^{i\mu\sigma} + 3R^{ij}_{\nu\sigma} R^{\mu\sigma}_{ij} - 2F_{\nu\sigma} F^{\mu\sigma}) = t^\mu_\nu \tag{6.13}$$

where the currents J^ν_i, $J^{ij\nu}$, $\bar{J}^\nu_i$, J^ν are the ones carried by the gauge fields themselves, and are defined by:

$$J^\nu_i \equiv b[\omega_\mu Q^{\mu\nu}_i + \omega^j_{i\mu} Q^{\mu\nu}_j - 4\bar{\omega}_{i\mu} F^{\mu\nu} + 6\omega_{j\mu} R^{j\mu\nu}_i] \tag{6.14a}$$

$$J^{ij\nu} \equiv b[6\omega^{[j}_{m\mu} R^{i]m\mu\nu} + \omega^{[j}_\mu P^{i]\mu\nu} + \omega^{[i}_\mu Q^{j]\mu\nu}] \tag{6.14b}$$

$$\bar{J}^\nu_i \equiv b[\omega^i_{j\mu} P^{j\mu\nu} - \omega_\mu P^{i\mu\nu} + 4\bar{\omega}^i_\mu F^{\mu\nu} - 6\bar{\omega}^j_\mu R^{i\mu\nu}_j) \tag{6.14c}$$

$$J^\nu \equiv b[\omega^i_\mu Q^{\mu\nu}_i - \bar{\omega}_{i\mu} P^{i\mu\nu}]. \tag{6.14d}$$

The total currents (matter plus gauge) clearly satisfy a true conservation law.

The system of equations (6.12)-(6.14) represent an example of a fully gauge conformal invariant field theory. The various gauge fields all enter on an equal footing with a source current for each one parameter subgroup determining the corresponding curvature component in a manner similar to Maxwell's equations. Furthermore, the simplicity of the group determines the exact proportion of the various contributions to the Lagrangian of (6.11), leaving only one coupling constant, γ to be chosen. The unfavorable aspects of these field equations include the presence of "intrinsic" translational currents, and the apparent lack of correspondence with the Einstein field equations for gravity. The physical interpretation of the new "geometrical" fields enterring into the conformal gauge structure we leave for the moment as an open question.

REFERENCES

1. S. WEINBERG, Rev. Mod. Phys. 46, 255 (1974).
2. R. UTIYAMA, Phys. Rev. 101, 1597 (1956).
3. T.W.B. KIBBLE, J. Math. Phys. 2, 212 (1961).
4. D.W. SCIAMA, "On the analogy between charge and spin in general relativity" in the "Festschrift for Infeld" (Pergaman Press, New York 1962).
5. A. TRAUTMAN, Bull. Akad. Polon. Sci. 20, 184, 503, 895 (1972); ibid. 21, 345 (1973); Rep. Math. Phys. 1, 29 (1970).
6. J.P. HARNAD and R.B. PETTITT, J. Math. Phys. 17, 1827 (1976).
7. A. BREGMAN, Progr. Theor. Phys. 49, 667 (1973); J.M. CHARAP and W. TAIT, Proc. Roy. Soc. Lond. A349, 249 (1974); P.G.O. FREUND, Ann. of Phys. 84, 440 (1974); H. WEYL, "Space, Time, Matter" (Dover, N.Y. 1961).
8. S. KOBAYASHI, "Transformation Groups in Differential Geometry" (Springer-Verlag, Berlin 1972).
9. G. MACK and A. SALAM, Ann. of Phys. 53, 174 (1969); S. FERRARA, R. GATTO and A.F. GRILLO, Springer Tracts in Modern Physics 67 (Springer-Verlag, Berlin 1973).
10. S. KOBAYASHI and K. NOMIZU, "Foundations of Differential Geometry" (Interscience, N.Y. 1963), A. LICHNEROWICZ, "Théorie globale des connexions et des groupes d'holonomie", Ed. Cremonese, Rome, 1955; K. NOMIZU, "Lie Groups and Differential Geometry" (Math. Soc. Japan Publ., vol. 2, 1956).
11. M. MATHISSON, Acta. Phys. Polon. 6, 1937, 163; A. PAPAPETROU, Proc. Roy. Soc. A209, 248 (1951), A.H. TAUB, J. Math. Phys. 5, 112 (1964).

THEORY OF COHERENT QUARKS. ON THE DYNAMICAL ORIGIN OF THE OKUBO-ZWEIG-IIZUKA RULE

M. Hongoh

I. INTRODUCTION

Discoveries of the ψ particles have surprised many physicists because of their strikingly narrow widths. Their narrowness is due neither to unfavorable phase space, nor to conservation of quantum numbers in the usual sense. A possible explanation is provided by the Okubo-Zweig-Iizuka (OZI) rule [1] (see Figure 1).

Figure 1.

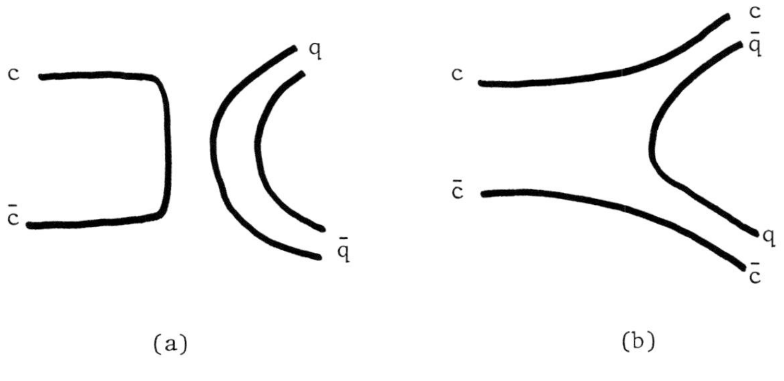

To be precise, let us consider the ϕ-meson of the vector nonet of SU(3). If SU(3) symmetry is not broken by mass, ϕ has the quark structure

$$\frac{1}{\sqrt{3}}\,(u\bar{u}+d\bar{d}+s\bar{s}),$$

rather than pure $s\bar{s}$. However the symmetry is broken and the *physical* ϕ is a mass eigenstate. In fact one can see ϕ has almost pure $s\bar{s}$ structure by diagonalizing the mass matrix. Or one can introduce a mixing angle between the octet and the singlet, and assuming ideal mixing, ϕ is also given a pure $s\bar{s}$ structure. Note, however, the mixing angle is a purely phenomenological quantity and has nothing to do with the angles of rotations defined in the manifold of the symmetry group. The purpose of the present work is to investigate a possible dynamical origin of the OZI rule by constructing new basis functions for SU(3), which are analogous to the SU(2) Bloch coherent states.

The SU(2) Bloch coherent states, $|\mu\rangle$, are well known in atomic physics [2]. They can be obtained via rotation of the ground state $|-J\rangle$ (the lowest Dicke state) in angular momentum space. $|\mu\rangle$ is related to the field coherent states, $|\alpha\rangle$ (Glauber coherent states) as the SU(2) algebra contracts to the Heisenberg algebra. We are interested in the generalization of the coherent states of this kind which we might call "Bloch type". Group theoretically, the generalized coherent states are vectors defined in the homogeneous factor space $\sim G/H$ [3], where H is a particular stability group, e.g. $|\mu\rangle \in SU(2)/O(2)$. The generalized coherent states defined in this way provide us with an attractive possibility of describing physical systems with spontaneously broken symmetry. In the case of $|\mu\rangle$, $SU(2) \sim O(3)$ is spontaneously broken by choosing the lowest angular momentum state $|-J\rangle$ for which, however, O(2) is the stability group.

The SU(3) symmetry of the strongly interacting particles is broken spontaneously by the difference in quark masses, and the direction of the symmetry breaking lies along the hypercharge axis, which is set by the weak and the electromagnetic interactions. This makes SU(2) × U(1) a good (internal) symmetry of the strong interaction in the limit in which electromagnetic and weak interactions are ignored.

II. SU(3) COHERENT STATES

In this section we consider coherent states ~ SU(3)/SU(2) × U(1). The method used in ref. [2] is not particularly convenient for SU(3), since an invariant manifold of SU(3) is not congruent to the defining point field. For this reason we shall introduce an algebraic approach [4]. Consider the SU(3) scalars,

$$A = (\eta_1\bar{\eta}_2+\xi_1\bar{\xi}_2+\zeta_1\bar{\zeta}_2)^K, \tag{2.1a}$$

$$B = (\bar{\eta}_1\eta_2+\bar{\xi}_1\xi_2+\bar{\zeta}_1\zeta_2)^L, \quad K,L\text{: nonnegative integers.} \tag{2.1b}$$

$\bar{\eta}$, $\bar{\xi}$, $\bar{\zeta}$ are complex variables *conjugate* (not complex conjugate!) to η, ξ, ζ respectively. We denote by η^*, ξ^*, ζ^* the variables *complex conjugate* to η, ξ, ζ. In (1) the suffixes refer to two distinct representation spaces. All the self-contragredient IR (SCIR) of SU(3) are orthogonal, and due to the Malcev-Dynkin theorem [5], corresponding bilinear forms are symmetric. We shall consider first the SU(3) coherent states associated with the quark state, since D(K,0) and D(0,L) separately form invariant subspaces. In each subspace a representation of SU(2) occurs only once. Let us consider an invariant form

$$F(u,v) = \sum_{s=0}^{k}\sum_{p=0}^{s} g_p^{s*}(u)\,g_p^s(v) \tag{2.2}$$

where

$$g_p^s(z) = \left[\frac{k!}{p!(s-p)!(k-s)!}\right]^{\frac{1}{2}} z_1^p z_2^{s-p} z_3^{k-s}. \tag{2.3}$$

$g_p^s(z)$ are homogeneous polynomials of degree k, and can be viewed as the orthonormal tensorial set defined over the $\frac{1}{2}(k+1)(k+2)$-dimensional space of the IR D(K,0). The comparison between (1a) and (2) leads to the correspondence,

$$\begin{pmatrix} \eta_1 \\ \xi_1 \\ \zeta_1 \end{pmatrix} \rightarrow \begin{pmatrix} v_1 \\ v_2 \\ v_3 \end{pmatrix}; \qquad \begin{pmatrix} \bar{\eta}_2 \\ \bar{\xi}_2 \\ \bar{\zeta}_3 \end{pmatrix} \rightarrow \begin{pmatrix} u_1^* \\ u_2^* \\ u_3^* \end{pmatrix}. \tag{2.4}$$

Without loss of generality we can assume,

$$F(v,v) = (|v_1|^2+|v_2|^2+|v_3|^2)^K = 1 \tag{2.5a}$$

and

$$|v_1|^2 + |v_2|^2 = \text{const. } (s). \tag{2.5b}$$

This amounts to a completeness relation for the orthonormal basis,

$$\sum_s \sum_p |sp\rangle_k \langle sp|_k = 1. \tag{2.6}$$

In (6) $|sp\rangle_k \equiv g_p^s(z)$, and the correspondence (4) is understood. Let us write $z(u\ d\ s)$ for $v(v_1 v_2 v_3)$, and $\bar{\omega}(\bar{\alpha}\ \bar{\beta}\ \bar{\gamma})$ for $u^*(u_1^* u_2^* u_3^*)$, where u, d, s, denote ordinary triplet quarks. Then from (2) and (4) we obtain

$$|\bar{\alpha}\bar{\beta}\bar{\gamma}\rangle = \sum_{s=0}^{k} \sum_{p=0}^{s} \left[\frac{k!}{p!(s-p)!(k-s)!}\right]^{\frac{1}{2}} \bar{\alpha}^p \bar{\beta}^{s-p} \bar{\gamma}^{k-s} |sp\rangle_k. \tag{2.7}$$

Assuming that SU(2) × U(1) be the stability group of the strong interaction, we may parametrize,

$$|u|^2 + |d|^2 + |s|^2 = 1$$

$$|s|^2 \neq |u|^2 = |d|^2 : \text{ in z space} \tag{2.8a}$$

and

$$|\bar{\alpha}| = |\bar{\beta}| = 1, \quad |\bar{\gamma}| = \varepsilon; \text{ in } \bar{w} \text{ space.} \tag{2.8b}$$

Thus, dividing by the square root of the norm, we shall write the SU(3) coherent states in the following form,

$$|Q\rangle_k = \frac{1}{(2+|\varepsilon|^2)^{k/2}} \sum_{s=0}^{k} \sum_{p=0}^{s} \times \left[\frac{k!}{p!(s-p)!(k-s)!}\right]^{\frac{1}{2}} \bar{\alpha}^p \bar{\beta}^{s-p} \bar{\gamma}^{k-s} \, |sp\rangle_k \Big|_{|\bar{\alpha}|=|\bar{\beta}|=1} . \tag{2.9a}$$

Similarly

$$|\bar{Q}\rangle_L = \frac{1}{(2+|\varepsilon|^2)^{\ell/2}} \sum_{s=0}^{\ell} \sum_{p=0}^{s} \times \left[\frac{\ell!}{p!(s-p)!(L-s)!}\right]^{\frac{1}{2}} \alpha^p (-\beta)^{s-p} \gamma^{\ell-s} \, |\bar{s}\bar{p}\rangle_\ell \Big|_{|\alpha|=|\beta|=1} \tag{2.9b}$$

where $|\bar{\alpha}|$, $|\bar{\beta}|$ and $|\alpha|$, $|\beta|$ are taken to be unity after the action of the group representation operators. We might call $|Q\rangle_k$ and $|\bar{Q}\rangle_L$ quark and antiquark coherent states of SU(3) with respect to the subgroup SU(2) × U(1). These states are not orthogonal as can easily be seen. Note that SU(2) × U(1) is the stability group for the lowest (highest) state of the triplet (antitriplet). In (9), $|\varepsilon| = |\gamma| = |\bar{\gamma}|$ may be regarded as the angle through which these states are rotated from the z_3 axis. This may be proved by a method similar to that used for the Bloch coherent states.

III. THE FAMILY OF PARTICLES AND THE COHERENT STATES

If we assume that the physical hadrons are states in $W = SU(3)/SU(2) \times U(1)$, they can be written as a product of $|Q\rangle_k$ and $|\bar{Q}\rangle_L$, e.g. the vector nonet mesons are,

$$f(0)\{|u\,\bar{u}\rangle + |u\,\bar{d}\rangle + |\bar{u}\,d\rangle + |d\,\bar{d}\rangle\} \sim \rho,\omega$$

$$f(\gamma)\{|u\,\bar{s}\rangle + |d\,\bar{s}\rangle\} \sim K^*,$$

$$f(\bar{\gamma})\{|s\,\bar{u}\rangle + |s\,\bar{d}\rangle\} \sim \bar{K}^*,$$

$$f(\gamma\bar{\gamma})\{|s\,\bar{s}\rangle\} \sim \phi. \tag{3.1}$$

SU(3) quantum numbers are carried by the "mathematical" basis of the symmetry in the curly brackets. Note the *physical* ϕ meson has the pure $s\bar{s}$ structure in this basis. The factor f in (1) is then strongly suggestive of the dynamical origin of the OZI rule. In general, the rotation in the manifold depends on three parameters α, β, and γ, and it is quite possible that the anomaly for pseudoscalar mesons is due to the dynamics which depend on α and β as well as γ.

This result can be immediately carried over to the ψ particles, where SU(3) symmetry breaking is considerably smaller compared to the badly broken SU(4). The SU(4) coherent states $\sim$ SU(4)/SU(3) $\times$ U(1) suggests,

$$\begin{aligned}\psi\text{-family} = {} & g(0)\{D(10) \otimes D(01)\} \\ & + g(\bar{w})\{D(10) \otimes \bar{c}\} \\ & + g(w)\{c \otimes D(01)\} \\ & + g(w\bar{w})\{c\bar{c}\}.\end{aligned}$$

Again the ψ particle has a pure $c\bar{c}$ structure and the OZI rule applies in the same manner.

IV. THE DYNAMICAL ORIGIN OF THE OZI RULE

It is easy to see that the SU(3) algebra contracts to the direct sum of the 1 and 2-dimensional Heisenberg algebra, $H_2 \oplus H_1$. Letting $\sqrt{s}\,\bar{\alpha} = X$ and $\sqrt{k}\,\bar{\gamma} = Z$ in $|\bar{Q}\rangle$, we obtain

$$|\bar{Q}\rangle\Big|_{|\bar{\beta}|=1} \underset{k\geq s\to\infty}{\longrightarrow} |X;Z\rangle = e^{-\frac{|X|^2}{2}} \sum_p \frac{X^p}{[p!]^{\frac{1}{2}}}|p\rangle$$

$$\times\, e^{-\frac{|z|^2}{2}} \sum_s \frac{Z^{k-s}}{[(k-s)!]^{\frac{1}{2}}}|k-s\rangle . \quad (4.1)$$

Since $|\bar{Q}\rangle$ is not an eigenstate of the lowering operator, the Hamiltonian analogy in the following holds only in the limit of the contraction of the algebra. If we let $\sqrt{\varepsilon} \sim 1/\sqrt{s}$ and $\varepsilon \sim 1/\sqrt{k}$, it can be seen that $|X;Z\rangle$ is a product of eigenstates of the annihilation operators w^- and $z_1^- + z_2^-$, where

$$A_1^{\pm} = E_1^{\pm} + E_2^{\pm} \to z_1^{\pm}, \qquad A_2^{\pm} = E_1^{\pm} - E_2^{\pm} \to z_2^{\pm}$$

$$M_1 + M_2 \to z_0$$

$$C^{\pm} = E_3^{\pm} \to w^{\pm}, \qquad M_3 \to w_0 \quad (4.2)$$

and

$$[z_i^{\pm}, z_j^-] = \delta_{ij} z_0, \qquad [z_0, z_i^{\pm}] = 0$$

$$[w^+, w^-] = w_0, \qquad [w_0, w^{\pm}] = 0. \quad (4.3)$$

Consider a simple Hamiltonian

$$H_{12}^0 = \varepsilon^2 \cdot \{\Sigma_i A_i^+ A_i^- - 2(M_1 + M_2)\}$$

$$H_3^0 = \varepsilon \cdot (c^+ c^- - M_3) \quad (4.4)$$

and c-number forces coupled to $A_i^{\pm}$ and $c^{\pm}$, i.e.

$$H_{12}' = \frac{1}{2k}\{\Sigma_i A_i^+ A_i^- - 2(M_1+M_2)\} + \frac{1}{\sqrt{k}} \Sigma_i \{p_i(A_i^+ - A_i^-) + q_i(A_i^+ + A_i^-)\}$$

$$H_3' = \frac{1}{2s}\{c^+ c^- - M_3\} + \frac{1}{\sqrt{s}} p_3(c^+ - c^-) + \frac{1}{\sqrt{s}} q_3(c^+ + c^-). \quad (4.5)$$

The total Hamiltonian reduces to that for the driven harmonic oscillators when the algebra contracts.

$$\tfrac{1}{2}\{\Sigma_i [P_i-p_i)^2+(Q_i-q_i)^2]-2\} - \tfrac{1}{2}\Sigma_i (p_i^2+q_i^2)$$

and

$$\tfrac{1}{2}\{(P_3-p_3)^2+(Q_3-q_3)^2-1\} - \tfrac{1}{2}(p_3^2+q_3^2), \tag{4.6}$$

where

$$Q_i = \frac{1}{\sqrt{2}} (z_i^- + z_i^+)$$

and

$$P_i = \frac{1}{\sqrt{2}} (z_i^- - z_i^+),$$

and

$$Q_3 = \frac{1}{\sqrt{2}} (w^- + w^+), \qquad P_3 = \frac{1}{\sqrt{2}} (w^- - w^+)$$

are canonical position and momentum operators. It can be seen easily that $|X;Z\rangle$ is the product of eigenvectors of these Hamiltonians with the eigenvalue

$$-\tfrac{1}{2}(p_3^2+q_3^2)\cdot\{-\tfrac{1}{2}\Sigma_i (p_i^2+q_i^2)\}. \tag{4.7}$$

V. CONCLUDING REMARKS

The SU(3) scalar forces rotate the lowest (highest) state of the triplet (antitriplet) in the base manifold of the SU(3) group and they are responsible for the OZI rule. However the coherent states do not fix the amount of the rotation. There should be no confusion about the role of such forces. Clearly they are not directly responsible for the spontaneous breaking of the SU(3) symmetry, for the symmetry is spontaneously broken as soon as we choose the lowest (highest) state to start with. The SU(3) coherent states considered in this paper may play a role similar to that of the Bloch coherent states in atomic physics.

REFERENCES

1. S. OKUBO, Phys. Lett. 5, 165 (1963); G. ZWEIG, CERN report, 8419/TH 412 (1964), unpublished; J. IIZUKA, Supplement to Prog. Theor. Phys. 37-38, 21 (1966).
2. J.M. RADCLIFFE, J. Phys. (London) A4, 313 (1971); F.I. ARECCHI, E. COURTENS, R. GILMORE and H. THOMAS, Phys. Rev. A6, 2211 (1972).
3. A.M. PERELOMOV, Commun. Math. Phys. 26, 222 (1972).
4. M. HONGOH, University of Montreal, preprint.
5. A.I. MALCEV, Am. Math. Soc. Transl. Ser. 1, 9, 172 (1962); E.B. DYNKIN, Dokl. Akad. Nank. SSSR (N.S.) 71, 221 (1950); E.B. DYNKIN, ibid. 76, 629 (1951).

FIGURE CAPTION

Fig. 1: The disconnected quark diagram (a) corresponds to the suppressed decay of ψ into ordinary hadrons, $\psi \to M\bar{M}$; while the connected diagram (b) corresponds to the normal decay of ψ into charmed particles, $\psi \to D\bar{D}$.

RELATIVISTIC COHERENT-STATE REPRESENTATIONS

Gerald Kaiser

1. INTRODUCTION

The coherent-state representation and its variants [1-3] have found many applications in quantum physics, in particular as a tool for the study of the classical limit [4-6]. For finite degrees of freedom, such representations are usually confined to non-relativistic systems. The purpose of this paper is to construct similar representations which are applicable to relativistic particles. In section 2 we develop a family of representations for the dynamics of a free non-relativistic particle which is closely related to the coherent-state representation. This family is extended in section 3 to include relativistic particles. In section 4 we summarize some properties of the new wave packets.

2. NON-RELATIVISTIC PARTICLE

The wave function $f(\vec{x},t)$ for a non-relativistic free particle in R^n evolves under the Schrödinger equation

$$i \frac{\partial f}{\partial t} = Hf, \qquad H = -\frac{1}{2m} \Delta. \tag{2.1}$$

The solutions are given by

$$f(\vec{x},t) = (e^{-itH}f)(\vec{x}) = (2\pi)^{-n/2} \int_{R^n} e^{-itp^2/2m+i\vec{x}\cdot\vec{p}}\hat{f}(\vec{p})d^np \quad (2.2)$$

where $\hat{f}(\vec{p})$ is the Fourier transform of the initial function $f(\vec{x},0)$. Now let $\vec{z} = \vec{x}-i\vec{y} \in C^n$ and let $\tau = t-i\beta$ be in the lower-half plane C^- (i.e., $\beta > 0$). Then $\exp(-i\tau p^2/2m + i\vec{z}\cdot\vec{p})$ decays rapidly as $|\vec{p}| \to \infty$ and eq. (2.2) defines a function $f(\vec{z},\tau)$ holomorphic in $\mathcal{D} = C^n \times C^-$. Let $G = \{f(\vec{z},\tau): \hat{f} \in L^2(R^n)\}$ be the vector space of all such functions. Then for each $\beta > 0$ the function $f_\beta(\vec{z},t) = f(\vec{x}-i\vec{y},t-i\beta)$ satisfies (2.1) in $\vec{x}$ and t. Let G_β be the space of all such functions $f_\beta(\vec{z},t)$. On G_β define the map $(e^{-itH}f_\beta)(\vec{z},s) = f_\beta(\vec{z},s+t)$. We are going to make G_β into a Hilbert space such that e^{-itH} is unitary for every real t, giving us a unitary representation of dynamics on G_β for every $\beta > 0$. Although these representations are all unitarily equivalent, the spaces G_β have some interesting properties, as we shall see.

Thus let $\beta > 0$ and $\vec{z} = \vec{x}-i\vec{y} \in C^n$. Then

$$f_\beta(\vec{z},0) = (e^{-\beta H}f)(\vec{z}) = (2\pi)^{-n/2} \int_{R^n} e^{-\beta p^2/2m+i\vec{z}\cdot\vec{p}}\hat{f}(\vec{p})d^np$$

$$\equiv \langle e^\beta_{\vec{z}} | f \rangle \quad (2.3)$$

where

$$\langle e^\beta_{\vec{z}} | p \rangle = (2\pi)^{-n/2}\exp(-\beta p^2/2m+i\vec{z}\cdot\vec{p}) \quad (2.4)$$

with Fourier transform

$$\langle e^\beta_{\vec{z}} | \vec{x}' \rangle = \left(\frac{2\pi\beta}{m}\right)^{-n/2}\exp\left(-\frac{m(\vec{z}-\vec{x}')^2}{2\beta}\right). \quad (2.5)$$

The e^β_z are minimum-uncertainty spherical wave packets with $\langle X_k \rangle = x_k = \mathrm{Re}(z_k)$, $\langle P_k \rangle = (m/\beta)y_k$ and diameter $\Delta X_k = \sqrt{\beta/2m}$ $(k = 1,2,\ldots,n)$. For f_β in G_β define

$$\|f\|_\beta^2 = \int_{C^n} |f_\beta(\vec{z},0)|^2 d\mu_\beta(\vec{z}), \tag{2.6}$$

where

$$d\mu_\beta(\vec{z}) = (\frac{m}{\pi\beta})^{n/2} \exp(-\frac{my^2}{\beta}) d^n x d^n y. \tag{2.7}$$

Theorem 1. Let $t \in R$, $\beta > 0$, $f \in L^2(R^n)$ and $f_\beta = e^{-\beta H} f$. Then

$$\|f\|_\beta = \|f\|. \tag{2.8}$$

In particular,

(a) $\|\cdot\|_\beta$ is a norm on G_β under which G_β is a Hilbert space.

(b) The map $e^{-\beta H}$ is unitary from $L^2(R^n)$ onto G_β.

(c) The map e^{-itH} is unitary on G_β.

Remarks. 1. (2.8) can of course be polarized to give a resolution of the identity: for f, g in $L^2(R^n)$,

$$\langle f|g\rangle_\beta \equiv \int_{C^n} \langle f|e^\beta_{\vec{z}}\rangle\langle e^\beta_{\vec{z}}|g\rangle d\mu_\beta(\vec{z}) = \langle f|g\rangle. \tag{2.9}$$

2. $e^{-\beta H}$ intertwines [7] the dynamics on $L^2(R^n)$ with the dynamics on G_β.

Proof. Let $f \in S(R^n)$. By (2.3), $f_\beta(\vec{x}-i\vec{y},0) = \check{g}_{\beta,\vec{y}}(\vec{x})$ where $g_{\beta,\vec{y}}(\vec{p}) = \exp(-\beta p^2/2m + \vec{y}\cdot\vec{p})\, \hat{f}(\vec{p})$ and $\check{g}$ denotes the inverse Fourier transform of g. Thus by Plancherel's theorem (and Fubini's),

$$\|f\|_\beta^2 = (\frac{m}{\pi\beta})^{n/2} \int_{R^n} e^{-my^2/\beta} d^n y \int_{R^n} d^n p\, e^{-\beta p^2/m+2\vec{y}\cdot\vec{p}} |\hat{f}(\vec{p})|^2$$

$$= \int_{R^n} |\hat{f}(\vec{p})|^2 d^n p = \|f\|^2,$$

which proves (2.8) for f in $S(R^n)$, hence also in $L^2(R^n)$ by continuity. (a)-(c) are obvious.

3. RELATIVISTIC PARTICLE

We sketch a generalization of the results of section 2 to relativistic particles. We confine ourselves to n=3.

The evolution of a free scalar relativistic particle of mass $m > 0$ is given by the Klein-Gordon equation

$$(-\frac{1}{c^2}\frac{\partial^2}{\partial t^2} + \Delta - m^2)f(\vec{x},t) = 0. \tag{3.1}$$

We consider only positive-energy solutions. These are given by

$$f(\vec{x},x_o) = (e^{-ix_oH}f)(\vec{x}) = (2\pi)^{-3/2}\int_{R^3} e^{-ixp}\hat{f}(\vec{p})\,d\Omega(\vec{p}), \tag{3.2}$$

where $x_o = ct$, $H = \sqrt{(mc)^2-\Delta}$, $xp = x_o\omega-\vec{x}\cdot\vec{p}$ with $\omega = \sqrt{(mc)^2+\vec{p}^2}$, and $d\Omega(p) = d^3p/\omega$ is the Lorentz-invariant measure in momentum space. $\hat{f}$ is the ordinary Fourier transform on R^3. For every $\hat{f}$ in $L^2(\Omega)$ the solution $f(\vec{x},x_o)$ is the boundary-value of a function $f(\vec{z},z_o) = f(z)$ holomorphic in the forward tube [8]

$$\mathcal{T} = \{x-iy \in C^4: x \in R^4,\ y \in V_+\},$$

where

$$V_+ = \{y \in R^4: y_o > |\vec{y}|\}$$

is the open forward light cone. This is so because

$$|e^{-izp}| = e^{-yp} < \exp(-(y_o-|\vec{y}|)|\vec{p}|)$$

decays rapidly as $|\vec{p}| \to \infty$ for fixed z in $\mathcal{T}$. $\mathcal{T}$ will replace $\mathcal{D} = C^3 \times C^-$ of section 2 and is strictly contained in $\mathcal{D}$. The analogue of G is the space $K = \{f(\vec{z},z_o):\hat{f} \in L^2(\Omega)\}$. To obtain counterparts of the G_β we need a phase space. In section 2 that was the set $\{(\vec{z},\tau) \in \mathcal{D}: \tau = -i\beta\} \approx C^3$. This will not do since it is not contained in T. Thus we deform it: let

$$P_\lambda = \{z = x-iy \in C^4: z_o = -i\sqrt{\lambda^2+\vec{y}^2}\}, \qquad \lambda \geq 0.$$

The functions

$$f_\lambda(\vec{z},x_o) = f(\vec{x}-i\vec{y},x_o-i\sqrt{\lambda^2+\vec{y}^2})$$

satisfy (3.1) in $\vec{x}$ and $x_o = ct$. Let $K_\lambda = \{f_\lambda(\vec{z},x_o): \hat{f} \in L^2(\Omega)\}$ and denote the map $\hat{f}(\vec{p}) \to f_\lambda(\vec{z},x_o)$ by U_λ. Define dynamics on K_λ by

$$(e^{-ix_o'H}f_\lambda)(\vec{z},x_o) = f_\lambda(\vec{z},x_o+x_o').$$

For $\lambda > 0$,

$$\begin{aligned} f_\lambda(\vec{z},0) &= (2\pi)^{-3/2}\int \exp(-\sqrt{\lambda^2+\vec{y}^2}\,\omega+i\vec{z}\cdot\vec{p})\hat{f}(\vec{p})d\Omega(\vec{p}) \\ &\equiv \langle e_{\vec{z}}^\lambda | f\rangle \end{aligned} \tag{3.3}$$

where

$$\langle e_{\vec{z}}^\lambda | p\rangle = (2\pi)^{-3/2}\exp(-\sqrt{\lambda^2+\vec{y}^2}\,\omega+i\vec{z}\cdot\vec{p}) \tag{3.4}$$

and all inner products are in $L^2(\Omega)$ until further notice. The e_z^λ are in $L^2(\Omega)$: for $z = x-iy$ in P_λ and $z' = x'-iy'$ in $P_{\lambda'}$, (where $\lambda,\lambda' > 0$),

$$\begin{aligned} \langle e_{\vec{z}}^\lambda | e_{\vec{z}'}^{\lambda'}\rangle &= (2\pi)^{-3}\int \exp\{-(y_o+y_o')\omega+i(\vec{z}-\bar{\vec{z}}')\cdot\vec{p}\}d\Omega(\vec{p}) \\ &= -2i\Delta_+(z-\bar{z}'). \\ &= \frac{mc}{4\pi^2\eta}K_1(2\eta mc), \end{aligned} \tag{3.5}$$

where y_o denotes $\sqrt{\lambda^2+\vec{y}^2}$, Δ_+ is the two-point function for the free scalar field of mass m [8] and $2\eta = [-(z-\bar{z}')^2]^{\frac{1}{2}}$ is defined by analytic continuation from $[-(z-\bar{z})^2]^{\frac{1}{2}} = [4y^2]^{\frac{1}{2}} = 2\lambda$ for $z = z' = x-iy$ in P_λ. K_n $(n = 0,1,2,\ldots)$ denotes a modified Bessel function. For $\lambda=0$, (3.3) still gives $f_o(\vec{z},0)$ and the functions $e^o_{\vec{z}}$ are still defined, but are no longer in $L^2(\Omega)$, as (3.5) shows. For $f_\lambda \in K_\lambda$ $(\lambda \geq 0)$ define

$$\|f\|^2_\lambda = \int_{C^3} |f_\lambda(\vec{z},0)|^2 d\mu_\lambda(\vec{z}) \tag{3.6}$$

where

$$d\mu_\lambda(\vec{z}) = C_\lambda d^3x d^3y \tag{3.7}$$

with $C_\lambda = [2\pi(\lambda/mc)^2 K_2(2\lambda mc)]^{-1}$ for $\lambda > 0$ and $C_o = (mc)^4/\pi$. Then our main result is the following

Theorem 2. Let $\lambda \geq 0$ and $\hat{f} \in L^2(\Omega)$. Then

$$\|f\|_\lambda = \|\hat{f}\|. \tag{3.8}$$

In particular,

(a) $\|\cdot\|_\lambda$ is a Lorentz-invariant norm on K_λ under which K_λ is a Hilbert space.

(b) The map U_λ is unitary from $L^2(\Omega)$ onto K_λ.

(c) e^{-ix_oH} is unitary on K_λ.

The remarks following Theorem 1 apply here as well.

Comparing the measures (2.7) and (3.7), note that $d\mu_\lambda$ has no weight function. This is a consequence of the curvature of the phase space P_λ. The "weight" has been absorbed into the functions f_λ themselves, which are consequently bounded:

$$|f_\lambda(\vec{z},0)|^2 = |\langle e^\lambda_{\vec{z}}|f\rangle|^2 \leq \|e^\lambda_{\vec{z}}\|^2\|\hat{f}\|^2 = \frac{mc}{4\pi^2\lambda} K_1(2\lambda mc)\|\hat{f}\|^2. \tag{3.9}$$

Finally note that Theorem 2 gives us a unitary, irreducible

representation of the restricted Poincaré group on K_λ. Define the action on K by $(U(g)f)(z) = f(g^{-1}z)$, $g \in P_+^\uparrow$. This induces an action on K_λ with the desired properties.

4. CONCLUSION

The $e^\lambda_{\vec{z}}$ have other interesting properties which we can only mention here for lack of space. In the state $e^\lambda_{\vec{z}}$, the particle appears as a wave packet centered about $\vec{x} = \mathrm{Re}(\vec{z})$ with expected momentum proportional to $\vec{y} = -\mathrm{Im}(\vec{z})$. The wave packet, which is spherical in the rest frame, shows contraction in the direction of motion and has minimal undertainties in a natural sense. It diameter increases from zero (when $\lambda mc \to 0$) to $\sim \sqrt{\lambda/2mc}$ (when $\lambda mc \to \infty$). Thus $e^\lambda_{\vec{z}}$ describes an extended, relativistic particle.

ACKNOWLEDGMENTS

I thank Lon Rosen for many helpful comments and suggestions. I have also benefited from a number of conversations with Alan Cooper, Zbigniew Haba and Ira Herbst at various stages of progress.

REFERENCES

1. J.R. KLAUDER, Ann. Phys. 11, 123 (1960).
2. V. BARGMANN, Communs. Pure Appl. Math. 14, 187 (1961).
3. I.E. SEGAL, Illinois J. Math. 6, 500 (1962).
4. I. BIALYNICKI-BIRULA, Ann. Phys. 67, 252 (1971).
5. E.H. LIEB, Commun. Math. Phys. 31, 327 (1973).
6. K. HEPP, Commun. Math. Phys. 35, 265 (1973).
7. I.M. GELFAND et. al., *Generalized Functions*, vol. 5 (Academic Press, New York, 1966).
8. R.F. STREATER and A.S. WIGHTMAN, *PCT, Spin and Statistics and All That* (Benjamin, New York, 1964).

GAUGE FIELDS OVER THE SUPER-SYMMETRY SPACE

Richard Kerner

In some recently published papers Arnowitt, Nath, Zumino, Salam and others have considered an extension of the super-symmetry formalism which includes quite naturally both the gravitational and the non-abelian gauge fields. The technique used by these authors is essentially an extension of the Cartan differential calculus onto a space of generalized coordinates containing the four space-time variables x^i (i,j=0,1,2;3) and the four anti-commuting spinorial variables θ^α (α,β=1,2,3,4). A point in this generalized manifold (called hereafter the super-symmetry space) is labeled by means of the generalized coordinates z^A (A,B=1,2,...,8).

The differential forms dz^A can be easily introduced, having the following properties:

	x^i	θ^α	dx^i	$d\theta^\alpha$
x^j	c	c	c	c
θ^β	c	a	c	a
dx^j	c	c	a	c
$d\theta^\beta$	c	a	c	c

Here a means anticommutation, and c means commutation. We say that the z^A form a graded Lie algebra with the multiplication law given by

$$z^A z^B = (-1)^{p(A)p(B)} z^B z^A$$

where

$$p(A) = \begin{cases} 0 & \text{if } A=i,j,\ldots \\ 1 & \text{if } A=\alpha,\beta,\ldots \end{cases}$$

A generalized differential geometry can be introduced on the super-symmetry space. It is defined by a collection of the fundamental 1-forms:

$$\xi^A = dz^M E^A_M(z)$$

and the connection 1-forms:

$$\omega^B_A = dz^M \tilde{\omega}^B_{MA} = \xi^C \omega^B_{CA}(z).$$

We use the convention in which the differentials dz^A are always shifted to the left.

We define the flat metric over the super-symmetry space as follows:

$$\eta_{AB} = \begin{pmatrix} \eta_{ij} & 0 \\ 0 & \eta_{\alpha\beta} \end{pmatrix}$$

where η_{ij} is the usual Minkowskian metric, and $\eta_{\alpha\beta}$ is the spinorial metric, such that $\eta_{\alpha\beta} = -\eta_{\beta\alpha}$, $\det \eta_{\alpha\beta} = 1$. In what follows we assume that the super-symmetry space is a direct product of the four-dimensional space-time manifold and the linear space of the anticommuting variables θ^α.

The *flat* super-symmetry space has the following collection of fundamental 1-forms:

$$E^A_M = \begin{pmatrix} E^i_k = \delta^i_k & \vdots & E^\alpha_k = 0 \\ \hline E^k_\beta = i(\gamma^k)_{\beta\delta}\theta^\delta & \vdots & E^\alpha_\beta = \delta^\alpha_\beta \end{pmatrix}$$

Remark that

$$E^A_i \partial_A = \partial_i$$

and

$$E^\alpha_\alpha \partial_A = \partial_\alpha - i(\bar{\theta}\gamma^k)_\alpha \partial_k.$$

Furthermore, we can introduce the torsion and the curvature 2-forms:

$$\Theta^A = D\xi^A = d\xi^A - \xi^B\omega^A_B = \tfrac{1}{2}\xi^C\xi^B\Theta^A_{BC}$$

$$\Omega^B_A = D\omega^B_A = d\omega^B_A - \omega^C_A\omega^B_C = \tfrac{1}{2}\xi^D\xi^C R^B_{CDA}.$$

The explicit expressions are rather lengthy; here we give one of them as an example:

$$\Theta^A_{BC} = (-1)^{p(B)p(C)+p(B)p(M)}E^M_C E^N_B \partial_N E^A_M$$

$$- (-1)^{p(C)p(M)}E^M_B E^N_C \partial_N E^A_M - \omega^A_{BC} + (-1)^{p(B)p(C)}\omega^A_{CB}.$$

In order to generalize the Riemannian geometry we have to put

$$\omega^C_{AB} = \begin{Bmatrix} C \\ AB \end{Bmatrix} = \tfrac{1}{2}(-1)^{p(B)p(C)}[(-1)^{p(B)p(D)}\partial_B g_{AD}$$

$$+ (-1)^{p(A)[p(D)+p(B)+1]+p(B)}\partial_A g_{BD} - \partial_D g_{AB}]g^{DC}.$$

Then it is easy to generalize the notions of the Ricci tensor and the scalar curvature; the same is valid for the notion of the determinant. Finally, introducing formal integration of anticommuting forms θ^α as follows:

$$\int \theta^\alpha d\theta^\alpha = 1, \quad \int d\theta^\alpha = 0,$$

we can write the generalized variational principle for the super-symmetry space:

$$\delta \int d^8z \sqrt{-g}\, R = 0$$

which gives of course the generalized Einstein equations:

$$R_{BA} - \tfrac{1}{2} g_{BA} R = 0.$$

It is important to note that every component of the generalized metric $g_{AB}(z)$ is now a super-field, i.e. a polynomial in θ's with the coefficients being fields of different spin. We write symbolically:

$$g_{AB}(z) = \{g_{ij}(z), g_{i\beta}(z), g_{\alpha j}(z), g_{\alpha\beta}(z)\}$$

where the components have the following properties:

$$[g_{ij}, g_{k\ell}] = 0, \quad [g_{\alpha\beta}, g_{\gamma\delta}] = 0$$

and

$$\{g_{i\beta}, g_{j\delta}\}_+ = 0.$$

Therefore, the $g_{ij}(z)$ expand as follows:

$$g_{ij}(z) = \overset{\circ}{g}_{ij}(x) + \bar{\theta}\gamma_i \psi(x)\bar{\psi}(x)\gamma_j\theta + \bar{\theta}\theta p_{ij}(x) + \ldots, \text{ etc.},$$

and the $g_{i\beta}(z)$ can be written as:

$$g_{i\beta}(z) = \overset{\circ}{\psi}_{i\beta}(x) + V_i(x)\theta_\beta + \bar{\theta}_\beta(\bar{\chi}(x)\gamma_i\theta) + \ldots, \text{ etc.}$$

The next problem arising naturally is to combine in some way the internal symmetries of the type SU(3), SU(2), etc., with the super-symmetry formalism. This can be done in two different ways: one, proposed by Arnowitt, Nath and Zumino, is to multiply the number of anticommuting variables by N, labeling them with a supplementary index, and assume that some irreducible representation of the internal symmetry group acts in the Cartesian product thus obtained. Here we propose a slightly different way, which is to construct a fibre bundle over the super-symmetry space with the structural group SU(n). The group SU(n) admits a natural metric which is the Cartan-Killing form; therefore, we can construct a Riemannian metric over the fibre bundle. Introducing the connection in the fibre bundle compatible with this metric, we obtain the analog of the Yang-Mills theory.

Let the metric in the flat super-symmetry space be

$$g_{AB} = \begin{pmatrix} g_{ij} & 0 \\ 0 & \eta_{\alpha\beta} \end{pmatrix}$$

and the Killing-Cartan form be given by $g_{ab} = C^{d}_{ac}C^{c}_{db}$ where C^{a}_{bc} are the structure constants of SU(n), a,b,... = 1,2,...,N, N = dim SU(n).

Let the 1-form of the connection in the fibre bundle over the super-symmetry space be

$$\omega^{a}_{\Gamma} = (\delta^{a}_{b},\omega^{a}_{A}) = (\delta^{a}_{b},\omega^{a}_{j},\omega^{a}_{\alpha})$$

where

$i,j = 0,1,2,3$ (space-time indices)

$\alpha,\beta = 1,2,3,4$ (spinor indices)

$A,B = 1,2,\ldots,8$ (super-symmetry space indices)

$a,b = 1,2,\ldots,N$ (group indices)

$\Gamma,\Delta = 1,2,\ldots,N+8$ (fibre bundle indices).

The curvature 2-form of the bundle is written locally as

$$F^a_{AB} = \partial_A \omega^a_B - (-1)^{p(B)} \partial_B \omega^a_A + \tfrac{1}{2} C^a_{bc} [\omega^b_A \omega^c_B - (-1)^{p(A)p(B)} \omega^b_B \omega^c_A].$$

There exists one and only one canonical metric over the fibre bundle, compatible with the connection, i.e. such that the horizontal and the vertical subspaces of the tangent space are orthogonal to each other. This metric is:

$$g_{\Gamma\Delta} = \left(\begin{array}{c|c} g_{AB} + \omega^a_A g_{ab} \omega^b_B & \omega^a_A g_{ab} \\ \hline g_{ab} \omega^b_B & g_{ab} \end{array}\right)$$

and

$$g^{\Delta\Pi} = \left(\begin{array}{c|c} g^{AB} & -g^{AB} \omega^b_B \\ \hline -\omega^a_A g^{AB} & g^{ab} + \omega^a_A g^{AB} \omega^b_B \end{array}\right).$$

Here g_{AB} is the flat metric given above, whereas the connection coefficients ω^a_A are supposed to be the superfields. The variational principle analogous to the Einsteinian one will yield the following equations:

$$(-1)^{p(A)} g^{AB} \nabla_A F^a_{CB} + (-1)^{p(A)} C^a_{bc} g^{AB} \omega^b_A F^c_{CB} = 0$$

and

$$T_{AB} = g_{ab} (-1)^{p(C)p(D)} F^a_{AC} g^{CD} F^b_{BD}$$

$$+ \tfrac{1}{4} g_{ab} (-1)^{p(C)p(D)+p(E)p(F)} g_{AB} F^a_{CD} g^{CE} g^{FD} F^b_{EF}.$$

The first set of equations is an analog to the equations of motions in the usual Yang-Mills theory, whereas the second set is analogous to the definition of the momentum-energy tensor. Here, however, due to the super-symmetry, the content of these equations is even richer. For example, in the second set of equations, when

$A,B = i,j$, we obtain the definition of the energy-momentum tensor as usual; but when $A,B = i,\beta$ or $A,B = \alpha,\beta$, we obtain some supplementary equations defining the sources and the currents, as well as some gauge conditions eliminating some degrees of freedom.

In the simplest cases, e.g. when we put

$$\omega^a_A = \{\omega^a_i = 0, \omega^a_\alpha = \varphi^a(x)\theta_\alpha\}$$

we obtain the usual equations for the massless scalar multiplet:

$$F^a_{ij} = 0, \quad F^a_{i\beta} = -F^a_{\beta i} = \partial_i\varphi^a\theta_\beta,$$

$$F^a_{\alpha\beta} = F^a_{\beta\alpha} = 2\delta_{\alpha\beta}\varphi^a,$$

and

$$\square\,\varphi^a = 0, \quad T_{ij} = \tfrac{1}{4}g_{ab}\partial_i\varphi^a\partial_j\varphi^b.$$

The equations for the classical Yang-Mills field interacting with the spinor massless field are obtained with the following connection:

$$\omega^a_A = \{\omega^a_i = \bar{\psi}^a\gamma_i\theta + \bar{\theta}\gamma_i\psi^a + A^a_i(x), \quad \omega^a_\alpha = \psi^a_\alpha + (\gamma^i A^a_i\theta)_\alpha\}.$$

Then we obtain

$$\gamma^i(\delta^b_c\partial_i - C^b_{ac}A^a_i)\psi^c = 0, \quad \bar{\psi}^c(\delta^b_c\overleftarrow{\partial}_i + C^b_{ac}A^a_i)\gamma^i = 0$$

and

$$\partial^i F^a_{ij} + C^a_{bc}A^{bi}F^c_{ij} = C^a{}_c(\bar{\psi}^b\partial_j\psi^c - \partial_j\bar{\psi}^b\psi^c)$$

where

$$F^a_{ij} = \partial_i A^a_j - \partial_j A^a_i + C^a_{bc}A^b_i A^c_j$$

etc.

One can easily see that all the super-fields corresponding to the connections are massless, just as in the classical Yang-

Mills theory. The invariant interactions with the massive fields can be introduced by admitting that not only the connection but also the underlying metric g_{AB} is a super-field (see (3)). It has been also shown recently by Fayet (7) that the Higgs-Kibble mechanism of building up the mass due to the broken symmetry can work as well in the super-field formalism.

REFERENCES

1. J. WESS and B. ZUMINO, Nucl. Phys. B70, 39 (1974).
2. A. SALAM and J. STRATHDEE, Nucl. Phys. B76, 477 (1974).
3. R. ARNOWITT and P. NATH, Northeastern Univ. Preprint N.U.B. 2246 (1975).
4. R. ARNOWITT, P. NATH and B. ZUMINO, Phys. Lett., 56 B, No. 1, 81 (1975).
5. R. KERNER, Ann. Inst. H. Poincaré, Vol. IX, No. 3, 141 (1968).
6. A. KOTECKY, Rep. on Math. Phys., Vol. 8 (1975).
7. P. FAYET, Nuovo Cimento A, Vol. 31A, No. 4 (1976).

ON SUPERSYMMETRY AND THE HOPE FOR A NON-TRIVIAL FUSION BETWEEN INTERNAL AND SPACE-TIME SYMMETRIES

C.N. Ktorides

Supersymmetry is the name given to certain relativistic (in the sense that they contain the Poincaré generators) graded Lie algebras. We recall that these algebras assign a grade to each of their elements, which then fall into even and odd categories. The most basic properties of the graded Lie product ( ) are the following two

$$(V_\ell V_k) = V_{\ell+k}; \qquad (V_\ell V_k) = -(-1)^{\ell k}(V_k V_\ell) \tag{1}$$

where the subscript denotes the grade of the corresponding element. The second property shows that the graded Lie product is symmetric with respect to the exchange $V_\ell \leftrightarrow V_k$ only if the two elements are odd; otherwise it is antisymmetric.

For the supersymmetry case, the odd elements are identified as Lorentz (or SL(2,C)) Majorana spinors. This is the most striking and potentially significant feature of supersymmetry. In fact, for the first time one possesses in relativistic theories a non-trivial symmetry scheme which could possibly encompass all particles occurring in nature (bosons and fermions). These odd elements, to be referred to as spinorial charges from now on, were independently introduced by Volkov and Akulov [1] and by Wess and Zumino [2].

The original supersymmetry algebra proposed by Wess and Zumino had 8 spinorial charges $Q_\alpha^{(0)}$, $Q_\alpha^{(1)}$, $\alpha=1,\ldots,4$, the 15 generators of the conformal group SO(4,2) and the generator Π of γ_5-transformations. The graded structure is as follows: $K_\mu(+2)$, $Q_\alpha^{(1)}(+1)$, $M_{\mu\nu}$, D, $\Pi(0)$, $Q_\alpha^{(0)}(-1)$, $P_\mu(-2)$. Our notation adheres to the familiar one for the conformal group and needs no further explanation.

Some examples of supersymmetry (anti)commutators, exhibiting the graded structure, are the following

$$[Q_\alpha^{(0)},P_\mu] = [Q_\alpha^{(1)},K_\mu] = 0;$$

$$\{Q_\alpha^{(0)},Q_\beta^{(1)}\} = -2(\gamma^{\mu\nu}M_{\mu\nu}-D+2\gamma_5\Pi)_\alpha^{\ \gamma}(\gamma_0)_\gamma^{\ \beta};$$

$$[Q_\alpha^{(1)},P_\mu] = (\gamma_\mu)_\alpha^{\ \beta}Q_\beta^{(0)}; \quad [Q_\alpha^{(0)},K_\mu] = (\gamma_\mu)_\alpha^{\ \beta}Q_\beta^{(1)}; \text{ etc.} \tag{2}$$

Of the multitude of ideas which supersymmetry gave rise to, regarding its physical applicability, the only one which has received a thorough and satisfactory treatment up to now is the question of a possible non-trivial fusion between internal and space-time symmetries. We shall turn our attention to this particular problem by first recalling the motivations supersymmetry has given for its consideration. Then, we shall proceed with a constructive proof of the answer, originally given by Haag, Lopuszanski and Sohnius [3].

Let J_A be the generators of a certain internal symmetry group, say SU(n). The index A refers to an SU(n)-vector space, in particular the algebra su(n). Therefore, the J_A are completely insensitive to a Lorentz rotation occurring in Minkowski space. In short, the J_A are Lorentz *scalars* and we must have the corresponding transformation property

$$[M_{\mu\nu},J_A] = 0. \tag{3}$$

This commutation relation lies at the heart of the so-called no-go theorems which brought to a halt all attempts aiming at a non-trivial mixing of an internal symmetry group with the Poincaré group.

Supersymmetry, with its spinorial charges, leads to the following commutation relations

$$[M_{\mu\nu}, Q_\alpha] = (\sigma_{\mu\nu})_\alpha^{\ \beta} Q_\beta \tag{4}$$

where $\sigma_{\mu\nu} = 2i(\gamma_\mu\gamma_\nu - \gamma_\nu\gamma_\mu)$. Relation (4) simply expresses the transformation property of a Lorentz-spinorial object.

Comparing (3) with (4) we see that the spinorial charges open up a new dimension to the question whether internal symmetries can be non-trivially fused with the Poincaré group. The immediate problem being posed to us now is the following: Can the Q_α be non-trivially linked with the generators of an (internal) unitary group?

An affirmative answer to this question was given by Haag et al [3]. They found that a necessary condition is the inclusion of the *whole* Wess and Zumino supersymmetry algebra and not just the algebra spanned by $M_{\mu\nu}$, P_μ and $Q_\alpha^{(0)}$. The latter has been the one which has attracted the attention of most supersymmetry workers.

The approach of Haag et al has been purely algebraic. Explicitly, they inserted in the supersymmetry algebra additional generators while at the same time they expanded the spinorial sector by introducing an index L running from 1 to n. They systematically exploited (graded) Jacobi identities in order that the emerging algebra, spanned by $M_{\mu\nu}$, P_μ, K_μ, D, $Q_\alpha^{(0)L}$, $Q_\alpha^{(1)L}$ as well as the additional generators W_A, form a graded Lie algebraic structure in a consistent manner. The result is that, given n, the generators W_A span the algebra $\underset{\sim}{u(n)}$. Furthermore, the whole structure is non-trivially linked together, which means that the internal symmetry sector is inseparable from the rest of the algebra.

We consider this result as pointing to the direction which sypersymmetry should take for future progress. Our task [4], in collaboration with M. Daniel, has been to understand this result constructively, the scope being three-fold. First, to identify the tools and the building blocks of supersymmetry. Second, to better understand the structure of supersymmetry itself and to acquire a working familiarity with it. Finally, to prove by construction the result of Haag et al, thus enhancing its validity.

The answer to the first question had already been furnished by Ferrara [5], whose work came at the early stages of supersymmetry and was, therefore, preoccupied with its algebraic structure. Ferrara correctly identified SU(2,2) spinors (i.e., Penrose's twistors) as the building blocks of supersymmetry. For the kinds of insights we want to obtain we must go back to the basic work of Cartan on spinors [6]. Cartan's work is on a purely classical, and mostly geometric, level so we can employ his spinors without the reservations one has when dealing with their quantum counterparts.

Cartan informs us that given any $(2m+\varepsilon)$-dimensional (pseudo) Euclidean space $E_{2m+\varepsilon}$, m integer and $\varepsilon = 0,1$, there exists by construction a corresponding 2^m-dimensional spinor space which we shall denote by S_{2^m}. It can be readily shown how Cartan's construction yields a space that carries a double-valued representation of the rotation group in $E_{2m+\varepsilon}$; this is the unmistakable property which physicists know in conjunction with spinorial representations.

The same construction which furnishes S_{2^m} also gives the Clifford algebra $C_{2m+\varepsilon}$ associated with $E_{2m+\varepsilon}$. This is a $2^{2m+\varepsilon}$-dimensional algebra with a unique finite irreducible representation in terms of $2^m \times 2^m$ matrices. The one-vectors of $C_{2m+\varepsilon}$ are in a 1-1 correspondence with the tangent vectors of $E_{2m+\varepsilon}$. Thus, we have a natural representation in terms of $2^m \times 2^m$ matrices for every $E_{2m+\varepsilon}$-vector. In addition to one-vectors, $C_{2m+\varepsilon}$ possesses two-vectors (corresponding to second rank tensors), three-vectors, etc. up to an including the $(2m+\varepsilon)$-vector which is actually a pseudo-scalar quantity. Finally, $C_{2m+\varepsilon}$ also includes the unit scalar (identity matrix). The following theorem now is of extreme importance: The two-vectors of $C_{2m+\varepsilon}$ are in a 1-1 correspondence with the generators of the (pseudo) rotation group in $E_{2m+\varepsilon}$. By local isomorphism the same correspondence with the generators of its doubly-covering spin group is also valid.

The common language between tensors and spinors at our disposal can be exploited in order to give answers to our remaining two questions. We shall accordingly narrow our considerations to the pseudo-Euclidean space E(4,2). Cartan's analysis is equally valid for pseudo- as well as Euclidean spaces, modulo a number of subtleties. Here we just need to mention the presence of the conjugation matrix J for the pseudo-Euclidean case; J is essential for the definition of conjugate spinors and is intimately connected with the signature.

We characterize the supersymmetry algebra as the inhomogeneous algebra of rotations in the spinor space S_8, associated with E(4,2), plus an extra generator Π. The SU(2,2) generators define infinitesimal rotations in S_8 while the spinor charges $(Q_\alpha^{(0)}, Q_\alpha^{(1)})$, $\alpha=1,\ldots,4$, generate translations. The following lemma [4] summarizes the above result:

Lemma. There exist bilinear expressions in ξ_a, ξ'_b (elements of S_8), a,b=1,...,8, which, if symmetrized with respect to the exchange $\xi \leftrightarrow \xi'$, form exclusively the components of a two-vector or of the pseudo-scalar in $C_{2m+\varepsilon}$.

By setting $\xi \equiv \xi'$ the above symmetrization is automatically achieved. Therefore, if we solve for the symmetric combinations $\xi_a\xi_b + \xi_b\xi_a$ we must get a linear combination of bivector components as well as a pseudo-scalar part only:

$$\xi_a\xi_b + \xi_b\xi_a \equiv \{\xi_a,\xi_b\} = \eta^{AB}(\gamma_{AB}J)_{ab} + \eta(\beta_7J)_{ab} \tag{5}$$

where the γ_{AB} form a two-vector basis in $C_{4,2}$, β_7 is the unit pseudo-scalar, J is the conjugation matrix and η, η_{AB} are the expansion coefficients.

By raising (5) to a generator status - just as one goes from p_μ, the dual coordinate to x_μ, to the generators of translations P_μ - we obtain the most fundamental relations of the supersymmetry scheme

$$\{\Xi_a,\Xi_b\} = K^{AB}(\gamma_{AB}J)_{ab} + \Pi(\beta_7J)_{ab} \tag{6}$$

where Ξ is an 8-spinor which can be broken into two 4-spinors $Q^{(0)}$ and $Q^{(1)}$, K^{AB} are the fifteen generators of SU(2,2) and Π is the generator of (pseudo-scalar) γ_5-transformations. We remark that the appearance of Π has been forced on us by our lemma.

Two comments are now in order about Π. First, we note the consistency between its presence and that of the conformal group. Indeed, γ_5-invariance imposes, among other things, masslessness on the spinorial sector of a given Lagrangian density. Masslessness, on the other hand, is the "trade-mark" of conformal invariance. Second, and more important, is the fact that this extra generator Π contains the seeds of the inherent presence of unitary group generators within the supersymmetry framework. To see this one must expand the spinorial sector à la Haag et al by introducing the additional index L, running from 1 to n. What one now finds is that the conformal generators are still present but Π is replaced by a greater number of generators which depends on n. In fact, by following the same steps which led to (6) one now obtains [4]

$$\{\Xi^L_a,\Xi^M_b\} = \delta^{LM}(\gamma_{AB}J)_{ab}K^{AB} + B^{LM}(\beta_7J)_{ab} + B{*}^{LM}(J)_{ab} \tag{7}$$

where the last term on the right is a $C_{4,2}$ scalar contribution, which is non-vanishing because the indices L and M spoil the exchange symmetry $\Xi \leftrightarrow \Xi'$.

In Ref. 4 we have explicitly shown that the generators B^{LM}, B^{*LM} span the algebra $\underset{\sim}{u(n)}$, which can be split into $\{\Pi\} \oplus \underset{\sim}{su(n)}$ where $\{\Pi\}$ denotes the one-dimensional algebra of infinitesimal γ_5-transformations. Relation (7) constitutes the answer to our third question and it reduces to the relations of Haag et al once the Ξ^L's are split into their two 4-spinor parts $Q^{(0)L}$ and $Q^{(1)L}$.

REFERENCES

1. D.V. VOLKOV and V.P. AKULOV, Phys. Letters 46B, 109 (1973).
2. J. WESS and B. ZUMINO, Nucl. Phys. B70, 39 (1974).
3. R. HAAG, J. LOPUSZANSKI and M. SOHNIUS, Nucl. Phys. B88, 257 (1975).
4. M. DANIEL and C.N. KTORIDES, Univ. of Athens preprint, to appear in Nuclear Physics B.
5. S. FERRARA, Nucl. Phys. B72, 73 (1974).
6. E. CARTAN, *The Theory of Spinors*, MIT Press (1966).

PHYSICS AND LIE PSEUDOGROUPS

J.F. Pommaret

We shall first recall some basic definitions that can be found in the local theory of Lie pseudogroups (4,6). Then we shall focus on the definition and use of the normalizer of a Lie pseudogroup. Finally we shall indicate the physical motivations involved and propose a conjecture.

As our study will deal mainly with local problems, we shall use local coordinates and illustrate the definitions with basic examples, the computations of which are straightforward and left to the reader.

Let X be a C^∞ connected, paracompact manifold with dim X = n and Y be a copy of X.

Definition. By a *"finite transformation"* $f:X \to Y$ with *"source"* $x \in X$ and *"target"* $y \in Y$, we mean an invertible map written $y^k = f^k(x^i)$, using suitable local coordinates. By an *"infinitesimal transformation"* ξ, we understand the map $\exp(t\xi)$ where ξ is a section of the tangent bundle $T = T(X)$ and t is a small parameter. We have $y^k = x^k + t\xi^k(x) + \ldots$.

If $\mathcal{E}$ is a fibered manifold (fiber bundle) over X with local coordinates (x^i,y^k) we denote by $J_q(\mathcal{E})$ the fibered manifold (fiber bundle) of q-jets of $\mathcal{E}$, using local coordinates (x^i,y^k,p^k_μ) with

$\mu = (\mu_1,\ldots,\mu_n)$ and $1 \le |\mu| = \mu_1 + \ldots + \mu_n \le q$ or simply (x^i, y^k_μ) with $0 \le |\mu| \le q$ and $y^k_0 = y^k$ [2,6].

Definition. By a *"nonlinear system"* of partial differential equations of order q on $\mathcal{E}$ we mean a fibered submanifold $\mathcal{R}_q \subset J_q(\mathcal{E})$, that is to say a submanifold with surjective induced projection onto X. By a *"solution"* of $\mathcal{R}_q$ we mean a local section f of $\mathcal{E}$ over an open set $U \subset X$ such that $j_q(f)(x) \in \mathcal{R}_q$, $\forall\ x \in U$ [2].

Definition. In a similar way, if E is a vector bundle over X, by a *"linear system"* of p.d.e. we mean a vector subbundle $R_q \subset J_q(E)$ and by a solution of R_q we mean a section ξ of E over an open set $U \subset X$ such that $j_q(\xi)(x) \in R_q$, $\forall\ x \in U$ [5].

Now, the transformation groups of a manifold (in general $\mathbb{R}^4$) that can be found in physics are of the following types:

1) Action of a Lie group G on X noted: $G \times X \to Y$.

2) Invariance of one or a set of tensor fields, sometimes connected by some (generally algebraic) relations.

It has to be noticed that, in both cases, the finite (infinitesimal) transformations to be considered are solutions of a nonlinear (linear) system of p.d.e. called *"finite"* (*"infinitesimal"*). However, one generally looks for the infinitesimal generators of the action in case 1), rather than to the defining systems of p.d.e.

Examples. We write

$$y' = \frac{\partial y}{\partial x} = p.$$

1) Action on $\mathbb{R}$:

$$y = x + a, \qquad y = ax + b, \qquad y = \frac{ax+b}{cx+d}.$$

Finite equations:

$$y' - 1 = 0, \qquad y'' = 0, \qquad \frac{y'''}{y'} - \frac{3}{2}\left(\frac{y''}{y'}\right)^2 = 0.$$

Infinitesimal equations:

$$\xi' = 0, \qquad \xi'' = 0, \qquad \xi''' = 0.$$

2) Let $f: X \to Y$, $f_*: T(X) \to T(Y)$ and ω be a p-form defined on X. Then the finite equations are written $f^*(\omega) = \omega$ and it is well known that the corresponding infinitesimal equations are $\mathcal{L}(\xi)\omega = 0$ where we have introduced the classical Lie derivative.

3) The following type is different from the preceding ones:

$$X = \mathbb{R}^2.$$

Finite transformations:

$$y^1 = f(x^1), \qquad y^2 = x^2/f'(x^1).$$

Finite equations:

$$y^2 \frac{\partial y^1}{\partial x^1} - x^2 = 0, \qquad \frac{\partial y^1}{\partial x^2} = 0, \qquad x^2 \frac{\partial y^2}{\partial x^2} - y^2 = 0.$$

Infinitesimal equations:

$$x^2 \frac{\partial \xi^1}{\partial x^1} + \xi^2 = 0, \qquad \frac{\partial \xi^1}{\partial x^2} = 0, \qquad x^2 \frac{\partial \xi^2}{\partial x^2} - \xi^2 = 0.$$

Remark. In fact we can write

$$\begin{pmatrix} y^2 & dy^1 \\ \frac{1}{y^2} & dy^2 \end{pmatrix} = \begin{pmatrix} 1 & 0 \\ a & 1 \end{pmatrix} \begin{pmatrix} x^2 & dx^1 \\ \frac{1}{x^2} & dx^2 \end{pmatrix},$$

eliminating the constant a, and the reader will notice that this is a G-structure where $G \supset GL(2,\mathbb{R})$ is a one parameter Lie subgroup which is isomorphic to the additive group of $\mathbb{R}$.

From the above examples, identifying a finite transformation $f: X \to Y$ with its graph also denoted $f: X \to X \times Y$, we are led to the following definitions, which are not the most useful (unless in the analytic case) but which will be convenient for our purpose:

Definition. A Lie pseudogroup Γ is a group of transformations that are solutions of a nonlinear system $\mathcal{R}_q \subset J_q(X\times Y)$, that is to say:

• If f and g are two solutions, then $g \circ f$, when it is defined, is also a solution.

• If f is a solution, then f^{-1} exists and is a solution.

We shall write locally

$$\mathcal{R}_q \qquad \phi^\tau(x,y,p) = 0, \qquad \tau=1,\ldots,\operatorname{codim} \mathcal{R}_q.$$

From the group axioms we see that the identity transformation id must be a solution. Thus we can linearize the nonlinear system $\mathcal{R}_q$ in order to get the linear system R_q defining the infinitesimal transformations of the pseudogroup Γ:

$$R_q \qquad A_k^{\tau\mu}(x)\partial_\mu\xi^k \equiv \frac{\partial\phi^\tau}{\partial y_\mu^k}(j_q(id)(x))\partial_\mu\xi^k = 0.$$

Definition. Γ is said to be "*transitive*" if $\forall\, x \in X$ and $\forall\, y \in Y$, there exists a neighbourhood U of x and a solution f of $\mathcal{R}_q$ such that $y = f(x)$.

Remark. When Γ is transitive, it can be shown [4,6] that there is a nice way to write out the equations of $\mathcal{R}_q$ by pushing out the source in the right members. This will not be used explicitly in the sequel, though it is very important in order to prove the results.

From now on we shall be concerned only with transitive Lie pseudogroups.

Preliminary definition. The "*normalizer*" $\tilde{\Gamma}$ of a Lie pseudogroup Γ is the biggest Lie pseudogroup contained in the Lie pseudogroup Aut(X) of local automorphism of X, in which Γ is normal.

However, though it is intuitive, this definition is not a good one because it is difficult to work with it. If we denote by Θ the set of solutions of R_q, we shall adopt the following definition:

Definition. $\tilde{\Gamma}$ is the set of finite transformations $\tilde{f}: U \subset X \to V \subset Y$ such that $\tilde{f}_*(\Theta(X)|U) \subset \Theta(Y)|V$ with $\Theta = \Theta(X)$.

When R_q is formally integrable [2,5], that is to say when we cannot get other equations of order q by differentiations and eliminations, the former definition is equivalent to $\tilde{f}(R_q(X)|U) \subset R_q(Y)|V$ with $R_q = R_q(X) \subset J_q(T(X))$ and $R_q(Y) \subset J_q(T(Y))$.

The next example will make clear how R_q is transformed by $\tilde{f}$:

Example.

$$\Gamma: \mathbb{R} \to \mathbb{R}: y = ax, \qquad \mathcal{R}_1: \frac{1}{y}\frac{\partial y}{\partial x} = \frac{1}{x}, \qquad R_1: \frac{1}{x}\frac{\partial \xi}{\partial x} - \frac{\xi}{(x)^2} = 0.$$

Defining N(y) by

$$N(\tilde{f}(x)) = \frac{\partial \tilde{f}(x)}{\partial x} \cdot \xi(x)$$

we must have

$$\frac{1}{y}\frac{\partial N(y)}{\partial y} - \frac{N(y)}{(y)^2} = 0$$

when

$$\frac{1}{x}\frac{\partial \xi(x)}{\partial x} - \frac{\xi(x)}{(x)^2} = 0.$$

As

$$\frac{\partial N(\tilde{f}(x))}{\partial y} \cdot \frac{\partial \tilde{f}(x)}{\partial x} = \frac{\partial^2 \tilde{f}(x)}{\partial x^2} \cdot \xi(x) + \frac{\partial \tilde{f}(x)}{\partial x} \cdot \frac{\partial \xi(x)}{\partial x}$$

it follows that $\tilde{f}$ must be a solution of the nonlinear system:

$$\tilde{\mathcal{R}}_2 \qquad\qquad - \frac{y''}{y'} + \frac{1}{y} y' = \frac{1}{x}$$

and we get $\tilde{\Gamma}: \mathbb{R} \to \mathbb{R} : y = ax^b$ with a and b arbitrary constants.

Remark. Doing the same computation with $\Gamma : y = \frac{ax+b}{cx+d}$ we get $\tilde{\Gamma} = \Gamma$. Thus, if two pseudogroups Γ_1 and Γ_2 are such that $\Gamma_1 \subset \Gamma_2$, we have not in general $\tilde{\Gamma}_1 \subset \tilde{\Gamma}_2$.

Remark. As the reader can check for the G-structure given at the beginning, when Γ is defined by a G-structure, then the normalizer $\tilde{G}$ of G in $GL(n, \mathbb{R})$ has nothing to do in general with $\tilde{\Gamma}$. This is the reason why we believe it is not useful in general to try to generalize the idea of G-structure for an arbitrary pseudogroup.

Remark. $\forall\, x \in X$, we may find many components for the fibers $\tilde{\mathcal{R}}_{q+1,x}$ but we shall keep only the ones containing the point $j_q(id)(x)$, and their union for $x \in X$ is called the *"component of the identity"*.

Example. When $\Gamma : \mathbb{R}^2 \to \mathbb{R}^2$ is defined by

$$\mathcal{R}_1 : \frac{\partial y^1}{\partial x^2} = 0, \qquad \frac{\partial y^2}{\partial x^1} = 0$$

there are two components for $\tilde{\mathcal{R}}_2$, and one can pass from one to the other by the finite transformation $y^1 = x^2$, $y^2 = x^1$.

Thus we can get a chain of linear systems R_q, $\tilde{R}_{q+1}$, $\tilde{\tilde{R}}_{q+2}, \ldots$ increasing the order by one each time, or of pseudogroups $\Gamma, \tilde{\Gamma}, \tilde{\tilde{\Gamma}}, \ldots$.

Conjecture. The chain is stationary after a finite number of steps.

Examples. All the classical examples to be found in the literature (and in particular the so-called simple ones of the Cartan's classification) are such that $\tilde{\tilde{\Gamma}} = \tilde{\Gamma}$. The only example we know that stops one step further is given by the following action on $\tilde{f}$ with infinitesimal generators [3]:

$$\frac{\partial}{\partial x^1}, \qquad \frac{\partial}{\partial x^2} + x^1 \frac{\partial}{\partial x^3}, \qquad \frac{\partial}{\partial x^3}.$$

We may restrict the conjecture to the case, obtained as above, when Γ arises from the action of a Lie group G on X.

If the vector fields ξ_α are the infinitesimal generators of this action, we have $[\xi_\beta, \xi_\gamma] = C^\alpha_{\beta\gamma} \cdot \xi_\alpha$. If now $\tilde{\xi}$ is an infinitesimal transformation of $\tilde{\Gamma}$ there must exist a constant matrix $[B^\alpha_\beta]$ such that $[\tilde{\xi}, \xi_\beta] = B^\alpha_\beta \cdot \xi_\alpha$. It follows easily that $\tilde{\Gamma}$ is obtained by the action of a Lie group $\tilde{G}$ on X. When dim G = dim X, if $\mathcal{G}$ is the Lie algebra of G, the Lie algebra of $\tilde{G}$ is $\tilde{\mathcal{G}} = \mathcal{G} \oplus Z^1(\mathcal{G},\mathcal{G})$ with bracket $[(\lambda,\sigma),(\mu,\tau)] = ([\lambda,\mu]+\sigma(\mu)-\tau(\lambda), \sigma\circ\tau-\tau\circ\sigma)$ where $\lambda,\mu \in \mathcal{G}$ and $\sigma,\tau \in Z^1(\mathcal{G},\mathcal{G})$ (= derivations of $\mathcal{G}$). However, it is easy to see that the study of the chain $\mathcal{G}, \tilde{\mathcal{G}}, \tilde{\tilde{\mathcal{G}}}, \ldots$ is not a purely algebraic problem, because, in general, (unless dim G = dim X) dim $\tilde{\mathcal{G}}$ depends on dim $\mathcal{G}$ *and* dim X.

Remark. It would be very helpful to prove the above conjecture or at least to have a counter-example, because, as the computations are rational with respect to the jets of $\tilde{f}$ of order ≥ 1, it follows that $\tilde{\Gamma}, \tilde{\tilde{\Gamma}}, \ldots$ are algebraic pseudogroups in the sense of [4], whatever is Γ.

We may now state the main result we have proved:

Theorem. If the symbol G_q of R_q is 2-acyclic and if $R_{q+1} \to R_q$ is surjective, then $\tilde{G}_{q+r} = G_{q+r}$, $\forall\ r \geq 1$ and R_{q+1} is formally integrable.

Finally we give an example, taken from theoretical mechanics, that shows why to introduce the idea of the normalizer in physics.

Example. A canonical transformation is often defined by saying that it must transform any Hamiltonian flow into another one, and this seems to be exactly the definition of a certain normalizer.

In fact it is known by different methods [1,2] that if ω is a closed 2-form of maximum rank, from $\Gamma:\mathbb{R}^n \to \mathbb{R}^n$, $f^*(\omega) = \omega$ and $\mathcal{L}(\xi)\omega = 0$ we get $\tilde{\Gamma}:\tilde{f}^*(\omega) = a\omega$ and $\mathcal{L}(\tilde{\xi})\omega = A\omega$ with a,A constants.

We shall prove that, in the case of a *"homogeneous formulation"* of mechanics, we must have a = 1 (or A = 0). We use the following picture:

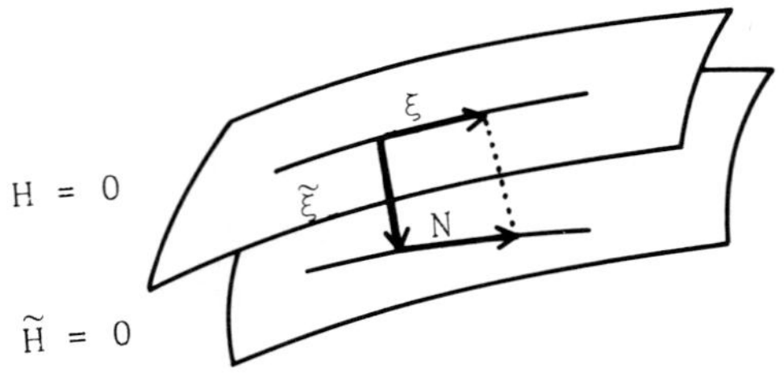

With classical notations, if $\mathcal{L}(\xi)\omega = i(\xi)d\omega + di(\xi)\omega = 0$, locally we have

$$\xi^i = \omega^{ij} \frac{\partial H}{\partial x^j}$$

where $[\omega^{ij}]$ is the matrix inverse to the one determined by the 2-form $\omega = \frac{1}{2}\omega_{ij}(x)dx^i \wedge dx^j$. Using the fact that $i(\tilde{\xi})\omega = A\omega$ and the well known transformation law of vector fields under infinitesimal transformations: $\xi \to N = \xi + t[\xi,\tilde{\xi}] + \dots$ a straightforward but tedious computation shows that:

$$N^i = \omega^{ij} \frac{\partial}{\partial x^j} (H - t\tilde{\xi}^r \frac{\partial H}{\partial x^r} - tAH) + \dots .$$

But H = 0 is the manifold of states and we do want that

$$N^i = \omega^{ij} \frac{\partial \tilde{H}}{\partial x^j}$$

with

$$\tilde{H} = H_0 \exp(-t\tilde{\xi}) = H - t\tilde{\xi}^r \frac{\partial H}{\partial x^r} + \dots .$$

Therefore we must have A = 0.

REFERENCES

1. A. AVEZ, A. LICHNEROWICZ, and A. DIAZ-MIRANDA, J. of Diff. Geometry, no. 1, 1-40 (1974).

2. H. GOLDSCHMIDT, J. of Diff. Geometry, 1, 269-307 (1967).

3. J.F. POMMARET, C.R. Acad. Sc. Paris, t. 280, A, 1495 (1975).

4. J.F. POMMARET, C.R. Acad. Sc. Paris, t. 280, A, 1693 (1975).

5. J.F. POMMARET, C.R. Acad. Sc. Paris, t. 282, A, 587 (1976).

6. J.F. POMMARET, C.R. Acad. Sc. Paris, t. 282, A, 635 (1976).

Part III

Classical and Quantum Mechanics

CLASSICAL FUNCTIONS ASSOCIATED WITH SOME GROUPS OF AUTOMORPHISMS OF THE WEYL GROUP

Ph. Combe, R. Rodriguez,
M. Sirugue-Collin and M. Sirugue

I. INTRODUCTION

It has been recently suggested to study Feynman path integrals in phase space, see [1,2]. An essential feature is the partition of the time interval and it is argued that, for small time, the Weyl quantization of the exponential of a classical hamiltonian function is the exponential of the quantized classical hamiltonian function. But, this is clearly not true since Weyl quantization procedure is not an algebraic homomorphism.

It is likely to be true in the limit $t \to 0$. Nevertheless, in the derivation of the Feynman path integral formula in phase space, one has to make this replacement n times and then to go to infinity. Hence, it is necessary to have better control of the above limits.

In this paper, we are interested in the first problem for the special situation where the hamiltonian is at most quadratic. This case is also of some interest in the study of the unitary representations of the metaplectic group [3].

In what follows we recall briefly the essential features of the quantization procedure and of the inversion formula, which connects an arbitrary bounded operator in the usual representation of canonical commutation relations to some functions on the classical phase space. Proposition (2.12), as a special case, connects the metaplectic group with the inhomogeneous symplectic group on phase space and allows a simplification of our results. Proposition (2.18) gives the correspondence between the one parameter subgroups of the inhomogeneous symplectic group to a family of classical functions on the classical phase space. The explicit calculations are carried out in the appendix.

Finally, in our case, we discuss the convergence of the exponential of the quantized hamiltonian to the quantized version of the exponential of the classical function as the time goes to zero.

II. CLASSICAL FUNCTIONS ASSOCIATED WITH THE METAPLECTIC GROUP

For the sake of simplicity, we shall restrict ourselves to systems with one degree of freedom, the general case only introduces notational complexity.

Let us consider R^2 the 2-dimensional vector space with a symplectic form

$$\sigma((x,p),(x',p')) = \frac{1}{4\pi}(px'-xp').^{*} \tag{2.1}$$

The Weyl correspondence allows us to associate with a rather general function f on the classical phase space an operator Q(f) [4,6].

$$Q(f) = \frac{1}{4\pi}\int \tilde{f}(v)\omega(-\frac{v}{2})dv \tag{2.2}$$

where dv is the Lebesgue measure on R^2,

$$\tilde{f}(v) = \frac{1}{4\pi}\int e^{2\pi i\sigma(v,v')}f(v')dv' \tag{2.3}$$

* $\hbar = 1$.

and $\omega(v)$ is the Weyl system, namely an application of R^2 to a group of unitary operators on some Hilbert space such that

$$\lambda \to \omega(v+\lambda v') \tag{2.4}$$

is continuous

$$\omega(v)\omega(v')^* = e^{-2\pi i\sigma(v,v')}\omega(v-v'). \tag{2.5}$$

Notice that the usual representation (x representation) on $L^2(R;dx)$ acts according to the formula

$$(\omega(\lambda,\mu)\varphi)(x) = e^{-\frac{i}{2}\lambda\mu} e^{i\mu x}\varphi(x-\lambda), \quad \varphi \text{ in } L^2(R;dx). \tag{2.6}$$

A formula equivalent to (2.2), see [7], uses the function f itself, namely,

$$Q(f) = \frac{1}{4\pi}\int f(\tfrac{v}{2})\omega(v)Mdv \tag{2.7}$$

where M is the parity operator defined by

$$(M\varphi)(x) = \varphi(-x), \quad \varphi \text{ in } L^2(R;dx). \tag{2.8}$$

Vice versa, given a bounded operator A on $L^2(R;dx)$, one can find a function f on the classical phase space such that

$$A = Q(f) \tag{2.9}$$

f being given explicitly by [8]

$$f(v) = 2\lim_{\beta\to 0} \mathrm{Tr}(e^{-\beta H}\omega(v)^*A\omega(v)M) \tag{2.10}$$

where Tr denotes the trace and H is the hamiltonian of a harmonic oscillator,

$$H = \tfrac{1}{2}(P^2+Q^2) \tag{2.11}$$

P and Q being the usual momentum and position operators. Our aim, in the following, is to use formula (2.10) to compute explicitly classical functions corresponding to some one parameter groups of unitary operators.

In order to simplify the notation, let us derive the following proposition:

Proposition (2.12). Let G be the inhomogeneous symplectic group, that is, the semidirect product of the symplectic group (which leaves invariant the symplectic form) with the translation group in R^2.

Let $g \to f^g$ be its action on a function on the phase space $f^g(u) = f(g^{-1}u)$. Let U_g be the unitary projective representation of G in $L^2(R;dx)$. Then

$$Q(f^g) = U_g Q(f) U_g^{-1}. \tag{2.12}$$

The unitary groups we have in mind are those which correspond to an infinitesimal generator K at most quadratic in P and Q. In this respect, proposition (2.12) is especially interesting in the sense that it allows us to consider the five typical cases [9].

$$K = P, \tag{2.13}$$

$$K = \frac{P^2}{2}, \tag{2.14}$$

$$K = \tfrac{1}{2}(P^2+Q^2), \tag{2.15}$$

$$K = \tfrac{1}{2}(P^2-Q^2), \tag{2.16}$$

$$K = \frac{P^2}{2} + Q. \tag{2.17}$$

The first case is trivial and we give in the appendix the calculation of exp(itK) in the other cases. Let us summarize the result as a proposition.

Proposition (2.18). We have the following correspondence

$$Q(f_t) = \exp it\,\frac{P^2}{2} \qquad f_t(x,p) = \exp it\,\frac{p^2}{2}$$

$$Q(f_t) = \exp \frac{it}{2}(P^2+Q^2) \qquad f_t(x,p) = \frac{1}{\cos\frac{t}{2}} \exp i(x^2+p^2)\,\mathrm{tg}\,\frac{t}{2}$$

$$Q(f_t) = \exp \frac{it}{2}(P^2-Q^2) \qquad f_t(x,p) = \frac{1}{\mathrm{ch}\,\frac{t}{2}} \exp i(p^2-x^2)\,\mathrm{th}\,\frac{t}{2}$$

$$Q(f_t) = \exp it\left(\frac{P^2}{2}+Q\right) \qquad f_t(x,p) = \exp i\left(\frac{p^2}{2}\,t+tx-\frac{t^3}{24}\right). \quad (2.18)$$

Using proposition (2.12), we get immediately the general case, namely the function f_t corresponding to the unitary group generated by the most general quadratic hamiltonian.

Notice that the classical function corresponding to some bounded operator A on $L^2(R;dx)$ can be used to write the action of A on a function in $L^2(R;dx)$ at least formally

$$(Q(f_t)\varphi)(x) = \frac{1}{2\pi}\iint f_t\left(\frac{\xi+x}{2},\zeta\right)e^{-i\zeta(\xi-x)}\varphi(\xi)\,d\xi\,d\zeta; \qquad (2.19)$$

this expression is a bit formal in the sense that in general f is not in the class that is considered for pseudodifferential operators [10]. Nevertheless, let us make some remarks.

i) Formula (2.19) is a computational tool for calculating the global solution of the Schrödinger equation. One can derive the Maslov index [11].

ii) Explicitly, in our examples,

$$\lim_{t\to 0} |f_t(x,p)-f_t^c(x,p)| = 0$$

uniformly on compact sets of the phase space R^2,

$$f_t^c(x,p) = \exp it\,h(x,p),$$

h(x,p) being the classical quadratic hamiltonian associated with K and this seems to be a step to the solution of the problem raised in section 1.

APPENDICES

From a theoretical point of view, it is possible to calculate the trace in (2.10) in different ways. Actually, for practical reasons, we shall use different methods in the three cases.

A. The classical function corresponding to $\exp(\frac{it}{2}(P^2+Q^2))$.

It is known that the trace of a trace class operator can be written

$$\mathrm{Tr}\, A = \frac{1}{2\pi} \int d^2\alpha (\alpha | A\alpha)$$

where $|\alpha\rangle$ is a coherent state defined as

$$\alpha = \omega(x,p)|0\rangle , \qquad \alpha = p - ix;$$

$|0\rangle$ is the vacuum state of the harmonic oscillator, $d^2\alpha = dxdp$. Thus

$$2\mathrm{Tr}(e^{-\beta H}\omega(\alpha')e^{\frac{it}{2}(P^2+Q^2)}\omega(\alpha')M)$$

$$= \frac{1}{\pi} \int d^2\alpha \langle\alpha| e^{-\beta H}\omega(\alpha')^* e^{it(P^2+Q^2)}\omega(\alpha')|-\alpha\rangle ,$$

$$= \frac{1}{\pi} \int d^2\alpha \langle\alpha| e^{-(\beta-it)H}\omega(e^{-it}\alpha')^*\omega(\alpha')|-\alpha\rangle$$

where we used explicitly that $M|\alpha\rangle = |-\alpha\rangle$.

Now after an obvious but tedious calculation using Weyl relations and the value of the scalar product of two coherent states

$$(\gamma|\gamma') = e^{-\frac{1}{4}(\gamma\bar{\gamma}'-\bar{\gamma}\gamma')}e^{-\frac{1}{4}|\gamma-\gamma'|^2}$$

we are led to calculate two Fourier transforms of Gaussians.

Finally

$$2\mathrm{Tr}(e^{-\beta H}\omega(\alpha')*e^{\frac{it}{2}(P^2+Q^2)}\omega(\alpha')M) = \frac{4e^{\frac{1}{2}(\beta+it)}e^{-\frac{1}{2}|\alpha'|^2(1-e^{it})}}{e^{2\beta}+1+2e^{\beta+it}}$$

$$\times \exp \frac{(e^{it}-1)^2[(\bar{\alpha}'-\alpha' e^{\beta})^2-(\bar{\alpha}'+\alpha' e^{\beta})^2]}{4(e^{2\beta}+1+2e^{\beta+it})}.$$

The limit $\beta \to 0$ can be performed without difficulty

$$\lim_{\beta\to 0} 2\mathrm{Tr}(e^{-\beta H}\omega(\alpha')*e^{\frac{it}{2}(P^2+Q^2)}\omega(\alpha')M) = \frac{1}{\cos\frac{t}{2}}\, e^{i(x'^2+p'^2)\operatorname{tg}\frac{t}{2}}.$$

B. Classical function corresponding to $\exp(\frac{it}{2}(P^2-Q^2))$.

By proposition (2.12), it is equivalent to study the unitary group generated by $\frac{1}{2}(PQ+QP)$ which is in fact the infinitesimal generator of the dilatation group, namely the unitary group in $L^2(R;dx)$, $t \to U_t$ such that

$$(U_t\psi)(x) = e^{-t/2}\psi(xe^{-t}). \tag{B.1}$$

Moreover, we calculate the trace in the form (2.10) using the orthogonal basis of eigenfunctions of the harmonic oscillator. Explicitly,

$$\psi_n(x) = \frac{1}{2^{n/2}\sqrt{n!}\,\pi^{\frac{1}{4}}} H_n(x)e^{-\frac{x^2}{2}}, \tag{B.2}$$

where $H_n(x)$ is the nth Hermite polynomial. Notice that

$$(M\psi_n)(x) = (-1)^n\psi_n(x). \tag{B.3}$$

Thus

$$\mathrm{Tr}(e^{-\beta H}\omega(v)^*U_t\omega(v)M) = e^{-\beta/2}\sum_{n=0}^{\infty}(-e^{-\beta})^n\frac{\pi^{-\frac12}e^{-t/2}}{2^n n!}$$

$$\times\int_{-\infty}^{+\infty} e^{-i\mu x}H_n(x-\lambda)e^{-\frac12(x-\lambda)^2}e^{i\mu(e^{-t}x)}H_n(e^{-t}x-\lambda)e^{-\frac12(e^{-t}x-\lambda)^2}dx, \tag{B.4}$$

where $v = (\lambda,\mu)$; we have explicitly used (2.6).

We can exchange the summation with the integration and use a classical formula about Hermite polynomials (see [12]). We obtain, with $z = -e^{-\beta}$, for (B.4)

$$(1-z^2)^{-\frac12}2\pi^{-\frac12}e^{-\beta/2}e^{-t/2}\int e^{i\mu x(e^{-t}-1)}e^{-\frac12(x-\lambda)^2}e^{-\frac12(e^{-t}x-\lambda)^2}$$

$$\times\exp(1-z^2)^{-1}\{2z(x-\lambda)(e^{-t}x-\lambda)-z^2[(x-\lambda)^2+(e^{-t}x-\lambda)^2]\}. \tag{B.5}$$

This last integral can be evaluated immediately:

$$2\mathrm{Tr}(e^{-\beta H}\omega(v)^*U_t\omega(v)M) = 2\sqrt2\,\{e^{\beta+t}[(1+z^2)(1+e^{-2t})-4ze^{-t}]\}^{-\frac12} \tag{B.6}$$

$$\times\exp\frac{-2i\mu\lambda(e^{-t}-1)(1-z)^2(e^{-t}+1)-\mu^2(e^{-t}-1)^2(1-z^2)-4\lambda^2(1-z)^2}{z[(1+z^2)(1+e^{-2t})-4ze^{-t}]}$$

$$\times\exp\frac{\lambda^2(1-z)^4(1+e^{-t})^2}{\{(1+z^2)(1+e^{-2t})-4ze^{-t}\}^2}. \tag{B.7}$$

The limit $\beta\to0$ gives the result

$$\frac{1}{\mathrm{ch}\,\frac{t}{2}}\exp 2i\lambda\mu\,\mathrm{th}\,\frac{t}{2}. \tag{B.8}$$

Coming back to the original generator $\frac12(P^2-Q^2)$, we have the desired result: the classical function is

$$f(\lambda,\mu) = \frac{1}{\operatorname{ch}\frac{t}{2}} \exp i(\lambda^2-\mu^2)\operatorname{th}\frac{t}{2}. \tag{B.9}$$

C. Classical function corresponding to $\exp it(\frac{P^2}{2}+Q)$.

In order to calculate the last case, let us derive the formula

$$e^{it(\frac{P^2}{2}+Q)} = \xi(t)\omega(\xi_t,\zeta_t)e^{it\frac{P^2}{2}}, \tag{C.1}$$

where $\xi(t)$ is a function of t whose modulus is one. Indeed, for fixed t, the transformations of $\mathcal{B}(\mathcal{H})$

$$A \to U_t A U_t^* \tag{C.2}$$

coincide, for U_t both members of (C.1). This can easily be checked on the $\omega(u)$, the Weyl operators, which are weakly dense in $\mathcal{B}(\mathcal{H})$. One has also to use for the r.h.s. the Dyson expansion:

$$e^{itH}Ae^{-itH} = \sum_n \frac{(it)^n}{n!}[H[\ldots[H,A]]\ldots]. \tag{C.3}$$

The phase factor $\xi(t)$ has to be determined by the group property of the ℓ.h.s. and one can find the solution to be

$$\xi(t) = e^{i\frac{t^3}{12}}. \tag{C.4}$$

Hence $\exp it(\frac{P^2}{2}+Q)$ is of the form

$$\exp it(\frac{P^2}{2}+Q) = \xi(t)Q(f_t)Q(g_t) \tag{C.5}$$

where

$$g_t = e^{ip^2t} \tag{C.6}$$

$$\tilde{f}_t(\lambda,\mu) = \delta(\frac{\mu}{2}+t)\delta(\frac{\lambda}{2}-\frac{t^2}{2}). \quad (C.7)$$

But now for general f and g

$$Q(f)Q(g) = Q(f*g)$$

(see [5]), where

$$(f*g)(v) = \int e^{-2\pi i\sigma(v,v')}\tilde{f}(v')g(v-\frac{v'}{4})dv'.$$

The result is now obvious:

$$f_t(\xi,\zeta) = \exp i(\frac{\zeta^2 t}{2} + \xi t - \frac{t^3}{24}). \quad (C.8)$$

ACKNOWLEDGEMENT

We are very indebted to A. Grossmann for fruitful discussions.

REFERENCES

1. C.P. KORTHALS ALTES, Internal report Marseille 73/P553.
2. M. MIZRAHI, J.M.P. 16, 2201 (1975).
3. V. BARGMANN, Conf. Indiana University Bloomington, June 26 (1968), in Analytic Methods in Mathematical Physics, ed. G. Newton, Gordon and Breach.
4. J.C.T. POOL, J.M.P. 7, 66 (1966).
5. A. GROSSMANN, G. LOUPIAS and E.M. STEIN, Ann. Inst. Fourier 18, 343 (1969).
6. K.B. WOLF, The Heisenberg-Weyl ring in quantum mechanics, in Group Theory and its Applications, Vol. III, ed. E.M. Loebl, Academic Press, 1975, other references therein.
7. A. GROSSMANN, Parity operator and quantization of δ function, Preprint Marseille 75/P763.
8. A. GROSSMANN and M. SIRUGUE, to be published.
9. G. BURDET and M. PERRIN, J.M.P. 16, 2172 (1975).
10. L. HÖRMANDER, Fourier integral operators, Act. Math. 127, 79 (1971).

11. J.M. SOURIAU, Construction explicite de l'indice de Maslov. Applications, Preprint Marseille 75/P773.

12. BATEMAN Project Vol. 2, ed. A. Erdélyi, page 193.

CYCLIC GROUPS AND ADAPTATION IN THE VISUAL SYSTEM OF THE CAT

Allan A. Harkavy

In the striate cortex one finds simple cells which respond most intensely to illumination at a specific retinal field which has a light or dark line or an edge at a specific angle, as the preferred visual stimulus (PVS). Hubel and Wiesel[1,2,3] believe that this is the result of the simple cell's connection to several lateral geniculate cells which together have this linear spatial relation as their combined retinal field. In addition to the simple cells they find in the same or neighboring columnar location, complex cells which respond to the same PVS as the nearby simple cells but for a much extended retinal field. Again, they have indirect evidence that this is due to connections from several simple cells to a complex cell.

Polymodal neurons have a primary response to specific light stimuli, but they also respond to one or several other modalities - sound, shock, vestibular, and altered light stimulation. These neurons are found in the lateral geniculate body[4], and in the striate[5], and the parastriate[6], visual cortex of the cat. Ten to fifteen per cent of polymodal neurons in the parastriate cortex of adult cats are plastic cells because if the PVS is coupled to one of the other modality stimuli, the response is altered. Upon subsequent stimulation with the PVS alone, the response remains altered for some time. This is observed in the histograms of unit responses versus time after stimulation. These are evoked response studies[6] and twenty response trials are summed in one post stimulus histogram (PSH).

The hypothetical analysis I offer is that the effect of the coupled non-visual stimulus is to perturb and thereby lower the information content in the altered response to the PVS. It is known pyschophysically that the central nervous system will undergo visual adaptation when there is a conflict between different modalities. This altered response will permit an ambiguous visual perception by another level of the cortex for this region of the visual field, one subject to structuring by tactile and/or auditory stimuli. The ambiguity is related to loss of resolution.

It will be assumed that there is a cyclic group symmetry associated with the response to the PVS of the cortical cell and that the perturbation by the coupled stimulus results in a lower symmetry. This lower symmetry, it will be shown, can be related geometrically to the loss of resolution in the angular orientation and spatial frequency of the cell. It is because cyclic groups can be geometrically related to angular orientation and spatial frequency, the area of concern of many vision physiologists, that justifies their consideration.

The PSH are replicable and are characterized by one or several peaks occurring at different latencies. The altered response consists of a change in the number of these peaks. Two approaches are taken; the first emphasizes the redundant and integrated activities of the brain; the second emphasizes a particular wired in approach. The second approach is only briefly considered in this article. Both treat evoked response PSH in a fashion somewhat analogous to group theoretical quantum mechanical treatment of atomic absorption spectra. Whereas the two approaches start from different premises, it is hoped that elements from both approaches can be brought together in a non-mutually exclusive synthesis.

THE FIRST APPROACH

The spatio-temporal pattern of post synaptic stimulation arriving at cells in an extended region of the brain is given by a wave, a time window of which can be spatially summated to give a function, F, that can be written as a Fourier series, the coefficients of which possess a phase factor, which in the evoked response case is phase locked to the stimulus. It is postulated that the interaction of F with a cell is analogous to an atom absorbing a photon from a broad spectrum of EM radiation. The interactions of F with a neuron is defined as a Hamiltonian. The Hamiltonian will consist of many terms, the main or preferred interaction being perturbable by other terms of the Hamiltonian.

The beginning of the response of the neuron is to select from the Fourier series the phase and amplitude of those temporal frequencies to which it is sensitive. Now assume the PVS corresponds to one or several of these temporal frequencies. The response consists of transitions between states. These states are not further defined physically and they are not observable. I assume there is a particular cyclic group symmetry associated with the preferred visual response, which can be altered by the Hamiltonian, and the states correspond to representations of the cyclic group. Transitions between states are controlled by the phase found in the temporal frequency selection as it is convenient to use phase-like functions for cyclic group eigenfunctions. In the second, wired in, approach phase is associated with a time delay which is related to spatial frequency.

One reason for considering cyclic groups is that Wiener and Paley[7] in 1933 demonstrated that the characters of a denumerable Abelian group, formed from cyclic group operators, also constituted a denumerable group. Now the transitions between cyclic group representations, which I postulate to be responsible for the peaks in the PSH, are computed from the products of such characters. Thus, the wave, F, for some other region of the brain, will contain this group symmetric coding.

The function F, is given by

$$F = \sum C_n e^{iw_n t}, \ldots, (1);$$

assuming no perturbation the initial interaction is

$$f(\theta_m) = \int_{-\infty}^{\infty} F e^{-iw_m t} dt = C_m = A_m e^{i\theta_m}, \ldots, (2),$$

where the neuron is sensitive to only one frequency, A_m is the amplitude, and $e^{i\theta_m}$ is the phase function associated with the interaction. Let us assume that the neuron is governed by cyclic group C4. A set of eigenfunctions for this group is $e^{4i\theta}$; $e^{i\theta}$, $e^{-i\theta}$; $e^{2i\theta}$; which define the three possible representations. Now we have no way of knowing which possible states and associated representations exist. Arbitrarily, let us assume that the lowest level for any cyclic group is an A1 level (the identity representation) and that it is then followed by one or more two-dimensional representations, beginning with E1. This completes our pictures of an odd cyclic group. An even cyclic group has one additional anti-symmetric one-dimensional representation, A2, to complete the group and we make it our highest level.

The cyclic groups are mathematically Abelian. However, because of the time reversal symmetry of quantum mechanics, complex conjugate representations are degenerate. The representations, as given by the characters of E1 (or E2) are complex conjugates of each other and the E levels are degenerate and two-dimensional.

Now we assume that the neuron is in a state so that effectively one or more levels are occupied. If we find that there is only one PSH peak, then only one level was occupied and there was only one transition. If we find two peaks then we assume that there were two levels occupied and two transitions. We do not know which levels were occupied and only if there is extensive data collected could we determine this and then our model would be predictive. Subsequent to a lowering of symmetry we find several PSH peaks which we attribute to transitions between essentially the same levels as before the removal of degeneracy but now these levels have separated sufficiently to create the new peaks.

Now if as in equation (2), $f(\theta)$ is a single phase function, θ_m can be any integral multiple of θ. If $f(\theta)$ is some linear function of $e^{i\theta}$, then transitions from one level to the next highest level are possible as

$$\frac{1}{2\pi}\int_0^{2\pi} e^{-ni\theta}e^{i\theta}e^{i(n-1)\theta}d\theta = 1.$$

Here $f(\theta)$ has the symmetry of E1, as an eigenfunction of E1 is $e^{i\theta}$.

When one lowers the symmetry of a group, the new symmetry is a subgroup of the original group and for cyclic groups the orders of the subgroups are divisors of the order of the group[8].

In figure 1(a) we see the group C12 representations and the associated representations in the subgroups. We see the removal of degeneracy, which is always from an E representation to two A1 or two A2 representations.

I developed an APL-360 computer program which generates the real part of the group characters of the cyclic groups and which then gives the representation in each group and the associated subgroup representations. The representations are numbered as indicated on the left of figure 1(a), so that for group C12 they are number from 1 to 7. For even numbered groups having 2N elements, number 1 is always the 1 dimensional identity representation. Then follow N-1 two dimensional representations (here number 2 to 6) and then a one dimensional anti-symmetric representation (here numbered 7). Odd numbered cyclic groups of order N,

FIGURE 1(a). Group C12 representations and the associated representations in the subgroups

	GROUP	SUBGROUPS			
	C12	C6	C4	C3	C2
1	A1	A1	A1	A1	A1
2	E1	E1	E1	E1	A2,A2
3	E2	E2	A2,A2	E1	A1,A1
4	E3	A2,A2	E1	A1,A1	A2,A2
5	E4	E2	A1,A1	E1	A1,A1
6	E5	E1	E1	E1	A2,A2
7	A2	A1	A2	A1	A1

FIGURE 1(b). Cyclic group representations electric dipole type transitions (E+A1)

GROUP - SUBGROUP													
	Representations							Transitions					
12	1	2	3	4	5	6	7	1	1	1	1	1	1
2	1	2	1	2	1	2	1	2	4	4	4	4	2
3	1	2	2	1	2	2	1	1	1	2	2	1	1
4	1	2	3	2	1	2	3	1	2	2	2	2	1
6	1	2	3	4	3	2	1	1	1	2	2	1	1

have only 1 one-dimensional representation and (N-1)/2 two-dimensional representations.

Because of computer space limitations only eleven cyclic groups, their associated subgroups and only certain transitions - those analogous to electric dipole transitions in atomic absorption spectra were examined on the computer. The symmetry of these transitions is E1 + A1 and were between adjacent numbered representations in the group. For each single transition in the group the number of transitions in the subgroup was found for the corresponding representations. In figure 1(b) we see the computer readout for group C12. Computer readouts for groups C4, C6, C8, C9, C10, C15, C16, C18, C20, and C24 were also done but show little additional information which is not redundant to that in figure 1(b) for our analytical purpose.

The data I examined consisted of 15 examples of plastic cells. I examined only the change in the number of peaks in the PSH for the primary visual response before and after a coupling with a perturbing stimulus. This I associate with the change in the number of transitions between adjacent representations in going from a group to a subgroup as the symmetry is lowered. The changes were all of the form 1 to 1; 1 to 2; 2 to 2; 1 to 3; 2 to 3; and 1 to 4. They are all consistent with the arbitrarily selected electric dipole-like transitions as shown in figure 1(b). One to 3 is not found as such on the computer printouts of figure 1(b) in going from some group to cyclic group C2. However, 1 to 4 is found and in some multiplex situations 1 to 4 becomes 1 to 3 transitions when the separation of the states is symmetrical. Magnetic dipole-like transitions, having the symmetry E1 + A2 were also computed and they likewise fit the data.

THE SECOND APPROACH

A wired in model of lateral inhibition at the lateral

FIGURE 1(c). Linear orientation and cyclic group representation

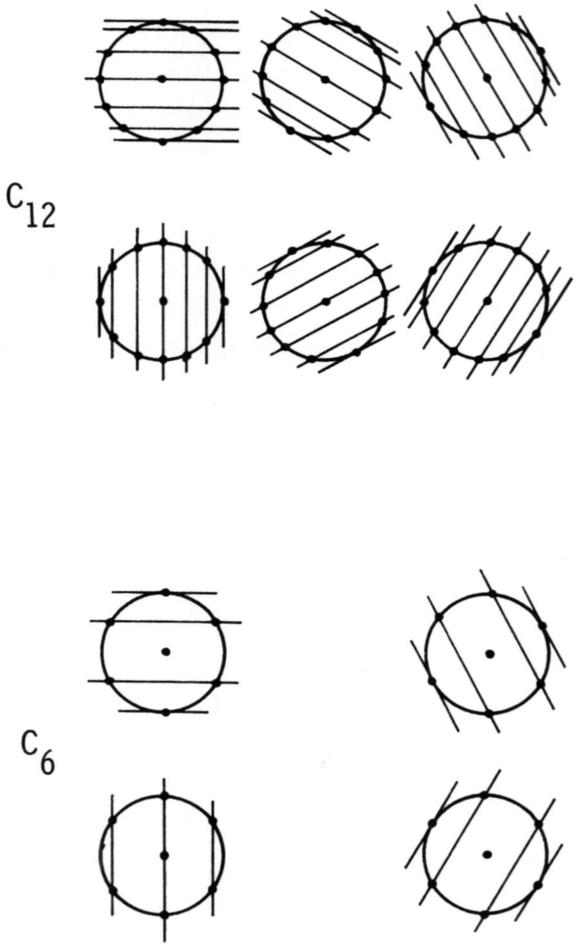

geniculate level of the visual system which assumes a circular perimeter geometry[9] for the receptive fields of the inhibiting geniculate cells suggests a mechanism for visual adaptation. In figure 1(c) there are 7 parallel lines drawn through the 12 cells on the perimeter of the circle. Two lines are tangent to the circle and strike only one cell while the remaining 5 lines strike two cells each. This is exactly analogous to the number and dimensionality of the representations of C12. However, this identification is not productive. Instead, each line defines a pair of operators or a single operator of C12 and physically each line defines a pair of cells or a single cell connected to an interneuron. The seven interneurons inhibit the geniculate cell at the center. The lowering of group symmetry results from the inactivation of interneurons which in turn results in a loss of spatial frequency resolution and angular orientation.

REFERENCES

1. D.H. HUBEL and T.N. WIESEL, Integration in the cat's lateral geniculate body. Journal of Physiology, London 155, 385-398, (1961).

2. D.H. HUBEL and T.N. WIESEL, Receptive fields, binocular interaction and functional architecture in the cat's visual cortex. Journal of Physiology, London 160, 106-154 (1962).

3. D.H. HUBEL and T.N. WIESEL, Receptive fields and functional architecture in two nonstriate visual areas (18 and 19) of the cat, Journal of Neurophysiology 28, 229-289 (1965).

4. K.L. CHOW, D.F. LINDSLEY and M. GOLLENDER, Modification of response patterns of lateral geniculate neurons after paired stimulation of contralateral and ipsilateral eyes, Journal of Neurophysiology 31, 729-739 (1968).

5. Private communication from Frank Morrell.

6. F. MORRELL, Electrical signs of sensory coding, in The Neurosciences - A Study Program, ed. by G.C. Quarton, T. Melnechuk and F.O. Schmitt, New York, Rockefeller Univ. Press 1967, pp. 452-469.

7. R.E.A.C. PALEY and N. WIENER, Characters of Abelian groups. Proceedings of the National Academy of Science 19, 253-257, (1933).

8. M. HAMERMESH, Group Theory, Reading, Mass., 1962, Addison-Wesley Publinshing Company, Inc. (See Chapter 6.)

9. A.A. HARKAVY, Spatial frequency response function of lateral geniculate neurons, Biol. Cybernetics 18, 15-18 (1975).

GEOMETRICAL MECHANICS AND DIRAC BRACKET

Kishore B. Marathe

1. INTRODUCTION

In studying generalized Hamiltonian dynamics, Dirac [1] introduced a bracket operation to replace the classical Poisson bracket when dealing with constrained systems. This bracket is then used to study the time evolution of the system in place of the Poisson bracket. Proof that this bracket operation defines a Lie algebra structure on the space of differentiable functions is quite long and does not bring out the geometrical significance of this new bracket operation. That this Dirac bracket is in fact the Poisson bracket with respect to an appropriate symplectic structure is the main result of the present paper. Those constraints that cannot be eliminated in this way will be discussed in a future paper.

Bergmann and Goldberg [2] and Mukunda and Sudarshan [3] have discussed the Dirac bracket from an algebraic point of view. We begin with the geometrical formulation of Hamiltonian dynamics by using the natural symplectic structure on phase space in section 2. Details of this formulation are discussed in the author's lecture notes [4], see also the excellent books by Abraham [5] and Souriau [6] for extensive discussion of geometrical mechanics.

In section 3 we discuss the generalized Poisson bracket and Lagrange bracket. Symmetry groups of Hamiltonian systems are discussed. Integrals of motion obtained by using the action of these groups may be regarded as constraints on the original system. The Dirac bracket formulation is closely related to the structure of the manifold of zeros of these constraints. This is discussed in section 4.

2. SYMPLECTIC MANIFOLDS AND HAMILTONIAN SYSTEMS

Let M be an m-dimensional manifold. A symplectic structure on M is a nondegenerate closed 2-form ω on M. Nondegeneracy implies that m must be even, say 2n. Requiring that ω be closed is equivalent to the existence of local canonical coordinates. Thus at each point $B \in M$ there exists a coordinate chart (x,U) such that

$$\omega|_U = \sum_{i=1}^{n} dx^i \wedge dx^{i+n} \tag{1}$$

This result is called Darboux's theorem and x^i are called canonical (or symplectic) coordinates at B.

Symplectic manifolds arise naturally in the study of dynamical systems as phase spaces. If N denotes the n-dimensional manifold of configuration space, then $M = T^*N$ the cotangent bundle of N can be identified with the phase space. A point of M can be written as (q,α) where $q \in N$ and $\alpha \in T_q^*(N)$. Thus in terms of the basis dq^i of $T_q^*(N)$ induced by the coordinate chart with coordinates q^i, $1 \leq i \leq n$, we have $\alpha = p_i dq^i$ (here and in what follows the summation convention is used). Thus $(q^1,\ldots,q^n,p_1,\ldots,p_n)$ can be taken as local coordinates at (q,α) on M. These coordinates become canonical coordinates with respect to a natural symplectic structure ω on M which can be expressed in these coordinates as

$$\omega = dq^i \wedge dp_i . \tag{2}$$

In what follows ω always denotes this natural symplectic structure on the phase space M.

Using ω we can associate a 1-form $f(X)$ with a vector field X on M by the definition

$$f(X)Y = \omega(X,Y), \tag{3}$$

where Y is any vector field on M. At each point $B \in M$, f induces an isomorphism of $T_B(M)$ and $T_B^*(M)$. If v_B is its inverse, then we can define a map v which associates with a 1-form θ on M a vector field $v(\theta)$ on M defined by

$$v(\theta)_B = v_B(\theta_B). \tag{4}$$

In particular, if H is a differentiable function on M, $v(dH)$ is a vector field on M which is denoted by X_H.

We say that a vector field X on M is locally Hamiltonian if the Lie derivative $L_X\omega$ of ω along X is zero. I.e.,

$$L_X\omega = 0. \tag{5}$$

The connection with the classical Hamilton's equations comes from the theorem that X is locally Hamiltonian if and only if for every $B \in M$, there exists a neighborhood U of B and a differentiable function H defined on U such that $X = X_H = v(dH)$ on U. If U = M then X is said to be a Hamiltonian vector field. We note that by definition of X_H we have

$$X_H = -\frac{\partial H}{\partial q^i} \cdot \frac{\partial}{\partial p_i} + \frac{\partial H}{\partial p_i} \cdot \frac{\partial}{\partial q^i}. \tag{6}$$

Thus the theorem implies that the integral curves of X_H are solutions of the system of differential equations

$$\frac{dq^i}{dt} = \frac{\partial H}{\partial p_i} \quad \text{and} \quad \frac{dp_i}{dt} = -\frac{\partial H}{\partial q^i}, \qquad 1 \le i \le n. \tag{7}$$

Equations (7) are the classical Hamilton's equations of motion.

A well known method of obtaining integrals of equations (7) is by the use of symmetry groups. We begin by considering the action of Lie groups on M. We say that a k-parameter Lie group G acts as a group of diffeomorphisms of manifold M if there is an injective homomorphism of G into the group of diffeomorphisms of M with itself. Using this homomorphism we identify each $g \in G$ as a diffeomorphism of M with itself. This action of G on M induces a Lie algebra homomorphism of L(G) (the Lie algebra of G) into $\mathcal{X}(M)$ (the Lie algebra of differentiable vector fields on M). The image $\tilde{A} \in \mathcal{X}(M)$ of $A \in L(G)$ under this homomorphism is called the infinitesimal transformation of M induced by A.

If M is the phase space with natural symplectic structure ω then we say that the action of G is (locally) Hamiltonian if $\tilde{A}$ is a (locally) Hamiltonian vector field for each $A \in L(G)$. G is called a symmetry group of a Hamiltonian function H on M if the action of G on M is Hamiltonian and if G preserves H, i.e.,

$$(g)^*H = H, \quad \text{for every} \quad g \in G. \tag{8}$$

If A_i, $1 \le i \le k$ form a basis of L(G) and if θ^i are functions on M such that $\tilde{A}_i = v(d\theta^i)$, $1 \le i \le k$, then $\theta^i = c^i$ are k integrals of equations (7). By redefining θ^i, if necessary we can absorb the constants c^i and write the integrals of motion as $\theta^i = 0$. Moreover, the k 1-forms $d\theta^i$ are linearly independent. Therefore, the zeros of θ^i determine a submanifold of M. The integrals θ^i may be regarded as constraints on the original system. Under certain conditions these constraints may be used to obtain a new Hamiltonian system with 2n-k degrees of freedom, whose time evolution is given by the Dirac bracket. Dirac bracket arises as the generalized Poisson bracket with respect to a symplectic structure induced by ω on the manifold determined by the constraints.

3. GENERALIZED POISSON AND LAGRANGE BRACKETS

In classical mechanics Poisson and Lagrange brackets are defined for functions, say f and g in terms of canonical coordinates and momenta. In using the symplectic structure ω on the phase space M to define these brackets we have to keep in mind the fact that the 2-form ω acts on vector fields. The corresponding definitions thus involve obtaining vector fields to correspond to functions f and g and then acting on them by the 2-form ω. The new definitions are given by

$$\{f,g\} = \omega(X_f, X_g), \quad \text{for the Poisson bracket,} \tag{9}$$

and

$$[f,g] = \omega\left(\frac{\partial}{\partial f}, \frac{\partial}{\partial g}\right), \quad \text{for the Lagrange bracket,} \tag{10}$$

In (10) we require that f and g form a 2-surface element so that they can be used as local coordinates to express q^i and p_i as their functions. Using canonical local coordinates it is easy to verify that (9) and (10) reduce in these coordinates to their well known classical expressions. In particular, if $u^1, u^2, \ldots, u^{2n}$ are the coordinates of a new coordinate system, then

$$\{u^j, u^k\} = \omega(X_{u^j}, X_{u^k}) = \omega^{jk}, \tag{11}$$

and

$$[u^j, u^k] = \omega\left(\frac{\partial}{\partial u^j}, \frac{\partial}{\partial u^k}\right) = \omega_{jk}, \quad 1 \le j,k \le 2n. \tag{12}$$

From our definitions the skew symmetry of the Poisson and Lagrange brackets is immediate. The identities connecting them follow from (11) and (12) as a consequence of the fact that the nondegeneracy of ω is equivalent to the following condition

$$\omega^{ij}\omega_{jk} = \delta^i_k, \quad \text{for } 1 \le i,j,k \le 2n. \tag{13}$$

However the important Jacobi identity satisfied by these brackets is not apparent from their definitions. It is equivalent to the fact that ω is closed. Thus the Lie algebra structures determined by these brackets are closely connected to the symplectic structure ω on the phase space M. But clearly these properties do not seem to depend on the special nature of ω. Thus we can define the generalized Poisson and Lagrange brackets by using any symplectic structure (i.e. a nondegenerate, closed 2 form) θ in place of ω in the definitions (9) and (10). We will show in the next section that the Dirac bracket is such a generalized Poisson bracket.

4. THE DIRAC BRACKET

Let us suppose that we have $2n-2r$ differentiable functions θ^a, $1 \le a \le 2n - 2r$, defined on the phase space M such that $\theta^a = 0$, express the constraints. We assume that $d\theta^a$ are linearly independent and thus the manifold M^* of the zeros of these functions is a submanifold of M of dimension 2r. Let $i: M^* \to M$ denote the inclusion map. Then the 2-form ω on M induces the 2-form $i^*\omega = \omega^*$ on M^*. Following Dirac we now assume that the matrix of Poisson brackets $\{\theta^a, \theta^b\}$ $1 \le a,b \le 2n-2r$ is non-singular. We show that this condition is equivalent to requiring that ω^* be a symplectic structure on M^*. We introduce local coordinates $\psi^1, \psi^2, \ldots, \psi^{2r}$ at a point $A \in M^*$ such that ψ^ℓ's and θ^a's become local coordinates at $A \in M$. In what follows we assume that a,b,c,d range over 1 to 2n-2r and ℓ, m, n range over 1 to 2r and correspond to coordinates ψ. We write

$$\{\theta^a, \theta^b\} = \omega^{ab}, \quad \{\theta^a, \psi^m\} = \omega^{am}, \quad \{\psi^m, \psi^n\} = \omega^{mn} \tag{14}$$

and

$$[\theta^a, \theta^b] = \omega_{ab}, \quad [\theta^a, \psi^m] = \omega_{am}, \quad [\psi^m, \psi^n] = \omega_{mn}. \tag{15}$$

By definition of ω^* we have

$$\omega^*\left(\frac{\partial}{\partial\psi^m}, \frac{\partial}{\partial\psi^n}\right) = \omega_{mn}. \tag{16}$$

Now by the Dirac assumption the matrix ω^{ab} is non-singular. Let α_{bc} denote its inverse matrix. I.e., $\omega^{ab}\alpha_{bc} = \delta^a_c$. Using this relation and the identities (13) where the coordinates u^j are replaced by θ^a, ψ^m we get

$$\omega_{\ell m}[\omega^{mn}-\omega^{mc}\alpha_{ca}\omega^{an}] = \delta^n_\ell. \tag{17}$$

Thus $\omega^{*mn} = \omega^{mn} - \omega^{mc}\alpha_{ca}\omega^{an}$ defines a matrix inverse to ω_{mn}. Hence by (16) ω^* is a nondegenerate 2-form on M^*. It is easy to see that the converse of this result is also true. We observe that the exterior differentiation d commutes with i^* and hence ω closed implies that ω^* is also closed. Thus ω^* defines a symplectic structure on M^*. Our definition of ω^* depends only on the manifold M^* defined by the constraints. Consider the generalized Poisson bracket defined by ω^*. If f, g are defined on M we denote their restriction to M^* by the same symbols. Thus we define

$$\{f,g\}^* = \omega^*(X^*_f, X^*_g). \tag{18}$$

Using equations (9) and (17) in (18) we get

$$\{f,g\}^* = \{f,g\} - \{f,\theta^a\}\alpha_{ab}\{\theta^b,g\}. \tag{19}$$

The expression on the right hand side of (19) is the original definition given by Dirac. Our definition shows that the Dirac bracket is in fact the Poisson bracket defined by the symplectic structure ω^* induced by ω on the manifold determined by the constraints. It therefore satisfies the Jacobi identity and thus defines a Lie algebra structure on the space of differentiable functions on M. From the definition it also follows that our bracket will treat the constraints θ^a as constants, i.e., $\{f,\theta^a\}^* = 0$. Thus if H is the Hamiltonian function on M, then

$$\{f,H\}^* = \{f,H\} = \frac{df}{dt}. \tag{20}$$

Equation (20) shows that the time evolution of the constrained system can now be studied by using the Dirac bracket in place of the Poisson bracket. In particular, we see under what conditions the integrals furnished by the symmetry groups can be used for a complete reduction of the degrees of freedom from 2n to 2r.

Constraints which do not satisfy these conditions have been called by Dirac first class constraints. We propose to take up their study in a later paper.

REFERENCES

1. P.A.M. DIRAC, Can. J. Math. 2, 129 (1950).
2. BERGMANN and GOLDBERG, Phys. Rev. 98, 531 (1955).
3. MUKUNDA and SUDARSHAN, J. Math. Phys. 9, 411 (1968); Classical Dynamics, Wiley, New York, 1974.
4. K.B. MARATHE, Symplectic manifolds and Hamiltonian mechanics, Séminaire de Mathématiques Supérieures, été 1976, Université de Montréal Press.
5. R. ABRAHAM, Foundations of Mechanics, Benjamin, New York, 1967.
6. SOURIAU, Structure des systèmes dynamique, Dunod, Paris, 1970.

OTHER SYMMETRIES AND CONSTANTS OF THE MOTION

G. Marmo and E.J. Saletan

We take a *dynamical system* to consist of a manifold Q (configuration space) and the set of trajectories on this manifold (that is, the $q_i(t)$). Through each point of Q, however, there pass many trajectories, and these are separated by going from Q to the tangent bundle TQ (the manifold of positions and velocities)[1].

On TQ the separated trajectories form a flow, one trajectory passing through each point. The tangent vectors to this flow form the dynamical vector field D. In local coordinates D may be written in the form

$$D = \dot{q}^k \frac{\partial}{\partial q^k} + f^k(q,\dot{q}) \frac{\partial}{\partial \dot{q}^k} , \qquad (1)$$

where the f^k are the "force functions" which give the "accelerations" according to $\ddot{q}^k = f^k(q,\dot{q})$. The vector field D is uniquely determined by the dynamical system, and its projection down on Q is just the dynamical system. If g is any function over TQ, then $D(g) \equiv dg(D) \equiv L_D g$ (where L_D is the Lie derivative along D) is also such a function and is in fact dg/dt.

A *symmetry of the dynamical system*, or briefly a *symmetry of* D is a diffeomorphism $\phi \in$ diff (TQ) which leaves D invariant;

that is $\phi^*D = D$. This means that if a curve $c : \mathbb{R} \to TQ$ is a solution for D through $m \in TQ$, then $\phi \circ c$ is a solution for D through $\phi(m)$. In particular, a Q-*symmetry* of D is a symmetry of D which is the lift of a diffeomorphism on Q. (In local coordinates, a symmetry may mix q's and $\dot{q}$'s in an arbitrary way, whereas a Q-symmetry arises from what is often called a point transformation.) Much of what we say below applies to symmetries of D in general, though we illustrate it with Q-symmetries.

Let $\phi(s)$ be a one-parameter group of symmetries of D, and let $X \in \mathcal{X}(TQ)$ be the associated vector field (or infinitesimal generator). Then $L_X D = 0$, or

$$[D,X] = 0. \tag{2}$$

Example. Let the dynamical system be the two-dimensional isotropic oscillator. According to Eq. (1) the dynamical field is

$$D = \dot{q}^k \frac{\partial}{\partial q^k} - q^k \frac{\partial}{\partial \dot{q}^k}, \qquad k = 1,2.$$

a) Dilation in q^1 is a Q-symmetry. The one-parameter group of diffeomorphisms on Q is given by

$$\Psi(s) : (q^1,q^2) \to (e^s q^1,q^2).$$

This lifts to the following diffeomorphism on TQ:

$$\phi(s) : (q^1,q^2,\dot{q}^1,\dot{q}^2) \to (e^s q^1,q^2,e^s \dot{q}^1,\dot{q}^2).$$

The associated vector field $\dot{X}_1$ (the dot indicates that it arises as a lift from Q) is

$$\dot{X}_1 = q^1 \frac{\partial}{\partial q^1} + \dot{q}^1 \frac{\partial}{\partial \dot{q}^1}.$$

A simple calculation then shows that $[D,\dot{X}]$ in fact vanishes, in agreement with (2).

b) Dilation in q^2 is similarly a Q-symmetry.

c) Dilation in both q^1 and q^2 is a Q-symmetry. The associated vector field in $X(TQ)$ is

$$\dot{\Delta} = q^k \frac{\partial}{\partial q^k} + \dot{q}^k \frac{\partial}{\partial \dot{q}^k} .$$

d) *Squeeze*, dilation in q^1 and contraction (by the same factor) in q^2, is also a Q-symmetry. The one-parameter group is given by

$$\phi(s) : (q^1, q^2, \dot{q}^1, \dot{q}^2) \to (e^s q^1, e^{-s} q^2, e^s \dot{q}^1, e^{-s} \dot{q}^2),$$

and the vector field by

$$\dot{S} = q^1 \frac{\partial}{\partial q^1} - q^2 \frac{\partial}{\partial q^2} + \dot{q}^1 \frac{\partial}{\partial \dot{q}^1} - \dot{q}^2 \frac{\partial}{\partial \dot{q}^2} .$$

e) Rotation is also a Q-symmetry. The vector field is

$$\dot{R} = q^2 \frac{\partial}{\partial q^1} - q^1 \frac{\partial}{\partial q^2} + \dot{q}^2 \frac{\partial}{\partial \dot{q}^1} - \dot{q}^1 \frac{\partial}{\partial \dot{q}^2} .$$

Of all these symmetries, however, only rotation leaves invariant the Lagrangian

$$\mathcal{L}_1 = \tfrac{1}{2}[(\dot{q}^1)^2 + (\dot{q}^2)^2 - (q^1)^2 - (q^2)^2]. \qquad (3)$$

For example, if one calculates $d\mathcal{L}_1/ds$ for squeeze, namely $L_{\dot{S}}\mathcal{L}_1 \equiv \dot{S}(\mathcal{L}_1)$, one obtains

$$L_{\dot{S}}\mathcal{L}_1 = -(q^1)^2 + (q^2)^2 + (\dot{q}^1)^2 - (\dot{q}^2)^2,$$

which fails to vanish in general. Thus a symmetry of D need not be a symmetry of the Lagrangian.

Now, one usually associates constants of the motion with symmetries of the Lagrangian through Noether's theorem, but this theorem does not apply to more general symmetries of D. Is there some other way to associate constants of the motion with symmetries of D? We show one way below.

In order to associate functions with vector fields one goes to the dual space. The first step (the first step also in making use of Noether's theorem) is to put a symplectic structure on TQ. This is done by pulling back the natural symplectic structure from the cotangent bundle T*Q (phase space)[1]. The connection between TQ and T*Q is through the fiber derivative $F\mathcal{L}$ of the Lagrangian, and so the symplectic structure on TQ depends on the Lagrangian, although that on T*Q does not.

Let the symplectic form on T*Q be ω_o. Then we define the symplectic form for TQ through

$$\omega_{\mathcal{L}} = F\mathcal{L}_* \omega_o .$$

The equations of motion can now be written in terms of $\omega_{\mathcal{L}}$ by pulling back Hamilton's canonical equations of motion from T*Q. One obtains

$$i_D \omega_{\mathcal{L}} = dE, \qquad (4)$$

where the *energy function* E is given by

$$E(m) = F\mathcal{L}(m)m - \mathcal{L}(m), \quad m \in TQ.$$

Let X be the infinitesimal generator of a one-parameter group of symmetries of D. Then the function

$$\omega_{\mathcal{L}}(D,X) = f \qquad (5)$$

is a constant of the motion:

$$L_D f = L_D L_X E = L_X L_D E = 0,$$

where we use Eq. (4), the commutation of X and D, and the fact that E is a constant of the motion.

Example. For the Lagrangian $\mathcal{L}_1$ of Eq. (3) we obtain

$$\omega_{\mathcal{L}_1} = dq^1 \wedge d\dot{q}^1 + dq^2 \wedge d\dot{q}^2.$$

Then with squeeze, we have

$$\omega_{\mathcal{L}_1}(D,\dot{S}) = (q^1)^2 + (\dot{q}^1)^2 - [(q^2)^2+(\dot{q}^2)^2] \tag{6}$$

which is the difference between the energies in the two modes of the oscillator. Thus in spite of the fact that $\mathcal{L}_1$ is not invariant under squeeze, one can use $\mathcal{L}_1$ to associate the constant of Eq. (6) with squeeze symmetry of D.

Usually one associates symmetry with a constant of the motion f by using the inverse of Noether's theorem, that is by solving for X in the differential equation

$$i_X\omega_{\mathcal{L}} = df. \tag{7}$$

But there is no f such that $\dot{S}$ is obtained in this way from $\omega_{\mathcal{L}_1}$. A necessary condition would be $d(i_{\dot{S}}\omega_{\mathcal{L}_1}) = 0$. Calculation yields

$$i_{\dot{S}}\omega_{\mathcal{L}_1} = q^1 d\dot{q}^1 - q^1 d\dot{q}^1 - q^2 d\dot{q}^2 + q^2 d\dot{q}^2,$$

and it is easily seen that this will not satisfy the condition.

There are, however, other Lagrangian functions which yield the same dynamics, for instance[2]

$$\mathcal{L}_2 = \dot{q}^1\dot{q}^2 - q^1 q^2.$$

(In the sense of the following paper, $\mathcal{L}_1$ and $\mathcal{L}_2$ are equivalent.) It is easily seen that

$$\omega_{\mathcal{L}_2} = dq^1 \wedge d\dot{q}^2 + dq^2 \wedge d\dot{q}^1$$

and that

(i) $$i_D\omega_{\mathcal{L}_2} = d(\dot{q}^1\dot{q}^2 + q^1q^2) \equiv dE_2:$$

$\mathcal{L}_2$ does indeed yield the same dynamics D, but the "energy" function is now E_2;

(ii) $$L_{\dot{S}}\mathcal{L}_2 = 0:$$

$\mathcal{L}_2$ is invariant under squeeze;

(iii) $$i_{\dot{S}}\omega_{\mathcal{L}_2} = d(q^1\dot{q}^2 - q^2\dot{q}^1) = -i_{\dot{R}}\omega_{\mathcal{L}_1}:$$

squeeze is associated with angular momentum through $\mathcal{L}_2$ (just as rotation is associated with angular momentum through $\mathcal{L}_1$). In fact one can use Noether's theorem, Eq. (7), to make this association.

In other words one constant of the motion can be associated with more than one symmetry, even by Noether's theorem.

Similarly, in the method of Eq. (5), one symmetry can be associated with more than one constant of the motion.

Example. Consider dilation of Example (c). One easily obtains

$$\omega_{\mathcal{L}_1}(D,\dot{\Delta}) = (q^1)^2 + (q^2)^2 + (\dot{q}^1)^2 + (\dot{q}^2)^2$$

and

$$\omega_{\mathcal{L}_2}(D,\dot{\Delta}) = 2(\dot{q}^1\dot{q}^2 + q^1q^2).$$

It is interesting that the constants obtained on the right-hand sides are twice the associated energy functions.

REFERENCES

1. R. ABRAHAM and J.E. MARSDEN, *Foundations of Mechanics* (W.A. Benjamin, New York, 1967). Terminology and notation is taken mostly from this work.

2. Y. GELMAN and E.J. SALETAN, Nuovo Cimento 18B, 53 (1972).

Q-SYMMETRIES AND EQUIVALENT LAGRANGIANS

G. Marmo and E.J. Saletan

We take configuration space Q and the trajectories on it to be physically relevant, and construct a vector field D on TQ whose flow projected on Q reproduces the trajectories. Let $\mathcal{L} \in \mathcal{F}(TQ)$ be a Lagrangian function from which D is constructed. We say $\mathcal{L}' \in \mathcal{F}(TQ)$ is *subordinate* to $\mathcal{L}$ (or $\mathcal{L}' < \mathcal{L}$) if

$$i_D\omega_{\mathcal{L}} = dE_{\mathcal{L}} \Rightarrow i_D\omega_{\mathcal{L}'} = dE_{\mathcal{L}'},$$

where $E_{\mathcal{L}}$ is defined by

$$E_{\mathcal{L}}(m) = F\mathcal{L}(m)m-\mathcal{L}(m), \qquad m \in TQ$$

and $\omega_{\mathcal{L}}$ is the pull-back of ω_o from T*Q through $F\mathcal{L}_*$. Locally $\mathcal{L}' < \mathcal{L}$ if for every set of solutions $q^k(t)$ of the Euler-Lagrange equations[1]

$$\frac{d}{dt}\frac{\partial\mathcal{L}}{\partial\dot{q}^k} - \frac{\partial\mathcal{L}}{\partial q^k} = 0$$

we have

$$\frac{d}{dt}\frac{\partial\mathcal{L}'(q(t),\dot{q}(t))}{\partial\dot{q}^k} - \frac{\partial\mathcal{L}'(q(t),\dot{q}(t))}{\partial q^k} = 0.$$

Note that $\mathcal{L}' < \mathcal{L} \Rightarrow L_D\omega_{\mathcal{L}'} = 0$. We say that $\mathcal{L} \approx \mathcal{L}'$ ($\mathcal{L}$ is *equivalent to* $\mathcal{L}'$) iff $\mathcal{L} < \mathcal{L}'$ and $\mathcal{L}' < \mathcal{L}$.[2] If $\mathcal{L}' < \mathcal{L}$, then $E_{\mathcal{L}'}$ is a constant of the motion.

It is possible to use Q-symmetries, defined in the previous paper, to generate subordinate (sometimes equivalent) Lagrangians. It can be shown[3] that lifts of Q-transformations are the only transformations on TQ which preserve the second-order nature of vector fields, that is the only ones that preserve the form of Newton's equations on TQ. The following theorem[3] states a property of Q-transformations: $\phi_*\omega_{\mathcal{L}} = \omega_{\phi_*\mathcal{L}}$ $\forall\ \mathcal{L} \in \mathcal{F}(TQ)$ iff ϕ is the lift of a Q-transformation. A useful corollary is: such ϕ is a Q-symmetry for D iff $\phi_*\mathcal{L} \approx \mathcal{L}$, where $\mathcal{L}$ is any regular Lagrangian such that $i_D\omega_{\mathcal{L}} = dE_{\mathcal{L}}$.

Let ϕ_s be a one-parameter group of transformations and X^ϕ the associated infinitesimal generator. If $X \in \mathcal{X}(Q)$, we call $\dot{X}$ the vector field associated with X by the tangent functor TX, that is

$$(X : Q \to TQ) \xrightarrow{T} (TX : TQ \to TTQ).$$

It follows that if $X = X^\phi$, $\phi \in \text{diff}\ (Q)$, then $\dot{X} = X^{T\phi}$. For lifted vector fields we then have the infinitesimal version of the previous theorem:

$$L_X\omega_{\mathcal{L}} = \omega_{L_X\mathcal{L}} \quad \forall\ \mathcal{L} \text{ iff } \exists :\ Y \in \mathcal{X}(Q)$$

such that

$$X = \dot{Y}.$$

A useful corollary is: $[\dot{X},D] = 0$ iff

$$(L_{\dot{X}})^{n+1}\mathcal{L} < (L_{\dot{X}})^n\mathcal{L} < \ldots < L_{\dot{X}}\mathcal{L} < \mathcal{L},$$

where $\mathcal{L}$ is any regular Lagrangian such that $i_D\omega_{\mathcal{L}} = dE_{\mathcal{L}}$. Thus we may obtain subordinate Lagrangians in this way.

Remark. The same considerations can be carried through in time dependent mechanics dealing with an exact contact manifold through the procedure of reduction to the autonomous case[1].

These results will appear elsewhere.

If one deals with symmetries which are not Q-symmetries, it is no longer true that $L_X\omega_{\mathcal{L}} = \omega_{L_X\mathcal{L}}$. In this case we introduce the notion of *kinematically subordinate* Lagrangians. We say that $\mathcal{L}'$ is kinematically subordinate to $\mathcal{L}$ if $i_D\omega_{\mathcal{L}} = dE_{\mathcal{L}} \Rightarrow L_D\omega_{\mathcal{L}'} = 0$. We say that $\mathcal{L}'$ is *globally kinematically subordinate* to $\mathcal{L}$ if $i_D\omega_{\mathcal{L}} = dE \Rightarrow i_D\omega_{\mathcal{L}'} = dE$, but $E \neq E_{\mathcal{L}'}$ and in fact cannot be obtained from $\mathcal{L}'$ alone.

If $\mathcal{L}'$ is globally kinematically subordinate to $\mathcal{L}$, it is possible to find $f \in F(Q)$ (a "potential energy") such that $\mathcal{L}'-f < \mathcal{L}$. The fact that $L_X\omega_{\mathcal{L}} \neq \omega_{L_X\mathcal{L}}$ makes f necessary. Indeed, it can be shown that

$$i_D(L_X\omega_{\mathcal{L}}-\omega_{L_X\mathcal{L}}) = d(f+L_{[X,V]}\mathcal{L}),$$

where V is the Liouville vector field (see Godbillon[1]). In a local chart,

$$V(q,\dot{q}) = \dot{q}^k \frac{\partial}{\partial \dot{q}^k} .$$

With these definitions it is then possible to recover many of the previous results, but in this weaker sense.

For simplicity we shall limit ourselves to Q-transformations on going to T*Q. The Euler-Lagrange equations can be pulled back to T*Q by means of any of the many diffeomorphisms $F\mathcal{L}_1, F\mathcal{L}_2, \ldots$ where $\mathcal{L}_1, \mathcal{L}_2, \ldots$ are all equivalent hyperregular[1] Lagrangians. The situation is illustrated by the following diagram:

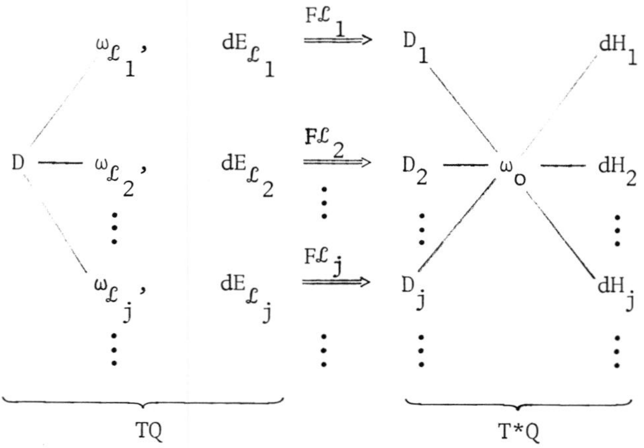

All of the $D_i \in \mathcal{X}(T^*Q)$ are q-*equivalent*[2,4]; that is, their projections on Q give the same trajectories. They are connected by *fouling* transformations. It can be shown that under suitable regularity conditions *any* two q-equivalent vector fileds on T*Q are connected by fouling transformations.

It is possible also to use just one Lagrangian and hence just one diffeomorphism $F\mathcal{L}_j$ to go to T*Q. The situation is then illustrated by the following diagram:

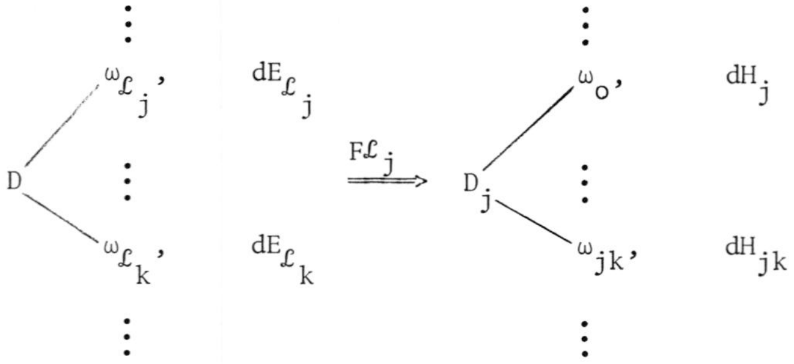

A one-parameter group of Q-symmetries ϕ_s for a given dynamics D on TQ yields a family of equivalent Lagrangians $\mathcal{L}_s = (\phi_s)_* \mathcal{L}$, where $i_D \omega_{\mathcal{L}} = dE_{\mathcal{L}}$, and a one-parameter group of fouling transformations on T*Q.

Example. Consider $\mathcal{L}_1$ of the preceding paper for the two-dimensional isotropic oscillator:

$$\mathcal{L}_1 = \tfrac{1}{2}[(\dot{q}^1)^2+(\dot{q}^2)^2-(q^1)^2-(q^2)^2].$$

With $\dot{R}$, $\dot{S}$, and $\mathcal{L}_2$ also as defined in that paper, we have

$$L_{\dot{R}} L_{\dot{S}} \mathcal{L}_1 = 2\mathcal{L}_2.$$

We have seen that

$$E_{\mathcal{L}_1} = \tfrac{1}{2}[(\dot{q}^1)^2+(\dot{q}^2)^2+(q^1)^2+(q^2)^2],$$

$$E_{\mathcal{L}_2} = \dot{q}^1\dot{q}^2 + q^1q^2.$$

The two diffeomorphisms to $T^*Q = T^*\mathbb{R}^2$ are given by

$$F\mathcal{L}_1 : (\dot{q}^1,\dot{q}^2) \to (p_1,p_2),$$

and

$$F\mathcal{L}_2 : (\dot{q}^1,\dot{q}^2) \to (p_2,p_1),$$

which can be calculated by writing $p_k = \partial\mathcal{L}/\partial\dot{q}^k$.

The vector fields and Hamiltonians on T^*Q are:

$$D_1 = (F\mathcal{L}_1)^*D = p_1 \frac{\partial}{\partial q^1} + p_2 \frac{\partial}{\partial q^2} - q^1 \frac{\partial}{\partial p_1} - q^2 \frac{\partial}{\partial p_2},$$

$$D_2 = (F\mathcal{L}_2)^*D = p_2 \frac{\partial}{\partial q^1} + p_1 \frac{\partial}{\partial q^2} - q^2 \frac{\partial}{\partial p_1} - q^1 \frac{\partial}{\partial p_2};$$

$$H_1 = \tfrac{1}{2}[p_1^2+p_2^2+(q^1)^2+(q^2)^2],$$

$$H_2 = p_1p_2 + q^1q^2.$$

Of course $(F\mathcal{L}_1)^*\omega_{\mathcal{L}_1} = (F\mathcal{L}_2)^*\omega_{\mathcal{L}_2} = \omega_o$.

If just $\mathcal{L}_1$, and hence only the diffeomorphism $F\mathcal{L}_1$ is used to go to T*Q, the only vector field obtained in T*Q is D_1, whereas the symplectic forms and Hamiltonians obtained are

$$\omega_{21} = dq^1 \wedge dp_2 + dq^2 \wedge dp_1, \quad H_{21} = p_1 p_2 + q^1 q^2;$$

$$\omega_1 = \omega_o, \quad H_1 = H_1 \text{ as before.}$$

(It seems to be just a coincidence that $H_{21} = H_2$.) We see that one dynamics can be described by two different Hamiltonians through two different symplectic structures.

Let us now study the dynamical group of symmetries for these two different symplectic structures. For H_1, as is well known, we have the usual SU(2). That is, with

$$f_1 = p_1 p_2 + q^1 q^2; \quad f_2 = q^2 p_1 - q^1 p_2;$$

$$f_3 = \tfrac{1}{2}[(q^1)^2 + p_1^2 - (q^2)^2 - p_2^2]$$

and Poisson bracket defined by ω_o, we obtain the Lie algebra su(2). As H_1 has compact surfaces, a theorem by R. Palais assures us that the algebra integrates to a group.

For H_{21} the Lie algebra of symmetries is generated by the functions

$$f_1 = q^2 p_1 - q^1 p_2; \quad f_2 = \tfrac{1}{2}[p_1^2 + (q^1)^2]$$

$$f_3 = \tfrac{1}{2}[p_2^2 + (q^2)^2],$$

and the Poisson bracket $\{\ \}_{21}$ is defined by ω_{21}. Then we have

$$\{f_1, f_2\}_{21} = 2f_2, \quad \{f_1, f_3\}_{21} = -2f_3, \quad \{f_2, f_3\}_{21} = -f_1,$$

which is su(1,1). In this case, too, it is possible to prove that the Lie algebra integrates to a group.

Thus with the same dynamics on the phase space it is possible to have two different groups of symmetries, a result which may be useful in the Kostant-Souriau geometrical quantization scheme. We are of the opinion that in this way we can obtain the same quantized dynamical system.

The so-called "prequantization scheme" suggested by Kirillov[5] consists of associating with any function F the operator

$$\hat{F} = i\hbar X_F + \theta(X_F).$$

In our simple case this gives

$$\hat{H}_{12} = i\hbar D = \hat{H}_1$$

where $\theta_1 = -p_k dq^k$ and $\theta_{12} = -(p_2 dq^1 + p_1 dq^2)$.

REFERENCES

1. R. ABRAHAM and J.E. MARSDEN, *Foundations of Mechanics* (W.A. Benjamin, New York, 1967). C. GODBILLON, *Géométrie différentielle et mécanique analytique* (Hermann, Paris, 1969).
2. D. CURRIE and E.J. SALETAN, J. Math. Phys. 7, 967 (1966). G. MARMO, Proceedings of the IVth International Colloquium on Group Theoretical Methods in Physics (Nijmegen, 1975).
3. G. MARMO, Tesi Scuola di Perfezionamento in Fisica Teorica e Nucleare, Marzo, 1975, Universitá di Napoli.
4. Y. GELMAN and E.J. SALETAN, Nuovo Cimento 18B, 53 (1972). G. CARATU et al., Nuovo Cimento 19B, 228 (1974).
5. A. KIRILLOV, Eléments de la théorie de représentations (Editions de Moscou, Moscow, 1974).

KINEMATICAL SYMMETRIES OF THE
NAVIER-STOKES EQUATION

U. Niederer

1. INTRODUCTION

A kinematical symmetry of an equation is a coordinate transformation which leaves invariant this equation. Examples of kinematical symmetries are provided by the Galilei or Poincaré invariance of the free particle wave-equations of nonrelativistic or
relativistic quantum mechanics. It may also happen that the
largest kinematical symmetry group is larger than the Galilei or
Poincaré group, as in the case of the conformal group [1] for the
Maxwell equations or the Schrödinger group [2,3] for the free
Schrödinger equation or the diffusion equation. The present paper
is a report on an attempt to apply the notion of kinematical symmetry to nonlinear equations, namely, to the Navier-Stokes equation of hydrodynamics.

The Navier-Stokes equation (NS) for the velocity field $\psi(t,\underset{\sim}{x})$
of a fluid is

(NS): $$\partial_t \underset{\sim}{\psi} + \underset{\sim}{\psi} \cdot \partial \underset{\sim}{\psi} - \nu \Delta \underset{\sim}{\psi} + \frac{1}{\rho} \underset{\sim}{\partial} p = 0, \qquad (1.1)$$

where ν is the kinematical viscosity, ρ the density, and p the
pressure. It is usually derived from a more general equation by
assuming constant density and then inferring $\underset{\sim}{\partial} \cdot \underset{\sim}{\psi} = 0$ from the

continuity equation. However, for (NS) to hold the slightly weaker requirement that the density is a function of t alone is actually sufficient and the continuity equation (C) then reads

$$\text{(C):} \qquad \underset{\sim}{\partial}\cdot\underset{\sim}{\psi} = -\frac{1}{\rho}\,\partial_t\rho, \quad \rho = \rho(t). \tag{1.2}$$

(NS) may be considered as a first-order differential equation for p and the corresponding integrability condition (I) is

$$\text{(I):} \qquad \partial_t\underset{\sim}{w} + \underset{\sim}{\psi}\cdot\underset{\sim}{\partial}\underset{\sim}{w} - \underset{\sim}{w}\cdot\underset{\sim}{\partial}\underset{\sim}{\psi} + \underset{\sim}{w}\underset{\sim}{\partial}\cdot\underset{\sim}{\psi} - \nu\Delta\underset{\sim}{w} = 0, \tag{1.3}$$

$$\underset{\sim}{w} := \underset{\sim}{\partial}\times\underset{\sim}{\psi}, \tag{1.4}$$

where $\underset{\sim}{w}$ is the vorticity vector. The content of (NS) is thus the condition (I) for $\underset{\sim}{\psi}$ and the definition (NS) of pressure. In the sequel the kinematical symmetries of one or more of the equations (NS), (C), (I) are investigated. Throughout, $\rho = \rho(t)$ and ν = const. is assumed.

We next give a precise definition of what is meant by a kinematical symmetry. Let there be given a coordinate transformation

$$g:(t,\underset{\sim}{x}) \rightarrow g(t,\underset{\sim}{x}). \tag{1.5}$$

At the same time the functions $\underset{\sim}{\psi}$, p, ρ are transformed into new functions according to

$$g:\underset{\sim}{\psi} \rightarrow T_g\underset{\sim}{\psi}, \quad p \rightarrow Q_g p, \quad \rho \rightarrow S_g\rho. \tag{1.6}$$

(The viscosity ν is being kept fixed all along.) Whether the transformation (1.5) leaves invariant, say, the NS-equation depends to some extent on what transformations T_g, Q_g, S_g are permitted. In the present paper we demand that T_g be of the form

$$(T_g\underset{\sim}{\psi})(t,\underset{\sim}{x}) = M_g[g^{-1}(t,\underset{\sim}{x})]\underset{\sim}{\psi}[g^{-1}(t,x)] + \underset{\sim}{h}_g[g^{-1}(t,\underset{\sim}{x})], \tag{1.7}$$

where M_g and $\underset{\sim}{h}_g$ are, respectively, a 3 × 3 matrix and a vector which do not depend on $\underset{\sim}{\psi}$. The definition (1.7) of T_g differs from the corresponding definition in the case of the Schrödinger equation [2] by the additional vector $\underset{\sim}{h}_g$. In the latter case the maps T_g form a representation of the kinematical symmetry group and this property would be destroyed if an inhomogeneous part $\underset{\sim}{h}_g$ is allowed. In the present case, however, the NS-equation is nonlinear and no (linear) representation can be carried by its solutions anyhow. The Ansatz (1.7) is the smallest possible deviation from a linear homogeneous transformation and in fact, as will be seen later, without the additional flexibility introduced by the term $\underset{\sim}{h}_g$, the NS-equation would not even exhibit invariance under Galilean boosts. We do not yet specify the transformations $Q_g p$ and $S_g \rho$; they will later either be determined automatically ($S_g \rho$) or have to be specified by some reasonable Ansatz ($Q_g p$) which in turn will affect the possible symmetries.

The transformation (1.5) is called a kinematical symmetry of the NS-equation if $(T_g \underset{\sim}{\psi}, Q_g p, S_g \rho)$ are a solution of (NS) whenever $(\underset{\sim}{\psi}, p, \rho)$ are a solution, and analogous definitions hold for the equations (C) and (I).

In section 2 the largest kinematical symmetry group of the NS-equation is determined up to a stage where a decision has to be taken on the permitted transformation behaviour of pressure. The result of the section is summarized in theorem 1. In section 2, two points of view are taken as concerns the transformation of pressure. On the one hand, if the transformation of pressure is left free entirely it is found (theorem 2) that the kinematical symmetry group of (NS) or, equivalently, of (I), is an infinite-dimensional Lie group. On the other hand, if pressure is required to transform in a simple and natural way then (theorem 3) the symmetry group is the Schrödinger group. Finally, in section 4, we briefly point out a quick way to verify Schrödinger invariance by tracing it back to Euclidean invariance and invariance under a single discrete transformation.

2. CONDITIONS FOR THE SYMMETRIES OF THE NS-EQUATION

In this section we want to determine the kinematical symmetry group of the NS-equation as far as is possible without specifying the transformation of pressure and density. The condition of invariance under (1.5) and (1.6,1.7) is

$$\{\partial'_t+[M_g(t,\underset{\sim}{x})\underset{\sim}{\psi}(t,\underset{\sim}{x})+\underset{\sim}{h}_g(t,\underset{\sim}{x})]\cdot\underset{\sim}{\partial}'-\nu\Delta'\}[M_g(t,\underset{\sim}{x})\underset{\sim}{\psi}(t,\underset{\sim}{x})+\underset{\sim}{h}_g(t,\underset{\sim}{x})] + \frac{1}{(S_g\rho)(t')}\underset{\sim}{\partial}'(Q_g p)(t',\underset{\sim}{x}') = 0, \tag{2.1}$$

where $(t',\underset{\sim}{x}') := g(t,\underset{\sim}{x})$ and $\underset{\sim}{\psi}$ is any solution of (NS). (2.1) is simply (NS) for $(T_g\underset{\sim}{\psi},Q_g p,S_g\rho)$ written at the point $(t',\underset{\sim}{x}')$. The analysis of (2.1) is rather technical and only the main points are given below. The result can be stated as follows:

Theorem 1: The largest kinematical symmetry group of the NS-equation is given by

$$g(t,\underset{\sim}{x}) = \left(\int dt\,\frac{1}{d^2},\ \frac{R\underset{\sim}{x}+\underset{\sim}{y}}{d}\right), \tag{2.2}$$

$$M_g(t,\underset{\sim}{x}) = dR,\quad \underset{\sim}{h}_g(t,\underset{\sim}{x}) = -\dot{d}R\underset{\sim}{x} - \dot{d}\underset{\sim}{y} + d\dot{\underset{\sim}{y}}, \tag{2.3}$$

where $R \in O(3)$, and $d = d(t)$ and $\underset{\sim}{y} = \underset{\sim}{y}(t)$ are real functions which have to be determined from the condition

$$d\ddot{d}(\underset{\sim}{x}+R^{-1}\underset{\sim}{y}) - d^2R^{-1}\ddot{\underset{\sim}{y}} = \underset{\sim}{\partial}\left\{\frac{(Q_g p)[g(t,\underset{\sim}{x})]}{(S_g\rho)[g(t)]} - d^2\,\frac{p(t,\underset{\sim}{x})}{\rho(t)}\right\}. \tag{2.4}$$

Proof. Defining the derivatives

$$d^2(t,\underset{\sim}{x}) = \partial t/\partial t',\quad b_i(t,\underset{\sim}{x}) = \partial x_i/\partial t',$$

$$c_i(t,\underset{\sim}{x}) = \partial t/\partial x'_i,\quad d_{ik}(t,\underset{\sim}{x}) = \partial x_i/\partial x'_k, \tag{2.5}$$

we first convert equation (2.1) into an equation with the differential operators ∂_t, $\underset{\sim}{\partial}$. Then, replacing $\partial_t \underset{\sim}{\psi}$ by space-derivatives by means of (NS) and noting that $\underset{\sim}{\psi}$ is an arbitrary solution of (NS) and hence the coefficients of different orders of derivatives must vanish separately, we obtain a number of conditions for the unknown quantities M_g, $\underset{\sim}{h}_g$ and (2.5).

The condition for the vanishing of the highest order derivative is $\underset{\sim}{c} = 0$, and this condition considerably simplifies the subsequent calculations. In particular, it implies that the transformations $g(t,\underset{\sim}{x})$ are of the Galilean type, i.e. that t' depends on t only, and hence that the transformed density $S_g\rho$ is again a function of t alone. The next condition is

$$d_{is}d_{ks} = d^2\delta_{ik}, \tag{2.6}$$

which implies that d_{ik}/d is a rotation:

$$d_{ik} = dR_{ki}, \quad R \in O(n). \tag{2.7}$$

The derivatives (2.5) are now inverted to

$$\partial t'/\partial t = \frac{1}{d^2}, \quad \partial t'/\partial x_i = 0,$$

$$\partial x_i'/\partial t = -\frac{1}{d^3} R_{ik}b_k, \quad \partial x_i'/\partial x_k = \frac{1}{d} R_{ik}, \tag{2.8}$$

and the integrability conditions for (2.8) are

$$d = d(t), \quad R = R(t), \quad \partial_i b_k = d\dot{d}\delta_{ik} - d^2 R_{sk}\dot{R}_{si}. \tag{2.9}$$

Integrating (2.8) we obtain the transformations (2.2) where the vector $\underset{\sim}{y}(t)$ is related to the vector $\underset{\sim}{b}$ by

$$\underset{\sim}{b} = d\dot{d}(\underset{\sim}{x}+R^{-1}\underset{\sim}{y}) - d^2(R^{-1}\dot{R}\underset{\sim}{x}+R^{-1}\dot{\underset{\sim}{y}}). \tag{2.10}$$

The remaining conditions from (2.1) turn out to be

$$M_g = dR, \quad \underset{\sim}{h}_g = -\frac{1}{d} R\underset{\sim}{b}, \quad \dot{R} = 0, \tag{2.11}$$

$$d\dot{\underset{\sim}{b}} - \dot{d}\underset{\sim}{b} = d\underset{\sim}{\partial}\left[\frac{(Q_g p)(t',\underset{\sim}{x}')}{(S_g \rho)(t')} - d^2 \frac{p(t,\underset{\sim}{x})}{\rho(t)}\right], \tag{2.12}$$

the latter implying (2.4).

3. KINEMATICAL SYMMETRIES AND THE TRANSFORMATION OF PRESSURE

In theorem 1 the general form of the kinematical symmetries was found. The actual form depends on the assumptions of the transformation Q_g in condition (2.4), and in this section we analyze condition (2.4) for two different assumptions. At the same time we also take into account the continuity equation (C).

The kinematical symmetry group of (I)

If the transformation Q_g of pressure is left free completely the condition (2.4) may be considered as definition of $Q_g p$ and there are then no restrictions on the functions d(t) and $\underset{\sim}{y}(t)$. Leaving free the pressure is equivalent to replacing (NS) by (I) and it is straightforward to verify that (I) is indeed left invariant by the full group of transformations (2.2), the transformation of the vorticity vector being given by

$$g:\underset{\sim}{w} \to V_g\underset{\sim}{w}: = \underset{\sim}{\partial} \times T_g\underset{\sim}{\psi}, \quad (V_g\underset{\sim}{w})[g(t,\underset{\sim}{x})] = d^2(t)R\underset{\sim}{w}(t,\underset{\sim}{x}). \tag{3.1}$$

Furthermore, the continuity equation (C) is also invariant under (2.2) and the transformation of density is determined to be

$$(S_g\rho)[g(t)] = d^3(t)\rho(t). \tag{3.2}$$

We may thus formulate as follows:

Theorem 2. The largest kinematical symmetry group of the combined equations (I) and (C) is the infinite-dimensional group of transformations (2.2) with arbitrary real functions d(t), $\underset{\sim}{y}(t)$

and with T_g, S_g given by (2.3) and (3.2). The same group is also a symmetry group pf (NS) provided pressure is transformed as

$$(Q_g p)[g(t,\underset{\sim}{x})] = d^5 p(t,\underset{\sim}{x}) + d^4 \rho(t)[\tfrac{1}{2}\ddot{d}\underset{\sim}{x}^2 + R\underset{\sim}{x}\cdot(\ddot{d}\underset{\sim}{y} - d\ddot{\underset{\sim}{y}})] + k(t), \quad (3.3)$$

where k(t) is arbitrary.

If we insist on constant density then the invariance of (C) requires that the function d be constant while the function $\underset{\sim}{y}$ is still arbitrary.

The Schrödinger group as symmetry group of (NS)

The simplest Ansatz for Q_g is the linear homogeneous transformation

$$(Q_g p)[g(t,\underset{\sim}{x})] = f_g(t,\underset{\sim}{x})p(t,\underset{\sim}{x}), \quad (3.4)$$

where f_g is some p-independent function. With (3.4) and (3.2) the condition (2.4) takes the form

$$\rho d^4[\ddot{d}(\underset{\sim}{x}+R^{-1}\underset{\sim}{y}) - dR^{-1}\ddot{\underset{\sim}{y}}] = (f_g - d^5)\underset{\sim}{\partial} p. \quad (3.5)$$

Since this condition is to hold for arbitrary solutions $(\underset{\sim}{\psi},p)$ of (NS) we conclude

$$\ddot{d} = 0, \quad \ddot{\underset{\sim}{y}} = 0, \quad f_g = d^5. \quad (3.6)$$

Writing $d = \gamma t + \delta$, $\underset{\sim}{y} = \underset{\sim}{v}t + \underset{\sim}{a}$, we obtain the following result:

Theorem 3. The largest kinematical symmetry group of the combined equations (NS) and (C), under the assumption that p transforms as in (3.4), is the *Schrödinger group* given by the transformations

$$g = [\begin{pmatrix}\alpha & \beta\\ \gamma & \delta\end{pmatrix}, \underset{\sim}{a}, \underset{\sim}{v}, R], \quad \begin{pmatrix}\alpha & \beta\\ \gamma & \delta\end{pmatrix} \in SL(2,\mathbb{R}), \quad \underset{\sim}{a},\underset{\sim}{v} \in \mathbb{R}^3, \; R \in O(3), \quad (3.7)$$

$$g(t,x) = \left(\frac{\alpha t+\beta}{\gamma t+\delta}, \frac{R\underset{\sim}{x}+\underset{\sim}{v}t+\underset{\sim}{a}}{\gamma t+\delta}\right), \quad (3.8)$$

$$(T_{g}\psi)(t,x) = \frac{1}{\alpha-\gamma t}(R\psi[g^{-1}(t,x)]-\gamma x+v), \tag{3.9}$$

$$(Q_g p)(t,x) = \frac{1}{(\alpha-\gamma t)^5} p[g^{-1}(t,x)], \tag{3.10}$$

$$(S_g \rho)(t) = \frac{1}{(\alpha-\gamma t)^3} \rho[g^{-1}(t)]. \tag{3.11}$$

If the density is required to be constant then $\gamma = 0$ and the symmetry group is the dilated Galilei group.

Note that the presence of the factors d^n in the transformations of ψ, p, ρ is due to the dilation-type transformations (γ,δ), and indeed these factors come out very naturally as can be seen by considering the dimensions of ψ, p, ρ and by observing that time and length are dilated into time/d^2 and length/d. The invariance under δ-dilations, of course, reflects the well-known hydrodynamic similarity law.

Finally we would like to point out that the operator T is a homomorphism of the Schrödinger group, i.e. it satisfies

$$T_{g_2} T_{g_1} = T_{g_2 g_1}. \tag{3.12}$$

This is in contrast to the case of the free Schrödinger equation [2] where the corresponding operator T is not a true but a projective representation.

4. A SIMPLE CRITERION FOR SCHRODINGER INVARIANCE

The following criterion may be useful for a quick verification that an equation is invariant under the Schrödinger group.

Criterion:

A system of equations is invariant under the Schrödinger group

$$g(t,\underset{\sim}{x}) = \left(\frac{\alpha t+\beta}{\gamma t+\delta}, \frac{R\underset{\sim}{x}+\underset{\sim}{v}t+\underset{\sim}{a}}{\gamma t+\delta}\right) \tag{4.1}$$

iff it is simultaneously invariant under the group $T_0 \otimes E(3)$ of time-translations and Euclidean transformations

$$g_0(t,\underset{\sim}{x}) = (t+\beta, R\underset{\sim}{x}+\underset{\sim}{a}) \tag{4.2}$$

and under the discrete transformation

$$\Sigma(t,\underset{\sim}{x}) = (-\frac{1}{t}, \frac{1}{t}\underset{\sim}{x}). \tag{4.3}$$

Proof. First we note that both transformations (4.2) and (4.3) are present among the transformations (4.1), in particular, Σ is obtained for $\alpha = \delta = 0$, $\beta = -1$, $\gamma = 1$, $\underset{\sim}{a} = \underset{\sim}{v} = 0$, $R = 1$. Conversely, we show that the Schrödinger group is generated by $T_0 \otimes E(3)$ and Σ by showing that boosts $B(\underset{\sim}{v})$, γ-dilations $\Gamma(\gamma)$, and δ-dilations $\Delta(\delta)$ can be written as products of translations and Σ. In fact, denoting space-translations by $T(\underset{\sim}{a})$ and time-translations by $T_0(\beta)$ one easily verifies the relations

$$B(\underset{\sim}{v}) = \Sigma^{-1}T(\underset{\sim}{v})\Sigma, \tag{4.4}$$

$$\Gamma(\gamma) = \Sigma^{-1}T_0(-\gamma)\Sigma, \tag{4.5}$$

$$\Delta(\delta) = \Gamma(\delta-1)T_0(1)\Gamma(\alpha-1)T_0(-\delta), \quad (\alpha\delta=1). \tag{4.6}$$

As an example we note that (NS) is certainly invariant under rotations and translations. Furthermore it is also invariant under Σ with

$$(T_\Sigma\underset{\sim}{\psi})(t,\underset{\sim}{x}) = -\frac{1}{t}[\underset{\sim}{\psi}(-\frac{1}{t},-\frac{1}{t}\underset{\sim}{x})-\underset{\sim}{x}],$$

$$(Q_\Sigma p)(t,\underset{\sim}{x}) = (-\frac{1}{t})^5 p(-\frac{1}{t},-\frac{1}{t}\underset{\sim}{x}), \quad (S_\Sigma\rho)(t) = (-\frac{1}{t})^3\rho(-\frac{1}{t}). \tag{4.7}$$

Hence it is invariant under the Schrödinger group and the

transformations T_g, Q_g, S_g of (3.9-3.11) may be calculated from (4.7) by writing g as a product of transformations (4.2) and Σ.

REFERENCES

1. M. FLATO, J. SIMON and D. STERNHEIMER, Ann. Phys. 61, 78 (1970).
2. U. NIEDERER, Helv. Phys. Acta 45, 802 (1972).
3. U. NIEDERER, Helv. Phys. Acta 47, 167 (1974).

NORMAL MODES OF NONLINEAR DISCRETE SYMMETRIC SYSTEMS BY GROUP REPRESENTATION THEORY

M.C. Singh and A.K. Mishra

INTRODUCTION

Non-linear modes of vibrations for non-linear springs connected with lumped masses have been analysed by Rosenberg [1]. The motion of a spring mass system, with n masses, in a three-dimensional physical space can, in general, be represented by a 3n-dimensional configuration space. The solution yields a set of 3n non-linear coupled equations of motion. For a symmetrical system group representation theory can be employed to determine generalized symmetry adapted coordinates. The equations of motion are then expressed in this coordinate system, and have, in general, a lower order of coupling. These equations are solved for non-linear modes by the method evolved by Rosenberg [1,2].

SYSTEM SYMMETRY AND SYMMETRY ADAPTED COORDINATES

Consider a spring mass network in equilibrium with n joints. Let $R_1, R_2, \ldots, R_g$ be the symmetry operations characterizing the symmetry of the system in equilibrium. The set of all the symmetry operations forms a group. Let this group be denoted by G. Construct a displacement vector $\bar{r}$ in 3n-dimensional vector space,

$$\bar{r} = \sum_{k=1}^{3n} x_k \bar{u}_k, \tag{1}$$

where the orthogonal set of unit vectors $\bar{u}_{3j-2}$, $\bar{u}_{3j-1}$ and $\bar{u}_{3j}$ and the corresponding set of displacement components x_{3j-2}, x_{3j-1} and x_{3j} are associated with the j-th joint. Operating by each symmetry operator R on the base vectors $\bar{u}_i$, a reducible representation D(R) is obtained as,

$$R\bar{u}_k = \sum_{j=1}^{3n} \bar{u}_j D(R)_{jk}, \tag{2}$$

where $D(R)_{jk}$ is the transformation matrix corresponding to the symmetry operator R.

The symmetry adapted basis vectors are obtained in terms of the base vectors $\bar{u}_i$ by making use of the projection operators [3] as,

$$\sum_j c_{kj}^{(\nu,r)} \bar{e}_j^{(\nu,r)} = \sum_i \chi^{(\nu,r)}(R_i)\cdot R_i(\bar{u}_k), \tag{3}$$

where $\bar{e}_j^{(\nu,r)}$ is the j-th normalized symmetry adapted basis vector which transforms according to the r-th row of the ν-th irreducible representation (IR). $c_{kj}^{(\nu,r)}$ are the constant coefficients and $\chi^{(\nu,r)}(R_i)$ is the character for the operator R_i corresponding to the ν-th IR. $r = 1,2,\ldots,n_\nu$ and $j = 1,2,\ldots,a_\nu$, where n_ν and a_ν are the dimensionality and the number of occurrences of the ν-th IR, respectively.

NON-LINEAR EQUATIONS OF MOTION

The displacements and other kinetic and kinematic quantities can be expressed in terms of components of the symmetry adapted basis vectors. A symmetry adapted component $q_j^{(\nu,r)}$ of the displacement vector $\bar{r}$ can be expressed as

$$q_j^{(\nu,r)} = \bar{r}\cdot\bar{e}_j^{(\nu,r)}. \tag{4}$$

The kinetic energy, in terms of the symmetry adapted coordinates is expressed as,

$$T = \frac{1}{2} \sum_{i,j,\nu,r} A_{ij}^{(\nu,r)} \dot{q}_i^{(\nu,r)} \dot{q}_j^{(\nu,r)}, \tag{5}$$

where $A_{ij}^{(\nu,r)}$ are the elements of the kinetic energy matrix.

Assuming that the non-linear system is homogeneous such that the springs resist deflection with a force $F_i^{(\nu,r)}$ which is proportional to its k-th power, the spring force can be expressed as,

$$F_i^{(\nu,r)} = \sum_{j_1,\ldots,j_k} \beta_{ij_1,\ldots,j_k}^{(\nu,r)} q_{j_1}^{(\nu,r)},\ldots,q_{j_k}^{(\nu,r)}, \tag{6}$$

$$i,j_1,\ldots,j_k = 1,\ldots,a_\nu; \quad k = 1,3,5,\ldots \text{ (odd integer)},$$

where

$$\beta_{ij_1,\ldots,j_k}^{(\nu,r)}$$

are constant coefficients corresponding to the ν-th IR of dimensionality r.

The strain energy can be written by using equation (6), as

$$V = \frac{1}{k+1} \sum_{\nu,r,i,j_1,\ldots,j_k} \beta_{ij_1,\ldots,j_k}^{(\nu,r)} q_i^{(\nu,r)} q_{j_1}^{(\nu,r)},\ldots,q_{j_k}^{(\nu,r)}. \tag{7}$$

Lagrangian equations of motion are obtained on the basis of Hamilton's principle by making use of the kinetic and strain energies as

$$\ddot{q}_i^{(\nu,r)} = -\sum_{j_1,\ldots,j_k} \beta_{ij_1,\ldots,j_k}^{(\nu,r)} q_{j_1}^{(\nu,r)},\ldots,q_{j_k}^{(\nu,r)}. \tag{8}$$

These are equations of motion for non-linear homogeneous systems of degree k. The equations for different IR's are independent. Coupling between a number of equations occurs in those cases in which the IR's involved are present more than once. A degeneracy is present in the case of an IR of dimension higher than one, the order of degeneracy being the same as the dimension of the IR.

NORMAL MODE SOLUTIONS OF THE EQUATIONS OF MOTION

Corresponding to an IR occurring once, the non-linear equation of motion is independent and thus can be solved by the classical approach [4]. When an IR occurs more than once, the equations of motion are coupled and the solutions can be obtained following Rosenberg [1]. For each invariant and irreducible subspace, one can write the strain energy as

$$V^{(\nu,r)} = V^{(\nu,r)}(q_1^{(\nu,r)},\ldots,q_{a_\nu}^{(\nu,r)}) = -V_o^{(\nu,r)}. \qquad (9)$$

where $V^{(\nu,r)}$ is the strain energy corresponding to the r-th dimension of the ν-th IR and $V_o^{(\nu,r)}$ is the maximum value of $V^{(\nu,r)}$. Equation (9) defines an equipotential surface. A modal line passes through the origin and intersects the equipotential surface orthogonally; this implies that the distance from the origin to the point where a modal line intersects the equipotential surface must be stationary [1]. Let the equation of the equipotential surface be written in terms of generalized polar coordinates s, $\Theta_1^{(\nu,r)},\Theta_2^{(\nu,r)},\ldots,\Theta_{a_\nu-1}^{(\nu,r)}$ as

$$V^{(\nu,r)}(s,\Theta_1^{(\nu,r)},\ldots,\Theta_{a_\nu-1}^{(\nu,r)}) = s^{k+1}\Theta^{(\nu,r)}(\Theta_1^{(\nu,r)},\ldots,\Theta_{a_\nu-1}^{(\nu,r)}). \qquad (10)$$

The locus of the stationary points is given by

$$\frac{\partial\Theta^{(\nu,r)}}{\partial\Theta_i^{(\nu,r)}} = 0, \qquad (i = 1,2,\ldots,a_\nu-1). \tag{11}$$

The roots of equations (11) lead to the relations between $q_i^{(\nu,r)}$ and $q_1^{(\nu,r)}$ as

$$q_i^{(\nu,r)} = a_i^{(\nu,r)} q_1^{(\nu,r)}, \qquad i = 2,3,\ldots,a_\nu, \tag{12}$$

where $a_i^{(\nu,r)}$ are constant coefficients. The normal mode vectors, in terms of the symmetry adapted basis vectors, can now be written in normalized form, with the aid of equations (4) and (12) as,

$$\bar{p}_i^{(\nu,r)} = \sum_{j=1}^{a_\nu} a_{ij}^{(\nu,r)} \bar{e}_j^{(\nu,r)}, \tag{13}$$

where $a_{ij}^{(\nu,r)}$ are constant coefficients, and $\bar{p}_i^{(\nu,r)}$ is the i-th normal mode vector corresponding to the ν-th IR.

The equations of motion can now be written as,

$$\ddot{p}_i^{(\nu,r)} + d_i^{(\nu,r)} [p_i^{(\nu,r)}]^k = 0; \quad i = 1,\ldots,a_\nu, \tag{14}$$

where $d_i^{(\nu,r)}$ are constant coefficients and k is the order of homogeneity. Equations (14) are uncoupled equations of motion. These equations are solved for the velocity-displacement and the frequency-amplitude relations. The velocity displacement relations can be written as,

$$\frac{1}{2} [\dot{p}_i^{(\nu,r)}]^2 + \frac{2}{h(k+1)} d_i^{(\nu,r)} [p_i^{(\nu,r)}]^{k+1} = 1; \quad i = 1,\ldots,a_\nu. \tag{15}$$

The frequency-amplitude relations are

$$\omega_i^{(\nu,r)} = \frac{\Pi}{2\int_0^{b_i^{(\nu,r)}} \frac{dP_i^{(\nu,r)}}{\sqrt{h - \frac{2d_i^{(\nu,r)}[P_i^{(\nu,r)}]^{k+1}}{k+1}}}}, \tag{16}$$

where $\omega_i^{(\nu,r)}$ is the frequency corresponding to $\bar{P}_i^{(\nu,r)}$, h is twice the maximum strain energy $V_o^{(\nu,r)}$ and is represented by

$$h = [\dot{P}_i^{(\nu,r)}]^2 + \frac{2}{k+1} d_i^{(\nu,r)} [P_i^{(\nu,r)}]^{(k+1)} \tag{17}$$

and $b_i^{(\nu,r)}$ is the amplitude of vibration and is equal to the maximum value of the normal mode displacement:

$$b_i^{(\nu,r)} = P_i^{(\nu,r)}\Big|_{max} = \left[\frac{(k+1)h}{2d_i^{(\nu,r)}}\right]^{\frac{1}{k+1}}. \tag{18}$$

All the frequencies in the numerical examples are evaluated by assuming the amplitude of vibration $b_i^{(\nu,r)}$ equal to unity.

NUMERICAL EXAMPLE

The spring mass system shown in Figs. (1-a,b) represents a simplified model of the ISISA Satellite [5]. The system consists of twelve mass points in three-dimensional physical space and thus forms a thirty-six dimensional configuration space.

The non-linearity of the system is due to material non-linearity of the springs, and the force-deflection relation is obtained from equation (6) for k=3. The spring displacements are assumed to be infinitesimal. The system remains invariant under the symmetry operations of the group D_{4h}. This group is of order sixteen and has ten IR's. The type and number of IR's contained in the reducible representation D are determined as

FIG. 1-a. Spring Mass System

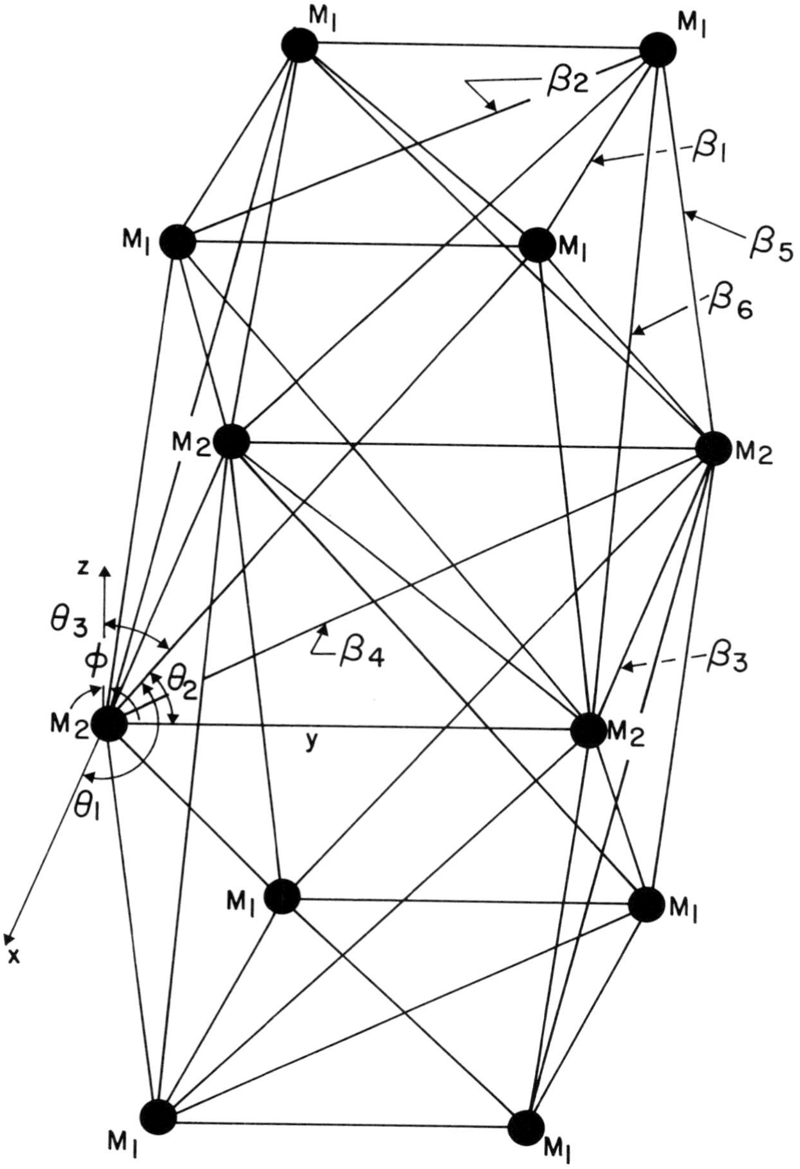

FIG. 1-b. Base Vectors and Symmetry Operations

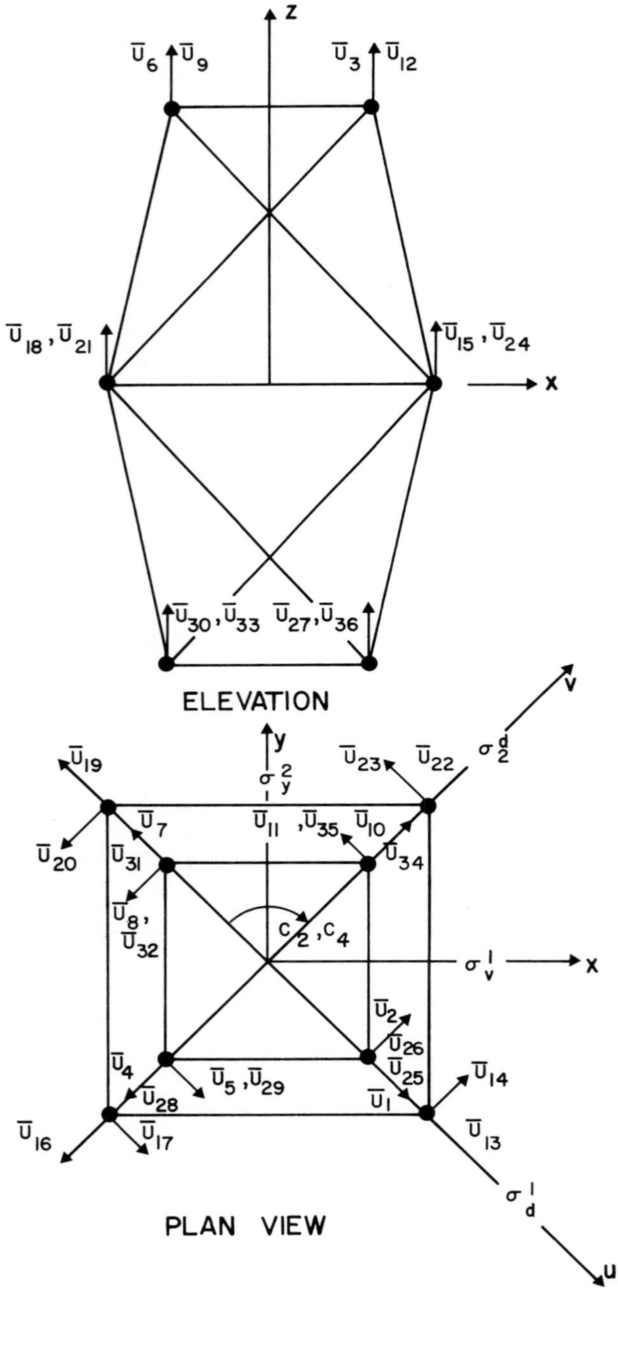

$$\Gamma = 3A_{1g}+2A_{2g}+2B_{1g}+3B_{2g}+4E_{g}+A_{1u}+3A_{2u}+3B_{1u}+B_{2u}+5E_{u}. \qquad (19)$$

Each mass M_i is assumed to be equal to unity; the spring constants, $\beta_i = \beta = 1000$; the angle $\phi = 15°$ and the mid base width equals the height above.

The symmetry adapted basis vectors corresponding to each IR are given in Table 1. The strain energy corresponding to each IR is written according to equation (7). The equations of motion along the normal coordinates are obtained as in equation (14). The velocity-displacement and the frequency-amplitude relations are determined according to equations (15) and (16), respectively.

The normal mode vectors, values of the constant coefficients $d_i^{(\nu,r)}$ for all normal coordinates, and the frequencies calculated by assuming unit amplitude of vibration are given in Table 2. The results tabulated are given for a set of typical IR's.

REFERENCES

1. R.M. ROSENBERG, On Non-linear Vibrations of Systems with Many Degrees of Freedom, Advances in Applied Mechanics, Vol. 9, Academic Press, New York, pp. 156-243, 1966.
2. A.K. MISHRA and M.C. SINGH, Int. J. Non-Linear Mechanics, Vol. 9, pp. 463-480.
3. J.F. CORNWELL, Group Theory and Electronic Energy Bands in Solids, North Holland Publication Co., Amsterdam, 1969.
4. N.V. BUTENIN, Elements of the Theory of Non-linear Oscillations, Blaisdell, New York, 1965.
5. N.R.C. of Canada, "Balloons, Rockets and Satellites to Study Earth's Atmosphere", Science Division, Vol. 1, No. 5, 1969.

TABLE 1. Symmetry Adapted Basis Vectors

Representation	Dimension	Number of Occurrence	Symbol	Symmetry Adapted Basis Vectors
A_{1g}	1	1	$\bar{e}_1^{(A_{1g},1)}$	$(2\sqrt{2})^{-1}(\bar{u}_1+\bar{u}_4+\bar{u}_7+\bar{u}_{10}+\bar{u}_{25}+\bar{u}_{28}+\bar{u}_{31}+\bar{u}_{34})$
		2	$\bar{e}_2^{(A_{1g},1)}$	$0.5\ (\bar{u}_{13}+\bar{u}_{16}+\bar{u}_{19}+\bar{u}_{22})$
		3	$\bar{e}_3^{(A_{1g},1)}$	$(2\sqrt{2})^{-1}(\bar{u}_3+\bar{u}_6+\bar{u}_9+\bar{u}_{12}-\bar{u}_{27}-\bar{u}_{30}-\bar{u}_{33}-\bar{u}_{36})$
A_{2g}	1	1	$\bar{e}_1^{(A_{2g},1)}$	$(2\sqrt{2})^{-1}(\bar{u}_2+\bar{u}_5+\bar{u}_8+\bar{u}_{11}+\bar{u}_{26}+\bar{u}_{32}+\bar{u}_{35}+\bar{u}_{29})$
		2	$\bar{e}_2^{(A_{2g},1)}$	$0.5\ (\bar{u}_{14}+\bar{u}_{17}+\bar{u}_{20}+\bar{u}_{23})$
E_g	2	1	$\bar{e}_1^{(E_g,1)}$	$0.5\ (\bar{u}_1-\bar{u}_7-\bar{u}_{25}+\bar{u}_{31})$
			$\bar{e}_1^{(E_g,2)}$	$0.5\ (\bar{u}_4-\bar{u}_{10}-\bar{u}_{28}+\bar{u}_{34})$
		2	$\bar{e}_2^{(E_g,1)}$	$0.5\ (\bar{u}_2-\bar{u}_8-\bar{u}_{26}+\bar{u}_{32})$
			$\bar{e}_2^{(E_g,2)}$	$0.5\ (\bar{u}_5-\bar{u}_{11}-\bar{u}_{29}+\bar{u}_{35})$

TABLE 1. (Cont.)

Representation	Dimension	Number of Occurrence	Symbol	Symmetry Adapted Basis Vectors
E_g	2	3	$\bar{e}_3^{(E_g,1)}$	$0.5\ (\bar{u}_3-\bar{u}_9+\bar{u}_{27}-\bar{u}_{33})$
			$\bar{e}_3^{(E_g,2)}$	$0.5\ (\bar{u}_6-\bar{u}_{12}+\bar{u}_{30}-\bar{u}_{36})$
		4	$\bar{e}_4^{(E_g,1)}$	$0.7071\ (\bar{u}_{15}-\bar{u}_{21})$
			$\bar{e}_4^{(E_g,2)}$	$0.7071\ (\bar{u}_{18}-\bar{u}_{24})$
A_{1u}	1	1	$\bar{e}_1^{(A_{1u},1)}$	$(2\sqrt{2})^{-1}(\bar{u}_2+\bar{u}_5+\bar{u}_8+\bar{u}_{11}-\bar{u}_{26}-\bar{u}_{29}-\bar{u}_{32}-\bar{u}_{35})$

TABLE 2. Normal Mode Vectors and Frequencies

Normal Mode Symbol	Normal Mode	Constant $d_i^{(\nu,r)}$	Frequency
$\bar{p}_1^{(A_{1g},1)}$	$.8309\ \bar{e}_1^{(A_{1g},1)} + 0.5507\ \bar{e}_2^{(A_{1g},1)} + 0.0799\ \bar{e}_3^{(A_{1g},1)}$	1562.7	33.49
$\bar{p}_2^{(A_{1g},1)}$	$.8176\ \bar{e}_1^{(A_{1g},1)} - 0.5757\ \bar{e}_2^{(A_{1g},1)} - 0.0119\ \bar{e}_3^{(A_{1g},1)}$	1503.25	32.85
$\bar{p}_3^{(A_{1g},1)}$	$.2359\ \bar{e}_1^{(A_{1g},1)} - 0.2898\ \bar{e}_2^{(A_{1g},1)} + 0.9276\ \bar{e}_3^{(A_{1g},1)}$	1470.77	32.49
$\bar{p}_1^{(A_{2g},1)}$	$.7929\ \bar{e}_1^{(A_{2g},1)} - 0.6096\ \bar{e}_2^{(A_{2g},1)}$	111.38	8.94
$\bar{p}_2^{(A_{2g},1)}$	$.6601\ \bar{e}_1^{(A_{2g},1)} + 0.7512\ \bar{e}_2^{(A_{2g},1)}$	0.0	0.0
$\bar{p}_1^{(E_g,1)}$	$0.0708\ \bar{e}_1^{(E_g,1)} - 0.152\ \bar{e}_2^{(E_g,2)} - 0.4514\ \bar{e}_3^{(E_g,1)} + 0.8764\ \bar{e}_4^{(E_g,1)}$	421,234.0	549.9
$\bar{p}_1^{(E_g,2)}$	$0.0708\ \bar{e}_1^{(E_g,2)} - 0.152\ \bar{e}_2^{(E_g,1)} - 0.4514\ \bar{e}_3^{(E_g,2)} + 0.8464\ \bar{e}_4^{(E_g,2)}$	421,234.0	549.9
$\bar{p}_2^{(E_g,1)}$	$0.4559\ \bar{e}_1^{(E_g,1)} + 0.1411\ \bar{e}_2^{(E_g,2)} + 0.7639\ \bar{e}_3^{(E_g,1)} + 0.4345\ \bar{e}_4^{(E_g,1)}$	802.53	24.0

TABLE 2. (Cont.)

Normal Mode Symbol	Normal Mode	Constant $d_i^{(\nu,r)}$	Frequency
$\bar{P}_2^{(E_g,2)}$	$0.4559\ \bar{e}_1^{(E_g,2)} + 0.1411\ \bar{e}_2^{(E_g,1)} + 0.7639\ \bar{e}_3^{(E_g,2)} + 0.4345\ \bar{e}_4^{(E_g,2)}$	802.53	24.0
$\bar{P}_3^{(E_g,1)}$	$0.6605\ \bar{e}_1^{(E_g,1)} - 0.1978\ \bar{e}_2^{(E_g,2)} + 0.4798\ \bar{e}_3^{(E_g,1)} + 0.5427\ \bar{e}_4^{(E_g,1)}$	394.81	16.84
$\bar{P}_3^{(E_g,2)}$	$0.6605\ \bar{e}_1^{(E_g,2)} - 0.1978\ \bar{e}_2^{(E_g,1)} + 0.4798\ \bar{e}_3^{(E_g,2)} + 0.5427\ \bar{e}_4^{(E_g,2)}$	394.81	16.84
$\bar{P}_4^{(E_g,1)}$	$0.6393\ \bar{e}_1^{(E_g,1)} + 0.6393\ \bar{e}_2^{(E_g,2)} - 0.2824\ \bar{e}_3^{(E_g,1)} - 0.3209\ \bar{e}_4^{(E_g,1)}$	0.0	0.0
$\bar{P}_4^{(E_g,2)}$	$0.6393\ \bar{e}_1^{(E_g,2)} + 0.6393\ \bar{e}_2^{(E_g,1)} - 0.2824\ \bar{e}_3^{(E_g,2)} - 0.3209\ \bar{e}_4^{(E_g,2)}$	0.0	0.0

Part IV

Relativistic Quantum Physics

THE CONCEPT OF A KINEMATICAL STABILITY GROUP IN IMPLEMENTING RELATIVISTIC SYMMETRY*

L.C. Biedenharn** and H. van Dam

I. INTRODUCTION

The concept of an elementary particle, considered as a structureless entity, received a definitive categorization in the classic work of Wigner; such an elementary particle is an irrep of the Poincaré group, which Wigner classified exhaustively. The prototype for this work was the electron, and its associated Dirac equation which - even for the very highest currently available energies - has yielded no compelling evidence for internal structure[1].

By contrast the strongly interacting systems, the hadrons, are far too numerous all to be elementary and exhibit, moreover, clear evidence of internal structure, not only in terms of hypothetical quark structure, but even more directly in that all hadrons have a finite size. Empirically all hadrons appear to belong to families characterized by (approximately) linear Regge trajectories: $(\text{Mass})^2$ = linear function of spin. It is reasonable to idealize these empirical facts and declare that hadrons

* Research supported in part by the U.S. National Science Foundation.

** Alexander von Humboldt Foundation Senior U.S. Scientist Award 1976; on leave from Duke University, Durham, N.C., U.S.A.

are to be characterized by (infinite) linear trajectories. Such a system, though composite, would have no continuum and could not fragment. Clearly such a system would necessarily belong to an infinitely reducible Poincaré representation.

This basic concept is by no means new; the problem has always been to implement it consistently. Attempts such as infinite-component wave equations[2] have always foundered on difficulties such as unphysical solutions. The symmetry, or group-theoretic approach (organizing the set of Poincaré irreps into larger groups containing $\mathcal{P}$) have been strongly discouraged[3] by the "no-go" theorems of McGlinn, O'Raifeartaigh, Jost, and Segal[4].

Actually, as clearly stated by Bacry[5], these theorems simply invite one to consider infinite dimensional "Lie" groups. Such structures are inherent for example in dual resonance models. The difficulty here is that such structures are largely unexplored mathematically, so the physical ideas must be the guide[6].

The purpose of the present paper is to discuss the concept of a *kinematical stability group*, which, for an arbitrary four-momentum, organizes the set of $\mathcal{P}$ irreps coherently into an irrep of a larger symmetry group. In terms of the no-go theorem, such a larger symmetry is implementable only as an infinite dimensional group.

We were led to these ideas in a round-about way, through our attempt to understand and interpret Dirac's positive energy relativistic wave equation[8]. We found[9,10] that Dirac's construction could in fact be viewed as a form of a positive energy relativistically covariant oscillator; this led to the concept of a kinematic stability group.

Before turning to the detailed discussion of these ideas let us remark that the brief note[11] by Bacry and Nuyts in 1967, and a related earlier paper by Finkelstein in 1955[12] contained models of the explicit structure we shall develop; the actual realizations, however, are quite different and the stability concept was not explicitly introduced.

II. AN INTERPRETATION OF DIRAC'S NEW RELATIVISTIC WAVE EQUATION

Dirac's constructions - and the generalization to all spins - are based on the well-known, but nonetheless remarkable, properties of a special realization over two (degenerate) harmonic oscillators of the Lie algebra of the deSitter group, $SO(3,2) \cong Sp(2,2)$. (A very complete discussion of this structure has been given by Böhm[13].)

The ten generators of this group have the form:

$$
\begin{aligned}
\{J\}:\quad J_1 &= \tfrac{1}{2}(a_1\bar{a}_2+a_2\bar{a}_1), & \{K\}:\quad K_1 &= \tfrac{1}{4}(a_1^2-a_2^2+\bar{a}_1^2-\bar{a}_2^2),\\
J_2 &= \tfrac{1}{2}(a_1\bar{a}_1-a_2\bar{a}_2), & K_2 &= -\tfrac{1}{2}(a_1a_2+\bar{a}_1\bar{a}_2),\\
J_3 &= \tfrac{1}{2}i(a_1\bar{a}_2-a_2\bar{a}_1), & K_3 &= \tfrac{1}{4}i(a_1^2+a_2^2-\bar{a}_1^2-\bar{a}_2^2);
\end{aligned}
\tag{1}
$$

and the Lorentz four-vector generators:

$$
\begin{aligned}
\{V\}:\quad V_1 &= \tfrac{1}{4}i(\bar{a}_1^2-\bar{a}_2^2+a_2^2-a_1^2),\\
V_2 &= \tfrac{1}{2}i(a_1a_2-\bar{a}_1\bar{a}_2),\\
V_3 &= \tfrac{1}{4}(a_1^2+a_2^2+\bar{a}_2^2+\bar{a}_1^2),\\
V_0 &= \tfrac{1}{2}(a_1\bar{a}_1+a_2\bar{a}_2+1).
\end{aligned}
\tag{2}
$$

(The special choices in this realization (J_2 diagonal) stem from compatibility requirements with front dynamics[10].)

The existence of the Hermitian four-vector operator is exploited in writing the Majorana equation[14]: $(\vec{V}\cdot\vec{P} + m)\ \psi = 0$. This equation (and the many closely related modifications[2]) suffers from the defect that it permits space-like solutions[15]. Dirac's construction avoids this problem in the following way:

First one introduces finite-dimensional non-unitary representations, (n,0), of the deSitter group. The simplest non-trivial case is (1,0); this is the four-dimensional basis, denoted by Q:

$$Q \equiv \text{column } (a_1, a_2, \bar{a}_2, -\bar{a}_1). \tag{3}$$

One then maps the generators, $\mathcal{O}$, into the matrices $\tilde{\mathcal{O}}$ by the association:

$$[\mathcal{O}, Q] \equiv \tilde{\mathcal{O}} Q. \tag{4}$$

This map preserves commutators, but *not* Hermiticity[9].

Next, using the (matrix) $\tilde{V}$, which is the map of the four-vector generator $\vec{V}$, Dirac writes the wave equation:

$$(\tilde{V} \cdot \vec{P} + m) Q\psi = 0. \tag{5}$$

This wave equation is not Lorentz invariant, but rather covariant, such that if valid in one frame it is valid in all frames. [It bears only a superficial resemblance to the Dirac electron equation, since the four elements in the column vector Q represent four independent equations constraining the single wave function ψ.] It can be verified that the solutions to this equation have the properties:

a) positive, non-zero, energy only;

b) spinless;

c) electromagnetic interactions via the minimal substitution are not possible.

Generalizations[9] to arbitrary spin possess properties a) and c). (Note that space-like solutions are forbidden precisely because the map $V \to \tilde{V}$ did not preserve Hermiticity).

Let us consider now the rest-frame solution to equation (5).

Replacing $\vec{P}$ by (0,0,0,m) one finds the two equations:

$$\bar{a}_1\psi = 0, \qquad \bar{a}_2\psi = 0. \tag{6}$$

In other words, in the rest frame the solution ψ is simply the oscillator state of no quanta.

What we wish to demonstrate now is that *Dirac's new wave equation is simply an invariant way to assert that a (relativistic) oscillator structure is in its ground state of no quanta.*

In order to establish this interpretation let us consider the effect on Q of an arbitrary Lorentz transformation, Λ. We find:

$$\Lambda : Q_i \rightarrow Q_i' = U(\Lambda)Q_iU^{-1}(\Lambda), \tag{7}$$

so that:

$$a_i \rightarrow a_i(\Lambda) \quad \text{and} \quad \bar{a}_i \rightarrow \bar{a}_i(\Lambda).$$

This transformation preserves the boson operator commutation relations, but it *changes* the ground state ket$|0\rangle$. To see this consider the ket$|0\rangle$. This ket is defined by the two conditions: $\bar{a}_i|0\rangle = 0$. It is invariant for those Lorentz transformations which leave the time axis invariant; that is, the rotations leaving the unit 4-vector (0,0,0,1) invariant. We must include this information in the notation for the ket; accordingly we denote $|0\rangle$ by $|0;\mathcal{U}_0\rangle$, where $\mathcal{U}$ denotes a unit 4-vector (proper velocity) and $\mathcal{U}_0$ the vector (0,0,0,1). Under the Lorentz transformation Λ, this ket becomes

$$e^{i\Lambda\cdot s}|0;\mathcal{U}_0\rangle \equiv |0;\mathcal{U}_\Lambda\rangle,$$

where $\mathcal{U}_\Lambda$ is the unit vector $\Lambda\mathcal{U}_0$. Clearly one has

$$\bar{a}_i(\Lambda)|0;\mathcal{U}_\Lambda\rangle = 0.$$

To find the wave function $\langle \xi_1\xi_2|0;\mathcal{U}_\Lambda\rangle$ adapted to the internal space $\xi_1\xi_2$ (each over the real line) one solves the two differential equations:

$$\langle \xi_1\xi_2|\bar{a}_i(\Lambda)|0;\mathcal{U}_\Lambda\rangle = 0. \tag{8}$$

Let Λ denote a general boost, parametrized by $(\chi\theta\phi)$, where $\tanh\chi = v/c$ and $(\theta\phi)$ specify the direction of the three vector v. Solving eq. (8) leads to the desired wave function:

$$\begin{aligned}\langle \xi_1\xi_2|0;\mathcal{U}_\Lambda\rangle &= [\pi(\cosh\chi + \sinh\chi\sin\theta\sin\phi)]^{-\frac{1}{2}}\\ &\times \exp\{-2\cosh\chi + 2\sinh\chi\sin\theta\sin\phi)^{-1}\\ &\times [(\xi_1^2+\xi_2^2) + 2i\sinh\chi\cos\theta\xi_1\xi_2\\ &\quad + i\sinh\chi\sin\theta\cos\phi(\xi_2^2-\xi_1^2)]\}. \end{aligned}\tag{9}$$

Adjoining the momentum eigenfunction for $\vec{P}$ and identifying Λ with the unit vector $\vec{P}/m$ we obtain Dirac's wave function solving eq. (5):

$$\langle \xi_1\xi_2 0;p\rangle = \exp\left[\frac{i}{h}(p\cdot x)\right]\langle \xi_1\xi_2|0;p/m\rangle. \tag{10}$$

The content of this general solution is now clear: we have simply Lorentz-transformed the rest-frame harmonic oscillator solution

$$\langle \xi_1\xi_2|0;(0,0,0,m)\rangle = \pi^{-\frac{1}{2}}\exp(-imt/\hbar) \times \exp[-\tfrac{1}{2}(\xi_1^2+\xi_2^2)]. \tag{11}$$

It is quite easy now to write out the generalization to non-zero spin. In a general frame Λ one has for spin s:

$$|sm_s;\Lambda\rangle = [(s+m_s)!(s-m_s)!]^{-\frac{1}{2}}(a_1(\Lambda)^{s+m_s})(a_2(\Lambda)^{s-m_s})|0;\mathcal{U}_\Lambda\rangle. \tag{12}$$

Hence the corresponding general wave function is given by

$$\langle \xi_1 \xi_2 | sm_s ;p \rangle = \exp(\frac{i}{\hbar} p \cdot x) \langle \xi_1 \xi_2 | sm_s ;p/m \rangle . \tag{13}$$

This result clearly has the invariant $p \cdot p = m^2$; moreover the Pauli-Lubansky invariant corresponds to spin s, since the *spin stability group* is generated by J *using bosons adapted to the frame p/m.*

III. THE TRAJECTORY CONSTRAINT

The above construction has obtained, in a uniform presentation, the set of all Wigner-Poincaré (m,s) irreps realized on a harmonic oscillator basis $(\xi_1 \xi_2)$, but the mass parameter m is, so far, not related to the spin parameter s. In an earlier paper[9], we found a way to impose the trajectory constraint $m^2 = f(s) > 0$ by a method which we called "Galilean Sub-Dynamics".

The relevant ideas can be summarized this way. Hamiltonian dynamics may be put in three distinct relativistic forms[16]: the instant, point and front forms, *each determined by a particular Poincaré subgroup.* The maximal proper subgroup of the Poincaré group is, however, the eight parameter extended Galilean group (in 2 space-1 time coordinates) augmented by a dilation operator. Galilean sub-dynamics is the use of this subgroup in constructing Poincaré irreps. Quite remarkably, two *Galilean* mass points (in 2 + 1 space time) interacting harmonically lead *precisely* to the construction and results of section II, but augmented now by the trajectory function: m^2 = linear function of spin.

Subsequently we determined[17] the explicit Poincaré generators yielding identically these same results. Once one has obtained these generators, however, one can forget where they came from, since they stand on their own merits. These generators are:

$$P_i = i\partial/\partial X_i ,$$

$$P_0 = + [\underline{P}^2 + f(a_1\bar{a}_1 + a_2\bar{a}_2 + 1)]^{\frac{1}{2}}$$

$$M_{ij} = \varepsilon_{ijk}(\underset{\sim}{X}\times\underset{\sim}{P})_k + J_{ij},$$
$$M_{0i} = \tfrac{1}{2}(X_iP_0+P_0X_i) + tP_i/P_0 + [P_0+M]^{-1}(\varepsilon_{ijk}P_jJ_k), \tag{14}$$

where

$$J_{ij} = \varepsilon_{ijk}J_k \quad \text{and} \quad M = [f(a_1\bar{a}_1+a_2\bar{a}_2+1)]^{\frac{1}{2}}.$$

It is readily verified that these generators close on the commutation relations of the Poincaré group; the generators are Hermitian with respect to the inner product:

$$(\phi,\psi) = \int d^3x d\xi_1 d\xi_2 \phi^*(\underset{\sim}{X},t;\xi_1\xi_2)\psi(\underset{\sim}{X},t;\xi_1\xi_2). \tag{15}$$

The generators given by eq. (14) define completely the free motion of a relativistic composite object whose mass-spin quantum numbers lie on a discrete Regge band, $m^2 = f(s)$; *the Hamiltonian is given by* P_0.

Remarks. 1) The coordinates (X,t) in this realization are "quasi-Newtonian" coordinates, that is, the position X is a three vector, and the time t is a c-number. (There is a very large literature on such coordinates, with the work of Wigner and T.D. Newton[18], and Foldy-Wouthuysen[19] (spin ½) being the best known).

2) The fact that the realization uses quasi-Newtonian coordinates poses, as is well known, problems as to interactions; we may avoid this by giving the equivalent Minkowski form[17].

3) It will be observed that these generators are actually in Hamiltonian form, and the operator P_0 is indeed the Hamiltonian. The motion generated is, however, the free motion of the object and there is no essential dynamics involved. Thus the real content of these generators is kinematical in nature as emphasized in the title.

IV. THE CONCEPT OF A KINEMATICAL STABILITY GROUP[20]

Let us now reconsider the results of sections I and II with the aim of identifying the underlying general principles. In section II we saw that Dirac's construction, in effect, made assertions about a covariant oscillator system; in section III we saw that this relationship became even more literal with the covariant oscillator structure implying a mass-spin trajectory.

The key element in understanding these results lies in the way that the Lorentz-group action has been transferred to the oscillators themselves: $a_i \to a_i(\Lambda)$, and to the ground state ket: $|0;\mathcal{U}_\Lambda\rangle$. This structure allows one to realize for every fixed value of the unit 4-vector, p/m, the (spin) stability group of the Poincaré group. (Note that this unit four-vector is the same for all mass states.)

It is clear that we can easily generalize this statement *to include the deSitter group itself as a stability group.* (This follows since the group structure, and the state vectors, for arbitrary p/m are *structurally* the same as those for the rest frame.) Thus the set of (m,s) Poincaré irreps comprise two irreps (integer and half-integer) of the deSitter group[21].

In essence what the construction accomplishes is to 'covariantize' the familiar oscillator representation of the deSitter group. This result becomes even clearer if we consider Fig. 1. One sees that the set of (m,s) Poincaré irreps (with $m^2 = 2sm_0^2$ in the figure) form a set of hyperbolae. For any given value of the unit four-vector p/m, there are points of intersection with each mass hyperbola, thereby defining the *states* of the oscillator basis parametrically indexed by the unit four-vector. The action of the deSitter group is stable for this value of the unit four vector, and realizes the irrep by carrying this set of states into itself.

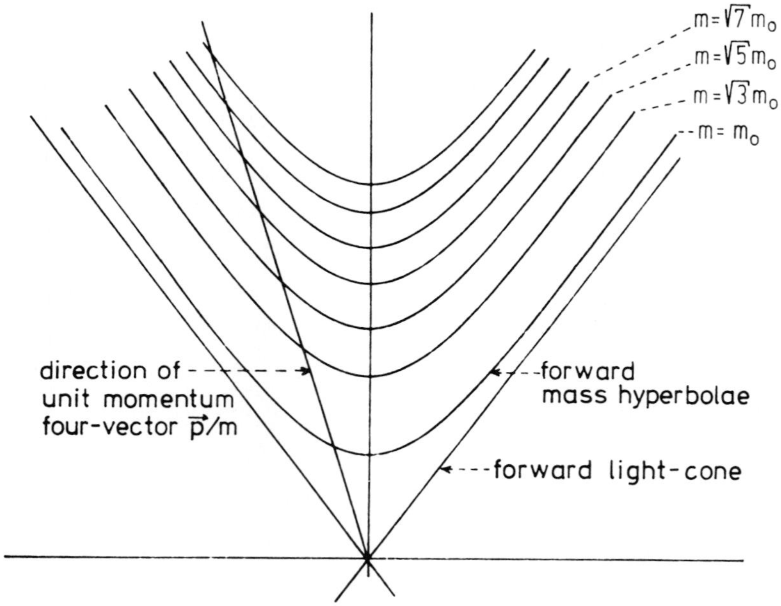

Fig. 1

The mass hyperbolae for the half-integer spin case with $m^2 = s \cdot (2m_o^2)$. For every momentum p the unit four-vector $\vec{p}/m$ intersects the set of mass hyperbolae each once and only once. The kinematical symmetry group leaves $\vec{p}/m$ invariant but transforms the set of state vectors -- of the different hyperbolae adapted to the direction $\vec{p}/m$ -- linearly among themselves.

From this picture one sees that the method is really quite general. For example, using this same basis, but augmenting the operator J by a quadrupole operator Q (see ref. 17) one can generate three irreps of the group $\overline{SL3R}$: the Regge bands (1/2,5/2,...); (0,2,4,...) and (1,3,5,...). (This group is interesting since it is the group of rotations and volume preserving deformations of three-space, and is consequently important as a symmetry group both in nuclear physics, and in the popular 'bag-models' of hadrons.)

One may generalize further by adjoining pairs of oscillators. For two pairs the group is that of the old strong coupling model; for three pairs relativistic SU(6)[22]; for unlimitedly many pairs, the 'rubber-sphere' dual resonance model[10].

V. CONCLUDING REMARKS

The essence of our construction is the picture presented in Fig. 1; the basic idea is a remarkably simple one when viewed in this way. This simplicity is to be contrasted with the complexity of the solution to the generators of section III, where one dealt with unlimitedly many Poincaré irreps of the trajectory function $M^2 = f(s)$.

We do not have time to present the Minkowski space generators, which are equivalent to the quasi-Newtonian generators of section III. (The relation between these two sets might be called a generalized "Foldy-Wouthuysen" transformation; in particular, the spin is no longer separately constant.)

One particular application is worth mentioning. If one chooses the trajectory function to be linear in the spin, the invariant P^2 becomes a quadratic form in eight internal and external variables, with one dimensional constant (the slope of the Regge band) setting the relative scales. If one seeks to factorize this form, *without* increasing the space-time structure beyond adjoining negative energies, one is led uniquely to the Cayley numbers. This procedure offers the hope of finding intrinsic superselection spaces[23] carried by these composite objects.

FOOTNOTES AND REFERENCES

1. This statement is not unanimously accepted, however, as indicated by the recent preprint of G.B. CVIJANOVICH and J.P. VIGIER, Uppsala College, East Orange, N.J.
2. E. ABERS, I. GRODSKY and R. NORTON, Phys. Rev. 159, 1222 (1967); G. FELDMAN and P.T. MATTHEWS, Phys. Rev. 154, 1241 (1967). Further references may be found in the conference report by Y. NAMBU, Proceedings of the 1967 International Conference on Particles and Fields, edited by Hagen, Guralnik and Mathur (Interscience, N.Y., 1967), and in the lectures by BÖHM (ref. (13) below).
3. The extent to which the over-reaction has occurred may be judged by reading the remarks of R.H. DALITZ, Proceedings of the Thirteenth International Conference on High Energy Physics (University of California Press, Berkeley, 1967) p. 215; and of S. WEINBERG (Phys. Rev. 139, B597, 1965).
4. L. O'RAIFEARTAIGH, Phys. Rev. Lett. 14, 575 (1965); R. JOST, Helv. Phys. Acta. 39, 369 (1966); I. SEGAL, J. Functional Analysis, 1, 1 (1967).
5. H. BACRY, Contribution to the Third International Colloquium on Group Theoretical Methods in Physics, Marseilles, 17-21 June 1974.
6. A qualitatively different way to avoid the no-go theorems is the use of supersymmetry (cf. ref. (7) below), but this suffers, probably fatally, from other difficulties.
7. J. WESS, "Supersymmetry", Fifteenth Schladming Winter School, 1976 (to be published).
8. P.A.M. DIRAC, Proc. Roy. Soc., London A322, 435 (1971); A328, 1 (1972); cf. also the spin-$\frac{1}{2}$ equation found by L.P. STAUNTON, Phys. Rev. D10, 1760 (1974).
9. L.C. BIEDENHARN, M.Y. HAN, and H. VAN DAM, Phys. Rev. D8, 11735 (1973).
10. L.C. BIEDENHARN and H. VAN DAM, Phys. Rev. D9, 471 (1974).
11. H. BACRY and J. NUYTS, Phys. Rev. 157, 1471 (1967).
12. D. FINKELSTEIN, Phys. Rev. 100, 924 (1955).
13. A. BÖHM, Lectures in Theoretical Physics (Gordon and Breach, N.Y., 1968) Vol. 10B, p. 483 ff.

15. The work of BACRY and CHANG, (ref. (5) above), also introduces a Hermitian four vector operator in their general construction. Space-like solutions are eliminated by explicitly restricting the Hilbert space (their $\mathcal{H}_u$).

16. P.A.M. DIRAC, Rev. Mod. Phys. 21, 392 (1949).

17. H. VAN DAM and L.C. BIEDENHARN, to appear in Phys. Rev. (July 1976).

18. E.P. WIGNER and T.D. Newton, Rev. Mod. Phys. 21, 400 (1949).

19. L.L. FOLDY and S.A. WOUTHUYSEN, Phys. Rev. 78, 29 (1950).

20. H. VAN DAM and L.C. BIEDENHARN, submitted to Physics Letters.

21. In terms of invariants of the deSitter group these are the irreps with the Casimir invariant = -3/8 and the second invariant zero.

22. L.P. STAUNTON and H. VAN DAM, Lett. N. Cimento 7, 371 (1973).

23. L.P. HORWITZ and L.C. BIEDENHARN, Helv. Phys. Acta 38, 385 (1965).

CAUSALITY AND SYMMETRY IN COSMOLOGY AND THE CONFORMAL GROUP

I.E. Segal

INTRODUCTION

I shall lead up to a new theoretical postulate in fundamental physics, which I have called the chronometric principle, because it deals primarily with the nature of time (or of its dual, or conjugate, the energy).

CONFORMAL GEOMETRY

Technically, the developments treated here are closely related to the conformal group and conformal space, or covering groups and covering spaces of them. Since these topics have been discussed in a course in the Summer School preceding this colloquium, I trust I may be fairly brief about them. From a fundamental physical standpoint, it is not the conformality which matters; in itself, conformality is a rather academic, quite mathematical, notion. But it is equivalent to the physically important notion of causality. Thus, the group of all local causality-preserving transformations in the vicinity of a point of Minkowski space is, as a local Lie group, identical with the conformal group. Essentially the same statement can be made globally on Minkowski space as follows: The set of all vector fields on Minkowski space which generate smooth local causality-preserving transformations is identical with the set of all conformal vector fields.

"Most" of but not all of these vector fields generate global one-parameter groups of conformal transformations. More precisely, the 11-dimensional subalgebra of infinitesimal Lorentz transformations, augmented by the infinitesimal scale transformation, of the 15-dimensional Lie algebra of all infinitesimal conformal transformations, consists of such vector fields. The classic Alexandrov-Ovchinnikova-Zeeman Theorem asserts conversely that every one-to-one causality preserving transformation of a Minkowski space of more than 2 space-time dimensions, is the product of a Lorentz transformation with a scale transformation. Complementary to this 11-dimensional subspace is a 4-dimensional subspace, consisting of the transforms of infinitesimal Minkowski space-time translations, by conformal inversion. These so-called "special" conformal transformations generate local one-parameter groups which are not global. That is, they develop singularities, which are similar to that of conformal inversion itself:

$$x_j \to x_j(x_0^2-x_1^2-x_2^2-x_3^2)^{-1} \qquad (j=0,1,2,3).$$

It is interesting and useful that the conformal group as a whole acts properly and globally on conformal space $\bar{M}$; or that suitable coverings of it act on coverings of conformal space. Conformal space may be defined as the set of all projective null spheres in Minkowski space, a projective sphere being one of the form

$$a(x_0^2-x_1^2-x_2^2-x_3^2) + 2(b_0x_0-b_1x_1-b_2x_2-b_3x_3) + c = 0,$$

where not all of the a, b_j (j=0,...,3), and c are zero, and a null sphere being one such that

$$b_0^2 - b_1^2 - b_2^2 - b_3^2 = ac.$$

Each point of Minkowski space M maps into the projective null sphere centered at the point, of zero radius in an evidently canonical way, and one obtains thereby an invariant imbedding of M into $\bar{M}$ (i.e. invariant with respect to the Lie algebra of all conformal vector fields.

There is an alternative way to describe the action of the conformal group on conformal space which is analytically simple although physically unfamiliar. The group SU(n,n) acts on the group U(n) in a way made familiar by C.L. Siegel: if $U \in U(n)$ and $T = \begin{pmatrix} A & B \\ C & D \end{pmatrix} \in SU(n,n)$, where A, B, C, and D are n×n matrices, then

$$T : U \to (AU+B)(CU+D)^{-1}.$$

In this action the center of SU(n,n), consisting of the Z_4 subgroup generated by $\begin{pmatrix} i & 0 \\ 0 & i \end{pmatrix}$, acts trivially on U(n); the quotient $SU(n,n)/Z_4$ acts "effectively" on U(n); and when n=2, one obtains a transformation group action precisely equivalent to that of the action of the conformal group $G_{\bar{M}}$ on conformal space $\bar{M}$. That is to say, the transformation group pair $(G_{\bar{M}},\bar{M})$ is abstractly the same as the transformation group pair $(SU(2,2)/Z_4,U(2))$. The equivalence can be implemented by the Cayley transform from the 2×2 hermitian matrices, to U(2), with the identification of the point (x_0,x_1,x_2,x_3) of M with the matrix

$$\begin{pmatrix} x_0-x_1 & x_2+ix_3 \\ x_2-ix_3 & x_0+x_1 \end{pmatrix},$$

and the extension of the Cayley transform to all of $\bar{M}$ by continuity.

There are two particular coverings of $\bar{M}$, or equivalently of U(2), which play a fundamental role. First there is the universal covering $\tilde{M}$, which for U(2) is the group $R^1 \times SU(2)$; this is globally causal, unlike the finite coverings, in the sense of having no closed time-like loops, or more cogently, being "globally hyperbolic" in the sense of Leray. It is this space $\tilde{M}$ which forms a conceivable alternative to M as a model for the cosmos, if one wants to retain global causality along with a local Minkowski structure. But photons and some other particles seem to "live" on the two-fold covering $T^1 \times SU(2)$ of U(2), in the sense that their wave functions on $\tilde{M}$ are obtained by lifting up associated wave functions on $T^1 \times SU(2)$. The latter space has the advantage of being compact, but more important, it is more easily parametrized than U(2) on the one hand, being a direct product of factors similar to "time" and "space"; and on the other hand, its conformal group is *linear*, unlike that of $\tilde{M}$. Any class of locally isomorphic Lie groups has a maximal element which is *linear*, in the sense of being globally isomorphic to a group of (finite-dimensional) matrices. In the case of the conformal groups of the coverings of $\bar{M}$, this so-called "linearizer" is SU(2,2). This is a two-fold covering of the conformal group of $T^1 \times SU(2)$, which is isomorphic to the connected component $SO_0(2,4)$ of SO(2,4), the group leaving invariant the quadratic form $F = b_0^2 - b_1^2 - b_2^2 - b_3^2 - ac$, which figured earlier. The element -I of $SO_0(2,4)$ acts trivially on the "projective quadric" obtained by setting F=0, in the 5-dimensional projective space of all $(a, b_0, b_1, b_2, b_3, c)$ not identically zero. For more details regarding conformal geometry, see Segal (1976) and the references given there; more recent developments particularly relevant to group-representation-theoretic aspects are given by Jakobsen (1976), Jakobsen and Vergne (1976) and Ørsted (1976).

REDSHIFT THEORY

The main physical validation for the chronometric principle, and the only one that can presently be correlated quantitatively with experiment is in the field of cosmology, i.e. ultramacroscopic physics. It therefore seems appropriate to begin with a derivation along elementary lines of the redshift, which serves to illustrate the principle in its simplest form.

Before commencing this derivation, however, let me suggest that you watch for the following important qualitative features, which are lacking in the conventional Riemannian geometric redshift theory:

a) the redness of the shift is automatic; it does not need to be supplied "by hand";

b) full Lorentz covariance is retained; this is no preferred time axis;

c) the main implications for the directly observable quantities (modulo small corrections which have no significant impact on the confrontation with theory, for astronomical samples which are reasonably large and systematic), such as apparent luminosity, apparent diameter, and the redshift itself, are independent of fundamental, unknown parameters, such as the "deceleration parameter q_0", or the "cosmic constant Λ", on which the predictions of the expanding-universe model depend.

I should also justify in physical terms my claim on your attention by reporting briefly on the cited confrontation. Suffice it here to suggest a quick glance at the figures given by Segal (1976a) which are striking in the excellent (for astronomical data, at least) fit of the chronometric predictions, and noteworthy also for the inadequacy of the expanding-universe predictions to provide a meaningful fit for many important observed relations. The only large sample which has become available since the cited work is the Uppsala General Catalogue, whose raw redshift-magnitude relation is shown in Figure 1, while Figure 2 shows the chronometric and expansion predictions, on the assumption that the sample is fair*. The variance of the deriva-

* I thank J.F. Nicoll for permission to reproduct these Figures from forthcoming joint work.

tions from the expansion prediction is more than twice that from the chronometric prediction, and of the same order of magnitude as the variance in the raw apparent magnitudes.

The case of a two-dimensional space-time is quite illustrative of the basic physics, and will be considered here for simplicity of the mathematics involved. The problem is to treat the propagation of photons, and especially of their energy, from the chronometric standpoint. With Minkowski coordinates x_0, x_1, the metric being $dx_0^2 - dx_1^2$, a scalar photon wave function φ satisfies the equation

$$\Box \varphi = 0 \qquad (\Box = \frac{\partial^2}{\partial x_0^2} - \frac{\partial^2}{\partial x_1^2}).$$

But there is nothing sacred about the coordinates x_0, x_1. The conformal compactification $\bar{M}$ is more easily described in terms of different coordinates τ and ρ. More specifically, $\bar{M}$ is obtained from the torus $T^1 \times T^1$ by identifying antipodal points. Letting τ and ρ denote the angles on the respective circles, then the metric $d\tau^2 - d\rho^2$ is conformally equivalent to the matric $dx_0^2 - dx_1^2$ (on the open dense subset M of $\bar{M}$ on which the latter is defined). That is,

$$d\tau^2 - d\rho^2 = k(dx_0^2 - dx_1^2),$$

for a certain non-vanishing function k on M. The parameters τ and ρ are respectively time and space-like, in a variety of ways, - the signature of the fundamental quadratic form; the facts that all points of the form (τ, ρ_0), ρ_0 fixed are mutually time-like while those of the form (τ_0, ρ), τ_0 fixed are mutually space-like, in the usual senses in Minkowski space; the displacement $\tau \to \tau + \tau_1$ is causality-preserving and into the future, if $\tau_1 > 0$; the displacement $\rho \to \rho + \rho_1$ is causality-preserving; etc. There is some apparent acausality in the circularity of τ, but this is easily removed by replacing $\bar{M}$ by its covering space $R^1 \times T^1$, in which τ

ranges over the infinite interval $-\infty < \tau < \infty$. Every solution of the wave equation on $R^1 \times T^1$ is automatically periodic in τ, of period 2π, τ and ρ being measured in radians. This means it is obtained in an obvious way from a solution of the wave equation on $T^1 \times T^1$, by "lifting" it up to $R^1 \times T^1$; as a consequence, there is no essential loss for present purposes in working with the compact space-time $T^1 \times T^1$ in place of the physical but non-compact space-time $R^1 \times T^1$.

If we take Minkowski space-time as given locally in terms of its causality features, there is no solid theoretical reason to prefer the "flat" (say) coordinates x_0, x_1 to the "curved" (say) coordinates τ, ρ. In terms of the latter coordinates, the wave equation takes the form

$$\Box' \varphi = 0, \qquad \Box' = \frac{\partial^2}{\partial \tau^2} - \frac{\partial^2}{\partial \rho^2},$$

even though $\Box$ and $\Box'$ are different.

For the flat equation $\Box \varphi = 0$, the energy E takes the form

$$E = \int [(\text{grad } \varphi)^2 + (\frac{\partial \varphi}{\partial x_0})^2] dx_1 .$$

This is invariant under (i.e. conserved by) the advance of time: $x_0 \to x_0 + x$, $x_1 \to x_1$. From a classical relativistic standpoint, E is the total energy observed when the photon interacts with matter. On the other hand, for the curved equation the energy is

$$E' = \int [(\frac{\partial \varphi}{\partial \tau})^2 + (\frac{\partial \varphi}{\partial \rho})^2] d\rho .$$

This is conserved by the advance of the curved time τ; if (τ, ρ) were the coordinates observed, E' would appear to be the observed energy. Now sufficiently locally, there is no *observable* difference between the flat coordinates (x_0, x_1) and the curved (τ, ρ) if the relation between them is normalized by requiring the maximal contact. Choosing the distance scale so that the radius of the circles T^1 is unity, this essentially unique relation is:

$$\tau = \tan^{-1} \frac{x_0}{1-\frac{1}{4}(x_0^2-x_1^2)}, \qquad \rho = \tan^{-1} \frac{x_1}{1+\frac{1}{4}(x_0^2-x_1^2)}.$$

Note that near the point of observation $x_0 = x_1 = \tau = \rho = 0$, where the relation between the flat and curved coordinates has been normalized,

$$\tau = x_0 + O(d^2), \qquad \rho = x_1 + O(d^2),$$

where d is the distance from the origin (measured by either metric). It follows similarly that for a photon of spatial extent d, - in the sense e.g. that at the time of observation, the corresponding electromagnetic field vanishes at distances greater than d from the point of observation, - the flat energy E differs from the curved energy E' by $O(d^2)$. This is unobservably small, if the photon wave function vanishes outside the solar system, and if, as is widely believed, the spatial universe extends to distances > 10^9 light years, for d is then of order 10^{-9}, leading to a difference in energy of less than 1 part in 10^{18}, which is far beyond observable limits even under the best optical conditions. But for a photon of very large spatial extent, such as one propagated from a cosmological distance, the two energies may well differ substantially, and the difference may grow rapidly with the distance (as its square, perhaps).

The question now arises of which of these energies is the "right" one, i.e. the one which is truly conserved, if such exists. Mathematically, one of the two energies must be right, if one believes in conservation of energy, Lorentz covariance, and similar quite rudimentary mathematical-physical principles. They cannot both be right, since they do not commute, and if conservation of energy is valid for one of them, it will be violated for the other.

In seeking to answer this question, it is helpful to make the following mathematical comparison between the "chronometric" (or "curved") energy, say H, and the "relativistic" (or "flat") energy, say H_0:

1) $H \geq H_0$; i.e. in every positive-energy representation of the conformal group (or of a covering group thereof), such as that defined by the Maxwell equations, the chronometric exceeds the relativistic energy.

2) H_0 is scale covariant, while H is not; i.e. if X denotes the infinitesimal generator of the scale transformations, $x_j \to e^{\lambda} x_j$ (j=0,1,2,3), then $[H_0,X] = H_0$, but $[H,X] \neq H$.

The non-scale-covariance of H can be understood physically as the result of a built-in fundamental length, the "radius of the spatial universe". This fixes a natural chronometric distance scale. In relativistic theory, such a fundamental length is entirely absent; the distance and time scales are entirely conventional, the present use of certain frequency standards being evidently a matter of convenience.

[It is interesting to note also that the excess $H_1 = H - H_0$ of the chronometric over the relativistic energy is "anti-scale covariant", i.e. $[H_1,X] = -H_1$. The decomposition $H = H_0 + H_1$ is the unique Lorentz-covariant decomposition of H into scale-covariant and anti-scale-covariant components.]

3) The expected relativistic energy $\langle H_0 \rangle$ of a photon is substantially unchanged by spatial cut-offs; i.e., it depends, within negligible (physically unobservably small) deviations only on the photon wave function in the immediate vicinity of the point of observation.

On the other hand, $\langle H \rangle$ depends quite materially on the spatial cut-off. For a spatially cut-off plane wave, $\langle H_1 \rangle$ varies roughly as the square of the number of oscillations, independently of the frequency. As a consequence, an extreme spatial cut-off yields only the $\langle H_0 \rangle$ constituent of $\langle H \rangle$. Detailed computations and rigorous bounds are given by Segal (1976b).

Now let us consider the question from a physical stand-point, while bearing in mind the foregoing mathematical results. First, measurement procedures for photon wave lengths which are based on conventional standards of length, such as those in actual use, would appear to be inherently scale covariant. For simply as a matter of definition, if the standard length is diminished by a factor k, the wave length is correspondingly increased, by the factor k^{-1}. On the other hand, there is no apparent physical reason for the total energy of a highly delocalized photon in a curved universe, say for example a spherical one, to be similarly scale covariant; indeed, scale transformations would not be applicable except to local coordinates.

Second, the energy of a photon state as measured in a laboratory would appear to be a highly localized quantity, in the sense that it seems inconceivable that the form of the wave function at large distances could significantly affect the laboratory measurement. Even its form at such cosmologically insignificant distances as those of the borders of the solar system seem irrelevant for the laboratory measurement of its energy.

These considerations lead rather definitely to the conclusion that actual measurements of photon energy, by homo sapiens, yield the relativistic energy $\langle H_0 \rangle$, and not the chronometric energy $\langle H \rangle$. On the other hand, the physical driving energy, on the assumptions of its Lorentz covariance and agreement with $\langle H_0 \rangle$ for localized photon states, etc. may be either H_0 or H; there is no apparent fundamental physical desideratum which excludes the latter possibility. Moreover, *if* the latter is indeed the case, then there follows immediately a prediction of a redshift for photons which progress from a localized to a delocalized state. For H is then conserved, while at all times τ,

$$H = H_0(\tau) + H_1(\tau);$$

since $\langle H \rangle \sim \langle H_0(0) \rangle$, and $H_0(\tau)$ and $H_1(\tau)$ are both positive, it follows that $\langle H_0(\tau) \rangle$ must decrease as τ increases from 0, at least initially. If however H_0 is both the observed and the driving energy, one must apparently look for an explanation other than non-conservation of observed energy to account for the redshift seen in the light from seemingly distant (as indicated e.g. by their angular diameters) galaxies.

On the tentative assumption that the driving energy is H, it remains to be seen whether there is quantitative agreement between the corresponding theoretical predictions and systematic observations of the light from distant objects. The more-or-less directly observable quantities associated with luminous objects in the sky are notably:

(1) The redshift (or blue shift) itself, normally from measurements of spectral lines. These lines are usually of the order of a few angstroms in width, reflecting apparent internal motions in the emitting object.

(2) The apparent luminosity (or magnitude), either in a visual band (usually but not always employed for optical frequencies) or monochromatically (employed for radio frequencies). The optical luminosities are somewhat sensitive to the apertures employed, and these are not always clearly defined, but for most galaxies as well as all quasars the aperture-dependence is not sufficient to strongly affect cosmological testing.

(3) The apparent angular diameter, which is again subject to some difficulties of precise definition, but with resulting ambiguities generally an order of magnitude less than that between large-sample predictions for qualitatively distinct cosmo-

logical theories.

Observations of the cited quantities have been published for a variety of large and/or systematic samples of galaxies or quasars (not to mention number counts for radio sources, which provide an additional cosmological test for theories postulating spatial homogeneity). There are a total of roughly 1000 such galaxy and 200 quasar observations.

Theoretical analysis shows that for the relativistic energy of a sharp-frequency photon, after elapsed time τ,

$$\langle H_0(\tau) \rangle = (1+z)^{-1} \langle H_0(0) \rangle ,$$

where $z \cong \tan^2 \frac{\tau}{2}$. Here $H_0(\tau) = e^{-i\tau H} H_0 e^{i\tau H}$, in accordance with quantum mechanics in the Heisenberg picture. The same result could equally be expressed in the Schrödinger picture, which is readily adaptable to a purely classical derivation, employing the classical energies E and E' earlier given. The quantity z is known as the redshift, and before the advent of quasars was largely limited to values < 0.1; but there are quasars now known with $z \sim 3$. The original galaxies treated by Hubble had redshifts < 0.01.

The distance $c\tau$ to a luminous object in the sky is not itself a directly observable quantity, but one can eliminate the distance from the various relations between redshift, apparent luminosity, and apparent diameter and distance, obtaining relations involving only such observable quantities. These predicted relations fit all large or statistically documented samples of galaxies and quasars quite satisfactorily (indeed, very well, in comparison with the expanding-universe model). There is however one relatively small sample, well known from its appearance in many textbooks, which fits the chronometric prediction less well than the expansion prediction. This is the sample developed over four decades by the successors to Hubble, consisting however of only 41 galaxies, and quite without published objective selection criteria.

THE CHRONOMETRIC PRINCIPLE

A variety of other observational indications could be cited for the "chronometric" hypothesis that the driving and observed hamiltonians are H and H_0. These indications relate notably to such matters as otherwise anomalous quasar energy outputs; apparent superluminal velocities; the so-called "Rubin-Ford" anomaly; etc. The observed apparent Planck law for the cosmic background radiation could also be related in a simple way to the Planck law predicted by conservation of energy and maximality of the entropy on the basis of the chronometric hypothesis. Segal (1976a) gives details concerning these matters.

It is natural at this point to raise the question of whether the chronometric hypothesis may not have non-trivial implications for physics at the other distance extreme, - the ultramicroscopic. At first glance the virtual identity of H and H_0 in relation to localized states would appear to foreclose this possibility. However the cosmological lifetimes of the stable particles, and the very large masses of the massive ones ($\sim 10^{40}$ for the proton, in the natural chronometric units in which $\hbar = c = R = 1$, where R is the "radius of the universe") make possible in principle the existence of non-trivial selection rules based on the chronometric hypothesis.

In particular it seems quite plausible that the observed fundamental particles are more accurately represented by vector bundles (scalar, spinor, vector, etc.) over the chronometric cosmos $\widetilde{M}$, than in the conventional fashion by trivial bundles on M itself, with corresponding transformation properties under the conformal group, and not merely the Poincaré subgroup.

The treatment of the corresponding representations of the

conformal group is of fundamental importance in this connection, as well as of much mathematical novelty and interest. Various aspects have been explored by E.G. Lee (1975; the scalar photon case); H.P. Jakobsen and M. Vergne (1976; the unitarity question for the fundamental positive-energy representations, etc.), H.P. Jakobsen (1976a), and in forthcoming work (1976b; further developments related to the cited paper); B. Speh (1976) (in a forthcoming paper, treating questions of non-trivial quasi-unitary composition series); B. Ørsted (1976) (in which, among other matters, group-representation-theoretic considerations directly related to the chronometric principle itself, - which we now proceed to define, - are initiated): M. Vergne (1976) (general methods for construction of massless particle representations and analogues).

Brevity permits only an indication of the chronometric principle. This is obtained from the chronometric hypothesis through its combination with Lorentz invariance. Lorentz invariance is the expression of local spatial and temporal isotropy; if we believe also in temporal invariance, - as we must for the pursuit of quantitative physics, - we must adjoin to the Lorentz group the group of displacements in time, to obtain the full applicable symmetry group. According as we use the chronometric energy H or the relativistic energy H_0, symmetry groups G_1 and G_0 locally isomorphic to SO(2,3) and the Poincaré group, are obtained. Together they generate the full conformal group; separately, they are each capable of describing particles of fixed non-vanishing mass - as the full conformal group is not.

The chronometric principle asserts that the group of symmetries of the objective non-anthropocentric physical world is G_1, while the group of symmetries of observation is G_0. These groups are simultaneously imbedded in the conformal group G (more exactly, respective covering groups are involved) in such a way that they osculate each other near the identity); the joint imbedding is then unique, within conjugacy, and may be specified by a point of observation together with a Lorentz frame and distance scale at the point.

The combination of the chronometric principle with general precepts of quantum field theory leads to a covariant and causal description of particle production in which non-linearities are supplanted by more sophisticated and comprehensive actions for the fundamental symmetry groups. This essentially convergent theory of interaction proposed by Segal (1976c) appears to provide an interesting point of departure for a possible more exact and scientifically economical description of fundamental particles.

REFERENCES

1. H.P. JAKOBSEN, Conformal harmonic analysis and intertwining operators, Ph.D. Thesis, M.I.T., Dept. of Mathematics (1976a).

2. H.P. JAKOBSEN, Intertwining differential operators for Mp(n,R) and SU(n,n), Preprint (1976b).

3. H.P. JAKOBSEN and M. VERGNE, Wave and Dirac operators and representations of the conformal group, to appear in Jour. Funct. Anal. (1976).

4. E.G. LEE, Conformal geometry and invariant wave equations, Ph.D. Thesis, M.I.T., Dept. of Mathematics (1975).

5. B. ØRSTED, Note on the conformal quasi-invariance of the Laplacian on a pseudo-Riemannian manifold, to appear in Letters in Math. Phys. (1976a).

6. B. ØRSTED, Wave equations, particles, and chronometric geometry, Ph.D. Thesis, M.I.T., Dept. of Mathematics (1976b).

7. I.E. SEGAL, Mathematical cosmology and extra-galactic astronomy, Academic Press (1976a).

8. I.E. SEGAL, Theoretical foundations of the chronometric cosmology, Proc. Nat. Acad. Sci. USA 73, 669-673 (1976b).

9. I.E. SEGAL, Interacting quantum fields and the chronometric principle, to appear in Proc. Nat. Acad. Sci. USA (1976c).

10. B. SPEH, Unitarity and composition series for certain incuded representations of the conformal group, Preprint (1976).

11. M. VERGNE, On the Weil representation of U(n,n), Preprint (1976).

THOMAS-BARGMANN-MICHEL-TELEGDI EQUATION FOR WIGNER PARTICLES

H. Bacry*

INTRODUCTION

The present contribution is an attempt towards a covariant Hamiltonian description for a quantum mechanical (as well as classical) system of elementary particles in interaction.

(i) It is an *attempt* because the only problem which is solved is the elementary particle in interaction with a homogeneous electromagnetic field through an electric charge and a magnetic moment.

(ii) It is *Hamiltonian* because the theory is given in the Heisenberg picture where observables are varying in time.

(iii) This Hamiltonian is *invariant* because the Hamiltonian is scalar with respect to Lorentz transformations (performed on the system combining both the particle and the external field).

(iv) The theory is at the same time *classical*, due to the one-to-one correspondence between the concepts of classical and quantum mechanical particle [1] [2].

(v) The present theory [3] is a generalization of Wigner theory [4] of elementary particle which is based on the Poincaré

* Université d'Aix-Marseille II and Centre de Physique Théorique, CNRS, Marseille.

group representation.

1. HEISENBERG AND SCHRODINGER PICTURES FOR THE WIGNER FREE PARTICLE

Let us denote by H, $\vec{P}$, $\vec{J}$, $\vec{K}$ the well known generators of the Poincaré group $\mathcal{P}$. We know that only H, $\vec{P}$, $\vec{J}$ commute with the Hamiltonian H and

$$[\vec{K},H] = i\vec{P}.$$

This has, as a consequence, that the energy-momentum $(H,\vec{P})$ of a *free* particle are constants of the motion. Instead, $\vec{K}$ (which is related with the position of the particle) does not. However, it is possible to get a whole set of generators of $\mathcal{P}$ as constants of the motion by replacing $\vec{K}$ by $\vec{K}'$

$$\vec{K}' = \vec{K} - \vec{P}t.$$

Such a substitution (an inner automorphism of $\mathcal{P}$) does not change the commutators but $\vec{K}'$ is a time-dependent operator and is a constant of the motion since

$$\frac{d\vec{K}'}{dt} = \frac{\partial\vec{K}'}{\partial t} + \frac{1}{i}\,[\vec{K}',H] = 0$$

Therefore, the set $\{H,\vec{P},\vec{J},\vec{K}\}$ (resp. $\{H,P,J,K'\}$) is the set of generators of $\mathcal{P}$ in the Schrödinger (resp. Heisenberg) picture.

2. COVARIANT HAMILTONIAN FOR THE FREE PARTICLE

The Hamiltonian H is the energy operator. In the rest frame, it coincides with the mass operator $\mathcal{H}_0 = m = \sqrt{P^\mu P_\mu}$. If we denote by τ the proper time parameter and by $M_{\mu\nu}$ and P_μ the generators of $\mathcal{P}$, we have, trivially:

$$\frac{dM_{\mu\nu}}{d\tau} = \frac{1}{i}\,[M_{\mu\nu},\mathcal{H}_0] = 0$$

$$\frac{dP_\mu}{d\tau} = \frac{1}{i}[P_\mu,\mathcal{H}_0] = 0.$$

$\mathcal{H}_0$ will be called the covariant Hamiltonian for the free elementary particle.

3. PARTICLE IN A HOMOGENEOUS EXTERNAL FIELD

We intend to find a covariant Hamiltonian $\mathcal{H}$ which generalizes the operator $\mathcal{H}_0$ of the preceding section and provides the right equations of motion for $M_{\mu\nu}$ and P_μ, when the particle is in a homogeneous electromagnetic field. Such a Hamiltonian is

$$\boxed{\mathcal{H} = \mathcal{H}_0 - \frac{e}{m} M_{\mu\nu}F^{\mu\nu} + \frac{\lambda}{m^2} W_\mu F^{*\mu\nu}P_\nu} \qquad (1)$$

where

$$F^{*\mu\nu} = \tfrac{1}{2}\varepsilon^{\mu\nu\rho\lambda}F_{\rho\lambda}$$

$$W^\mu = \tfrac{1}{2}\varepsilon^{\mu\nu\rho\lambda}M_{\rho\lambda}P_\nu \qquad \text{(Pauli-Lubanski vector).}$$

It is clear that $\mathcal{H}$ is invariant in the sense given in the introduction. The Heisenberg equations give:

$$\boxed{\frac{dP_\mu}{d\tau} = \frac{e}{m} P^\nu F_{\nu\mu}} \qquad (2)$$

which is the expected equation provided P_μ is interpreted as the energy-momentum at time τ.

The equation corresponding to $M_{\mu\nu}$ is somewhat complicated but the one associated with W_μ is

$$\frac{dW_\mu}{d\tau} = (\frac{e}{m} + \lambda)W^\nu F_{\nu\mu} - \frac{\lambda}{m^2} (W_\alpha F^{\alpha\beta} P_\beta) P_\mu \tag{3}$$

which is nothing else than the Thomas-Bargmann-Michel-Teledgi equation [5-6] now "proved" for all spinning particles with mass, both in quantum and classical mechanics. Obviously, we have taken $c = 1$ and $\lambda = (g-2) \frac{e}{2m}$.

The essential advantage of the Hamiltonian $\mathcal{H}$ is not only its existence (many physicists have failed in looking for a Hamiltonian formalism for the T-B-M-T equation and this one also gives the translational motion) but also the fact that the particle stays in a given representation of $\mathcal{P}$. In fact Eqs. (2) and (3) show that $P^\mu P_\mu$ and $W^\mu W_\mu$ are *both constants* of motion. In other words, the particle is always "tangent" to a fixed representation (a fact which is difficult to see in the Maxwell-Dirac field formalism, for instance).

4. CONCLUSION

We conclude with some remarks

a) Going to the rest frame gives for $\mathcal{H}$

$$\mathcal{H} \to m - e\vec{E}\cdot\vec{x} - \frac{ge}{2m} \vec{s}\cdot\vec{B}$$

which is the sum of m and the interaction energy.

b) The equivalent theory in Galilean case [3] is provided by the Hamiltonian

$$\mathcal{H}_{Gal.} = m + \frac{e}{m} (\vec{K}\cdot\vec{E}-\vec{J}\cdot\vec{B}) - \lambda\vec{S}\cdot\vec{B}$$

where

$$\vec{S} = \vec{J} + \frac{\vec{K}\times\vec{P}}{m}$$

one gets the expected equations

$$\frac{d\vec{P}}{d\tau} = e\left(\vec{E} + \frac{\vec{P}\times\vec{B}}{m}\right)$$

$$\frac{dS}{d\tau} = \left(\frac{e}{m} + \lambda\right)\vec{S} \times \vec{B}.$$

c) $\mathcal{H}$ is the sum of three invariant quantities since m, $M_{\mu\nu}F^{\mu\nu}$ and $W_{\mu}F^{*\mu\nu}P_{\nu}$ are commuting quantities. The last one has been found by L. Michel [7]; it is the magnetic energy in the rest frame.

d) The invariant $M_{\mu\nu}F^{\mu\nu}$ is an element of the Lie algebra of the Lorentz group. It is intimately related with the fact that the classical worldline of a charged particle is an orbit (in the group theoretical sense) of a subgroup of the Lorentz group [8-9].

REFERENCES

1. H. BACRY, Classical Hamiltonian for Spinning Particles, Princeton preprint, unpublished, 1966. H. BACRY, Commun. math. Phys. 5, 97 (1967). R. ARENS, Commun. math. Phys. 21, 139 (1971). R. ARENS, J. Math. Phys. 12, 2415 (1971).
2. J.M. SOURIAU, C.R. Acad. Sc. 263, B 1191 (1966). J.M. SOURIAU, Structure des systèmes dynamiques (Dunod, 1970).
3. H. BACRY, Wigner Elementary Particle in an External Homogeneous Field (to appear in Letters in Math. Phys.).
4. E.P. WIGNER, Ann. of Math. 40, 149 (1939).
5. L.H. THOMAS, Nature, 117, 514 (1926). L.H. THOMAS, Phil. Mag. 3, 1 (1927).
6. V. BARGMANN, L. MICHEL, and V. TELEGDI, Phys. Rev. Lett. 2, 435 (1959).
7. In H. BACRY, Ann. Phys. Paris 8, 197 (1963).
8. H. BACRY, Physics To Day 25, 15 (1972).
9. H. BACRY, Ph. COMBE, and P. SORBA, Rep. Math. Phys. 5, 145 (1974), Appendix C.

GENERATEURS DU GROUPE DE POINCARE
ASSOCIES AUX
REPRESENTATIONS IRREDUCTIBLES UNITAIRES
DU GENRE ESPACE

J. Beckers et M. Jaspers

Les représentations irréductibles unitaires du groupe de Poincaré ont été mises en évidence par Wigner[1] et, parmi elles, celles correspondant à des quadrivecteurs p du *genre espace* ont été associées à la description de particules de masse imaginaire et de spin s. Si ces représentations particulières ne trouvent par conséquent pas d'interprétation physique directe, elles restent toutefois d'un intérêt tout particulier pour diverses raisons: mentionnons seulement ici leur relation avec le cas physique de particules de masse nulle[2][3], leur intérêt dans les développements d'amplitudes associées à certains processus de diffusion[4][5], etc...

Un problème particulier relatif à *toute* représentation irréductible unitaire du groupe de Poincaré est de réaliser explicitement les dix générateurs associés afin de préciser leurs effets dans l'espace de représentation correspondant. Dans le cas de représentations du genre espace, un tel problème a été résolu par Shirokov[2] et d'autres propositions, plus spécifiquement "à caractère d'hélicité", ont été faites par Korff[6], Moses[7], Mukunda[8] et Chakrabarti[9] notamment.

La méthode développée par Chakrabarti[9] peut être exploitée plus avant en utilisant la transformation unitaire de Coester[10] diagonalisant l'opérateur d'hélicité. C'est l'objet de notre contribution: nous obtenons, en effet, de nouvelles formes des générateurs associés aux rotations ($\vec{J}$) et aux transformations pures de Lorentz ($\vec{K}$) (les générateurs P^{μ} associés aux translations étant diagonaux), formes qui sont simplement reliées aux autres via des transformations unitaires simples et qui ont un intérêt certain en liaison avec les opérateurs correspondants établis dans le cadre des représentations du genre lumière[11] - une étude complète des interrelations entre les différentes contributions relatives aux cas des genres-temps, - lumière et - espace sera publiée par ailleurs[12] -.

Rappelons que le petit groupe[1] relatif à des quadrivecteurs p du genre espace est le groupe SO(2,1) - le groupe homogène de Lorentz à trois dimensions - dont les générateurs R^1, R^2, S^3 vérifient les relations de commutation:

$$[S^3,R^1] = iR^2, \quad [S^3,R^2] = -iR^1, \quad [R^1,R^2] = -iS^3. \tag{1}$$

Les représentations irréductibles unitaires de SO(2,1) sont bien connues[14] et, dès lors, celles du groupe de Poincaré peuvent être obtenues par la méthode de Wigner[1].

Recherchons à présent la forme dite de Foldy[13]-Shirokov[2] des générateurs $\vec{J}$ et $\vec{K}$ associés à de telles représentations. Dans ce cas, la méthode de construction de Chakrabarti[8] conduit aux deux ensembles de générateurs $(\vec{J},\vec{K}_{(a)})$ et $(\vec{J},\vec{K}_{(b)})$ *,**:

$$\vec{J} = -i\vec{P}\wedge \frac{\partial}{\partial\vec{P}} + \vec{S} \tag{2}$$

---

* La métrique utilisée est celle définie par $g^{00} = 1$, $g^{ii} = -1$, $g^{\mu\nu} = 0$, $\mu \neq \nu$.

** Nos notations sont reliées à celles de Chakrabarti par $\vec{J} \equiv \vec{M}$, $\vec{K} \equiv -\vec{N}$.

$$\vec{K}_{(a)} = iP^0 \frac{\partial}{\partial\vec{P}} - \frac{1}{|\vec{P}|^2} (P^0\vec{S}+m\vec{R})\Lambda\vec{P} \tag{3}$$

et

$$\vec{K}_{(b)} = iP^0 \frac{\partial}{\partial\vec{P}} - \frac{1}{|\vec{P}|^2} \left(P^0\vec{S} + \frac{m(\vec{R}\Lambda\vec{P})}{|\vec{P}|}\right)\Lambda\vec{P}. \tag{4}$$

Dans les équations (2), (3), (4), les opérateurs $\vec{S}$ et $\vec{R}$ sont les générateurs du groupe O(3,1) caractérisés par les relations de commutation:

$$[S^i,S^j] = i\varepsilon^{ij}{}_k S^k, \quad [S^i,R^j] = i\varepsilon^{ij}{}_k R^k, \quad [R^i,R^j] = -i\varepsilon^{ij}{}_k S^k \tag{5}$$

incluant naturellement celles de SO(2,1). Les ensembles de générateurs $(\vec{J},\vec{K}_{(a)})$ et $(\vec{J},\vec{K}_{(b)})$ sont reliés par la transformation unitaire:

$$U = \exp(-i \frac{\pi}{2} \Lambda) \tag{6}$$

où Λ est l'opérateur d'hélicité:

$$\Lambda = \frac{\vec{J}\cdot\vec{P}}{|\vec{P}|} = \frac{\vec{S}\cdot\vec{P}}{|\vec{P}|}. \tag{7}$$

Chakrabarti[8] a appliqué la transformation unitaire:

$$U^{(2)} = \exp\left[i\omega \frac{P^1S^2-P^2S^1}{(P_1^2+P_2^2)^{\frac{1}{2}}}\right]$$

où

$$\omega = \text{arctg} \frac{(P_1^2+P_2^2)^{\frac{1}{2}}}{P^3} \tag{8}$$

aux ensembles de générateurs $(\vec{J},\vec{K}_{(a)})$ et $(\vec{J},\vec{K}_{(b)})$ donnés par (2), (3), (4) et a obtenu deux réalisations dans lesquelles *seuls* les générateurs de SO(2,1), c'est-à-dire R^1, R^2, S^3, apparaissent. Leurs formes explicites $(\vec{J}^{(2)},\vec{K}^{(2)}_{(a)})$ et $(\vec{J}^{(2)},\vec{K}^{(2)}_{(b)})$ sont:

$$\vec{J}^{(2)} = U^{(2)}\vec{J}U^{(2)-1} = -i\vec{P}\wedge \frac{\partial}{\partial\vec{P}} + S^3\vec{\xi}, \tag{9}$$

$$\vec{K}^{(2)}_{(a)} = U^{(2)}\vec{K}_{(a)}U^{(2)-1} = iP^0 \frac{\partial}{\partial\vec{P}} + S^3 \frac{P^0}{|\vec{P}|} (\vec{\xi}\wedge\vec{e}^3)$$

$$+ R^1 \frac{m}{|\vec{P}|^2} \left(- \frac{P^1P^2}{|\vec{P}|+P^3}, |\vec{P}| - \frac{P_2^2}{|\vec{P}|+P^3}, -P^2\right) \tag{10}$$

$$+ R^2 \frac{m}{|\vec{P}|^2} \left(\frac{P_1^2}{|\vec{P}|+P^3} - |\vec{P}|, \frac{P^1P^2}{|\vec{P}|+P^3}, P^1\right),$$

$$\vec{K}^{(2)}_{(b)} = U^{(2)}\vec{K}_{(b)}U^{(2)-1} = iP^0 \frac{\partial}{\partial\vec{P}} + S^3 \frac{P^0}{|\vec{P}|} (\vec{\xi}\wedge\vec{e}^3)$$

$$+ R^1 \frac{m}{|\vec{P}|^2} \left(|\vec{P}| - \frac{P_1^2}{|\vec{P}|+P^3}, \frac{-P^1P^2}{|\vec{P}|+P^3}, -P^1\right) \tag{11}$$

$$+ R^2 \frac{m}{|\vec{P}|^2} \left(- \frac{P^1P^2}{|\vec{P}|+P^3}, |\vec{P}| - \frac{P_2^2}{|\vec{P}|+P^3}, -P^2\right),$$

où

$$\vec{\xi} = \left(\frac{P^1}{|\vec{P}|+P^3}, \frac{P^2}{|\vec{P}|+P^3}, 1\right). \tag{12}$$

De la même manière, appliquons aux relations (2), (3), (4) la transformation unitaire de Coester[10]:

$$U^{(1)} = \exp(i\theta S^2)\exp(i\phi S^3) \tag{13}$$

où (θ,ϕ) sont les angles polaires de $\vec{P}$:

$$\theta = \arccos \frac{P^3}{P^0}, \quad \phi = \mathrm{arctg} \frac{P^2}{P^1} \tag{14}$$

et $\vec{S} \equiv (S^1,S^2,S^3)$ sont les opérateurs de spin engendrant la structure SU(2):

$$[S^i,S^j] = i\varepsilon^{ij}{}_k S^k. \tag{15}$$

Cette application conduit à nouveau à deux réalisations dans lesquelles *seuls* les générateurs de SO(2,1) apparaissent.

Leurs formes explicites $(\vec{J}^{(1)},\vec{K}^{(1)}_{(a)})$ et $(\vec{J}^{(1)},\vec{K}^{(1)}_{(b)})$ sont:

$$\vec{J}^{(1)} = U^{(1)}\vec{J}U^{(1)-1} = -i\vec{P}\wedge\frac{\partial}{\partial\vec{P}} + S^3\frac{|\vec{P}|}{P_1^2+P_2^2}(P^1,P^2,0), \tag{16}$$

$$\begin{aligned}\vec{K}^{(1)}_{(a)} = U^{(1)}\vec{K}_{(a)}U^{(1)-1} &= iP^0\frac{\partial}{\partial\vec{P}} + S^3\frac{P^0P^3}{|\vec{P}|(P_1^2+P_2^2)}(-P^2,P^1,0)\\ &+ R^1\frac{m}{|\vec{P}|(P_1^2+P_2^2)^{\frac{1}{2}}}(-P^2,P^1,0)\\ &+ R^2\frac{m}{|\vec{P}|^2(P_1^2+P_2^2)^{\frac{1}{2}}}(-P^1P^3,-P^2P^3,P_1^2+P_2^2),\end{aligned} \tag{17}$$

$$\begin{aligned}\vec{K}^{(1)}_{(b)} = U^{(1)}\vec{K}_{(b)}U^{(1)-1} &= iP^0\frac{\partial}{\partial\vec{P}} + S^3\frac{P^0P^3}{|\vec{P}|(P_1^2+P_2^2)}(-P^2,P^1,0)\\ &+ R^1\frac{m}{|\vec{P}|^2(P_1^2+P_2^2)^{\frac{1}{2}}}(P^1P^3,P^2P^3,-(P_1^2+P_2^2))\\ &+ R^2\frac{m}{|\vec{P}|(P_1^2+P_2^2)^{\frac{1}{2}}}(-P^2,P^1,0).\end{aligned} \tag{18}$$

Les transformations $U^{(1)}$ et $U^{(2)}$ jouissent toutes deux de la propriété de diagonaliser l'opérateur d'hélicité (7):

$$U^{(i)}\Lambda U^{(i)-1} = S^3 \qquad (i=1,2). \tag{19}$$

Elles ont été discutées et exploitées par ailleurs[15].

Les ensembles $(\vec{J}^{(i)},\vec{K}^{(i)}_{(a)})$ et $(\vec{J}^{(i)},\vec{K}^{(i)}_{(b)})$, d'une part, $(\vec{J}^{(1)},\vec{K}^{(1)}_{(\alpha)})$ et $(\vec{J}^{(2)},\vec{K}^{(2)}_{(\alpha)})$ $(\alpha=a,b)$, d'autre part, sont unitairement équivalents. Les premiers sont reliés pour chaque indice i par la transformation unitaire:

$$^{(i)}U = U^{(i)}U\, U^{(i)-1} \tag{20}$$

qui peut dans chaque cas prendre la forme simple:

$$^{(i)}U = \exp(-i\,\frac{\pi}{2}\,S^3) \tag{21}$$

et les seconds, pour chaque indice α, par la transformation unitaire:

$$U^{(0)} = \exp(-i\,\phi\,S^3) \tag{22}$$

où

$$\phi = \text{arctg}\,\frac{p^2}{p^1}.$$

Remarquons encore que si l'on pose $m = 0$ dans les ensembles de générateurs $(\vec{J},\vec{K}_{(\alpha)})$, $(\vec{J}^{(1)},\vec{K}^{(1)}_{(\alpha)})$, $(\vec{J}^{(2)},\vec{K}^{(2)}_{(\alpha)})$, on retrouve[11] les formes explicites des générateurs $\vec{J}$ et $\vec{K}$ dans le cadre des représentations du genre lumière "à spin discret". Ces formes sont respectivement celles de Foldy[13]-Shirokov[2], de Coester[10] et de Chakrabarti[16].

Notons enfin que ces résultats partiels seront inclus dans une étude complète[12] reliant l'ensemble des formes proposées pour les générateurs $\vec{J}$ et $\vec{K}$ tant dans le cas des genres temps et lumière que du genre espace.

REFERENCES

1. E.P. WIGNER, Ann. Math. 40, 149 (1939).
2. Iu.M. SHIROKOV, Sov. Phys. J.E.T.P. 6, 664, 919, 929 (1958).
3. D.W. ROBINSON, Helv. Phys. Acta 35, 98 (1962).
4. E.G. KALNINS, J. PATERA, R.T. SHARP et P. WINTERNITZ, Elementary Particle Reactions and the Lorentz and Galilei Groups, in Group Theory and its Applications, ed. E.M. Loebl, vol. III, p. 370 (Academic Press, 1975).
5. M. TOLLER, Nuovo Cim. 53A, 671 (1968).
6. D. KORFF, J. Math. Phys. 5, 869 (1964).
7. H.E. MOSES, J. Math. Phys. 9, 2039 (1968).
8. A. CHAKRABARTI, J. Math. Phys. 12, 1813 (1971).
9. N. MUKUNDA, Ann. Phys. (N.Y.) 61, 329 (1970).
10. F. COESTER, Phys. Rev. 129, 2816 (1963).
11. M. JASPERS, Nuovo Cim. (1976).
12. J. BECKERS et M. JASPERS, to be published (1977).
13. L.L. FOLDY, Phys. Rev. 102, 568 (1956).
14. V. BARGMANN, Ann. of Math. 48, 568 (1947).
15. J. BECKERS et C. PIROTTE, Nuovo Cim. 64, 439 (1969).
16. A. CHAKRABARTI, J. Math. Phys. 7, 949 (1966).

ON THE STRUCTURE OF 4-VECTOR OPERATORS, WITH AN APPLICATION TO THE POINT FORM OF RELATIVISTIC DYNAMICS

A.J. Bracken

Consider any representation of SL(2,C), with generators $J_{\lambda\mu}$ $(= -J_{\mu\lambda})$, $\lambda,\mu = 0,1,2,3$, where

$$i[J_{\lambda\mu},J_{\nu\rho}] = g_{\lambda\nu}J_{\mu\rho} + g_{\mu\rho}J_{\lambda\nu} - g_{\mu\nu}J_{\lambda\rho} - g_{\lambda\rho}J_{\mu\nu}. \qquad (1)$$

Take the metric tensor to be diagonal, with $g_{oo} = -g_{11} = -g_{22} = -g_{33} = 1$, and define $\tilde{J}_{\lambda\mu} = \frac{1}{2}\epsilon_{\lambda\mu\nu\rho}J^{\nu\rho}$, taking the alternating tensor to have $\epsilon_{0123} = -1$. Introduce the Casimir invariants $I_1 = \frac{1}{2}J_{\lambda\mu}J^{\lambda\mu}$ and $I_2 = \frac{1}{4}J_{\lambda\mu}\tilde{J}^{\lambda\mu}$.

Defining the 4×4 matrices of operators α and β with elements

$$\alpha^{\mu}_{\lambda} = -iJ^{\mu}_{\lambda} - \delta^{\mu}_{\lambda}E, \qquad \beta^{\mu}_{\lambda} = \tilde{J}^{\mu}_{\lambda}, \qquad (2)$$

where E is the identity operator in the representation space, we find in consequence of (1) the identities[1]

$$\alpha^4 - (I_1+E)\alpha^2 - (I_2)^2\epsilon = 0,$$

$$\beta^4 - (I_1+E)\beta^2 - (I_2)^2\epsilon = 0,$$

$$\alpha\beta = \beta\alpha = iI_2\epsilon; \quad \alpha^2+\beta^2 = (I_1+E)\epsilon. \qquad (3)$$

Here, for example, $\alpha\beta$ denotes the matrix of operators with elements $(\alpha\beta)^{\mu}_{\lambda} = \alpha^{\nu}_{\lambda}\beta^{\mu}_{\nu}$, and ϵ denotes the 4×4 unit matrix.

In the irreducible representation $[k_o,c]$, I_1 and I_2 are multiples of E by $(k_o^2+c^2-1)$ and ik_oc, respectively, where $2k_o$ is some integer and c is some complex number. More generally, in a fully-reducible representation

$$I_1 = K^2+C^2-E, \qquad I_2 = iKC, \tag{4}$$

where the operator 2K has certain integral eigenvalues and the spectrum of the operator C is some subset of the complex numbers. Using (4) with (3) we find

$$(\alpha-K\epsilon)(\alpha+K\epsilon)(\alpha-C\epsilon)(\alpha+C\epsilon) = 0$$

$$(\beta-K\epsilon)(\beta+K\epsilon)(\beta-C\epsilon)(\beta+C\epsilon) = 0 \tag{5}$$

$$\alpha\beta = \beta\alpha = -KC\epsilon; \qquad \alpha^2+\beta^2 = (K^2+C^2)\epsilon. \tag{6}$$

The identities (5) for the commuting matrices α and β are analogous to Cayley-Hamilton identities for matrices of numbers, but here the analogues of eigenvalues are multiples of the labelling operators K and C. In order to consider the analogues of eigenvectors, suppose now that there is at least one 4-vector operator V_λ acting in the representation space, so that

$$i[V_\lambda,J_{\mu\nu}] = g_{\lambda\nu}V_\mu - g_{\lambda\mu}V_\nu. \tag{7}$$

Construct, using (5), the 4×4 projection matrices $\theta^{(\pm 0)}$, $\theta^{(0\pm)}$, where

$$\theta^{(\pm 0)} = [\pm 2K(K^2-C^2)]^{-1}(\alpha\pm K\epsilon)(\alpha-C\epsilon)(\alpha+C\epsilon)$$

$$\theta^{(0\pm)} = [\pm 2C(C^2-K^2)]^{-1}(\alpha-K\epsilon)(\alpha+K\epsilon)(\alpha\pm C\epsilon) \tag{8}$$

so that

$$\theta^{(+0)} + \theta^{(-0)} + \theta^{(0+)} + \theta^{(0-)} = \epsilon$$

$$\theta^{(+0)}\theta^{(+0)} = \theta^{(+0)}, \quad \theta^{(+0)}\theta^{(0-)} = 0, \text{ etc.}$$

$$\alpha\theta^{(+0)} = K\theta^{(+0)}, \quad \beta\theta^{(+0)} = -C\theta^{(+0)}, \text{ etc.} \tag{9}$$

(Singular cases arise when the θ's apply to elements in the representation space on which one or more of C, K, C+K, C-K vanishes. These cases must be treated separately and carefully.)

Then we have

$$V_\lambda = V_\lambda^{(+0)} + V_\lambda^{(-0)} + V_\lambda^{(0+)} + V_\lambda^{(0-)} \tag{10}$$

where, for example $V_\lambda^{(+0)} = (\theta^{(+0)})_\lambda^\mu V_\mu$, so that

$$\alpha_\lambda^\mu V_\mu^{(+0)} = KV_\lambda^{(+0)}, \quad \beta_\lambda^\mu V_\mu^{(+0)} = -CV_\lambda^{(+0)}, \text{ etc.} \tag{11}$$

The remarkable feature of (10) is that it is a resolution of the 4-vector operator along 4-vector "Wigner operators". To see this note that since $V_\lambda^{(+0)}$, for example, is a 4-vector operator

$$[V_\lambda^{(+0)}, I_1] = 2iJ_\lambda^\mu V_\mu^{(+0)} + 3V_\lambda^{(+0)} = (1-2K)V_\lambda^{(+0)},$$

$$[V_\lambda^{(+0)}, I_2] = i\tilde{J}_\lambda^\mu V_\mu^{(+0)} = -iCV_\lambda^{(+0)} \tag{12}$$

using (11). Then from (4) we have

$$[(K-1)^2+C^2]V_\lambda^{(+0)} = V_\lambda^{(+0)}(K^2+C^2),$$

$$(K-1)CV_\lambda^{(+0)} = V_\lambda^{(+0)}KC \tag{13}$$

from which we conclude that, for suitable K and C satisfying (4),

$$KV_\lambda^{(+0)} = V_\lambda^{(+0)}(K+1), \quad CV_\lambda^{(+0)} = V_\lambda^{(+0)}C. \tag{14}$$

In a similar way we have

$$KV_\lambda^{(-0)} = V_\lambda^{(-0)}(K-1), \quad CV_\lambda^{(-0)} = V_\lambda^{(-0)}C$$

$$KV_\lambda^{(0\pm)} = V_\lambda^{(0\pm)}K \quad , \quad CV_\lambda^{(0\pm)} = V_\lambda^{(0\pm)}(C\pm 1). \tag{15}$$

In a representation which is a direct sum of irreducible representations, $V_\lambda^{(\pm 0)}$, $V_\lambda^{(0\pm)}$ are evidently 4-vector shift-operators for the labelling operators K and C. However, in a representation which is a direct integral of irreducible representations, so that C has a continuous spectrum, it may not be possible to maintain this interpretation, at least for the $V_\lambda^{(0\pm)}$. In any case, these V's must satisfy several relations, whatever the precise structure of the original operator V_λ - in particular, whether or not its components commute.

For example, since it follows from (14) that

$$KV_\lambda^{(+0)}V^{\lambda(+0)} = V_\lambda^{(+0)}V^{\lambda(+0)}(K+2),$$

and since no scalar operator can fail to commute with K, we must have $V_\lambda^{(+0)}V^{\lambda(+0)} = 0$. In this way one sees that all scalar products of $V_\lambda^{(\pm 0)}$, $V_\lambda^{(0\pm)}$ must vanish, excepting

$$V_\lambda^{(\pm 0)}V^{\lambda(\mp 0)} \quad \text{and} \quad V_\lambda^{(0\pm)}V^{\lambda(0\mp)}. \tag{16}$$

Again, since it follows from (14) that

$$K[V_\lambda^{(+0)},V_\mu^{(+0)}] = [V_\lambda^{(+0)},V_\mu^{(+0)}](K+2)$$

one must have $[V_\lambda^{(+0)},V_\mu^{(+0)}] = 0$, as it is easily seen that no antisymmetric second-rank tensor operator can shift the value of K (or C) by two units. In this way we deduce that

$$[V_\lambda^{(\pm 0)},V_\mu^{(\pm 0)}] = 0 = [V_\lambda^{(0\pm)},V_\mu^{(0\pm)}]. \tag{17}$$

Furthermore, it is not hard to see that any antisymmetric second-rank tensor operator which commutes with K and C, such as, for example,

$$V_{\lambda}^{(0+)}V_{\mu}^{(0-)} - V_{\mu}^{(0+)}V_{\lambda}^{(0-)}$$

must be of the form $AJ_{\lambda\mu} + B\tilde{J}_{\lambda\mu}$, where A and B are scalar operators determined by the structure of V_{λ}.

These techniques can in principle be generalized to solve the problem of resolving an irreducible tensor operator, in a representation of any semi-simple Lie group, into irreducible tensor operators, which are also shift-operators for the labelling operators in the representation[(2)]. The method is especially simple and direct in the case of finite-dimensional representations, and some applications have been given involving such cases[(3)]. In order to demonstrate the utility of the method in the treatment of algebraic problems posed in a space carrying an infinite-dimensional representation of a semi-simple group, we shall consider the problem of finding a 4-vector coordinate operator for a (positive energy) particle with rest mass m and spin 0.

Consider then the appropriate unitary representation (m,0,+) of the Poincaré group, with self-adjoint generators $J_{\lambda\mu}$, P_{λ}. Here P_{λ} is a 4-vector operator with commuting components, $P_o > 0$, $P_{\lambda}P^{\lambda} = m^2c^2E$ and $\beta_{\lambda}^{\mu}P_{\mu} = 0$ (the spin 0 condition). It is known that in this case the $J_{\lambda\mu}$ generate a direct integral representation of SL(2,C),

$$\int_{0}^{\infty} \oplus\ [0,i\rho]d\rho, \tag{18}$$

so that here K=0 (= I_2), and $I_1 = C^2-E$, where C may be taken to be skew-adjoint, with spectrum $i[0,\infty)$.

Since $\beta_{\lambda}^{\mu}P_{\mu} = 0$, we see at once from (3) that

$$(\alpha^2-C^2\epsilon)_{\lambda}^{\mu}P_{\mu} = 0, \tag{19}$$

so that in this case

$$P_\lambda = P_\lambda^{(0+)} + P_\lambda^{(0-)}. \tag{20}$$

Any 4-vector operator acting in the space has the form

$$Q_\lambda = A^{(+)}(C)P_\lambda^{(0+)} + A^{(-)}(C)P_\lambda^{(0-)} \tag{21}$$

where $A^{(\pm)}(C)$ is some function of C, there being no other non-trivial scalar operators acting in the space.

We seek a 4-vector operator satisfying the conditions

$$Q_\lambda^\dagger = Q_\lambda \quad , \quad [Q_\lambda, Q_\mu] = 0$$

$$Q_\lambda Q^\lambda = k^2 E, \quad k > 0, \quad Q_o > 0$$

$$\{Q_\lambda, P_\mu\} - \{Q_\mu, P_\lambda\} = 2J_{\lambda\mu}. \tag{22}$$

Such an operator then represents the analogue in quantum mechanics of the 4-vector coordinate of the particle where its world-line crosses the hyperboloid $x_\lambda x^\lambda = k^2$, $x_o > 0$ in Dirac's "point form" of relativistic classical dynamics(4). Alternatively, Q_λ may be viewed as the analogue for this space-like surface of the Newton-Wigner operator, which is appropriate to the surface $x_o = 0$.

In the classical case, Dirac has given the solution(4) - actually he gave the expression for P_λ in terms of Q_μ and $J_{\nu\rho}$, but there is an obvious symmetry between the roles of the coordinates and momenta in the point form of dynamics -

$$m^2c^2Q_\lambda = J_{\lambda\mu}P^\mu + P_\lambda(m^2k^2c^2 - \tfrac{1}{2} J_{\lambda\mu}J^{\lambda\mu})^{1/2}. \tag{23}$$

In the present case, we impose each of the requirements (22) on the operator (21), making use of [cf. (13-17)]

$$P_\lambda^{(0\pm)} P^{\lambda(0\pm)} = 0, \quad P_\lambda^{(0\pm)} P^{\lambda(0\mp)} = \frac{1}{2} m^2 c^2 (C \mp 1)/C,$$

$$[P_\lambda^{(0\pm)}, P_\mu^{(0\pm)}] = 0, \quad P_\lambda^{(0\pm)} P_\mu^{(0\mp)} - P_\mu^{(0\pm)} P_\lambda^{(0\mp)} = \mp \frac{1}{2} m^2 c^2 J_{\lambda\mu}/\hbar C,$$

$$CP_\lambda^{(0\pm)} = P_\lambda^{(0\pm)} (C \pm 1), \quad C^\dagger = -C. \tag{24}$$

We obtain thereby a number of non-linear relations (mostly difference equations) to be satisfied by the coefficients $A^{(\pm)}(C)$. These can be solved without great difficulty to yield

$$m^2c^2Q_\lambda = \frac{1}{2}\{J_{\lambda\mu}, P^\mu\}[1 - \frac{1}{2} i[B^{(+)}(C) - B^{(-)}(C)]/\hbar C]$$
$$+ P_\lambda[(C - \tfrac{1}{2})B^{(+)}(C) + (C + \tfrac{1}{2})B^{(-)}(C)]/2C, \tag{25}$$

where

$$B^{(\pm)}(C) = [m^2k^2c^2 - (C \pm \tfrac{1}{2})^2\hbar^2]^{1/2}. \tag{26}$$

We have introduced the appropriate multiples of $\hbar$ in (24-26) in order to be able to consider the classical limit, which is obtained when $\hbar \to 0$, $\hbar^2C^2 \to \frac{1}{2} J_{\lambda\mu}J^{\lambda\mu}$. Then

$$B^{(\pm)}(C) \to (m^2k^2c^2 - \tfrac{1}{2} J_{\lambda\mu}J^{\lambda\mu})^{1/2} \tag{27}$$

and (25) reduces to (23). Details of the calculations involved in obtaining (25) will be presented in a forthcoming note on the point form of dynamics.

It should be pointed out that although Q_λ has all the desirable properties (22), the commutator $[Q_\lambda, P_\mu]$ cannot be simply expressed, although it is easily shown that $[Q_\lambda, P^\lambda] = -3i\hbar$ and (trivially) that $[Q_\lambda, P_\mu P^\mu] = 0$. Hence, Q_λ is no more a "covariant position operator" than the Newton-Wigner operator. The latter transforms simply under transformations of the Euclidean subgroup of the Poincaré group, but not under Lorentz transformations or translations in time. This reflects the fact that the Euclidean group leaves invariant the surface $x_o = 0$. Quite analogously, since the invariance group of the hyperboloid is SL(2,C), the position operator in this case transforms simply under rotations and Lorentz transformations (as a 4-vector) but not under space-time translations generated by P_λ.

REFERENCES

1. A.J. BRACKEN and H.S. GREEN, J. Math. Phys. 12, 2099 (1971). A.J. BRACKEN, Ph.D. thesis, The University of Adelaide, 1970 (unpublished).
2. H.S. GREEN, J. Math. Phys. 12, 2106 (1971).
3. A.J. BRACKEN and H.S. GREEN, Nuovo Cim. 9A, 349 (1972); J. Math. Phys. 14, 1784 (1973). H.S. GREEN and A.J. BRACKEN, Int. J. Theor. Phys. 11, 157 (1974).
4. P.A.M. DIRAC, Rev. Mod. Phys. 21, 392 (1949).

ON THE LOCALIZABILITY OF MASSLESS PARTICLES

G. Burdet* and M. Perrin*

"We are forced to think of them as "particles" which are a little less "particle-like" than the others in that they have no position observables."(1)

I. The notion of localizability discussed here deals with states localized in space at a given time, and technically the above sentence means that there does not exist a transitive system of imprimitivity for the restriction on the 3-dimensional Euclidean group E(3) of the massless, discrete spin UIR U_0^s of the Poincaré group $\bar{P}_+^\uparrow$ based on $\mathbb{R}^3$, i.e. $U_0^s \downarrow E(3)$ is not equivalent to a representation of E(3) which is induced by a unitary representation D^s of SU(2):(2)

$$U_0^s \downarrow E(3) \neq D^s \uparrow E(3).$$

Let us introduce $\mathbb{R}^4 \,\square\, E(2) \subset \bar{P}_+^\uparrow$, the little group of a given light-like vector, obviously $E(3) \not\subset \mathbb{R}^4 \,\square\, E(2)$. However one can try to look for a system of imprimitivity based on the coset space $\mathbb{R}^3 \,\square\, E(2)/_{E(2)} \simeq \mathbb{R}^3$. Then it has been shown(3) that

---

* On leave of absence from Laboratoire de Physique Mathématique, Université de Dijon, France.

$U_0^s \downarrow \mathbb{R}^3 \square E(2)$ is never equivalent to a representation of $\mathbb{R}^3 \square E(2)$ induced by a unitary representation of E(2).

Faced with this situation, an enlarged Wightman's procedure has been used in Ref. (3), see also (4), and a non-transitive (each orbit being homeomorphic to $\mathbb{R}^2$) system of imprimitivity based on $\mathbb{R}^2 \otimes R^+$ was exhibited, i.e. there is an equivalence between $U_o^s \downarrow (\mathbb{R}^4 \square E(2) \square \mathbb{R}^+)$ and a representation of $R^4 \square E(2) \square R^+$, the stabilizer of a given light-like vector into Poincaré, induced by a representation of $\mathbb{R}^4 \square SO(2)$. Corresponding to this system, a position operator with two self-adjoint commuting components was proposed.

In this paper we want to show how the same operator also appears by using the formal association between a massless and a "non relativistic" particle which has been proposed in Ref. (5).

II. The representations of SU(2,2), the reduction of which to the Weyl $\bar{W}$ or to the Poincaré $\bar{P}$ groups remain irreducible, are the most degenerate discrete series, they lead to the zero mass, discrete spin s representations U_0^s of the Poincaré group[(6)]. Conversely, the extension to SU(2,2) of U_0^s is unique[(7)]. Hence massless particles can strictly be associated to UIR of the conformal group C_0 which is locally isomorphic to SU(2,2). Now the elements of the chain $P_+^\uparrow \subset W \subset C_0$ correspond to transformations in the Minkowski space M; then if we look for the stabilizers of a given light-like vector, in each case we get a maximal subgroup which, up to a dilation, corresponds to a group known as a group of transformations in a (2+1)-dimensional Newtonian space-time. More precisely one finds:

$$\begin{array}{ccccc} P_+^\uparrow & \subset & W & \subset & C_0 \\ \cup & & \cup & & \cup \\ \tilde{G}_2 \square \mathbb{R}^+ & \subset & \tilde{G}_2 \square (\mathbb{R}^+ \otimes \mathbb{R}^+) & \subset & \widetilde{Sch}_2 \square \mathbb{R}^+ \end{array}$$

where $\tilde{G}_2$ and $\widetilde{Sch}_2$ are the one parameter central extensions of the

Galilei and Schrödinger groups respectively.

The (2+1)-dimensional "Newtonian space-time" in which they act is a hyperplane $T(3) \subset M$, the elements of which are space-like vectors and the stabilized light-like direction $T(1)$. A representative of the class $T(3)/T(1)$ is usually called "the transversal plane" $T_{\perp}(2)$ to the considered light-like direction.

Then the repeated "non-relativistic" structures of the stabilizers(*) of a given light-like vector into the Poincaré, Weyl and conformal groups support the following formal analogy[5]:

In a two-dimensional space-like plane, transversal to its propagation direction, a massless particle behaves like a "non-relativistic" particle of variable mass.

By using this analogy, it is clear how a two-component "Galilean" position operator can be associated to a massless particle, moreover this operator is the same as the two-component one proposed in Ref. (3).

III. Here we want to illustrate the above discussion by considering the simplest non trivial case, namely the case of the neutrino described by the Weyl equation which, in the usual spinor representation is given by:

$$\partial_{\alpha\dot{\beta}}\psi^{\alpha}(x) = 0 \tag{1}$$

where

$$\partial_{\alpha\dot{\beta}} = \eta^{\mu\nu} \frac{\partial}{\partial x^{\mu}} (\sigma_{\nu})_{\alpha\dot{\beta}}.$$

In the Schrödinger-like form, Eq. (1) can be written:

(*) It is interesting to remark that the "non-relativistic" chain can be continued for instance by the embedding into $H_2 \,\square\, (Sp(4,\mathbb{R})\otimes\mathbb{R}^{+})$, but it is easy to show that a finite dimensional "relativistic" Lie algebra cannot be obtained which contains $H_2 \,\square\, (Sp(4,\mathbb{R})\otimes\mathbb{R}^{+})$ together with the conformal algebra.

$$H\psi(\underline{x},t) \equiv i(\sigma \cdot \frac{\partial}{\partial x})\psi(\underline{x},t) = i\sigma_0 \frac{\partial}{\partial t}\psi(\underline{x},t) \tag{2}$$

where H acts on a spinor-valued (two-component) function ψ of the space variable $\underline{x}$.

A representation of the conformal group acting on the solution space S_H of the Weyl equation can be obtained by allowing the wave function to co-transform according to

$$\hat{\psi}(\hat{x}) \equiv (T_g\psi)(g\cdot x) = S_g(g,x)\psi(x)$$

while $x \to \hat{x} = g\cdot x$ with $g \in C_0$.

Under a Poincaré transformation $x \to \hat{x} = \Lambda x+a$ we have

$$(T_g\psi)(\Lambda x+a) = S(\Lambda)\psi(x)$$

where $S(\Lambda)$ belongs to the representation $\mathcal{D}(0,\frac{1}{2})$ of SL(2,C). Then, as has been shown in Ref. (8), under a Weyl dilation and a special conformal transformation:

$$x \to \hat{x}_c = \lambda \frac{x+cx^2}{w(c,x)}, \quad w(c,x) = 1+2c\cdot x+c^2x^2$$

the wave function co-transforms according to:

$$\hat{\psi}(\hat{x}) = \lambda^{-3/2}w(c,x)^{3/2}S_0(c,x)\psi(x)$$

where $S_0(c,x)$ is implicitly determined as the unique solution of some matrix differential system, whose explicit solution is given by[9]:

$$S_0(c,x) = \exp\left\{[(c\cdot x)^2-c^2x^2]^{-\frac{1}{2}}\mathrm{Log}\left(\frac{1+c\cdot x+[(c\cdot x)^2-c^2x^2]^{\frac{1}{2}}}{1+c\cdot x-[(c\cdot x)^2-c^2x^2]^{\frac{1}{2}}}\right)a^\mu x^\nu M_{\mu\nu}\right\}$$

where the matrices $M_{\mu\nu}$ are the generators of the $\mathcal{D}(0,\frac{1}{2})$ representation of SL(2,C).

From these co-transformation laws, the following representation of the Lie algebra generators is deduced.

$$
\begin{aligned}
p_j &= i\,\frac{\partial}{\partial x^j} = i\partial_j \\
p_0 &= i\partial_t \\
M_{j0} &= -x^j p_0 + it\partial_j + \frac{i}{2}\sigma_j \\
M_{jk} &= -i(x^j\partial_k - x^k\partial_j + \frac{i}{2}\sigma_\ell) \qquad (3)\\
D &= tp_o + i(\underline{x}\cdot\underline{\partial} + \frac{3}{2}) \\
C_0 &= -(t^2+\underline{x}^2)\partial_t - 2it(\underline{x}\cdot\underline{\partial} + \frac{3}{2}) - i(\underline{\sigma}\cdot\underline{x}) \\
C_j &= 2x^j tp_0 + i(t^2-\underline{x}^2)\partial_j + 2ix^j(\underline{x}\cdot\underline{\partial} + \frac{3}{2}) + it\sigma_j + \varepsilon_{jk\ell}\sigma_k x^\ell .
\end{aligned}
$$

Let Z be a general element of the Lie algebra written as

$$Z = \rho^\mu p_\mu + \tau^\mu C_\mu + \delta D + \nu_j M_{j0} + \tfrac{1}{2}\varepsilon_{jk\ell}\mu_\ell M_{jk} \quad (\rho,\tau,\delta,\mu,\nu \in \mathbb{R});$$

it is easy to verify that

$$[H-i\partial_t, Z(\underline{x},t)] = i\Omega(\underline{x},t)(H-i\partial_t) \qquad (4)$$

with

$$\Omega(\underline{x},t) = (\underline{\sigma}\cdot\underline{\nu}) + \delta - 2\tau^0(t-(\underline{\sigma}\cdot\underline{x})) + 2(\underline{\tau}\cdot\underline{x}-(\underline{\tau}\cdot\underline{\sigma})t).$$

Other representations of the generators of the Lie algebra of C_0 can be deduced from (3); successively one constructs:

- The Schrödinger representation (obtained by setting $p_0 = H$ in (3)) which always acts on S_H. In this representation the relation (4) becomes:

$$[H, Z_s(\underline{x},t)] = i\,\frac{\partial Z_s}{\partial t}. \qquad (5)$$

- The Heisenberg representation, in which the generators no longer explicitly depend upon the time and which acts on $\mathcal{L}^2(\mathbb{R}^3,C^2)$.

- The p-representation obtained by a three-dimensional Fourier transform.

Let us suppose that the relativistic particle moves along the third axis, which usually is expressed by

$$p_j\psi(p) = 0, \quad (p_0+p_3)\psi(p) = 0 \tag{6}$$

while

$$(p_0-p_3)\psi(p) = c\psi(p) \tag{7}$$

where c is the colour.

The corresponding "Galilean" boosts are[5]:

$$K_1 = M^{10}-M^{31}, \quad K_2 = M^{20}+M^{23}$$

and the associated variable mass operator is:

$$M = p_0-p_3.$$

Then in the p-representation the two-component position operator is given by:

$$Q_1 = i\frac{\partial}{\partial p_1} + i\frac{p_1}{p_0-p_3}\frac{\partial}{\partial p_3} - \frac{1}{2(p_0-p_3)}(\sigma_2+i\sigma_1)$$

$$Q_2 = i\frac{\partial}{\partial p_2} + i\frac{p_2}{p_0-p_3}\frac{\partial}{\partial p_3} + \frac{1}{2(p_0-p_3)}(\sigma_1-i\sigma_2).$$

In "the transversal plane" the so-obtained Q_j fulfill the wanted commutation relations for a position operator, namely:

$$[Q_1,Q_2] = 0$$

$$[Q_j,p_k] = i\delta_{jk} \quad j,k = 1,2$$

$$[J_3, Q_j] = i\varepsilon_{jk3} Q_k \quad (J_3 \equiv M_{12}).$$

Moreover,

$$[Q_j, p_0 + p_3] = 0.$$

Hence the position operator and the colour can be simultaneously diagonalized.

From the position operator a velocity operator can be introduced whose components are given by:

$$v_j = p_j (p_0 + p_3)^{-1}$$

and satisfy

$$[v_j, p_\mu] = 0 \quad \forall\ \mu \in [0,1,2,3].$$

Hence the velocity operator can be simultaneously diagonalized with the four p_μ; then according to (6) and (7): $v_j \psi(p) = 0$. This clearly shows that the position and velocity operators obtained above are consistent with the usual notion of a light-like particle propagating along a given direction.

It is worth noticing that the position operator takes a diagonal matrix form in the "Foldy-Wouthuysen" representation, i.e. in the representation in which the Hamiltonian becomes $-\sigma_3 |p|$ and which is obtained by using the automorphism induced by:

$$\exp(\tfrac{i}{2}\,\sigma_3 \varphi)\exp(-\,\tfrac{i}{2}\,\sigma_1 \theta)\exp(-\,\tfrac{i}{2}\,\sigma_3 \varphi)$$

where

$$\varphi = \text{Arc tg}\,\frac{p_1}{p_2} \quad \text{and} \quad \theta = \text{Arc tg}\,\frac{\sqrt{p_1^2 + p_2^2}}{p_3}\,.$$

Ref. (3,4) and the above results lead us to conclude that the notion of localizability depends on the largest dimensional space-like region "orthogonal to the kind" of the considered particle, i.e.

- a three-dimensional hyperplane for a massive time-like particle;

- a two-dimensional transversal plane $T_{\perp}(2)$ for a massless light-like particle;

- a two-dimensional space-like plane for a tachyon!

Finally we want to note that a generalized four-component position operator in the sense of Ref. (10) can also be introduced for a massless particle. But these four components are not independent, they depend on the two-dimensional space-like plane $T_{\perp}(2)$ and describe the position on this plane, so they have to satisfy two constraints and two components reduce to a number.

REFERENCES

1. G.W. MACKEY, Induced Representations of Groups and Quantum Mechanics, W.A. Benjamin Inc., N.Y. and Ed. Boringhieri, Torino (1968).
2. This result is no longer valid for s=0 (not for the massless spin 1/2 particle with both helicities), and "so the phonon is localizable", A.S. WIGHTMAN, Rev. Mod. Phys. 34 845 (1962). In this paper it is also shown that in the massive case, and for any spin, there exists a system of imprimitivity based on the coset space $E(3)/_{SU(2)} \simeq \mathbb{R}^3$, i.e. $U^s_m \downarrow E(3) \approx D^s \uparrow E(3)$.
3. E. ANGELOPOULOS, F. BAYEN and M. FLATO, Physica Scripta 9, 173 (1974).
4. J. BERTRAND, Nuov. Cim. 15A, 281 (1973).
5. G. BURDET, M. PERRIN and P. SORBA, Comm. Math. Phys. 34, 85 (1973).
6. G. MACK and I. TODOROV, Jour. Math. Phys. 10, 2078 (1969). See also: D. STERNHEIMER, Jour. Math. Pure Appl. 47, 289 (1968).
7. J. MICKELSSON and J. NIEDERLE, Jour. Math. Phys. 13, 23 (1972).
8. M. FLATO, J. SIMON and D. STERNHEIMER, Ann. of Phys., N.Y. 61, 78 (1970).
9. G. ROSEN, Ann. of Phys., N.Y. 77, 452 (1973).
10. G.M. FLEMING, Phys. Rev. 137B, 188 (1965).

ON THE EXISTENCE OF CHARGES AND MASS-SPLITTING IN RELATIVISTIC QUANTUM FIELD THEORY*

E. Gal-Ezer and L.P. Horwitz†

The existence of charge operators associated with integrals of local densities in the (Wightman) framework of quantum field theory, in the presence of explicit symmetry breaking, can be demonstrated in certain cases[1]. Their construction, in terms of null-plane integrals, is rather delicate. The possibility that a finite number of null-plane charges, which includes the Poincaré generators, close on an algebra whose irreducible representations contain particles with different masses is considered[2]; domain problems are shown to invalidate the basic hypotheses of the O'Raifeartaigh theorem. Null-plane Fourier transforms, which enter into the discussion of current algebra at infinite momentum, are also studied. It is shown that slns behavior is the maximal growth of high energy off-mass-shell amplitudes consistent with the existence of null-plane charge and null-plane Fourier transforms. Under the assumption that asymptotic states exist, these results also hold in the case of spontaneously broken chiral

---

* Work supported in part by the U.S.-Israel Binational Science Foundation (BSF), Jerusalem, Israel.

† Work prepared while visiting the Centers for Particle Theory and Statistical Mechanics and Thermodynamics at The University of Texas, Austin, Texas. Work supported in part by the Energy Research and Development Administration under Contract E(40-1)3992.

symmetry, with massless pseudoscalar Goldstone bosons.

Let us first consider the integrated local density

$$j(\eta,\theta_R) = \int d^4x j(x^0,\vec{x})\eta(x^0)\theta_R(\vec{x}), \tag{1}$$

where $\theta_R(x) = \theta\left(\frac{|\vec{x}|}{R}\right)$ is a family of infinitely differentiable functions with compact support, and $\eta_T(x^0) = \frac{1}{T}\eta\left(\frac{x^0}{T}\right)$ is a family of infinitely differentiable fast decreasing functions. The sesquilinear form (s.q.f.) (for example, for j^0 the fourth component of a four-vector)

$$\lim_{T\to 0, R\to\infty} (A\Omega, j^0(\eta_T,\theta_R)B\Omega) \tag{2}$$

where Ω is the cyclic vacuum state and $A\Omega \in D_{ql}$ (quasi local states) defines an operator Q only if vacuum annihilation holds:

$$\lim_{T\to 0}\lim_{R\to\infty} (\Omega,[j^0(\eta_T,\theta_R),A]\Omega) = 0 \tag{3}$$

which, by Coleman's theorem, implies that $\partial_\mu j^\mu = 0$. For the case of broken symmetry, where $\partial_\mu j^\mu \neq 0$ or (3) does not hold, we study the null-plane integrals (m is a set of tensor indices)

$$\hat{t}^m(\eta_T,\theta_R) = \int dx^+ dx^- d^2x^\perp t^m(x)\eta_T(x^+)\theta_R(x^-,x^\perp), \tag{4}$$

on a domain $D_\pm$ dense in D_{in}, D_{out} (we shall assume a mass gap unless otherwise stated; $D_\pm$ have three-momentum wave functions which are infinitely differentiable, with compact support, and every pair of particles is disjoint in velocity space). Vacuum annihilation follows easily from the spectral representation. For ϕ, ψ many-particle states ($\tilde{p}=(p^+,p^\perp)$), consider

$$(\phi,t^m(x)\psi) = \int \frac{d^3\tilde{p}_1\ldots d^3\tilde{p}_n d^3\tilde{p}_{n+1}\ldots d^3\tilde{p}_{n+k}}{2p_1^+\ldots 2p_{n+k}^+}\phi^*(\tilde{p}_1\ldots\tilde{p}_n)\psi(\tilde{p}_{n+1}\ldots\tilde{p}_{n+k})$$

$$\times\ {}_{out}\langle\tilde{p}_1\ldots\tilde{p}_n|t^m(x)|\tilde{p}_{n+1}\ldots\tilde{p}_{n+k}\rangle_{in}, \tag{5}$$

and define the form factor

$$ {}_{out}\langle \tilde{p}_1 \ldots \tilde{p}_n | t^m(x) | \tilde{p}_{n+1} \ldots \tilde{p}_{n+k} \rangle_{in} = \frac{1}{(2\pi)^3} e^{iq\cdot x} F^m(q, \tilde{p}_2 \ldots \tilde{p}_{n+k}), \quad (6) $$

where

$$ q = \sum_{i=1}^{n} p_i - \sum_{j=n+1}^{n+k} p_j . $$

The limits of the matrix elements of (4) can then be taken using the smoothness of F^m. The s.q.f. $\lim(\phi, \hat{t}^m\psi)$ then defines a charge if it is continuous in ϕ. It is sufficient to show that $\lim\|\hat{t}^m(\eta_T, \theta_R)\psi\| = b(\psi) < \infty$. To do this, one defines the absorptive part ($q = \frac{1}{2}(q_1+q_2)$)

$$ V^{mn} = (2\pi)^2 \int d^4x e^{iq\cdot x} {}_{in}\langle \tilde{p}_1 \ldots \tilde{p}_n | t^m\left(\tfrac{x}{2}\right) t^n\left(-\tfrac{x}{2}\right) | \tilde{p}_{n+1} \ldots \tilde{p}_{2n} \rangle_{in}, \quad (7) $$

and, with the polarization tensors $\varepsilon_r^n(\lambda, q)$, assumes the asymptotic form

$$ V(s\lambda, r\tau) = \sum_{m,n} \varepsilon_s^m(\lambda, q_1) \varepsilon_r^n(\tau, q_2) V^{mn} \sim \beta(q^-)^{\alpha(Q^2)} (\ln q^-)^{\gamma(Q^2)}, \quad (8) $$

where $Q = \frac{1}{\varepsilon}(q_2 - q_1)$ and $\alpha \le 1$, $\gamma \le 2$ at $Q^2 = 0$ (unitarity bounds). We may then invert (8) with the result

$$ V^{mn} = \sum_{s,r} c_{sr} \varepsilon_s^m(\lambda, q_1) \varepsilon_r^n(\tau, q_2) V(s\lambda, r\tau) \quad (9) $$

and take the (off-shell) limits $Q^2 \to 0$, q_1^2, q_2^2, $q_1^\perp$, $q_2^\perp \to 0$ $(R \to \infty)$. To summarize our conclusions, it is convenient to express m in terms of null-plane indices, +, -, 1, 2, and suppose $\beta \neq 0$. Then, null-plane charges exist for "very good" tensors (two or more + indices in excess of - indices) for $\alpha \le 1$, and for "good" (one + index in excess) tensors for $\alpha < 1$. For $\alpha = 1$, good tensors do not lead to a charge if $\gamma(0) > 1$ (this is shown by considering a sequence $\chi_n = \phi(np_1^+, p_1^\perp)\phi(\tilde{p}_2)$ of two particle states, for which

$(\chi_n,\hat{j}^+\psi) \to \infty$ as $n \to \infty$ if $\gamma(0) > 1$). Hence, we obtain the bound slns for the existence of charges associated to non-conserved currents. For "bad" tensors (no + excess) or "very bad" tensors (excess of - indices), null-plane charges do not exist even if $\alpha < 1$ (exceptions: free field bilinears, + components of conserved currents, tensors with explicit x^+ factor). The null-plane charges defined in this way are symmetric on $D_+ \cup D_-$ (given the usual properties of the local densities).

O'Raifeartaigh[3] proved the following result:

Let $\{Q_i\}$ represent the finite Lie algebra, which includes the Poincaré and an internal symmetry algebra, $[q_i,q_j] = iC_{ijk}q_k \equiv D_{jk}(q_i)q_k$ in the sense $[Q_i,Q_j] \subset iC_{ijk}Q_k$ on $\mathcal{H}$, ψ_{m^2} (one particle state) satisfy $(P^2-m^2)\psi_{m^2} = 0$, and for all i, $B(P)Q_iC(P)\psi_{m^2}$ be defined, where B, C are any polynomials in P^μ. Then, $(P^2-m^2)Q_i\psi_{m^2} = 0$. If such ψ_{m^2} are dense in $\mathcal{H}_{m^2}$, then $\mathcal{H}_{m^2}$ is invariant under $\{Q_i\}$.

We have seen that the construction of charges in quantum field theory is rather delicate. The proof of O'Raifeartaigh utilizes nilpotence to show that $(P^2-m^2)^N Q_i\psi_{m^2} = 0$. For $\phi \in D((P^-)^N)$, $\psi_{m^2} \in \mathcal{H}_{m^2} \cap D_\pm$, $\lim((P^-)^N\phi,\ \hat{t}^m(\eta_T,\theta_R)\psi_{m^2}) \le \|\phi\|K$ is violated for finite N (take $\phi_n = \hat{t}^m(\eta_{T_n},\theta_{R_n})\psi_{m^2}$ as a bounded sequence for t^m very good, and $\phi(np_1^+,p_1^\perp)\psi_{m^2}(p_2^+,p_2^\perp)$ for t^m good). One can attempt to extend $\hat{Q}_i$ to a dense invariant domain D which does not contain $D_+ \cup D_-$ in the following way. Suppose $\cap D(\hat{Q}_i^\dagger) \supset D_+ \cup D_-$. Let $\hat{Q}_i$ be symmetric extensions of the null-plane charges to $D \cup (D_+ \cup D_-)$, where $D \subset \cap D(\hat{Q}_i^\dagger)\backslash D_+ \cup D_-$. Then, for $(P^2-m^2)^N\phi \in D$, nilpotence implies

$$((P^2-m^2)^N\phi,\hat{Q}_i\psi_{m^2}) = 0. \qquad (10)$$

Since, however, the restriction of P^2 to D is not necessarily essentially self-adjoint, and the limit $\phi \to \psi_{m'^2}$, leading to a negative conclusion on mass-splitting for discrete eigenstates, may not be possible to carry out ($\psi_{m'^2}$ may not be in the closure of the restricted operator). The Flato-Sternheimer[4] counter-examples satisfy (10) for N = 2. Translation invariance of D would imply, by a theorem of Segal, that $(P^2-m^2)^N$ is essentially self-adjoint[5].

The assumption of algebraic structures of the following form (not equivalent to the usual definition of representations of Lie algebras)

$$\lim_u(\phi,[Q_{i,u},Q_{j,u}]\psi) = iC_{ijk} \lim_u(\phi,Q_{k,u}\psi), \qquad (11)$$

for $\phi,\psi \in D_{ql}$ or $D_+ \cup D_-$, or

$$(\hat{Q}_i\psi,\hat{Q}_j\psi') - (\hat{Q}_j\psi,\hat{Q}_i\psi') = iC_{ijk}(\psi,\hat{Q}_k\psi'), \qquad (12)$$

for $\psi,\psi' \in D_+ \cup D_-$, lead to a negative conclusion for mass-splitting on discrete eigenstates.

We finally turn to a consideration of Fourier transforms of the form

$$\lim(\chi,\int d^4x\eta_T(x^+)\theta_R(x^-,x^\perp)e^{ik^\perp\cdot x^\perp}j_a^+(x)\psi); \qquad (13)$$

the associated generators have application to current algebra[6]. Using the techniques obtained above, one finds that $k_\perp \neq 0$ replaces a mass gap in the convergence of the s.q.f. and vacuum annihilation (assuming D_{in}, D_{out} can be constructed in the absence of a mass gap). In the sufficient condition for constructing a charge when $k^\perp \to 0$ in the absence of a mass gap, involving examination of an amplitude, there is a possible divergence at t = 0 due to the t-channel cut, unless the double spectral function $\rho(s,t=0) = 0$ for all s (we just consider currents j^+ here), since $k^\perp \to 0$ permits a singular denominator. It has been shown[7], however, using dispersion relations and two-body unitarity, that the Adler zeros in the case of chiral current conservation with massless pions imply that such a line of zeros exists. Hence the chiral charge can be defined in the presence of spontaneous symmetry breaking. The technique that we have followed is in agreement with the conjecture of Carlitz et al.[8].

REFERENCES

1. E. GAL-EZER and L.P. HORWITZ, Letters in Mathematical Physics 1, 147 (1976). Further details are given in E. GAL-EZER and L.P. HORWITZ, Physical Review, to be published.
2. E. GAL-EZER and L.P. HORWITZ, Letters in Mathematical Physics, to be published.
3. L. O'RAIFEARTAIGH, Phys. Rev. Lett. 14, 575 (1965).
4. L. O'RAIFEARTAIGH, Phys. Rev. 164, 2000 (1967).
5. M. FLATO and D. STERNHEIMER, Phys. Rev. Lett. 16, 1185 (1966); M. FLATO and D. STERNHEIMER, Comm. Math. Phys. 12, 296 (1969).
6. S.-J. CHANG, R. DASHEN, and L. O'RAIFEARTAIGH, Phys. Rev. 182, 1805 (1969).
7. S.P. AUERBACH, C. ROSENZWEIG, and M.R. PENNINGTON, Ann. Phys. 85, 214 (1974).
8. R. CARLITZ, D. HECKATHORN, J. KAUR, and W.-K. TUNG, Phys. Rev. D11, 1234 (1975).

PARTICLES, MASS SPECTRA AND INTERFERENCE IN RELATIVISTIC QUANTUM THEORY*

L.P. Horwitz† and Y. Rabin

In 1941, Stueckelberg(1) discussed a covariant formalism of classical relativistic mechanics in which the motion of a particle is described by a curve in space-time, parametrized by a continuous variable τ. He admitted the possibility of non-Einsteinian motion, involving curves that are locally light-like and space-like, and hence the existence of pair production and annihilation(2). He extended this idea to a covariant quantum theory with wave functions defined in an $L_2(R^4)$; the full space-time manifold is necessary to represent a theory in which x^μ, p^μ are dynamical variables. In 1950 Feynman(3), using his path integral formalism, derived a Schrödinger type equation of the form (metric -, +, +, +)

$$i \frac{\partial \psi}{\partial \tau} = \left(\frac{p^\mu p_\mu}{2M} + V(x)\right)\psi \equiv K\psi \tag{1}$$

where, in the presence of electromagnetic interaction, p^μ is

* Supported in part by the U.S.-Israel Binational Science Foundation (BSF), Jerusalem, Israel.

† This work was prepared while the first author was visiting the Centers for Particle Theory and Statistical Mechanics and Thermodynamics, The University of Texas at Austin, Austin, Texas, 78712, and supported in part by the U.S. Energy Research and Development Administration, Contract No. E(40-1) 3992.

replaced by $\pi^\mu = p^\mu - eA^\mu$ (and V may be zero). This equation was also stated by Stueckelberg[1] in the latter form.

Horwitz and Piron[4] more recently derived the canonical equations for the classical theory from a Cartan type variational principle:

$$\frac{dq^\mu}{d\tau} = \frac{\partial K}{\partial p_\mu}, \qquad \frac{dp^\mu}{d\tau} = -\frac{\partial K}{\partial q_\mu}. \tag{2}$$

The parameter τ was regarded as the historical evolution parameter for the entire system. Only one such parameter is introduced for a many body system as well. In this respect, their interpretation of τ differs from that of Feynman[3]. For the case of a free particle, $K = p^\mu p_\mu/2M$, p^μ is a constant of the motion, and $d\vec{q}/dq^0 = \vec{p}/p^0$ (the correct equation for the velocity is also obtained in the presence of an external electromagnetic field). In the corresponding space-time quantum theory, they found that the operator

$$\vec{x}_r = \vec{x} - \tfrac{1}{2}\{\frac{\vec{p}}{p^0}, t\}$$

is exactly the direct sum over mass shells of Newton-Wigner position operators defined for each mass. The extension of this theory to describe systems with spin was carried out by Horwitz, Piron and Reuse[5], and the structure of the kinematical group, $SO(3,1) \wedge (\mathbb{R}^4 \oplus \mathbb{R}^4)$, of the theory was clarified. The imprimitivity system is defined by

$$\begin{aligned}
(U(a)\psi)_n(x) &= \psi_n(x-a) \\
(U(\omega)\psi)_n(x) &= e^{i\omega x}\psi_n(x) \\
(U(\Lambda)\psi)_n(x) &= L^{-1}(n)\Lambda L(\Lambda^{-1}n)\psi_{\Lambda^{-1}n}(\Lambda^{-1}x),
\end{aligned} \tag{3}$$

where the family of Hilbert spaces forming the representation of this group is labelled by the superselection rule n^μ, the direction of time, and $L(n)n^0 = n$, $n^0 = (0,0,0,1)$. One may then induce a representation of the spin, analogous to that of Wigner, using the little group which leaves n^μ invariant. Such a representation is consistent with the unitary invariant scalar product $(\psi_1,\psi_2) = \int d^4p\psi_1^*(p)\psi_2(p)$ and $(\psi_1,x^\mu\psi_2)$ is covariant. In the following, we turn to a study of particle properties, mass spectra, and interference phenomena.

The condition that $\psi \in L_2(\mathbb{R}^4)$ implies that wave functions over space-time go to zero as $|t| \to \infty$. This behavior is easy to understand if we interpret $|\psi_\tau(x)|^2$ as the probability density for finding an *event* at the point $(\vec{x},t)$, at a value τ of the historical evolution parameter of the system. At increased τ, the wave packet of a free system will move in space-time to a position covering a later t. Hence, such a theory is fundamentally a theory of *events*, and the evolution of sequences of events to form particles with familiar properties is a consequence of dynamical laws. A condition* that can be imposed on a classical one-body system which is implied by particle-like behavior is

$$\frac{d}{d\tau}(p^\mu p_\mu) = 0. \tag{4}$$

Using the canonical equations (2) and the form (1) for K, this is equivalent to

$$p^\mu \frac{\partial V}{\partial x_\mu} = \frac{\partial}{\partial x_\mu}(p^\mu V(x)) = 0. \tag{5}$$

One may therefore interpret (-)V as a mass source density, and $p^\mu V$ as a mass current which does not transfer mass to the "particle" if it is divergenceless. The potential V is invariant, and if we exclude dependence on n^μ, it must be of the form $V(x^2)$. Then, (5) corresponds to

---

* We shall be concerned primarily with asymptotic properties in this paragraph.

$$p^{\mu} x_{\mu} V'(x^2) = 0. \tag{6}$$

The time-like motion of a free particle can carry it on a straight line asymptotically far from the light cone, so that condition (6) can be satisfied if $V'(x^2) \to 0$, $x^2 \to -\infty$. There is, however, a solution for $V' \neq 0$. Since $\frac{d}{d\tau} x^2 = 2p^{\mu}x_{\mu}/M$, $x^2 = \text{const.} > 0$, with p^{μ} time-like, we may have $x^{\mu}p_{\mu} = 0$ along the trajectory. For $V' = \text{const.}$, the equations of motion have a simple solution:

$$x^{\mu}(\tau) = x^{\mu}(0)\cosh \tau \sqrt{-\frac{V'}{M}} + \frac{1}{\sqrt{-\frac{V'}{M}}} \dot{x}^{\mu}(0)\sinh(\tau \sqrt{-\frac{V'}{M}}). \tag{7}$$

For $x^2 = \text{const.}$, $x^2(0) = \frac{1}{MV'} p^2(0) = -M/V'$; if x^2 is to be positive, $V' < 0$.* This hyperbolic motion then corresponds to the particle falling towards the light cone on a mass source density surface that dips towards the light cone from the space-like region; its transverse velocity results in a hyperbolic path (analogous to the "falling" in Kepler motion).

For the quantum case, we impose the condition

$$\frac{d}{d\tau} (\psi_{\tau}, p^2\psi_{\tau}) = \int d^4p' d^4p'' \psi^*_{\tau}(p')\psi_{\tau}(p'') \langle p'|V|p''\rangle (m'^2 - m''^2) = 0, \tag{8}$$

as characteristic of particle-like solutions. For V diagonal in x, it is convenient to introduce the Klein-Gordon functions

$$\phi_{\tau}(x,m^2) = \int \frac{d^3p}{2p^0} \langle x|p\rangle \psi_{\tau}(p)\Big|_{p^0=\sqrt{\vec{p}^2+m^2}}. \tag{9}$$

The condition (8) can then be written as

$$\int d^4x \int dm^2 \int dm'^2 \phi^*_{\tau}(x,m^2)\overleftrightarrow{\partial}_{\mu}\phi_2(x,m'^2)\,\partial^{\mu}V(x) = 0, \tag{10}$$

reminiscent of the classical condition (5). It follows from (8)

---

* There is also a light-like solution $x^2(\tau) = p^2(\tau) = 0$.

that the mass stability condition is satisfied if $\psi_\tau(p)$ is sharply peaked around a single mass (integration by parts of (10) yields the same result), if the Klein-Gordon mass transition current is orthogonal to $\partial_\mu V$, or has vanishing overlap with it.

Consider now the two body problem. If $V = V(x_1 - x_2)$, the center of mass momentum $P^\mu_{cm} = p^\mu_1 + p^\mu_2$ is constant, and we may separate the Hilbert space into a tensor product of wave functions of the center of mass and of the relative motion. For $p = (m_1 p_2 - m_2 p_1)/(m_1 + m_2)$, $x = x_1 - x_2$, $m = m_1 + m_2$ and $\mu = m_1 m_2/(m_1 + m_2)$, the total evolution operator is

$$K = \frac{p^2_{cm}}{2M} + \frac{p^2}{2\mu} + V(x) = \frac{p^2_{cm}}{2M} + K_{rel}. \tag{11}$$

Let us suppose that, as in the case of the relativistic harmonic oscillator[6], K_{rel} has only a discrete spectrum $\mathcal{E}_n$. In a spectral decomposition over K and K_{rel}, one finds

$$p^2_{cm}(k, \mathcal{E}_n) = 2M(k - \mathcal{E}_n). \tag{12}$$

In contrast to the non-relativistic situation, where the total Hamiltonian can take on any value through a variation of the momentum of the center of mass, the value of K is invariant. If we consider the preparation of the two body state S, resulting, for example, from a collision $A + B \to C + S$, where A, B, and C are one-particle asymptotic states, then $K_A + K_B = K_C + K_S$. However, K_A, K_B, K_C are determined by the particle masses (on shell) to be $-\frac{m_A}{2}$, $-\frac{m_B}{2}$, $-\frac{m_C}{2}$. Hence K_S cannot be arbitrary, and a relation of the form (12) may be associated with mass spectra for composite systems.

We finally turn to the use of interference phenomena[7] to affirm the applicability of Stueckelberg type wave functions. The wave function for a free particle propagates according to

$$\psi_\tau(\vec{r},t) = \int cd^3r't' \langle \vec{r}t|e^{-iK\tau/\hbar}|\vec{r}'t'\rangle \psi_0(r',t') \tag{13}$$

where

$$\langle \vec{r}t|e^{-iK\tau/\hbar}|\vec{r}'t'\rangle = -i\left(\frac{M}{2\pi\hbar\tau}\right)^2 e^{i\frac{M}{2\pi\tau\hbar}[(\vec{r}-\vec{r}')^2-c^2(t-t')^2]} . \tag{14}$$

The wave functions (13) are coherent over t as well as $\vec{r}$ (we do not expect this to be true in the non-relativistic Schrödinger theory). Integrating over a yz plane at x = 0 across the path of a wave packet moving with average momentum p_x, one may open the plane to transmission for two short times and look on a yz detector plane at x = L. The signal detected, due to this "double slit in time", will be proportional to

$$\left| e^{-i\frac{Mc^2}{2\tau\hbar}(t-t_1)^2} + e^{-i\frac{Mc^2}{2\tau\hbar}(t-t_2)^2}\right|^2 = 2\left(1+\cos\frac{Mc^2}{\tau\hbar}\varepsilon(t-\bar{t})\right), \tag{15}$$

where $\bar{t}$ is the mean time of the two pulses and ε is the interval between them. The period of the detected signal is

$$T = \frac{2\pi\hbar L}{\varepsilon p_x c^2}. \tag{16}$$

Although ε and T in this idealized experiment are very small for particles under normal conditions, the use of solid state current controlled devices to form the beam might lead to values of ε and T of the order of microseconds.

REFERENCES

1. E.C.G. STUECKELBERG, Helv. Phys. Acta 14, 588 (1941), Helv. Phys. Acta 15, 23 (1942).

2. P. HAVAS, Acta Phys. Austr. 3, 342 (1949) resolved an apparent causal anomaly paradox of this formulation. For a more recent review, see P. HAVAS, Synthese 18, 75 (1968). We thank Prof. W. Schieve for bringing these references to our attention.

3. R.P. FEYNMAN, Phys. Rev. 80, 440 (1950), App. A (with 2M ≡ 1). See also C. MORETTE, Phys. Rev. 81, 848 (1950).

4. L.P. HORWITZ and C. PIRON, Helv. Phys. Acta 46, 316 (1973). There has been a significant literature on this subject which we cannot review here, for example, C. GARROD, Rev. Mod. Phys. 38, 483 (1966), J.H. COOKE, Phys. Rev. 166, 1293 (1968), M.B. MENSKY, Comm. Math. Phys. 47, 97 (1976); see also references in P.L. HUDDLESTON, M. LORENTE and P. ROMAN, Found. of Physics 5, 75 (1975), Y. RABIN, Tel Aviv University preprint TAUP 544-76 and L.P. HORWITZ and Y. RABIN, to be published.

5. L.P. HORWITZ, C. PIRON, and F. REUSE, Helv. Phys. Acta 48, 546 (1975).

6. T.J. KARR, Univ. Maryland Tech. Rep. #76-085 (1976) and references therein.

7. L.P. HORWITZ and Y. RABIN, Tel Aviv University preprint TAUP 506-75, to be published in Nuovo Cimento Letters.

SU(4) BREAKING AND THE NEW PARTICLES: SOME APPLICATIONS

Paul Sorba*

The ψ particles, discovered in e^+e^- production and annihilation, are now interpreted as compound states of a charmed quark and its charge conjugate.

The existence of a fourth quark, already conjectured many years ago by Bjorken and Glashow in analogy with leptons, is not unexpected since it has been advocated by Glashow, Iliopoulos and Maiani in connection with the current-current Lagrangian of weak interactions: they have shown the usefulness of introducing a new quark, with the same charge as the proton quark u. The introduction of this fourth quark enlarges the internal symmetry group SU(3) of hadrons into SU(4) and therefore completes the list of well-known hadrons with the presumably existing "charmed" mesons and baryons. Note that very recently SPEAR obtained good evidence for the particles D^0 and D^+, the pseudoscalar mesons supposedly constituted with $c\bar{u}$ and $c\bar{d}$ respectively (we will denote u,d,s,c the up, down, strange and charmed quark respectively). During the writing of this talk we just learned that physicists at Fermilab (Experiment E87) observed a peak at 2250 MeV/c^2 in the effective mass distribution of $\bar{\Lambda}\pi^-\pi^-\pi^+$ produced in the reac-

* On leave of absence from Centre de Physique Théorique, CNRS, Marseille (France).

tion γ+Be $\rightarrow \bar{\Lambda}$ + pions +..., and an indication of a state near 2500 MeV/c^2 decaying into $\pi^{\pm}$ + ($\bar{\Lambda}\pi^-\pi^-\pi^+$). These results can be interpreted following B.W. Lee, Quigg and Rosner as decays involving charmed 1/2 and 3/2 antibaryons $\bar{C}$ and $\bar{C}$* (constituted with one c and 2 u,d quarks).

Moreover, the high value of the mass of the ψ particles compared to the 1^- vector mesons (or of the mass of D^0(1865) or D^+ (1870) compared to the mass of the pseudoscalar mesons) is a clear hint that the SU(4) symmetry is strongly broken.

Here we will present, in the context of the breaking of SU(4) two different exercises, the corresponding preprints of which already appeared and will be published in Lettere al Nuovo Cim. The first one, entitled "The Weak Charges of the Charmed Particles" is a result of a collaboration with F. Buccella, A. Pugliese and A. Sciarrino, while the second one, entitled "Relativistic Bound States: A Mass Formula for Vector Mesons" has been done with J.L. Richard. The reader could find in these two papers all the necessary references that we will not list here, being limited in the number of pages.

A. THE WEAK CHARGES OF CHARMED PARTICLES

If the internal symmetry group SU(3) is enlarged into the group SU(4), then the classification group $SU(6)_W$ is enlarged into $SU(8)_W \supset SU(4) \times SU(2)_W$. More precisely, adding the orbital part $O(3)_L$ we could consider the classification group of hadrons $SU(8)_W \times O(3)_L$ or, as we proposed last year during the IVth Colloquium in Nijmegen the classification group $SU(8)_W \times SU(3)_{GOM}$ (GOM = generalized orbital momentum).

We propose to give an expression for the vector and axial vector charges of hadrons. The vector charges Q^i (i=1,...,15) and the axial vector ones Q_5^i are supposed connected, at $P_z \rightarrow \infty$, to the corresponding generators of the classification group

$$A\left(\frac{\mu_i}{2}\right) \text{ and } A\left(\frac{\mu_i \sigma_2}{2}\right)$$

by a unitary transformation V.

Here we assume that the operator V acts on a system of quarks as the product of a set of commuting unitary operators acting on a single quark. As we had the opportunity to explain during the IIIrd Colloquium in Marseilles in 1974, the Melosh transformation V can be seen as the Wigner rotation between the quark rest frame and the hadron rest frame. Following this approach the V operator can be easily written:

$$\begin{cases} Vq_i^{\uparrow}(\vec{k},x) = \cos\,\hat{\theta}_i(\vec{k},x)q_i^{\uparrow} + \dfrac{k_x+ik_y}{k_\perp}\sin\,\hat{\theta}_i(\vec{k},x)q_i^{\downarrow} \\ \\ Vq_i^{\downarrow}(\vec{k},x) = \cos\,\hat{\theta}_i(\vec{k},x)q_i^{\downarrow} - \dfrac{k_x-ik_y}{k_\perp}\sin\,\hat{\theta}_i(\vec{k},x)q_i^{\uparrow} \end{cases}.$$

It follows that the matrix elements of the charges concerning the lowest meson and baryon multiplets of SU(8) are proportional to the corresponding generators of the same algebra by an amount characteristic of each charge. This factor is $\cos(\hat{\theta}_i \mp \hat{\theta}_j)$ for the nondiagonal vector and axial vector charges respectively, $\cos(2\hat{\theta}_i)$ for the diagonal axial charges, while the diagonal vector charges (T_3, Y, C) are not renormalized, since V does not change the internal quantum numbers of the quarks.

As already checked by F. Buccella, F. Nicolo and C.A. Savoy, this ansatz has worked reasonably well in connecting the different effects of SU(3) breaking on semi-leptonic decays of the baryon octet; note in particular that a different normalization factor for the strangeness changing and strangeness conserving axial charges has practically the same effect as introducing two different angles for the Cabibbo rotation of the vector and axial vector weak currents.

Among the predictions which can be drawn from this study, let us note that for the strong decays of the charmed particles into a pion plus another charmed particle, one predicts the same coupling constants obtained with the strange quark in place of the charmed one:

$$\langle D^{*+} | Q_5^3 | D^+ \rangle = \langle \overline{K^{*0}} | Q_5^3 | \bar{K}^0 \rangle .$$

In a similar way all the matrix elements of the charge Q_5^3 between the 20, $1/2^+$ and 20, $3/2^+$ states of the 120 of SU(8) (baryonic states) are all given in terms of one parameter.

An extension of PCAC to kaons allows one to obtain from $\langle F^{*+} | Q_5^{6-i7} | D^+ \rangle = - \langle D^{*+} | Q_5^{6+i7} | F^+ \rangle$ the corresponding equality for the amplitudes for $F^{*+} \to D^+ + K^0$ and $D^{*+} \to F^+ + \overline{K^0}$. More generally all the emissions of kaons within states of the 63 and 120 are obtained in terms of only one parameter given by $K^* \to K + \pi$. To get a quantitative idea of the renormalization parameters involved it is appropriate to use the phenomenological information available about the V operator. In the formalism adopted, one is led to write:

$$\begin{cases} V q_i^{\uparrow} | \text{ground state} \rangle = \cos \theta_i q_i^{\uparrow} | \text{ground state} \rangle + \sin \theta_i q_i^{\downarrow} | L_z = +1 \rangle \\ V q_i^{\downarrow} | \text{ground state} \rangle = -\sin \theta_i q_i^{\uparrow} | L_z = -1 \rangle + \cos \theta_i q_i^{\downarrow} | \text{ground state} \rangle \end{cases}$$

where now the θ_i's are numbers and the state $| L_z = \pm 1 \rangle$ is the same for each quark.

The mixing angle of the ordinary quarks can be obtained from the strangeness conserving and changing beta decays:

$$\theta_u = \theta_d \simeq 20^\circ \qquad \theta_\lambda \simeq 28^\circ .$$

To determine θ_c, one may assume that also for charmed particles the matrix elements of the magnetic moment are proportional to those of the axial charges (which was a successful hypothesis for the baryon octet). Then we obtain the following general form for the magnetic moment operator:

$$M = \mu_0\left(\frac{2}{3}\cos 2\theta_u \vec{S}_u - \frac{1}{3}\cos 2\theta_d \vec{S}_d - \frac{1}{3}\cos 2\theta_\lambda \vec{S}_\lambda + \frac{2}{3}\cos 2\theta_c \vec{S}_c\right).$$

This last operation relates the angle θ_c to the decay rate $\Gamma(\psi \to \eta_c + \gamma)$. Despite the fact that there is not even yet definite evidence for the existence of η_c particle, the measured branching ratio for the chain

$$\psi \to \eta_c(2800) + \gamma$$

$$\hookrightarrow \gamma + \gamma$$

requires, with reasonable assumptions on the η_c decays, that

$$\theta_c \simeq 45°.$$

From this value one can compute all the weak charges with $\Delta c = 1$.

$$\langle K^- | Q^{13+i14} | D^0 \rangle = -1.12$$

$$\langle \eta | Q^{13+i14} | F^+ \rangle = .96$$

$$\langle \overline{K^{*0}} | Q_5^{13+i14} | D^+ \rangle = \langle \phi | Q_5^{13+i14} | F^+ \rangle = .29$$

$$\langle \eta^- | Q^{11+i12} | D^0 \rangle = \langle K^0 | Q^{11-i12} | F^+ \rangle = .91$$

$$-\sqrt{2}\,\langle \rho^0 | Q_5^{11+i12} | D^+ \rangle = \langle K^{*0} | Q_5^{11+i12} | F^+ \rangle = .42.$$

B. A RELATIVISTIC MASS FORMULA FOR VECTOR MESONS

In order to obtain a relativistic description of interacting particles, we use the approach of Bakamjian and Thomas, itself reconsidered by Coester. We will deal with wave functions defined in momentum space. As we all know from the work of Currie, Jordan and Sudarshan, a relativistic description of interacting particles in terms of coordinates in space-time has been unsuccessful.

So let us consider two noninteracting particles of mass m_1 and m_2 and spin s_1 and s_2, and let us define the variables:

$$p = p_1+p_2 \text{ with } p_1^2 = m_1^2, \quad p_2^2 = m_2^2$$

$$k = 1/2L(p)^{-1}(p_1-p_2)$$

L(p) being the relativistic boost mapping $p^0 = (\sqrt{p^2},\vec{0})$ into p and leaving invariant the vectors orthogonal to p^0 and p.

One can rewrite the generators of the Poincaré group as:

i) Translations:

$$\vec{P} = \vec{p}, \quad M_0 = \sqrt{\vec{p}^2+M_0^2}$$

with

$$M_0 = \sqrt{\vec{k}^2+m_1^2} + \sqrt{\vec{k}^2+m_2^2}.$$

ii) Rotations:

$$\vec{J} = \vec{x}\times\vec{p} + \vec{y}\times\vec{k} + \vec{S}$$

with

$$\vec{x} = i\frac{\partial}{\partial\vec{p}}, \quad \vec{y} = i\frac{\partial}{\partial\vec{k}}$$

and $\vec{S} = \vec{S}_1+\vec{S}_2$ where $\vec{S}_1$ and $\vec{S}_2$ are the spin representations for particles (1) and (2) respectively.

iii) Boosts:

$$K = 1/2[\vec{x}H_0+H_0\vec{x}] - (\vec{y}\times\vec{k}+\vec{S}) \times \vec{p}(M_0+H_0)^{-1}.$$

As one can see, if we replace in the above expressions the mass M_0 by a "mass" M depending only on scalars made with $\vec{y},\vec{k}$ and the spin variables, the commutation relations of the new generators are still the same. In doing so, we introduce some kind of an interaction between the particles compatible with Poincaré symmetries.

In particular, we can exhibit a harmonic-oscillator-like spectrum in order to describe mesons as quark-antiquark bound systems. With simplicity as our only guide, we shall define the mass operator M in the very naïve form:

$$M = (\vec{k}^2+\omega^2\vec{y}^2-E_0+m_1^2)^{1/2} + (\vec{k}^2+\omega^2\vec{y}^2-E_0+m_2^2)^{1/2}.$$

As we see ω plays the role of a coupling constant, and M becomes M_0 when the coupling constant of the "interaction" is put to be zero. The constant E_0 is set to be the energy of the ground state for the harmonic oscillator while m_1 and m_2 are to be connected to the quark masses. The eigenvalues of M are then:

$$M_u = (2n\omega+m_1^2)^{1/2} + (2n\omega+m_2^2)^{1/2} \quad n=0,1,2,\ldots$$

As we see, we did not introduce terms connected with spin-spin and spin-orbit couplings. This could be done in order to obtain a more elaborate and more complete mass formula. However, we will see that such a crude mass formula gives results in good agreement with experimental values for the vector mesons.

We shall assume that a quark and the corresponding antiquark have the same mass, and we will suppose $m_u = m_d$ as usual.

For isospin one mesons, this leads to a linear law in the $(\text{mass})^2$ which has been frequently observed. More precisely, we get:

$$M^2_{n,I=1} = 4(2n\omega+m_u^2).$$

From the mass values of the $J^{PC} = 1^{--}$ $\rho(770)$ and $\rho(1600)$ mesons, corresponding to n=0 and n=2 respectively, and of the

$J^{PC} = 1^{+-}$ $\phi(1020)$ and $\psi(3100)$ corresponding to n=0, we deduce:

$$\omega = 123\times10^3 \text{ MeV}^2;\ m_u = 385 \text{ MeV};\ m_\lambda = 510 \text{ MeV};\ m_c = 1550 \text{ MeV}.$$

So, we find for the 1^{+-}, n=1, B(1220) meson the mass of 1255 MeV, for the 1^{--}, n=0, K(892) meson the mass of 895 MeV and for the 1^{+-}, n=1, K(1320) the mass of 1338 MeV.

We note that in this approach the excited state $\psi'(3700)$ corresponds to n=4; for n=2, we obtain the corresponding mass of 3400 MeV.

We have in this way the masses of the charmed vector mesons $D^*(c\bar{u})$ and $F^*(c\bar{\lambda})$:

$$m_{D^*(n=0)} \simeq 1935 \text{ MeV} \qquad m_{D^*(n=1)} \simeq 2230 \text{ MeV}$$

$$m_{F^*(n=0)} \simeq 2060 \text{ MeV} \qquad m_{F^*(n=1)} \simeq 2345 \text{ MeV}.$$

Thus this simple mass formula for vector mesons predicts masses in fairly good agreement with the experimental values. Let us emphasize the elementary nature of the expression we have chosen for the mass operator. One can hope to obtain on one hand a more general formula for mesons by introducing some kind of spin-spin or spin-orbit couplings, and on the other hand, a generalization of such a mass formula for baryons. We can remark we find again here the advantages of the Gell-Mann Okubo formula and its generalization to SU(4): SU(4) and SU(3) breakings appear here by the quark mass difference: $m_u = m_d < m_\lambda << m_c$.

ON A GROUP THEORETIC TREATMENT OF THE NUMBER OPERATOR OF PARTICLES

Aleksander Strasburger and Igor Szczyrba

We shall begin by recalling some background information concerning symmetry types of tensors. This is a standard and well known subject in the finite dimensional case, however the case of infinite dimensional ground space is perhaps less known, although it is in many respects similar to the former (cf. [1], [2]).

For a Hilbert space H (complex, separable) we denote by $\overset{n}{\otimes} H$, or simply H_n, the Hilbert space of the n-fold tensor product of H with itself (suitably completed if H is infinite dimensional). The canonical action of the unitary group Aut(H) of the space H in $\overset{n}{\otimes} H$ is defined by the following formula

$$U_v(x_1 \otimes \ldots \otimes x_n) := (vx_1) \otimes \ldots \otimes (vx_n), \qquad v \in \mathrm{Aut}(H),$$

and $v \to U_v$ is a unitary representation of Aut H in $\overset{n}{\otimes} H$. The infinitesimal generator (differential) of a one parameter unitary group $t \to U_{\exp(itA)}$, where A is a hermitian (bounded) linear operator on H, will be denoted by $N(A)$, i.e.

$$N(A) := \frac{1}{i} \frac{d}{dt} U_{\exp(itA)} \Big|_{t=0}.$$

The norm continuity of the one parameter group $t \to U_{\exp(itA)}$ for bounded hermitian A implies that $N(A)$ is a bounded operator on H_n. We extend by linearity N to a mapping $N: B(H) \to B(H_n)$, where $B(H)$ ($B(H_n)$ resp.) stands for the space of bounded linear operators on H (H_n resp.). It is easily seen that this mapping satisfies

$$N(vAv^{-1}) = U_v N(A) U_v^{-1}, \quad \text{for} \quad v \in \mathrm{Aut}(H).$$

The symmetric group S_n (i.e. the permutation group of n letters) acts on H_n by $S_n \ni \sigma \to \pi(\sigma) \in \mathrm{Aut}(H_n): \pi(\sigma)(x_1 \otimes \ldots \otimes x_n) := x_{\sigma(1)} \otimes \ldots \otimes x_{\sigma(n)}$. This action commutes with the representation of the group Aut(H), hence also with $N(A)$ for an arbitrary $A \in B(H)$. Denoting by $A(S_n)$ the algebra of bounded operators on H_n generated by the set of all $\pi(\sigma)$, $\sigma \in S_n$, we have in the case of $\dim H = \infty$ the following result of I. Segal [1]:

$$A(S_n) = \{U_v : v \in \mathrm{Aut}(H)\}',$$

where ' means the commuting algebra.

It follows that if $P \in A(S_n)$ is a projection with the range M then by restricting all the U_v to M we obtain an irreducible representation of Aut(H) if and only if P is a minimal projection in $A(S_n)$. Elements of the range of some minimal projection in $A(S_n)$ are called tensors of maximal symmetry type.

The same procedure as in the finite dimensional case can be used to assign a minimal projection in $A(S_n)$ to every Young diagram and every minimal projection may be obtained this way. We note in passing that, contrary to the finite dimensional case, not all the irreducible representations of the group Aut(H) arise in this fashion, cf. [7].

Two maximal symmetry types of tensors most frequently encountered in physical applications are, of course, (totally) symmetrical or antisymmetrical tensors describing bosons and fermions respectively. Their Young diagrams consist of one row or one column respectively.

Proposition 1. Let H be infinite dimensional and $K \subset H_n$ be the range of a minimal projection in $A(S_n)$ determined by a rectangular Young diagram (i.e. all columns of the same length) and let (abusing notation slightly) $U:\mathrm{Aut}(H) \to \mathrm{Aut}(K)$ denote the restriction of U to K. Also let $N:B(H) \to B(K)$ denote the differential of this representation.

Then N is the only mapping from B(H) to B(K) satisfying the following conditions:

i) positivity preserving, i.e. for a positive $A \in B(H)$ $N(A)$ is also positive,

ii) normalization, i.e. $N(I) = nI$, where I is the identity on H or K resp.,

iii) intertwining property, i.e. for every $v \in \mathrm{Aut}(H)$ and every $A \in B(H)$

$$N(vAv^{-1}) = U_v N(A) U_v^{-1},$$

iv) continuity with respect to the ultra-weak topology on B(H) and the weak topology on B(K).

We shall indicate here only the main lines of the proof referring the interested reader to [3] for the missing details.

The result depends heavily on the detailed study of the property iii) in the finite dimensional case given in [4] and [5]. Let $\dim H = p < \infty$ and let $v \to U_v$ be an irreducible unitary representation of Aut(H) in a finite dimensional space K. Let also the signature α of U have $S(\alpha)$ distinct components (we recall that the signature of a representation of the unitary group in p dimensions is a sequence $(m_{\hat{1}},\ldots,m_{\hat{p}})$ of p integers - components, with $m_{\hat{1}} \geq m_{\hat{2}} \geq \ldots \geq m_{\hat{p}}$, and that it determines the representation uniquely. Prescribing the signature of a representation is the same as prescribing the Young diagram corresponding to it. For all this and other relevant information see [6]).

Then the space of all intertwining mappings $T:B(H) \to B(K)$, i.e. linear mappings satisfying

$$T(vAv^{-1}) = U_v T(A) U_v^{-1} \quad \text{for all} \quad v \in \mathrm{Aut}(H)$$

has the dimension $S(\alpha)$.

Two linearly independent elements of this space can be easily pointed out, they are resp. the differential N of the representation U and the mapping T defined by $T(A) = \mathrm{tr}(A)I$, where tr is the trace and I is the identity on K. They span the space of all intertwining mappings when $S(\alpha) = 2$ which is the case precisely when the Young diagram determining U is rectangular.

Now, the actual proof of the proposition 1 proceeds by reduction to the case of finite dimension and then applying the preceding results. This reduction rests on the observation that when $A \in B(H)$ is reduced by a subspace $F \subset H$ (that is to say, both F and its orthogonal complement are invariant under A) then the condition iii) forces T(A) to be reduced by a suitably chosen subspace of K. It then follows that, roughly speaking, every mapping $T: B(H) \to B(K)$ satisfying the property iii) of the proposition 1, when restricted to a certain dense subspace of B(H) is a linear combination of the differential N and the trace mapping T. We then use the remaining assumptions on T to get rid of the summand involving T and fix the numerical factor of N.

Now, physics enters here (hopefully) via the following definition.

Definition. Let H and K be two Hilbert spaces (called one- resp. multi-particle space) and denote by S the unit sphere in H, i.e. $S = \{\varphi \in H : \|\varphi\| = 1\}$.

A mapping $S \ni \varphi \to N(\varphi) \in B(K)$ is called a number operator of particles if

I) for every $\varphi \in S$ $N(\varphi)$ is hermitian positive,

II) for a certain constant $N > 0$ and for every orthonormal basis $\{\varphi_i\}$ of H

$$\sum_i (\Phi | N(\varphi_i)\Phi) = N(\Phi|\Phi) \quad \text{for all} \quad \Phi \in K,$$

III) for every $v \in \mathrm{Aut}(H)$ the transformation $N(\varphi) \to N(v\varphi)$ is unitarily implemented, that is there exists an operator $U_v \in \mathrm{Aut}(K)$ such that

$$N(v\varphi) = U_v N(\varphi) U_v^{-1},$$

IV) the set $\{N(\varphi):\varphi \in S\}$ is irreducible, i.e. only scalars commute with all the $N(\varphi)$.

The reader might have observed that there is a strong resemblance between the conditions in the definition above and those in the proposition 1. In fact the two objects are identical as the following shows.

Proposition 2. Suppose $N: B(H) \to B(K)$ is a mapping satisfying, in addition to the assumptions i)-iv) of the proposition 1 the condition

v) $N(B(H))$ is an irreducible set of operators.

For $\varphi \in S$ let us denote by $P_{\hat{\varphi}}$ the projection in H with the range $\mathbb{C}\varphi$.

Then the mapping $S \ni \varphi \to N(P_{\hat{\varphi}}) \in B(K)$ is a number operator of particles in the sense of the definition above.

Conversely, every number operator extends to a linear mapping $N: B(H) \to B(K)$ satisfying properties i)-v). The only adjustment which one needs to make is to allow for a projective representation $v \to U_v$ in iii), since the condition III) of the preceding definition fails in general to define a proper representation of the group Aut(H).

For the proof see [5].

Such an approach to the number operator was proposed in [5].

It aims at introducing the number operator without using the creation and annihilation operators and also at placing a stronger emphasis at the part played by the unitary group of the one particle space rather than that of the symmetric group. The latter point is well illustrated by the crucial role of the condition iii) in the proof of the proposition 1.

Concerning the physical significance of our results it is perhaps worth remarking that the classical cases of the Fermi and Bose particles are clearly covered by the assumptions of the proposition 1 and hence they possess the unique number operator. The results of [5] in the finite dimensional case and also some further work on the other types of symmetries in the infinite dimensional case indicate that this may not be the case in general, hence substantiating the basic difference between the classical statistics on one side and the so-called parastatistics on the other.

REFERENCES

1. I.E. SEGAL, Trans. Amer. Math. Soc. 81 (1956), 106-134.
2. I.E. SEGAL, Proc. Amer. Math. Soc. 8 (1957), 197-203.
3. A. STRASBURGER and I. SZCZYRBA, Bull. Acad. Pol. Sci. (forthcoming).
4. I. SZCZYRBA, Rep. Math. Phys. 6 (1974), 199-212.
5. I. SZCZYRBA, Rep. Math. Phys. 7 (1975), 251-274.
6. H. WEYL, The classical groups, Princeton, 1946.
7. A.A. KIRILLOV, Dokl. Akad. Nauk SSSR 212 (1973), 288-290.

POINCARÉ GROUP AND MAGNETIC CHARGE

D. Villarroel

Our aim is to give a canonical Hamiltonian formulation for a system of classical electric and magnetic charges without the introduction of potentials. Let us begin by splitting the electric and magnetic fields into their longitudinal and transverse parts, namely

$$\underset{\sim}{E} = \underset{\sim}{E}^{\perp} + \underset{\sim}{E}^{\parallel}; \qquad \nabla\cdot\underset{\sim}{E}^{\perp} = 0; \qquad \nabla\times\underset{\sim}{E}^{\parallel} = 0. \tag{1}$$

In the present case we find

$$E_r^{\parallel} = -\frac{\partial}{\partial x_r}\sum_{j=1}^{N}\frac{e_j}{|\underset{\sim}{x}-\underset{\sim}{q}^j|}; \qquad B_r^{\parallel} = -\frac{\partial}{\partial x_r}\sum_{j=1}^{\hat{N}}\frac{\hat{e}_j}{|\underset{\sim}{x}-\underset{\sim}{\hat{q}}^j|}. \tag{2}$$

(Notation: the electric and magnetic charges are denoted by e_j and $\hat{e}_j$ respectively; their masses are m_j and $\hat{m}_j$. The position of the electric charge e_j is $\underset{\sim}{q}^j$, similarly $\underset{\sim}{\hat{q}}^j$ is the position of the magnetic charge $\hat{e}_j$. We use c=1).

The Maxwell equations are then

$$\begin{aligned} \nabla\times\underset{\sim}{B}^{\perp} - \frac{\partial \underset{\sim}{E}^{\perp}}{\partial t} &= 4\pi\underset{\sim}{j}^{\perp} \\ \nabla\times\underset{\sim}{E}^{\perp} + \frac{\partial \underset{\sim}{B}^{\perp}}{\partial t} &= -4\pi\underset{\sim}{\hat{j}}^{\perp} \end{aligned} \tag{3}$$

where

$$j_r^{\perp} = j_r + \partial_r \partial_s \sum_{i=1}^{N} \frac{e_i V_s^i}{4\pi |\underset{\sim}{x} - \underset{\sim}{q}^i|} ,$$

$$\hat{j}_r^{\perp} = \hat{j}_r + \partial_r \partial_s \sum_{i=1}^{\hat{N}} \frac{\hat{e}_i \hat{V}_s^i}{4\pi |\underset{\sim}{x} - \underset{\sim}{\hat{q}}^i|} ; \qquad (4)$$

$\underset{\sim}{j}$ and $\underset{\sim}{\hat{j}}$ are the electric and magnetic currents of the system.

Let us introduce the Fourier components of the transverse electric and magnetic fields. For this purpose we associate, as usual, with each propagation vector $\underset{\sim}{k}$ two polarization vectors $\underset{\sim}{\varepsilon}^{(1)}(\underset{\sim}{k})$ and $\underset{\sim}{\varepsilon}^{(2)}(\underset{\sim}{k})$ in such a way that

$$\underset{\sim}{\varepsilon}^{(\alpha)}(\underset{\sim}{k})\cdot\underset{\sim}{\varepsilon}^{(\beta)}(\underset{\sim}{k}) = \delta_{\alpha\beta}, \quad \underset{\sim}{\varepsilon}^{(\alpha)}(\underset{\sim}{k})\cdot\underset{\sim}{k} = 0 \qquad \alpha,\beta = 1,2. \qquad (5)$$

We denote the polarization indices by α,β. The vectors $\underset{\sim}{k}$, $\underset{\sim}{\varepsilon}^{(1)}$, and $\underset{\sim}{\varepsilon}^{(2)}$ form a three-dimensional orthogonal right hand system. We also adopt the following convention: $\underset{\sim}{\varepsilon}^{(1)}(-\underset{\sim}{k}) = \underset{\sim}{\varepsilon}^{(2)}(\underset{\sim}{k})$; $\underset{\sim}{\varepsilon}^{(2)}(-\underset{\sim}{k}) = \underset{\sim}{\varepsilon}^{(1)}(\underset{\sim}{k})$.

With the help of these polarization vectors we can write

$$\underset{\sim}{E}^{\perp}(\underset{\sim}{x}) = \int d\underset{\sim}{k}\; e^{i\underset{\sim}{k}\cdot\underset{\sim}{x}}\underset{\sim}{\varepsilon}^{(\alpha)}(\underset{\sim}{k})E_{\alpha}(\underset{\sim}{k}), \qquad \alpha = 1,2. \qquad (6)$$

Here and in what follows repeated polarization indices are to be summed over. We use the abbreviations $E_{\alpha} \equiv E_{\alpha}(\underset{\sim}{k})$, $\varepsilon^{(\alpha)} \equiv \varepsilon^{(\alpha)}(\underset{\sim}{k})$. In a similar manner the quantities B_{α} are associated with the magnetic transverse field $\underset{\sim}{B}^{\perp}$, and j_{α}, $\hat{j}_{\alpha}$; $\alpha = 1,2$ with $\underset{\sim}{j}$ and $\underset{\sim}{\hat{j}}$. Then it is easy to see that the Maxwell equations (3) are equivalent to the system

$$\dot{E}_1 + ikB_2 = -4\pi j_1,$$

$$\dot{E}_2 - ikB_1 = -4\pi j_2, \qquad (7)$$

$$\dot{B}_1 - ikE_2 = -4\pi\hat{j}_1,$$
$$\dot{B}_2 + ikE_1 = -4\pi\hat{j}_2. \tag{7}$$

Here and in the following we put $k \equiv |\underset{\sim}{k}| = (k_r k_r)^{\frac{1}{2}}$. A dot over a symbol means time derivative.

However, the quantities E_α, B_α, $\alpha = 1,2$, are not adequate for a canonical formalism; instead we choose as the basic dynamical variables for the description of the Maxwell equations the set $a^{\pm}_\alpha$ defined by

$$a_1^-(\underset{\sim}{k}) = \Pi k^{-\frac{1}{2}}[E_1(\underset{\sim}{k})+B_2(\underset{\sim}{k})],$$
$$a_2^-(\underset{\sim}{k}) = \Pi k^{-\frac{1}{2}}[E_2(\underset{\sim}{k})-B_1(\underset{\sim}{k})],$$
$$a_1^+(\underset{\sim}{k}) = \Pi k^{-\frac{1}{2}}[E_2(-\underset{\sim}{k})+B_1(-\underset{\sim}{k})], \tag{8}$$
$$a_2^+(\underset{\sim}{k}) = \Pi k^{-\frac{1}{2}}[E_1(-\underset{\sim}{k})-B_2(-\underset{\sim}{k})],$$

In terms of $a^{\pm}_\alpha$, the equations (7) are equivalent to:

$$\dot{a}_1^- + ika_1^- = -\frac{1}{2\pi}k^{-\frac{1}{2}}\sum_j e_j\underset{\sim}{v}^j\cdot\underset{\sim}{\varepsilon}^{(1)}(\underset{\sim}{k})e^{-i\underset{\sim}{k}\cdot\underset{\sim}{q}^j}$$
$$-\frac{1}{2\pi}k^{-\frac{1}{2}}\sum_j \hat{e}_j\hat{\underset{\sim}{v}}^j\cdot\underset{\sim}{\varepsilon}^{(2)}(\underset{\sim}{k})e^{-i\underset{\sim}{k}\cdot\hat{\underset{\sim}{q}}^j}$$
$$\dot{a}_2^- + ika_2^- = -\frac{1}{2\pi}k^{-\frac{1}{2}}\sum_j e_j\underset{\sim}{v}^j\cdot\underset{\sim}{\varepsilon}^{(2)}(\underset{\sim}{k})e^{-i\underset{\sim}{k}\cdot\underset{\sim}{q}^j}$$
$$+\frac{1}{2\pi}k^{-\frac{1}{2}}\sum_j \hat{e}_j\hat{\underset{\sim}{v}}^j\cdot\underset{\sim}{\varepsilon}^{(1)}(\underset{\sim}{k})e^{-i\underset{\sim}{k}\cdot\hat{\underset{\sim}{q}}^j} \tag{9}$$
$$\dot{a}_1^+ - ika_1^+ = -\frac{1}{2\pi}k^{-\frac{1}{2}}\sum_j e_j\underset{\sim}{v}^j\cdot\underset{\sim}{\varepsilon}^{(1)}(\underset{\sim}{k})e^{i\underset{\sim}{k}\cdot\underset{\sim}{q}^j}$$
$$-\frac{1}{2\pi}k^{-\frac{1}{2}}\sum_j \hat{e}_j\hat{\underset{\sim}{v}}^j\cdot\underset{\sim}{\varepsilon}^{(2)}(\underset{\sim}{k})e^{i\underset{\sim}{k}\cdot\hat{\underset{\sim}{q}}^j}$$

$$\dot{a}_2^+ - ika_2^+ = -\frac{1}{2\pi} k^{-\frac{1}{2}} \sum_j e_j \underset{\sim}{v}^j \cdot \underset{\sim}{\varepsilon}^{(2)}(\underset{\sim}{k}) e^{i\underset{\sim}{k}\cdot\underset{\sim}{q}^j} + \frac{1}{2\pi} k^{-\frac{1}{2}} \sum_j \hat{e}_j \hat{\underset{\sim}{v}}^j \cdot \underset{\sim}{\varepsilon}^{(1)}(\underset{\sim}{k}) e^{i\underset{\sim}{k}\cdot\hat{\underset{\sim}{q}}^j}. \tag{9}$$

Where we have used the explicit form of the currents.

We introduced the quantities $a_\alpha^\pm$ for the description of the electromagnetic field from our experience with quantum electrodynamics. The particle-coordinates $\underset{\sim}{q}^i$ and $\hat{\underset{\sim}{q}}^i$ are of course also fundamental dynamical variables of our canonical formalism. We need in addition some kind of momenta $\underset{\sim}{p}^i$ and $\hat{\underset{\sim}{p}}^i$ associated with the electric and magnetic charges, which of course cannot be the mechanical ones. For the moment, we do not need to be more explicit about $\underset{\sim}{p}^i$ and $\hat{\underset{\sim}{p}}^i$. As we are looking for a canonical formulation, we must impose the following Lie-bracket realization:

$$F,G = \sum_{j=1}^{N} \left(\frac{\partial F}{\partial q_r^j}\frac{\partial G}{\partial p_r^j} - \frac{\partial F}{\partial p_r^j}\frac{\partial G}{\partial q_r^j}\right) + \sum_{j=1}^{\hat{N}} \left(\frac{\partial F}{\partial \hat{q}_r^j}\frac{\partial G}{\partial \hat{p}_r^j} - \frac{\partial F}{\partial \hat{p}_r^j}\frac{\partial G}{\partial \hat{q}_r^j}\right) + i\int d\underset{\sim}{k}\left(\frac{\delta F}{\delta a^+(\underset{\sim}{k})}\frac{\delta G}{\delta a^-(\underset{\sim}{k})} - \frac{\delta F}{\delta a^-(\underset{\sim}{k})}\frac{\delta G}{\delta a^+(\underset{\sim}{k})}\right). \tag{10}$$

Let us introduce the following auxiliary momenta:

$$p_r^{\parallel i} \equiv (4\pi)^{-1}\varepsilon_{rsn}\int d\underset{\sim}{x} E_s^{\parallel i} B_n^{\perp}, \quad \hat{p}_r^{\parallel i} \equiv (4\pi)^{-1}\varepsilon_{rsn}\int d\underset{\sim}{x} E_s^{\perp} B_n^{\parallel i}.$$

Here $E_s^{\parallel i}$ and $B_n^{\parallel i}$ are the contributions of the i-th term in Eq. (2). We can easily write $p_r^{\parallel i}$ and $\hat{p}_r^{\parallel i}$ in terms of $a_\alpha^\pm$ and $\underset{\sim}{q}^i$, $\hat{\underset{\sim}{q}}^i$ to obtain

$$p_r^{\parallel j} = -\frac{ie_j}{\pi}\int d\underset{\sim}{k}\, k^{-\frac{1}{2}}\varepsilon_r^{(\alpha)}[a_\alpha^-(\underset{\sim}{k})e^{i\underset{\sim}{k}\cdot\underset{\sim}{q}^j} - a^+(\underset{\sim}{k})e^{-i\underset{\sim}{k}\cdot\underset{\sim}{q}^j}],$$
$$\hat{p}_r^{\parallel j} = \frac{i\hat{e}_j}{2\pi}\varepsilon^{\alpha\beta}\int d\underset{\sim}{k}\, k^{-\frac{1}{2}}\varepsilon_r^{(\alpha)}[a_\beta^-(\underset{\sim}{k})e^{i\underset{\sim}{k}\cdot\hat{\underset{\sim}{q}}^j} - a_\beta^+(\underset{\sim}{k})e^{-i\underset{\sim}{k}\cdot\hat{\underset{\sim}{q}}^j}]. \tag{12}$$

where $\varepsilon^{\alpha\beta}$ is the antisymmetric two-dimensional tensor with $\varepsilon^{12} = 1$.

Let us now introduce the following tentative Hamiltonian for the system of electric and magnetic charges:

$$H' = \sum_{j=1}^{N} \{(\underset{\sim}{p}^j - \underset{\sim}{p}^{\parallel j})^2 + m_j^2\}^{\frac{1}{2}} + \sum_{j=1}^{\hat{N}} \{(\hat{\underset{\sim}{p}}^j - \hat{\underset{\sim}{p}}^{\parallel j})^2 + \hat{m}_j^2\}^{\frac{1}{2}}$$
$$+ \int d\underset{\sim}{k}\; k a_\alpha^+ a_\alpha^- + \frac{1}{2} \sum_{i \neq j} e_i e_j |\underset{\sim}{q}^i - \underset{\sim}{q}^j|^{-1} + \frac{1}{2} \sum_{i \neq j} \hat{e}_i \hat{e}_j |\hat{\underset{\sim}{q}}^i - \hat{\underset{\sim}{q}}^j|^{-1}. \tag{13}$$

This Hamiltonian implies that the canonical and mechanical momenta are related by

$$\underset{\sim}{p}^i = \underset{\sim}{\Pi}^i + \underset{\sim}{p}^{\parallel i}.$$

It is easy to prove that this Hamiltonian together with the Lie-bracket (10) gives the field equations (9), but it fails to give the right equation of motion for the particles. In fact, we obtain:

$$\dot{\Pi}_n^i = e_i [E_n(\underset{\sim}{q}^i) + \varepsilon_{nsr} v_s^i B_r(\underset{\sim}{q}^i)] - e_i \varepsilon_{nsr} \sum_j^{\hat{N}} \hat{e}_j (v_s^i - \hat{v}_s^j) \frac{q_r^i - \hat{q}_r^j}{|\underset{\sim}{q}^i - \hat{\underset{\sim}{q}}^j|^3}$$
$$\dot{\hat{\Pi}}_n^i = \hat{e}_i [B_n(\hat{\underset{\sim}{q}}^i) - \varepsilon_{nsr} \hat{v}_s^i E_r(\hat{\underset{\sim}{q}}^i)] - \hat{e}_i \varepsilon_{nsr} \sum_{j=1}^{N} e_j (\hat{v}_s^i - v_s^j) \frac{\hat{q}_r^i - q_r^j}{|\hat{\underset{\sim}{q}}^i - \underset{\sim}{q}^j|^3}. \tag{14}$$

The failure to obtain the equations of motion for the particles is rather obvious from the structure of the Hamiltonian H'. Indeed, it does not contain the interaction of Coulombian type between electric and magnetic charges. Due to the complicated form of the additional terms in Eqs. (14) it follows that the static e-ê interaction cannot be so simple as the static e-e or ê-ê interaction. It is easy to see that the only change in H' that still gives the field equations (9) is:

$$\underset{\sim}{p}^{\parallel i} \to \underset{\sim}{p}^{\parallel i} + \underset{\sim}{F}(\underset{\sim}{q}, \hat{\underset{\sim}{q}}), \qquad \hat{\underset{\sim}{p}}^{\parallel i} \to \hat{\underset{\sim}{p}}^{\parallel i} + \hat{\underset{\sim}{F}}(\underset{\sim}{q}, \hat{\underset{\sim}{q}}). \tag{15}$$

With this change, the relation between the mechanical and canonical momentum is

$$\underset{\sim}{\pi}^i = \underset{\sim}{p}^i - \underset{\sim}{p}^{\|i} - \underset{\sim}{F}(q,\hat{q}). \tag{16}$$

The new Hamiltonian is

$$H = \sum_{j=1}^{N} \{(\underset{\sim}{p}^j - \underset{\sim}{p}^{\|j} - \underset{\sim}{F})^2 + m_j^2\}^{\frac{1}{2}} + \sum_{j=1}^{\hat{N}} \{(\hat{\underset{\sim}{p}}^j - \hat{\underset{\sim}{p}}^{\|j} - \hat{\underset{\sim}{F}})^2 + \hat{m}_j^2\}^{\frac{1}{2}}$$

$$+ \int d\underset{\sim}{k}\, k a_\alpha^+ a_\alpha^- + \frac{1}{2}\sum_{i \neq j} e_i e_j |q^i - q^j|^{-1} + \frac{1}{2}\sum_{i \neq j} \hat{e}_i \hat{e}_j |\hat{q}^i - \hat{q}^j|^{-1}. \tag{17}$$

This Hamiltonian gives the equations of motion for electric and
magnetic charges if and only if

$$\sum_{j=1}^{N} v_r^j \left(\frac{\partial F_r^j}{\partial q_n^i} - \frac{\partial F_n^i}{\partial q_r^j}\right) - \sum_{j=1}^{\hat{N}} \hat{v}_r^j \left(\frac{\partial F_n^i}{\partial \hat{q}_r^j} - \frac{\partial \hat{F}_r^j}{\partial q_n^i}\right) =$$

$$= e_i \epsilon_{nsr} \sum_{j=1}^{\hat{N}} \hat{e}_j (v_s^i - \hat{v}_s^j) \frac{q_r^i - \hat{q}_r^j}{|q^i - \hat{q}^j|^3}$$

$$\sum_{j=1}^{\hat{N}} \hat{v}_r^j \left(\frac{\partial \hat{F}_r^j}{\partial \hat{q}_n^i} - \frac{\partial \hat{F}_n^i}{\partial \hat{q}_r^j}\right) - \sum_{j=1}^{N} v_r^j \left(\frac{\partial \hat{F}_n^i}{\partial q_r^j} - \frac{\partial F_r^j}{\partial \hat{q}_n^i}\right) = \tag{18}$$

$$= -\hat{e}_i \epsilon_{nsr} \sum_{j=1}^{N} e_j (\hat{v}_s^i - v_s^j) \frac{\hat{q}_r^i - q_r^j}{|\hat{q}^i - q^j|^3}.$$

The right hand sides of these equations tell us that $\underset{\sim}{F}^i(q,\hat{q})$
and $\hat{F}^i(q,\hat{q})$ are necessarily of the form

$$F_r^i = \sum_{j=1}^{\hat{N}} f_r(q^i - \hat{q}^j) = f_r(q^i - \hat{q}^1) + f_r(q^i - \hat{q}^2) + \ldots$$

$$\hat{F}_r^i = -\sum_{j=1}^{N} f_r(\hat{q}^i - q^j) = -f_r(\hat{q}^i - q^1) - f_r(\hat{q}^i - q^2) - \ldots \tag{19}$$

The equation for $f_r(q^i - \hat{q}^j)$, implied by Eqs. (18) is

$$\varepsilon_{nsr} \frac{\partial f_r(p^{ij})}{\partial p_s^{ij}} = e_i \hat{e}_j \rho_n^{ij}/|\underset{\sim}{\rho}^{ij}|^3.$$

(20)

where $\underset{\sim}{\rho}^{ij} = q^i - \hat{q}^j$. This equation shows clearly all the pathologies inherent to a canonical formulation, and therefore to the quantum version, of electric and magnetic charges. In fact, there is no regular solution, but there are a lot of solutions that are valid "almost everywhere". The solution considered by Dirac[1]

$$\underset{\sim}{F}^i = -e_i \sum_j^{\hat{N}} \frac{\hat{e}_j}{|\underset{\sim}{\rho}^{ij}|} \frac{\hat{n} \times \underset{\sim}{\rho}^{ij}}{|\underset{\sim}{\rho}^{ij}| - \hat{n} \cdot \underset{\sim}{\rho}^{ij}} ;$$

$$\hat{\underset{\sim}{F}}^i = \hat{e}_i \sum_j^N \frac{e_j}{|\underset{\sim}{\rho}^{ji}|} \frac{\hat{n} \times \underset{\sim}{\rho}^{ji}}{|\underset{\sim}{\rho}^{ji}| - \hat{n} \cdot \underset{\sim}{\rho}^{ji}} ,$$

(21)

where $\underset{\sim}{\rho}^{ij} = q^i - \hat{q}^j$, and $\hat{\underset{\sim}{n}}$ an arbitrary unit vector, is the most simple one. Unfortunately, the physical predictions, like magnetic charge quantization, of different solutions do not seem to be completely equivalent. Calling "string" the domain where the solution of Eq. (20) fails, we see that the correct Lorentz force is acting on the particles out of the string.

If $f(L,x)$ is any particular solution of Eq. (20) (here we denote the string by L), then the functions $f(L,x)+\nabla g(x)$, are also solutions. In particular, we obtain the gauge of usual electrodynamics for $f=0$ and g an arbitrary good function. But in the present case, even g may depend on another string L'.

We now exhibit the other nine Poincaré group generators

$$P_t = \sum_{j=1}^N p_t^j + \sum_{j=1}^{\hat{N}} \hat{p}_t^j + \int d\underset{\sim}{k} k_t a_\alpha^+(k) a_\alpha^-(k),$$

$$J_t = \varepsilon_{tnm} \sum_{i=1}^N q_n^i (p_m^i - F_m^i) + \varepsilon_{tnm} \sum_{i=1}^{\hat{N}} \hat{q}_n^i (\hat{p}_m^i - \hat{F}_m^i) - \sum_{i=1}^N \sum_{j=1}^{\hat{N}} e_i \hat{e}_j \frac{q_t^i - \hat{q}_t^j}{|\underset{\sim}{q}^i - \hat{\underset{\sim}{q}}^j|}$$

$$- \frac{i}{2} \varepsilon_{trm} \int d\underset{\sim}{k} (\varepsilon_s^{(\alpha)} a_\alpha^+)(k_r \partial_m - k_m \partial_r)(\varepsilon_s^{(\beta)} a_\beta^-) + i\varepsilon^{\alpha\beta} \int d\underset{\sim}{k} k^{-1} k_t a_\alpha^- a_\beta^+ ,$$

$$K_t = \sum_{j=1}^{N} q_t^j \{(\underset{\sim}{p}^j - \underset{\sim}{p}^{\parallel j} - \underset{\sim}{F})^2 + m_j^2\}^{\frac{1}{2}} + \sum_{j=1}^{\hat{N}} \hat{q}_t^j \{(\hat{\underset{\sim}{p}}^j - \hat{\underset{\sim}{p}}^{\parallel j} - \hat{\underset{\sim}{F}})^2 + \hat{m}_j^2\}^{\frac{1}{2}}$$

$$+ \frac{1}{2\Pi} \sum_{j=1}^{N} e_j \int d\underset{\sim}{k} k^{-3/2} \varepsilon_t^{(\alpha)} [a_\alpha^- e^{i\underset{\sim}{k}\cdot\underset{\sim}{q}^j} + a_\alpha^+ e^{-i\underset{\sim}{k}\cdot\underset{\sim}{q}^j}]$$

$$- \frac{1}{2\Pi} \sum_{j=1}^{\hat{N}} \hat{e}_j \int d\underset{\sim}{k} k^{-3/2} \varepsilon^{\alpha\beta} \varepsilon_t^{(\alpha)} [a_\beta^- e^{i\underset{\sim}{k}\cdot\hat{\underset{\sim}{q}}^j} + a_\beta^+ e^{-i\underset{\sim}{k}\cdot\hat{\underset{\sim}{q}}^j}]$$

$$+ \frac{i}{2} \int d\underset{\sim}{k} k^{-1} k_t a_\alpha^- a_\alpha^+ + i \int d\underset{\sim}{k} k (\varepsilon_r^{(\alpha)} \partial_t \varepsilon_r^{(\beta)}) a_\alpha^+ a_\beta^- + i \int d\underset{\sim}{k} k a_\alpha^+ \frac{\partial a_\alpha^-}{\partial k_t}$$

$$+ \frac{1}{2} \sum_{i \neq j} e_i e_j q_t^i |\underset{\sim}{q}^i - \underset{\sim}{q}^j|^{-1} + \frac{1}{2} \sum_{i \neq j} \hat{e}_i \hat{e}_j \hat{q}_t^i |\hat{\underset{\sim}{q}}^i - \hat{\underset{\sim}{q}}^j|^{-1} .$$

Laborious but straightforward computations show that in fact these quantities satisfy the right algebra of the Poincaré group. The proof of this fact is, however, based on the equation (20), which as is well-known, is rather pathological. The troubles introduced by the string are even more complicated in an operator field theory[2].

ACKNOWLEDGMENTS

The author would like to thank Prof. A. Daigneault for his kind hospitality at the Département de Mathématiques, Université de Montréal.

REFERENCES

1. P.A.M. DIRAC, Proc. Roy. Soc. (London) A133, 60 (1931); Phys. Rev. 74, 817 (1948).
2. C.R. HAGEN, "Noncovariance of Schwinger Monopole", preprint, University of Rochester (1976).

Part V

Mathematical Physics

SOLVABILITY QUESTIONS FOR INVARIANT DIFFERENTIAL OPERATORS

Sigurdur Helgason

§1. INTRODUCTION

Consider an arbitrary differential operator

$$D = \sum_{(k)} a_{k_1,\dots,k_n}(x) \frac{\partial^{k_1}}{\partial x_1^{k_1}},\dots,\frac{\partial^{k_n}}{\partial x_n^{k_n}} \tag{1}$$

on $\mathbb{R}^n$, the

$$a_{k_1,\dots,k_n}$$

being C^∞ functions and the $(x_1,\dots,x_n)$ the ordinary Cartesian coordinates. Since each translation τ of $\mathbb{R}^n$ satisfies

$$\frac{\partial}{\partial x_i}(f(\tau(x)) = (\frac{\partial f}{\partial x_i})(\tau(x))$$

it is clear that the coefficients

$$a_{k_1,\dots,k_n}$$

of the operator D above are constants if and only if D is invariant under all translations τ, i.e.

$$D(f \circ \tau) = (Df) \circ \tau \qquad f \in C^\infty(\mathbb{R}^n), \tag{2}$$

$\circ$ denoting composition of mappings. While the property of having constant coefficients depends on the coordinate system used, property (2) is, of course, intrinsic.

A less trivial example is the wave operator

$$\Box = \frac{\partial^2}{\partial x_1^2} + \frac{\partial^2}{\partial x_2^2} + \frac{\partial^2}{\partial x_3^2} - \frac{\partial^2}{\partial x_4^2}.$$

This operator, or rather the polynomials $P(\Box)$ in it, can be characterized among all the operators (1) (for n=4) by their invariance under the Poincaré group.

In order to generalize this invariance notion consider a manifold M and G a separable Lie transformation group of M. We assume that G acts transitively on M. Then if $p \in M$ is a fixed point, H the subgroup of G leaving p fixed, it is a standard result in differential geometry that the mapping

$$gH \to g \cdot p$$

is a diffeomorphism of the coset space G/H onto M. (Here $g \cdot p$ denotes the image of p under the group element g.)

A differential operator D on the manifold G/H is called *invariant under* G if

$$D(\varphi \circ \tau(g)) = (D\varphi) \circ \tau(g) \qquad \varphi \in C^\infty(G/H) \tag{3}$$

for all $g \in G$, the mapping $\tau(g)$ denoting the translation $xH \to gxH$ of G/H onto itself. Let o denote the origin H in G/H and let $\mathbb{D}(G/H)$ denote the set of all differential operators on G/H invariant under G.

Of the many natural questions one can ask for the operators $D \in \mathbb{D}(G/H)$ (see [16], [17] for the connection with representation theory) we shall in this lecture focus attention on the following solvability questions:

A. *Does there exist a fundamental solution for* D, *i.e., a distribution* T *on* G/H *such that*

$$DT = \delta,$$

δ *denoting the delta-distribution at* o *on* G/H?

B. *Is* D *locally solvable, i.e. is*

$$DC^{\infty}(V) \supset C_c^{\infty}(V)$$

for some neighborhood V *of* o *in* G/H?

C. *Is* D *globally solvable, i.e., is*

$$DC^{\infty}(G/H) = C^{\infty}(G/H)?$$

In questions B and C one can of course replace the space of C^{∞} functions by other function spaces or distribution spaces.

If G is the group of translations of $\mathbb{R}^n$ and H = {0} then $\mathbb{D}(G/H)$ consists of all differential operators on $\mathbb{R}^n$ of constant coefficients; by well-known results of Ehrenpreis and Malgrange, questions A, B, and C all have positive answers for this case.

We shall now give a survey of some of the principal results which have been obtained for the questions above for general differential operators D.

Having quoted positive results, I will now quote a negative result which shows that the generality of G/H should be severely restricted. It was proved by H. Lewy [20] that the operator

$$\frac{\partial}{\partial x} + i\frac{\partial}{\partial y} + ix\frac{\partial}{\partial z}$$

in $\mathbb{R}^3$ is not locally solvable. It was noticed by several people (cf. [3]) that this operator is a left invariant differential operator on the Heisenberg group

$$\begin{pmatrix} 1 & x & z \\ 0 & 1 & y \\ 0 & 0 & 1 \end{pmatrix} \qquad x,y,z \in \mathbb{R}.$$

Thus question B has in general a negative solution and we shall accordingly restrict the generality of G/H.

§2. SOLVABILITY RESULTS FOR SYMMETRIC COSET SPACES. HUYGENS' PRINCIPLE

We now assume that G/H is a *symmetric coset space*, that is G has an involutive automorphism σ with fixed point set H.

Example. Let $G = \mathbf{SL}(n,\mathbb{R})$, $K = SO(n)$. Then the automorphism $\sigma: g \to ({}^tg)^{-1}$ of G (${}^tg$ denoting transpose) turns G/K into a symmetric coset space. Moreover, the mapping

$$gK \to g\,{}^tg \tag{4}$$

is a diffeomorphism of G/K onto the space $\mathbb{P}$ of positive definite $n\times n$ matrices of determinant one. If $h \in G$, the mapping $\tau(h)$ of G/K corresponds, via (4), to the mapping $Y \to hY\,{}^th$ of $\mathbb{P}$ onto itself. The corresponding G-invariant differential operators on $\mathbb{P}$ are known (cf. [21], [25]) to be generated by the operators

$$\mathrm{Trace}[(Y\frac{\partial}{\partial Y})^k] \qquad k = 2,3,\ldots,n,$$

where Y and $\partial/\partial Y$ denote the symmetric matrices $Y = (y_{rs})$,

$$\partial/\partial Y = \tfrac{1}{2}(1+\delta_{rs})\,\frac{\partial}{\partial y_{rs}}, \quad (1 \le r,s \le n).$$

Moreover $\mathrm{Tr}[(Y^{-1}dY)^2]$ is a G-invariant Riemannian metric on $\mathbb{P}$.

More generally, let G/K be a coset space where G is a connected noncompact semisimple Lie group with finite center and K a maximal compact subgroup. Such a space is called a *Riemannian symmetric space of the noncompact type.* By a result of É. Cartan such a space is indeed a symmetric coset space; "Riemannian" refers to the fact that as a result of the compactness of K such a space has a Riemannian metric invariant under the action of G. The Killing form of the Lie algebra of G provides such a metric.

Theorem 2.1. Each G-*invariant differential operator* D *on the symmetric space* $X = G/K$ *has a fundamental solution* T, *i.e.*

$$DT = \delta.$$

This was proved in [11] using Harish-Chandra's work [9] on zonal spherical functions on G. Since K is compact, Theorem 2.1 gives also a positive answer to question B; in fact if $f \in C_c^\infty(G/K)$ we get a solution u to the equation $Du = f$ by convolving f with the fundamental solution T.

If the group G is complex (and also if G is compact) the fundamental solution T can be constructed quite explicitly and from [11], p. 583 we have the following result.

Theorem 2.2. Let G *be a simple, odd-dimensional complex Lie group and let* Δ *denote the Laplace-Beltrami operator on the symmetric space* $X = G/K$. *For the constant* $c_0 = (\dim X)/12$ *the damped wave operator*

$$\Delta + c_0 - \frac{\partial^2}{\partial t^2} = 0$$

satisfies Huyghens' principle. For no constant $c \neq c_0$ *is this the case.*

The same proof is valid for the wave equation on a compact simple Lie group (the symmetric space "dual" to G/K). Noting that the metric is then given by the negative of the Killing form we have

Corollary. Let K *be a simple compact Lie group of odd dimension,* Δ_K *the Laplacian on* K *and* $c_o = (\dim K)/12$. *Then the damped wave operator*

$$\Delta_K - c_o - \frac{\partial^2}{\partial t^2}$$

satisfies Huyghens' principle.

Remark. At the colloquium P. Lax showed me a simple geometric proof of this for the case when K = SO(3). The cases where Theorem 2.2 and its corollary apply are K = SU(2n), SO(n), SP(n) and $\mathbb{E}_7$.

A modified version of Theorem 2.2 (and its corollary) holds for all odd-dimensional symmetric spaces X = G/K of the noncompact type. Let X_o denote the tangent space to X at the origin $o = \{K\}$, let Δ_o denote the Laplacian on X_o and Exp: $X_o \to X$ the usual exponential mapping which maps lines through the origin onto geodesics through o in X. Let ω^2 denote the Jacobian of Exp, i.e.

$$\int_X f(x)\,dx = \int_{X_o} f(\mathrm{Exp}\ Y)\omega^2(Y)\,dY \qquad f \in C_c^\infty(X)$$

for the volume elements dx and dY. Then by [13] p. 30,

$$(\Delta f)\circ\mathrm{Exp} + (df)\circ\mathrm{Exp} = \omega^{-1}\Delta_o((f\circ\mathrm{Exp})\omega),$$

for all K-invariant functions f on X, $\circ$ denoting composition of mappings and d the function defined by

$$d(\mathrm{Exp}\ Y) = \omega^{-1}(Y)\Delta_o(\omega)(Y). \qquad (5)$$

This gives the following new result generalizing Theorem 2.2.

Theorem 2.3. Let G/K *be an odd-dimensional Riemannian symmetric space of the noncompact type. Then the differential operator*

$$\Delta + d - \frac{\partial^2}{\partial t^2},$$

where the function d *is defined by* (5), *satisfies Huygens' principle.*

Remark. In analogy with the Corollary of Theorem 2.2, Theorem 2.3 has a counterpart for odd-dimensional symmetric spaces of the compact type.

Finally we consider Problem C for the symmetric space G/K.

Theorem 2.4. Let D *be a* G-*invariant differential operator on a symmetric space* X = G/K *of the noncompact type. Then*

$$D\, C^{\infty}(G/K) = C^{\infty}(G/K). \tag{6}$$

The proof (Helgason [15]) is based on a characterization of the image of $C_c^{\infty}(G/K)$ either under the Radon transform of under the Fourier transform on G/K introduced in [12]. This Paley-Wiener type theorem can be sharpened further, giving instead of (6) a global solvability in the space $\mathcal{D}'_0(X)$ of K-finite distributions on X (cf. [16])

$$D\mathcal{D}'_0(X) = \mathcal{D}'_0(X). \tag{7}$$

The global solvability holds also in the space $\mathcal{S}'(X)$ of tempered distributions on X (cf. [14])

$$D\mathcal{S}'(X) = \mathcal{S}'(X). \tag{8}$$

The proof is based on the characterization of the image of the space $\mathcal{S}(X)$ of rapidly decreasing functions on X under the Fourier transform (cf. [7], [8]).

§3. LIE GROUPS AS SYMMETRIC SPACES

Since left invariant differential operators on Lie groups are not in general locally solvable we consider now a Lie group L as a homogeneous space under left and right translations. In fact, L×L acts on L by the translation

$$(g_1,g_2): g \to g_1 g g_2^{-1} \qquad g, g_1, g_2 \in L$$

and the subgroup of L×L leaving the identity element e fixed is the diagonal L* in L×L so we have the coset space representation

$$L = (L\times L)/L^*,$$

which the involutive automorphism $\sigma: (g_1,g_2) \to (g_2,g_1)$ turns into a symmetric coset space. The algebra $\mathbb{D}((L\times L)/L^*)$ is canonically isomorphic to the algebra Z(L) of bi-invariant differential operators on L and for L the natural questions become: *Given* $Z \in Z(L)$ *is it locally (globally) solvable on* L?

For solvable Lie groups a positive answer was given by Duflo and Raïs [6].

Theorem 3.1. Each bi-invariant differential operator on a solvable Lie group G *is locally solvable.*

The proof uses harmonic analysis on L extending the method which Raïs had used earlier for the nilpotent case [22]. This latter proof uses the theory of characters on G (cf. [5], [19]) and the method used in [1], [2] to construct a tempered fundamental solution T for constant-coefficient operators

$$D = P\left(\frac{\partial}{\partial x_1},\ldots,\frac{\partial}{\partial x_n}\right),$$

(originally done by Hörmander and Lojasiewicz). According to this method, T can be chosen as the value at s = -1 of the Fourier transform of the distribution P^s on $\mathbb{R}^n$.

On the other hand, assuming G simply connected, there is for each f in the dual $\mathcal{G}^*$ of the Lie algebra $\mathcal{G}$ of G associated a unitary irreducible representation of G; its character T_f is an eigendistribution of Z, $ZT_f = \hat{Z}(f)T$, and the eigenvalue $\hat{Z}(f)$ is a polynomial in f. A fundamental solution for Z is given by the value at s = -1 of the Fourier transform of the distribution $(\hat{Z})^s$ on $\mathcal{G}^*$. This gives Theorem 3.1 for G nilpotent. A simpler proof of Theorem 3.1 was given by Rouvière [24].

The analog of Theorem 3.1 for semisimple Lie groups was proved in the meantime ([15], p. 477).

Theorem 3.2. Each bi-invariant differential operator Z *on a semisimple Lie group* G *is locally solvable.*

In contrast to Theorem 3.1 this does not require any harmonic analysis on G but is a consequence of a structure theorem of Harish-Chandra [10] which shows that on functions on G, which are locally invariant under inner automorphisms of G, the operator Z is reducible to a constant coefficient differential operator on the Lie algebra of G.

While the operator d/dθ on the circle group shows that the global version of Theorem 3.1 does not hold, Theorems 2.4 and 3.2 raise the question of global solvability of the bi-invariant operator Z on the noncompact semisimple Lie group G. It was pointed out by Cérèzo and Rouvière [4] that when SL(2,C) is considered as a *real* semisimple Lie group G, the imaginary part C_I of the complex Casimir operator on SL(2,C) is a bi-invariant differential operator on G but it annihilates all functions on SL(2,C) which are invariant under right translations by elements of the subgroup SU(2). Hence the image of $C^\infty(G)$ under C_I contains no such functions so C_I is not globally solvable. Thus the global version of Theorem 3.2 does not in general hold for a noncompact semisimple Lie group G.

Nevertheless, J. Rauch and D. Wigner proved [23] the

following positive result which for complex G had been proved by Cérèzo and Rouvière [4].

Theorem 3.3. Let G *be a noncompact semisimple Lie group with finite center. Then the Casimir operator* C *on* G *is globally solvable.*

A central step in the proof is showing that no null bicharacteristics of C lie over a compact subset of G. This, together with applications of the Holmgren uniqueness theorem, reduce the proof to a general result of Hörmander [18].

Since an arbitrary connected Lie group L is a semidirect product of a semisimple one and a solvable one, Theorems 3.1 and 3.2 raise the question of local solvability for L. More generally, in view of Theorem 2.4 we can ask:

Is an invariant differential operator on a symmetric coset space always locally solvable?

REFERENCES

1. M. ATIYAH, Resolution of singularities and division of distributions, Comm. Pure Appl. Math. 23 (1970), 145-150.

2. I.N. BERNSTEIN and S.I. GELFAND, The meromorphic function P^{λ}, Funkcional Anal., Vol. 3 (1969), 84-85.

3. A. CEREZO and F. ROUVIERE, Résolubilité locale d'un opérateur différentiel invariant du premier ordre, Ann. Sic. Ecole Norm. Sup. 4 (1971), 21-30.

4. A. CEREZO and F. ROUVIERE, Opérateurs différentiels invariants sur un groupe de Lie, Séminaire Goulaouic-Schwartz, Ecole Polytechnique, 1972-73.

5. J. DIXMIER, Sur les représentations unitaires des groupes de Lie nilpotents V, Bull. Soc. Math. France 87 (1959), 65-79.

6. M. DUFLO and M. RAIS, Sur l'analyse harmonique sur les groupes de Lie résolubles (preprint).

7. M. EGUCHI and K. OKAMOTO, The Fourier transform of the Schwartz space on a symmetric space (preprint).

8. M. EGUCHI, Asymptotic expansions of Eisenstein integrals and Fourier transform on symmetric spaces (preprint).

9. HARISH-CHANDRA, Spherical functions on a simisimple Lie group I, II, Amer. J. Math. 80 (1958), 241-310; 550-613.

10. HARISH-CHANDRA, Invariant eigendistributions on a semisimple Lie group, Trans. Amer. Math. Soc. 119 (1965), 457-508.

11. S. HELGASON, Fundamental solutions of invariant differential operators on symmetric spaces, Amer. J. Math. 86 (1964), 565-601.

12. S. HELGASON, Radon-Fourier transforms on symmetric spaces and related group representations, Bull. Amer. Math. Soc. 71 (1965), 757-763.

13. S. HELGASON, Analysis on Lie groups and homogeneous spaces CBMS Monograph No. 14, Amer. Math. Soc., Providence, R.I., 1972.

14. S. HELGASON, Paley-Wiener theorems and surjectivity of invariant differential operators on symmetric spaces and Lie groups, Bull. Amer. Math. Soc. 79 (1973), 129-132.

15. S. HELGASON, The surjectivity of invariant differential operators on symmetric spaces I, Ann. of Math. 98 (1973), 451-479.

16. S. HELGASON, A duality for symmetric spaces with applications to group representations II, Advan. Math. (to appear).

17. S. HELGASON, Invariant differential operators and eigenspace representations, Lecture notes from a Seminar on "Group theoretical methods in physics", Univ. de Montréal, 1976.

18. L. HORMANDER, On the existence and regularity of solutions of linear pseudo-differential equations, l'Enseignement Math. 17 (1971), 99-163.

19. A.A. KIRILLOV, Unitary representations of nilpotent Lie groups, Uspehi Mat. Nauk 17 (1962), 57-110.

20. H. LEWY, An example of a smooth linear partial differential equation without solution, Ann. of Math. (2) 66 (1957), 155-158.

21. H. MAASS, Spherical functions and quadratic forms, J. Indian Math. Soc. 20 (1956), 117-162.

22. M. RAIS, Solutions élémentaires des opérateurs différentiels bi-invariants sur un groupe de Lie nilpotent, C.R. Acad. Sci. Paris 273 (1971), 495-498.

23. J. RAUCH and D. WIGNER, Global solvability of the Casimir operator, Ann. of Math. 103 (1976), 229-236.

24. F. ROUVIERE, Sur la résolubilité locale des opérateurs bi-invariants (preprint).

25. A. SELBERG, Harmonic analysis and discontinuous groups in weakly symmetric Riemannian spaces with applications to Dirichlet series, J. Ind. Math. Soc. 20 (1956), 47-87.

SYMMETRY AND SEPARATION OF VARIABLES FOR LINEAR PARTIAL DIFFERENTIAL AND HAMILTON-JACOBI EQUATIONS*

Willard Miller, Jr.

INTRODUCTION

This work is concerned with the interplay between the symmetries of a partial differential equation, the coordinate systems in which the equation admits solutions via separation of variables, and the properties of the separated solutions. We present the theory by means of some very important examples although the basic ideas have general validity. The first two sections are concerned with the wave equation (1.1) in three-dimensional space time. Most of this material has appeared recently in papers by E.G. Kalnins and the author, [1]. Sections three and four are devoted to the (nonlinear) equation of the characteristics of (1.1). This work by C.P. Boyer and the author appears here for the first time and illustrates the intimate connection between the theory of linear and Hamilton-Jacobi differential equations in terms of symmetry and separation of variables.

A detailed bibliography for this field can be found in [5].

* Research partially supported by NSF Grant MC S76-04838.

1. THE WAVE EQUATION

In this section we will confine our attention to the wave equation in three-dimensional space time,

$$\Psi_{x^0x^0} - \Psi_{x^1x^1} - \Psi_{x^2x^2} = 0, \tag{1.1}$$

or for short, $\Box\Psi = 0$ where $\Box$ is the D'Alembertian. A *first order symmetry operator* for this equation is an operator

$$\underset{\sim}{L} = \sum_{\mu=0}^{2} a_\mu(x^\nu)\partial_{x^\mu} + b(x^\nu) \tag{1.2}$$

such that the function $\underset{\sim}{L}\Psi$ is (locally) a solution of (1.1) for any solution Ψ of (1.1). The set of such operators is easily seen to form a Lie algebra isomorphic to o(3,2) under the operations of addition and scalar multiplication of Lie derivatives and the commutator bracket

$$[\underset{\sim}{L}_1,\underset{\sim}{L}_2] = \underset{\sim}{L}_1\underset{\sim}{L}_2 - \underset{\sim}{L}_2\underset{\sim}{L}_1. \tag{1.3}$$

A basis for the Lie algebra is provided by the elements

$$\underset{\sim}{M}_{\mu\nu} = x_\mu\partial_{x^\nu} - x_\nu\partial_{x^\mu}, \qquad \mu \neq \nu$$

$$\underset{\sim}{P}_\mu = \partial_{x^\mu}, \quad \underset{\sim}{D} = x\cdot\partial_x + \tfrac{1}{2} = x^\mu\partial_{x^\mu} + \tfrac{1}{2}$$

$$\underset{\sim}{K}_\mu = 2x_\mu x\cdot\partial_x - x\cdot x\partial_{x^\mu} + x_\mu, \quad 0 \le \mu,\nu \le 2 \tag{1.4}$$

where $x_0 = x^0$, $x_j = -x^j$ for $j=1,2$, $x\cdot x = x^\mu x_\mu$ and the Einstein summation convention for repeated indices is adopted. (We have omitted the trivial symmetry of multiplication by 1.) These operators can be exponentiated to yield a local Lie group O(3,2) of symmetry operators whose physical significance has been discussed in many references, e.g. [1], [2].

We now introduce another basis for the symmetry algebra. The matrix algebra o(3,2) is conventionally defined as the ten-dimensional Lie algebra of 5 × 5 real matrices A such that $AG + GA^t = 0$ where 0 is the zero matrix and

$$G = \begin{pmatrix} 1 & & & & 0 \\ & 1 & & & \\ & & 1 & & \\ & & & -1 & \\ 0 & & & & -1 \end{pmatrix}.$$

Let E_{ij} be the 5 × 5 matrix with a 1 in row i, column j and zeros elsewhere. Then the matrices

$$\Gamma_{ab} = E_{ab} - E_{ba} = -\Gamma_{ba}, \quad a \neq b$$

$$\Gamma_{aB} = E_{aB} + E_{Ba} = -\Gamma_{Ba}, \quad 1 \le a,b \le 3, \quad B = 4,5$$

$$\Gamma_{45} = E_{54} - E_{45} = -\Gamma_{54}, \tag{1.5}$$

form a basis for o(3,2) with commutation relations

$$[\Gamma_{\alpha\beta}, \Gamma_{\gamma\varphi}] = G_{\beta\gamma}\Gamma_{\alpha\varphi} + G_{\alpha\varphi}\Gamma_{\beta\gamma} + G_{\gamma\alpha}\Gamma_{\varphi\beta} + G_{\varphi\beta}\Gamma_{\gamma\alpha}. \tag{1.6}$$

This basis is related to (1.4) via the identifications

$$\underset{\sim}{P}_0 = \Gamma_{14} + \Gamma_{45}, \quad \underset{\sim}{P}_1 = \Gamma_{25} + \Gamma_{12}, \quad \underset{\sim}{P}_2 = \Gamma_{35} + \Gamma_{13}$$

$$\underset{\sim}{K}_0 = \Gamma_{45} - \Gamma_{14}, \quad \underset{\sim}{K}_1 = \Gamma_{25} - \Gamma_{12}, \quad \underset{\sim}{K}_2 = \Gamma_{35} - \Gamma_{13}$$

$$\underset{\sim}{M}_{12} = \Gamma_{32}, \quad \underset{\sim}{M}_{01} = \Gamma_{24}, \quad \underset{\sim}{M}_{02} = \Gamma_{34}, \quad \underset{\sim}{D} = \Gamma_{51}. \tag{1.7}$$

In analogous manner to the computation of first order symmetries $\underset{\sim}{L}$ one can also determine the vector space S' of second order operators

$$\underset{\sim}{S}' = \sum_{\mu,\mu'} a_{\mu\mu'}(x^\nu)\partial_{x^\mu}\partial_{x^{\mu'}} + \sum_\mu b_\mu(x^\nu)\partial_{x^\mu} + c(x^\nu) \tag{1.8}$$

which (locally) map solutions of (1.1) to solutions. Among these
operators are the trivial symmetries $f(x^\nu)\square$ where f is an arbitrary function, which act (locally) as the zero operator on the
solution space of (1.1). Let η be the vector space of trivial
symmetries. One can show by a tedious computation that the factor
space $S \cong S'/\eta$ is spanned by the first and second order elements
in the enveloping algebra of $o(3,2)$. Indeed, if $\{\underset{\sim}{\Gamma}_j, 1 \le j \le 10\}$ is a
basis for the symmetry algebra $o(3,2)$ then elements of the form
$\underset{\sim}{\Gamma}_j$, $\{\underset{\sim}{\Gamma}_j, \underset{\sim}{\Gamma}_k\} = \underset{\sim}{\Gamma}_j\underset{\sim}{\Gamma}_k + \underset{\sim}{\Gamma}_k\underset{\sim}{\Gamma}_j$, $1 \le j \le k \le 10$, span S.

A second-order linear partial differential equation, such as
(1.1), for which the non-trivial second-order symmetries lie in
the enveloping algebra of the first-order symmetry algebra is
called *class I*. In all cases known to the author the separable
coordinate systems admitted by a class I equation are describable
in terms of the symmetry algebra, see [1], [3]. If an equation
admits non-trivial second-order symmetries which do not belong to
the enveloping algebra it is called *class II*. The best known example of a class II equation is the Schrödinger equation for the
hydrogen atom. However, even for these less symmetric equations
the separable coordinate systems are determined by elements of S
in all examples known to the author, e.g. [3], [4].

When acting on solutions Ψ of (1.1) the elements of S are not
all independent. For example, we have the identities

$$\underset{\sim}{P}_0^2 - \underset{\sim}{P}_1^2 - \underset{\sim}{P}_2^2 = \underset{\sim}{K}_0^2 - \underset{\sim}{K}_1^2 - \underset{\sim}{K}_2^2 = 0$$

$$\underset{\sim}{\Gamma}_{12}^2 + \underset{\sim}{\Gamma}_{13}^2 + \underset{\sim}{\Gamma}_{23}^2 = \underset{\sim}{\Gamma}_{45}^2 + \tfrac{1}{4}, \quad \underset{\sim}{M}_{12}^2 - \underset{\sim}{M}_{01}^2 - \underset{\sim}{M}_{02}^2 = -\underset{\sim}{D}^2 + \tfrac{1}{4} \quad (1.9)$$

valid on the solution space of (1.1).

We now describe explicitly the relationship between the above
and separation of variables for the wave equation. By an
R-separable coordinate system $\{u,v,w\}$ for (1.1) we mean a coordinate system such that under the substitution

$\Psi = R(u,v,w)U(u)V(v)W(w)$ the wave equation is equivalent to a triplet of ordinary differential equations, one for each of the functions U, V, W. Either $R \equiv 1$ or the fixed function R cannot be factored as a product $R_1(u)R_2(v)R_3(w)$. All known separable systems $\{u,v,w\}$ are characterized as follows: There is a pair of operators $\tilde{S}_1, \tilde{S}_2 \in \mathcal{S}$ such that $[\tilde{S}_1, \tilde{S}_2] = \tilde{0}$ and the separable solutions Ψ corresponding to $\{u,v,w\}$ are eigenfunctions of the operators $\tilde{S}_j$,

$$\tilde{S}_j \Psi = \lambda_i \Psi, \quad j=1,2. \tag{1.10}$$

The eigenvalues λ_j are the separation constants for the separable solutions, [1], [3], [5]. In a similar manner, equations in n variables have separable solutions associated with n - 1 commuting operators in $\mathcal{S}$. For class I equations these operators belong to the enveloping algebra of the first-order symmetry algebra $\mathcal{G}$. The elements of the local Lie symmetry group G of an equation map separable coordinates to separable coordinates in an obvious fashion and partition the coordinate systems into G-equivalence classes. The associated symmetry operators $\{\tilde{S}_1, \ldots, \tilde{S}_{n-1}\}$ generate orbits of commuting operators under the adjoint action of $\mathcal{G}$ and one obtains a correspondence between equivalence classes of separable coordinates and orbits of commuting symmetry operators. This correspondence also holds for class II equations even though the operators $\tilde{S}_j$ may not belong to the enveloping algebra of $\mathcal{G}$, e.g. [4].

To exploit this relationship between operators and separable solutions we make use of the Hilbert space of positive energy solutions of (1.1), [1], [2]. Let $\mathcal{H}$ be the Hilbert space of all functions f on R_2 (variables k_1, k_2) such that f is Lebesgue square integrable with respect to the measure $d\mu(k) = dk^1 dk^2/k^0$, $k^0 = [(k^1)^2 + (k^2)^2]^{\frac{1}{2}}$. The inner product is given by

$$\langle f,g \rangle = \int_{-\infty}^{\infty}\int_{-\infty}^{\infty} f(k^1,k^2)\bar{g}(k^1,k^2)\,d\mu(\tilde{k}), \quad f,g \in \mathcal{H}. \tag{1.11}$$

It is well-known that the functions $\Psi(x)$,

$$\Psi(x) = \frac{1}{4\pi}\int_{-\infty}^{\infty}\int_{-\infty}^{\infty} \exp(ik\cdot x) f(\underset{\sim}{k}) d\mu(k) = I(f)$$

$$f \in \mathcal{H}, \quad k = (k^0, k^1, k^2)$$

are (weak) solutions of (1.1). Furthermore, the set of all such solutions I(f) as f ranger over $\mathcal{H}$ is a Hilbert space with inner-product

$$\langle \Psi, \Phi \rangle \equiv \langle f, g \rangle = 4i \iint_{x^0=t} \Psi(x)\partial_{x^0}\bar{\Phi}(x)dx^1dx^2$$

$$\Psi = I(f), \quad \Phi = I(g), \tag{1.13}$$

independent of t.

The action of operators (1.4) on the solution space of (1.1) induces a corresponding action on $\mathcal{H}$ by operators

$$\underset{\sim}{P}_\mu = ik_\mu, \quad \underset{\sim}{M}_{12} = k^1\partial_{k^2} - k^2\partial_{k^1}, \quad \underset{\sim}{M}_{01} = k^0\partial_{k^1},$$

$$\underset{\sim}{M}_{02} = k^0\partial_{k^2}, \quad \underset{\sim}{D} = -(\tfrac{1}{2}+k^1\partial_{k^1}+k^2\partial_{k^2}),$$

$$\underset{\sim}{K}_0 = -ik^0(\partial_{k^1k^1}+\partial_{k^2k^2}), \quad \underset{\sim}{K}_1 = -i(k^1\partial_{k^1k^1}-k^1\partial_{k^2k^2}+2k^2\partial_{k^1k^2}+\partial_{k^1}),$$

$$\underset{\sim}{K}_2 = -i(-k^2\partial_{k^1k^1}+k^2\partial_{k^2k^2}+2k^1\partial_{k^1k^2}+\partial_{k^2}). \tag{1.14}$$

It is well-known, [2], that operators (1.14) induce a unitary irreducible representation of the universal covering group of the identity component of SO(3,2). Thus, the elements (1.14) of o(3,2) can be extended to domains on which they are skew-adjoint. Now the purely second-order members of S are real linear combinations of elements of the form $\{\underset{\sim}{L}_1, \underset{\sim}{L}_2\} = \underset{\sim}{L}_1\underset{\sim}{L}_2 + \underset{\sim}{L}_2\underset{\sim}{L}_1$, $\underset{\sim}{L}_j \in o(3,2)$. It is easy to see from this that, say restricted to C^∞ functions on $\mathcal{H}$ with compact support bounded away from $\underset{\sim}{k} = \underset{\sim}{0}$, these elements of S are symmetric operators. In many cases the operators are essentially self-adjoint; in any case one can always construct (possibly non-unique) self-adjoint extensions.

References [1] contain lists of all R-separable systems for (1.1) which E.G. Kalnins and I have so far been able to discover. For each of these 90-odd systems we have provided the associated pair of commuting operators $\underset{\sim}{S}_1, \underset{\sim}{S}_2 \in S$. As follows from the above remarks, one can define a pair of (possibly non-unique) commuting self-adjoint operators on $\mathcal{H}$. Thus, via the spectral theorem, there exists a generalized eigenbasis $\{f_{\lambda_1\lambda_2}\}$ for $\mathcal{H}$ such that

$$\underset{\sim}{S}_j f_{\lambda_1\lambda_2} = \lambda_j f_{\lambda_1\lambda_2}, \quad j=1,2,$$

$$\langle f_{\lambda_1\lambda_2}, f_{\lambda_1'\lambda_2'} \rangle = \delta(\lambda_1-\lambda_1')\delta(\lambda_2-\lambda_2'). \tag{1.15}$$

It follows that the functions $\Psi_{\lambda_1\lambda_2} = I(f_{\lambda_1\lambda_2})$ are solutions of (1.1) which satisfy the equations

$$\underset{\sim}{S}_j \Psi_{\lambda_1\lambda_2} = \lambda_j \Psi_{\lambda_1\lambda_2}, \quad j=1,2,$$

$$\langle \Psi_{\lambda_1\lambda_2}, \Psi_{\lambda_1'\lambda_2'} \rangle = \delta(\lambda_1-\lambda_1')\delta(\lambda_2-\lambda_2') \tag{1.16}$$

where here the $\underset{\sim}{S}_j$ are expressed in terms of the operators (1.4).

Note that the expression $\Psi_{\lambda_1\lambda_2} = I(f_{\lambda_1\lambda_2})$ determines an integral representation for the solutions $\Psi_{\lambda_1\lambda_2}$. In general the integral appears difficult to evaluate. However, one knows in advance that the $\Psi_{\lambda_1\lambda_2}$ satisfy (1.16), hence these solutions must be R-separable in the appropriate coordinate system:

$$\Psi_{\lambda_1\lambda_2} = R(u,v,w)U_{\lambda_1\lambda_2}(u)V_{\lambda_1\lambda_2}(v)W_{\lambda_1\lambda_2}(w),$$

(or at worst a linear combination of 8 separable terms). Since U, V, W satisfy known second-order ordinary differential equations one can immediately write down the general solution in terms of at most 8 arbitrary constants and these constants can be determined by computing the integral for special values of the parameters. (A large percentage of the integral representations for special functions have their origin in this interplay between separation of variables and models of group representations.)

If $f \in \mathcal{H}$ we have the Hilbert space expansion

$$f = \iint \langle f, f_{\lambda_1\lambda_2} \rangle f_{\lambda_1\lambda_2} d\lambda_1 d\lambda_2, \tag{1.17}$$

hence the expansion

$$\Phi = \iint \langle f, f_{\lambda_1\lambda_2} \rangle \Psi_{\lambda_1\lambda_2} d\lambda_1 d\lambda_2 \tag{1.18}$$

where $\Phi = I(f)$ is a solution of (1.1). Expression (1.18) shows how one can expand an arbitrary Hilbert-space solution of (1.1) in terms of an eigenbasis of R-separable solution $\{\Psi_{\lambda_1\lambda_2}\}$. The expansion coefficients $\langle f, f_{\lambda_1\lambda_2} \rangle$ and the spectral resolutions are all computed in the more tractable Hilbert space $\mathcal{H}$.

2. EXAMPLES OF SEPARABLE SYSTEMS

First we consider the coordinate system corresponding to the commuting operators $\underset{\sim}{\Gamma}_{45}^2$, $\underset{\sim}{\Gamma}_{23}^2$. To obtain the spectral resolutions of these operators we diagonalize the operators $\underset{\sim}{\Gamma}_{45}$, $\underset{\sim}{\Gamma}_{23}$. (Note that $\underset{\sim}{\Gamma}_{45}$ generates the subalgebra o(2) from the maximal compact subalgebra o(3) × o(2) of o(3,2).)

On $\mathcal{H}$ the eigenvalue equations are

$$\underset{\sim}{\Gamma}_{45} f = i\lambda f, \quad \underset{\sim}{\Gamma}_{23} f = imf, \quad \underset{\sim}{\Gamma}_{45} = \frac{i}{2} k^0 (-\partial_{k^1 k^1} - \partial_{k^2 k^2} + 1). \tag{2.1}$$

Setting $k^1 = k^0 \cos\theta$, $k^2 = k^0 \sin\theta$ we find that the orthonormal basis of eigenvectors is

$$F^{(\ell,m)}(\underset{\sim}{k}) = [(\ell-m)!/\pi(\ell+m)!]^{\frac{1}{2}} (2k^0)^m e^{-k^0} L_{\ell-m}^{(2m)}(2k^0) e^{im\theta},$$

$$\lambda = \ell + \tfrac{1}{2}, \quad \ell = 0,1,2,\ldots, m = -\ell, \quad -\ell+1,\ldots,\ell. \tag{2.2}$$

Here $L_n^{(\alpha)}(z)$ is a generalized Laguerre polynomial and the $\{f^{(\ell,m)}: m=-\ell,\ldots,\ell\}$ for fixed ℓ form a basis for the $(2\ell+1)$-dimensional irreducible representation D_ℓ of $o(3)$. On restriction from $o(3,2)$ to $o(3)$ our representation decomposes as $\sum_{\ell=0}^{\infty} \oplus D_\ell$

There is a very close connection between the eigenvalue equation $\Gamma_{45}f = i\lambda f$ on $\mathcal{H}$ and the quantum Kepler problem in two-dimensional space:

$$\underset{\sim}{H}g = \mu g, \quad \underset{\sim}{H} = -\partial_{xx} - \partial_{yy} + e/r, \quad r = (x^2+y^2)^{\frac{1}{2}}$$

$$\iint |g|^2 dxdy < \infty. \tag{2.3}$$

The two eigenvalue equations are identifiable provided $k^1 = x\sqrt{-\mu}$, $k^2 = y\sqrt{-\mu}$, $\mu = -e^2/4\lambda^2$. Although these eigenvalue problems are defined on Hilbert spaces with different inner products, the virial theorem shows that the eigenvectors of $\underset{\sim}{H}$ coincide with eigenvectors of $\underset{\sim}{\Gamma}_{45}$. Since the eigenvalues λ of $\underset{\sim}{\Gamma}_{45}$ are $\lambda = \ell + \frac{1}{2}$, $\ell=0,1,2,\ldots$, it follows that the point eigenvalues of $\underset{\sim}{H}$ are $\mu_\ell = -e^2/4(\ell+\frac{1}{2})^2$.

The corresponding solutions of (1.1) are given by the double integral $\psi^{(\ell,m)} = I(f^{(\ell,m)})$. It is easy to compute the coordinates in which the $\psi^{(\ell,m)}$ become R-separable. One merely employs elementary Lie theory to construct new coordinates such that each of the commuting operators $\underset{\sim}{\Gamma}_{45}$, $\underset{\sim}{\Gamma}_{23}$ becomes differentiation with respect to a single variable. The results are

$$x^0 = \frac{\sin \varphi}{y^0-\cos \varphi}, \quad x^1 = \frac{y^1}{y^0-\cos \varphi}, \quad x^2 = \frac{y^2}{y^0-\cos \varphi}$$

$$y^0 = \cos \sigma, \quad y^1 = \sin \sigma \cos \alpha, \quad y^2 = \sin \sigma \sin \alpha. \tag{2.4}$$

Note that $(y^0)^2 + (y^1)^2 + (y^2)^2 = 1$ so the vector $\underset{\sim}{y} = (y^0,y^1,y^2)$ ranges over the sphere S_2 while φ ranges over the circle S_1. We find

$$\underset{\sim}{\Gamma}_{45} = -\partial_\varphi + \tfrac{1}{2} \sin \varphi/(\cos \sigma - \cos \varphi), \quad \underset{\sim}{\Gamma}_{23} = \partial_\alpha. \qquad (2.5)$$

Our separation of variables problem has led us naturally to a compactification $S_1 \times S_2$ of Minkowski space M_3 such that each ordinary point x of M_3 is covered by two points in the compact space $S_1 \times S_2$. It is this compactification (in connection with the wave equation in four-dimensional space time) which is used by Segal in his model of the cosmos, [6]. The operator $\underset{\sim}{\Gamma}_{45}$ becomes the generator for unitime translations in Segal's theory.

Once the R-separable coordinates $\{\alpha, \sigma, \varphi\}$ have been determined it is a simple matter to determine the functional form of the solutions $\psi^{(\ell, m)}$. Computing the constants in the expansion by evaluating $I(f^{(\ell, m)})$ in special cases, we find

$$\psi^{(\ell, m)} = [(-i)^{m-1}/\sqrt{4\ell+2}][\cos \sigma - \cos \varphi]^{\frac{1}{2}} \quad e^{-i\varphi(\ell+\frac{1}{2})} Y_\ell^m(\sigma, \alpha) \qquad (2.6)$$

where $Y_\ell^m(\sigma, \alpha)$ is a spherical harmonic. Note that here the variables R-separate with $R = [\cos \sigma - \cos \varphi]^{\frac{1}{2}}$.

From a more general point of view the effect of diagonalizing $\underset{\sim}{\Gamma}_{45}$ is to reduce (1.1) to the eigenvalue equation for the Laplace operator on the sphere S_2. Indeed, if $\underset{\sim}{\Gamma}_{45}\Psi = i\lambda\Psi$ then $\Psi = R\Phi(\underset{\sim}{y})e^{-i\varphi(\ell+\frac{1}{2})}$ where $\underset{\sim}{y} \in S_2$. From one of the identities (1.9) we find

$$(\underset{\sim}{\Gamma}_{12}^2 + \underset{\sim}{\Gamma}_{13}^2 + \underset{\sim}{\Gamma}_{23}^2)\Phi = (\tfrac{1}{4} - \lambda^2)\Phi \qquad (2.7)$$

where $\underset{\sim}{\Gamma}_{12}, \underset{\sim}{\Gamma}_{13}, \underset{\sim}{\Gamma}_{23}$ are the usual angular momentum operators on S_2. Here (2.7) is the reduced equation obtained by splitting off the coordinate φ from (1.1). The symmetry algebra of the class I equation (2.7) is o(3) and this equation can be in turn separated using the methods discussed above. As shown in [7] there are two orbits of second order symmetries, associated with the two distinct separable coordinate systems for (2.7). One orbit has the representative $\underset{\sim}{\Gamma}_{23}^2$ and leads to spherical harmonic solutions (2.6).

The other orbit type has representative $\underset{\sim}{\Gamma}_{12}^2 + a^2\underset{\sim}{\Gamma}_{13}^2$, $1 > a > 0$ and yields separable solutions of (2.7) which are products of Lamé polynomials.

We call a coordinate system *split* if the corresponding symmetry operators can be chosen as $\underset{\sim}{A}^2$, $\underset{\sim}{B}^2$, $[\underset{\sim}{A},\underset{\sim}{B}] = 0$, $\underset{\sim}{A},\underset{\sim}{B} \in o(3,2)$. These coordinates, such as (2.4) are the simplest to study. More generally, a system is *semi-split* if the symmetry operators can be chosen as $\underset{\sim}{A}^2$, $\underset{\sim}{S}_2$, $[\underset{\sim}{A},\underset{\sim}{S}_2] = 0$, $\underset{\sim}{A} \in o(3,2)$. Semi-split systems are in general less tractable than split systems but much easier to handle than *non-split* systems in which neither symmetry operator can be chosen as a square of a Lie algebra symmetry. The R-separable solutions of (1.1) which are exponentials times products of Lamé polynomials are semi-split.

In [1] many other semi-split systems are derived through the simple expedient of splitting off a single variable from (1.1) and then studying the separable coordinate systems for the reduced equation so obtained. Thus if Ψ is an eigenfunction of $\underset{\sim}{P}_0 + \underset{\sim}{P}_1$, $(\underset{\sim}{P}_0+\underset{\sim}{P}_1)\Psi = i\lambda\Psi$ we can set $\Psi(x) = \exp(is\lambda)\Phi(t,x^2)$ where $2s = x^0 + x^1$, $2t = x^1 - x^0$. Here Φ satisfies the free particle Schrödinger equation

$$(i\lambda\partial_t+\partial_{x^2x^2})\Phi(t,x^2) = 0. \tag{2.8}$$

This equation admits the Schrödinger symmetry algebra with basis $\{\underset{\sim}{P}_2,\underset{\sim}{P}_1-\underset{\sim}{P}_0,\underset{\sim}{P}_1+\underset{\sim}{P}_0,\underset{\sim}{M}_{02}-\underset{\sim}{M}_{12},\underset{\sim}{D}+\underset{\sim}{M}_{01},\underset{\sim}{K}_0+\underset{\sim}{K}_1\}$ and R-separates in four coordinate systems corresponding to four orbits in this algebra, [8].

There are additional examples of semi-split coordinates in [1] in which the reduced equations are those of Helmholtz, Klein-Gordon, Euler-Poisson-Darboux, the Laplace-Beltrami operator on a hyperboloid and other equations. Moreover, in [1] there is a classification of all orthogonal R-separable coordinates for (1.1) for which the coordinate surfaces are families of confocal cyclides. The most general such coordinates are non-split. All known orthogonal systems are limiting forms of these general cyclidic systems.

3. HAMILTON-JACOBI EQUATIONS

Next we consider the separation of variables problem for Hamilton-Jacobi equations. To illustrate the intimate relationship between this problem and the corresponding linear theory we consider the example

$$W^2_{x^0} - W^2_{x^1} - W^2_{x^2} = 0, \quad W_{x^\mu} = \partial_{x^\mu} W(x), \tag{3.1}$$

the equation obeyed by the characteristics of (1.1). It is well-known, [9], that the space-time symmetry algebra of (3.1) is o(3,2). That is, the set of Lie derivatives

$$\underset{\sim}{L} = \sum_{\alpha=0}^{2} a_\alpha(x)\partial_{x^\mu}$$

such that $\underset{\sim}{L}W$ is a solution of (3.1) whenever W is a solution, forms a Lie algebra isomorphic to o(3,2). A basis is provided by the elements

$$\underset{\sim}{M}_{\mu\nu} = x_\mu \partial_{x^\nu} - x_\nu \partial_{x^\mu}, \quad \mu \neq \nu$$

$$\underset{\sim}{P}_\mu = \partial_{x^\mu}, \quad \underset{\sim}{D} = x\cdot\partial_x, \quad \underset{\sim}{K}_\mu = 2x_\mu x\cdot\partial_x - x\cdot x\partial_{x^\mu}, \quad 0 \leq \mu,\nu \leq 2. \tag{3.2}$$

Another realization of the symmetry algebra is provided by the functions

$$M_{\mu\nu} = x_\mu p_\nu - x_\nu p_\mu, \quad P_\mu = p_\mu, \quad D = x^\mu p_\mu,$$
$$K_\mu = 2x_\mu(x^\nu p_\nu) - x\cdot xp_\mu \tag{3.3}$$

where $p_\mu = W_{x^\mu}$ and the commutator is the Poisson bracket

$$\{F(x,p),G(x,p)\}_P = \sum_{\mu=0}^{2} (G_{x^\mu}F_{p_\mu} - F_{x^\mu}G_{p_\mu}). \tag{3.4}$$

(One can easily check that the appropriate commutation relations for o(3,2) are satisfied.)

By taking all possible products of operators (3.2) we can generate an enveloping algebra of o(3,2). Furthermore, we can identify the subspace S'_k of homogeneous symmetric kth-order elements in the enveloping algebra with the space K_k of kth-order polynomials in the basis functions (3.3). That is, the two subspaces are isomorphic as vector spaces and the adjoint action of o(3,2) on S'_k induced by the commutator $[\cdot,\cdot]$ agrees with the adjoint action on K_k induced by the Poisson bracket. In particular, S'_2 is spanned by elements of the form $\{\underset{\sim}{L}_1,\underset{\sim}{L}_2\} = \underset{\sim}{L}_1\underset{\sim}{L}_2 + \underset{\sim}{L}_2\underset{\sim}{L}_1$ where the Lie derivatives $\underset{\sim}{L}_j$ belong to o(3,2). If L_1, L_2 are the corresponding functions in the Lie algebra (3.3), the correspondence

$$\{\underset{\sim}{L}_1,\underset{\sim}{L}_2\} \leftrightarrow 2L_1L_2 \tag{3.5}$$

extended by linearity, provides the stated isomorphism between S'_2 and K_2.

We briefly review the well-known relationship between a first-order partial differential equation

$$H(x^\mu, p_\mu) = 0, \quad p_\mu = \frac{\partial W}{\partial x^\mu}, \quad 0 \le \mu \le n \tag{3.6}$$

and the Hamiltonian system of ordinary differential equations, [10],

$$\frac{dp_\mu}{d\tau} = -H_{x^\mu}, \quad \frac{dx^\mu}{d\tau} = H_{p_\mu}, \quad 0 \le \mu \le n. \tag{3.7}$$

Consider the n-dimensional surface $x_\mu = x_\mu(t_1,\ldots,t_n)$ and prescribe initial data on this surface

$$W = W(t_1,\ldots,t_n), \quad p_\mu = p_\mu(t_1,\ldots,t_n) \tag{3.8}$$

subject to the requirements

$$\frac{\partial W}{\partial t_j} = p_\mu \frac{\partial x^\mu}{\partial t_j}, \quad j=1,\ldots,n, \quad H(x^\mu(t_\ell),p_\mu(t_\ell)) = 0.$$

Then provided

$$\det\begin{pmatrix} H_{p_\alpha} \\ \dfrac{\partial x_\alpha}{\partial t_j} \end{pmatrix} \neq 0, \quad \alpha=0,\ldots,n, \quad j=1,\ldots,n$$

on the surface, the solutions of (3.7) with initial data (3.8) generate a local solution of (3.6).

Conversely, if

$$W = f(x^\mu,a_1,\ldots,a_n) + a_0 \tag{3.9}$$

is a complete integral of (3.6), i.e., if W is a solution of (3.6) for each choice of the n + 1 real constants a_α and the n × (n+1) matrix $(\partial_{a_\alpha}\partial_{x^\mu}W)$ has rank n, then (3.9) and relations

$$f_{a_j}(x^\mu,a_\ell) = \lambda_j, \quad j=1,\ldots,n$$

$$p_\nu = f_{x^\nu}(x^\mu,a_\ell), \quad \nu=0,\ldots,n \tag{3.10}$$

with $a_0,\ldots,a_n,\lambda_1,\ldots,\lambda_n$ fixed define a solution of the characteristic system (1.12).

Furthermore, the canonical transformation generated by (3.7) preserves Poisson brackets, [10]. Thus if $F_j^\tau(x^\mu,p_\nu) = F_j(x^\mu(\tau),p_\nu(\tau))$, j=1,2, where

$$x^\mu(\tau) = x^\mu(\tau,x^{\mu'},p_{\nu'}), \quad p_\nu(\tau) = p_\nu(\tau,x^{\mu'},p_{\nu'}) \tag{3.11}$$

are solutions of (3.7) such that $x^\mu(0) = x^\mu$, $p_\nu(0) = p_\nu$ then

$$\{F_1^\tau,F_2^\tau\}_P = \{F_1,F_2\}_P. \tag{3.12}$$

Furthermore,

$$\frac{d}{d\tau} F^\tau = F_{x^\mu} H_{p_\mu} - F_{p_\mu} H_{x^\mu} = \{H^\tau,F^\tau\}$$

so $F^\tau \equiv F$ if F commutes with H.

Applying this theory to (3.1) we find

$$H = p_0^2 - p_1^2 - p_2^2 \tag{3.13}$$

so the associated Hamiltonian system is

$$\frac{dp_\mu}{d\tau} = 0, \quad \frac{dx^\mu}{d\tau} = g^{\mu\nu} p_\nu, \quad 0 \le \mu,\nu \le 2. \tag{3.14}$$

Thus we can obtain a solution of (3.1) by prescribing initial data for W, x^μ and p_μ on a 2-dimensional surface in x-space and solving (3.14). For some computations we take this data in the form

$$x^0 = 0,\ p_0 = \sqrt{p_1^2+p_2^2},\ x^1 = t_1,\ x^2 = t_2,\ p_j = p_j(t_1,t_2). \tag{3.15}$$

Note that the basis functions (3.3) restricted to this surface are

$$P_\alpha = p_\alpha,\quad M_{ij} = x_i p_j - x_j p_i,\quad M_{0j} = x^j p_0,$$

$$K_0 = ((x_1)^2+(x_2)^2)p_0,\quad D = x^i p_i$$

$$K_j = 2x_j(x^i p_i) + ((x_1)^2+(x_2)^2)p_j,\quad i,j=1,2. \tag{3.16}$$

Model (3.15) and its relationship to (3.1) via integration of (3.14) is an analogy of the model $\mathcal{H}$ for (1.1).

Each separated system for (1.1) was characterized by a pair

of commuting second order symmetries $\underset{\sim}{S}_1$, $\underset{\sim}{S}_2$. Similarly, each separated system for (3.1) is characterized by functions $F_1, F_2 \in K_2$ such that $\{F_1, F_2\}_P = 0$. A separable system for (1.1) yields additive R-separation for (3.1) where the $\underset{\sim}{S}_j$ and F_j are related by (3.5). The eigenvalue equations (1.10) are here replaced by

$$p \cdot p = 0, \quad F_j(x^\mu, p_\nu) = \lambda_j, \quad j=1,2 \tag{3.17}$$

where the λ_j are separation constants.

4. SEPARABLE SYSTEMS FOR HAMILTON-JACOBI EQUATIONS

To illustrate the analogy between separation of variables for (1.1) and (3.1) we consider semi-split systems characterized by operators $\underset{\sim}{\Gamma}_{45}^2$, $\underset{\sim}{S}$. (Here the $\underset{\sim}{\Gamma}_{\alpha\beta}$ are related to the basis functions (3.3) by (1.7). Setting $\Gamma_{45} = \lambda$ we find that (3.1) reduces to

$$\Gamma_{12}^2 + \Gamma_{13}^2 + \Gamma_{23}^2 = \lambda^2. \tag{4.1}$$

It is easy to see that o(3) is the symmetry algebra for (4.1). This equation can be viewed as the result of splitting off one variable φ in W. Indeed, we choose new coordinates such that $\underset{\sim}{\Gamma}_{45} = -\partial_\varphi$. Lie theory yields equations (2.4) where $\underset{\sim}{y}$ ranges over S_2. Choosing any parametrization $y_j(\sigma,\alpha)$ for S_2 we obtain a new set of coordinates for space time. The equation $\Gamma_{45} = \lambda$, i.e., $\underset{\sim}{\Gamma}_{45}W = \lambda$, implies

$$W = -\lambda\varphi + U(\sigma,\alpha). \tag{4.2}$$

Substituting (4.2) into (3.1) we obtain the reduced equation (4.1) for U.

As mentioned in §2 there are two possibilities for F: 1) $F = \Gamma_{23}^2$ or 2) $F = \Gamma_{12}^2 + a^2\Gamma_{13}^2$. For the orbit of type 1) we introduce spherical coordinates on S_2:

$$y^0 = \cos\sigma, \quad y^1 = \sin\sigma\cos\alpha, \quad y^2 = \sin\sigma\sin\alpha. \tag{4.3}$$

Then (4.1) becomes

$$\csc^2\sigma\, p_\alpha^2 + p_\sigma^2 = \lambda^2, \quad p_\alpha = \frac{\partial U}{\partial \alpha}, \quad p_\sigma = \frac{\partial U}{\partial \sigma}, \tag{4.4}$$

and the requirement $\Gamma_{23} = p_\alpha = m$ yields the separated solution

$$U = m\alpha + \int \sqrt{m^2\csc^2\sigma - \lambda^2}\, d\sigma.$$

For the orbit of type 2) we introduce elliptic coordinates on S_2:

$$y^0 = a'^{-1}\mathrm{dn}(\alpha,a)\mathrm{dn}(\sigma,a), \quad y' = iaa'^{-1}\mathrm{cn}(\alpha,a)\mathrm{cn}(\sigma,a)$$

$$y^2 = a\,\mathrm{sn}(\alpha,a)\mathrm{sn}(\sigma,a) \tag{4.5}$$

where $a' = \sqrt{1-a^2}$ and dn, cn, sn are Jacobi elliptic functions, [11]. Then (4.1) becomes

$$p_\alpha^2 - p_\sigma^2 = -\lambda^2 a^2(\mathrm{sn}^2\alpha - \mathrm{sn}^2\sigma) \tag{4.6}$$

and the condition $\Gamma_{12}^2 + a^2\Gamma_{13}^2 = (\mathrm{sn}^2\alpha - \mathrm{sn}^2\sigma)^{-1}(-\mathrm{sn}^2\sigma p_\alpha^2 + \mathrm{sn}^2\alpha p_\sigma^2) = \mu^2$ leads to the separated solution

$$U = \int \sqrt{\mu^2 - \lambda^2 a^2 \mathrm{sn}^2\alpha}\, d\alpha + \int \sqrt{\mu^2 - \lambda^2 a^2 \mathrm{sn}^2\sigma}\, d\sigma. \tag{4.7}$$

On the surface (3.15) the condition $\Gamma_{45} = \lambda$ for a solution of (3.1) becomes

$$(x^1)^2 + (x^2)^2 - \frac{2\lambda}{p_0} = -1, \quad p_0 = \sqrt{p_1^2 + p_2^2}. \tag{4.8}$$

Performing the canonical transformation $p_j \to x^j$, $x^j \to -p_j$, which preserves Poisson brackets, we transform (4.8) to the Hamilton-Jacobi equation for the Kepler problem:

$$p_1^2 + p_2^2 - \frac{2\lambda}{r} = -1, \quad r = \sqrt{(x^1)^2 + (x^2)^2}. \tag{4.9}$$

The o(3) symmetry algebra for (4.8), generated by Γ_{12}, Γ_{13}, Γ_{23} is mapped to an o(3) symmetry algebra for (4.9). If $\varphi(\xi_1,\xi_2)$ is a solution of (4.9) (with $x^1 = \xi_1$, $x^2 = \xi_2$) then by prescribing the initial data $x^j = \varphi_{\xi_j}$, $p_j = \xi$ on the surface (3.15) and integrating along characteristics, we find a solution of (3.1) with $\Gamma_{45} = \lambda$. Conversely, if $W(x^\mu)$ is a solution of (3.1) with $\Gamma_{45} = \lambda$ then a function $\varphi(\xi_1,\xi_2)$ such that

$$x^j = -\varphi_{\xi_j}, \quad W_{x^j}(0,x^\ell) = \xi_j, \quad j,\ell = 1,2 \tag{4.10}$$

with $\det(W_{x^j x^\ell}) \neq 0$ is a solution of (4.9). This relationship is a classical analogy of the Fock treatment of the quantum mechanical hydrogen atom.

We have obtained the reduced equation (4.1) from (3.1) by additively separating off the variable φ. However, (4.1) can also be viewed as the equation for the graph of (3.1). Indeed, set $\lambda = 1$ for simplicity and let $U(\sigma,\alpha)$ be a solution of (4.1). A *graph* of U is a function $W(\sigma,\alpha,U)$ of three variables such that $W(\sigma,\alpha,U(\sigma,\alpha)) = 0$. Since $W_\sigma + W_U U_\sigma = 0$, $W_\sigma + W_U U_\sigma = 0$ it follows from (4.1)-(4.4) that W satisfies (3.1) with φ replaced by U. In this sense equations (3.1) and (4.1) are equivalent.

Since (3.1) admits the symmetry algebra o(3,2), so does (4.1). Now however, o(3,2) acts not only on σ,α but also on U. We have used only an o(3) subalgebra of o(3,2) to explain the two systems in which (4.1) admits an additive separation of variables. However, we can use commuting pairs of second-order elements in the enveloping algebra of o(3,2) to distinguish many other symmetry adapted solutions of (4.1). Our restriction to the subalgebra o(3) merely picks out those solutions which are additively separable for (2.1).

Similarly, each of the semi-split systems discussed in §2 permits separation of variables in (3.1). Each reduced equation for (2.1) has its Hamilton-Jacobi counterpart. For example the requirement $P_0 + P_1 = \lambda$ in (3.1) leads to the Hamilton-Jacobi equation for the free particle

$$\lambda p_t + p_x^2 = 0 \tag{4.11}$$

in analogy with (2.8). Here (4.11) is additively R-separable in the same four coordinate systems for which (2.8) permits R-separation and these systems are related to four orbits in the Schrödinger algebra, a symmetry algebra of (4.11). Moreover, the Hamilton-Jacobi equations for the harmonic oscillator, repulsive oscillator and linear potential are all equivalent to (4.11), [9]. It is easy to check that the equation of the graph of (4.11) is (2.1). Thus (4.11) admits $o(3,2)$ as a symmetry algebra; the Schrödinger algebra is just the subalgebra of $o(3,2)$ which commutes with $P_0 + P_1$. All symmetry adapted solutions of (2.1), such as (4.4) and (4.7), thus yields complete integrals for (4.11). However, only those solutions characterized in terms of the Schrödinger algebra are additively separable.

Each of the non-split cyclidic coordinate systems for (1.1) classified in [1] also yields additively separable solutions of (2.1). For example, substitution of the coordinates [311](i) from [1],

$$x^0 = -\tfrac{1}{2}(\cos^2\alpha+\cos^2\beta+\cos^2\gamma), \quad x^1 = \sin\alpha \sin\beta \sin\gamma,$$
$$x^2 = \cos\alpha \cos\beta \cos\gamma \tag{4.12}$$

into (2.1) yields

$$(\sin^2\beta-\sin^2\gamma)p_\alpha^2 + (\sin^2\gamma-\sin^2\alpha)p_\beta^2 + (\sin^2\alpha-\sin^2\beta)p_\gamma^2 = 0, \tag{4.13}$$

$p_\alpha = \frac{\partial W}{\partial \alpha}$, etc. It is not possible to additively separate one of these variables from the other two. However, use of the defining elements

$$2(P_0-P_1)M_{02} + P_0^2 + P_1^2 = \mu, \quad 2P_2M_{02} - M_{12}^2 + P_2^2 = \nu \tag{4.14}$$

leads to the separated solution

$$W = \int \sqrt{\mu \sin^2\alpha+\nu}\, d\alpha + \int \sqrt{\mu \sin^2\beta+\nu}\, d\beta + \int \sqrt{\mu \sin^2\gamma+\nu}\, d\gamma. \quad (4.15)$$

The corresponding separated solutions of (1.1) in these coordinates are products of Lamé-type functions. The ordinary differential equation for each factor involves both separation constants, thus leading to a multiparameter eigenvalue problem.

REFERENCES

1. E.G. KALNINS and W. MILLER, Jr., J. Math. Phys. 16, 2507-2516 (1975); 17, 331-355 (1976); 17, 356-368 (1976); 17, 369-377 (1976).
2. L. GROSS, J. Math. Phys. 5, 687-695 (1964). H. KASTRUP, Phys. Rev. 140, B183-186 (1965). G. POST, J. Math. Phys. 17, 24-32 (1976).
3. C.P. BOYER, E.G. KALNINS and W. MILLER, Jr., J. Math. Phys. 16, 499-511 (1975). C.P. BOYER, E.G. KALNINS and W. MILLER, Jr., Nagoya Math. J. 60, 35-80 (1976). E.G. KALNINS, W. MILLER, Jr. and P. WINTERNITZ, SIAM J. Appl. Math. 30, 630-664 (1976).
4. C.P. BOYER, SIAM J. Math. Anal. 7, 230-263 (1976).
5. W. MILLER, Jr., "Symmetry, separation of variables and special functions", in *Theory and Application of Special Functions*, R. Askey, ed., Academic Press, New York, 1975.
6. I.E. SEGAL, *Mathematical Cosmology and Extragalactic Astronomy*, Academic Press, New York, 1976.
7. J. PATERA and P. WINTERNITZ, J. Math. Phys. 14, 1130-1139 (1973).
8. E.G. KALNINS and W. MILLER, Jr., J. Math. Phys. 15, 1728-1737 (1974).
9. C.P. BOYER and E.G. KALNINS, Symmetries of the Hamilton-Jacobi Equation, J. Math. Phys. (to appear).
10. R. COURANT and D. HILBERT, *Methods of Mathematical Physics*, Vol. 2, Wiley, New York, 1962.
11. A. ERDELYI et al., *Higher Transcendental Functions*, Vol. 2, McGraw-Hill, New York, 1951.

SUBGROUPS OF LIE GROUPS
AND SYMMETRY BREAKING

P. Winternitz

1. INTRODUCTION

The subject of symmetry breaking is a very old one, occurring in various forms and manifestations in all fields of physics. In elementary particle theory in particular, spontaneously broken symmetries are enjoying a period of great popularity in connection with the breaking of gauge invariance, the Higgs phenomenon, etc.

Our aim here is certainly not to review this vast field. What we plan to do is to show how a certain program of research, presently being conducted here in Montreal, basically of mathematical nature, has its applications in a systematic study of symmetry breaking. The system under consideration can essentially be arbitrary, as long as it is described by an equation (or set of equations) that has a Lie group as a symmetry group. The symmetry can be broken "explicitly" by including additional terms of various types in the equation (switching on external fields, taking additional interactions into account, etc.), or "spontaneously" by imposing some supplementary conditions with a lower symmetry (e.g. boundary conditions, nonsymmetric ground states, etc.).

The program under discussion (involving mainly the work of

J. Patera, R.T. Sharp, H. Zassenhaus and myself) is one of classifying all closed subgroups of a given Lie group G into conjugacy classes under some group of automorphisms of G and finding a representative of each class. So far we have mainly (but not exclusively) concentrated on Lie subgroups of Lie groups. General classification methods have been developed and applied to find all Lie subgroups of the Poincaré group, the similitude group (Poincaré extended by dilations), the de Sitter groups O(4,1) and O(3,2), the Euclidean group E(3), the Schrödinger group S_1 (invariance group of the one dimensional time dependent Schrödinger equation) and other groups[1-4].

If the group G is the invariance group of a certain equation and G_k (k runs through a discrete or continuous set of values) are its subgroups then a classification of all G_k provides a classification of all symmetry breaking "influences". Indeed, we can add further terms (or boundary conditions, etc.) to the equation in such a manner that the new equation is invariant under $G_k \subset G$, rather than under G. The remaining invariance group G_k (as long as it is larger than the identity transformation) can be used to solve the new modified equation, or at least to find some properties of the solutions. We now proceed to illustrate the above discussion by several specific examples.

2. A NONRELATIVISTIC PARTICLE IN AN EXTERNAL POTENTIAL AND SUBGROUPS OF E(3).

The work to be discussed in this section was performed by J. Beckers, J. Patera, M. Perroud and myself[5]. Consider the stationary Schrödinger equation for a free particle of spin 0:

$$-\tfrac{1}{2}\Delta\psi(\vec{r}) = E\psi(\vec{r}). \qquad (2.1)$$

Its kinematical invariance group is the Euclidean group E(3), generated by the linear momenta (translations) P_i and angular momenta L_i (i=1,2,3), satisfying

$$[H,P_i] = [H,L_i] = 0$$

$$[L_i,L_k] = i\varepsilon_{ik\ell}L_\ell, \quad [L_i,P_k] = i\varepsilon_{ik\ell}P_\ell, \quad [P_i,P_k] = 0. \quad (2.2)$$

It is a simple application of our general algorithms[2] to find representatives of all subalgebras of the Lie algebra LE(3) of E(3). Once they are found we consider a particle in a field described by an arbitrary scalar and vector potential. This leads to a Schrödinger equation of the form

$$\{-\tfrac{1}{2}\Delta+V(\vec{r})+\vec{A}(\vec{r})\vec{P}\}\psi(\vec{r}) = E\psi(\vec{r}) \quad (2.3)$$

where self-adjointness of the Hamiltonian $H = H^+$ implies

$$A_i(\vec{r}) = A_i^*(\vec{r}), \quad \text{Im } V(\vec{r}) = -\tfrac{1}{2} \text{ div } \vec{A}(\vec{r}). \quad (2.4)$$

We now run through the list of representatives of all subalgebras of LE(3) and for each of them find the most general potentials $V(\vec{r})$ and $\vec{A}(\vec{r})$ that are left invariant by the corresponding subgroup of E(3). These will then represent the interaction that breaks the E(3) symmetry of the free equation (2.1) to that of the corresponding subgroup.

Let us consider just one subgroup (the least obvious one) as an example, namely the subgroup $\bar{E}(2)$ (universal covering group of the Euclidean group E(2)), generated by

$$L_3 + aP_3, \quad P_1, \quad P_2, \qquad a \neq 0. \quad (2.5)$$

Introduce helical coordinates ρ, u and v:

$$x = \rho\cos(u-v), \quad y = \rho\sin(u-v), \quad z = a(u+v). \quad (2.6)$$

In these coordinates it is easy to check that the condition $[H,X] = 0$, where $H = -(\tfrac{1}{2})\Delta + \vec{V}(\vec{r}) + \vec{A}(\vec{r})\vec{P}$ and X runs through the three generators (2.5) implies

$$V = \text{const.}$$
$$\vec{A} = (f\cos(u+v)+g\sin(u+v), f\sin(u+v)-g\cos(u+v), h). \qquad (2.7)$$

Thus, the interaction is completely specified up to the four constants V, f, g and h. To solve the Schrödinger equation (2.3) with the potentials (2.7) we make use of the symmetry group $\bar{E}(2)$ and add the equations

$$P_1\psi = k_1\psi, \quad P_2\psi = k_2\psi \qquad (2.8)$$

to (2.3). This leads to a separation of variables in cartesian coordinates and provides us with the wave function

$$\psi(x,y,z) = e^{-ihz} Z\left(\frac{z}{2a} + \frac{\alpha}{2}\right) e^{ik_1x+ik_2y} \qquad (2.9)$$

where $Z(\xi)$ satisfies the Mathieu equation

$$Z''(\xi) + 4a^2(A-D+2D\cos^2\xi)Z(\xi) = 0$$
$$A = 2(E-V) - k_1^2-k_2^2,\ D\cos\alpha = -fk_1 + gk_2,\ D\sin\alpha = fk_2 + gk_1. \qquad (2.10)$$

Thus, the requirement that the symmetry E(3) be broken to $\bar{E}(2)$ has led to a quite definite and nontrivial interaction term and the Schrödinger equation turned out to be separable and solvable analytically.

For a general consideration of all subgroups of E(3) along the above lines we refer to our article[5]; here we just make several comments.

1. We have found a one-to-one correspondence between subgroups and symmetry breaking interactions.

2. The one-dimensional subalgebras $\{P_3\}$, $\{L_3\}$ and $\{L_3+aP_3\}$

lead to interactions involving arbitrary functions of two variables, to a partial separation of variables in the Schrödinger equation and to an explicitly known dependence of the wave function on one variable. The subalgebras $\{P_1,P_2\}$, $\{L_3,P_3\}$, $\{L_1,L_2,L_3\}$ and $\{L_3,P_1,P_2\}$ lead to interactions involving arbitrary functions of one variable, to a complete separation of variables and to an explicit dependence of the wave functions on two variables. The subalgebras $\{L_3+aP_3,P_1,P_2\}$, $\{P_1,P_2,P_3\}$, $\{L_3,P_1,P_2,P_3\}$ and $\{L_1,L_2,L_3,P_1,P_2,P_3\}$ determine the interaction completely (up to constants), lead to a complete separation of variables (sometimes in several coordinate systems) and to an explicit known form of the wave functions (and the energy spectrum).

3. The approach can be generalized to particles of arbitrary spin and to more general Ansatzes on the form of the interaction. Spin $\frac{1}{2}$ has already been treated[5].

3. THE TIME DEPENDENT LINEAR AND NONLINEAR SCHRODINGER EQUATIONS

The results of this section were obtained in collaboration with C.P. Boyer and R.T. Sharp[6]. Let us consider the one dimensional time dependent Schrödinger equation for a free particle:

$$\hat{S}\psi = \frac{\partial^2\psi}{\partial x^2} + i\,\frac{\partial\psi}{\partial t} = 0. \tag{3.1}$$

Its invariance group S_1 (or S_n in n space dimensions) is well known (in the guise of the heat equation since the times of S. Lie), is of dimension 6 and its Lie algebra LS_1 can be represented by the generators

$$X_1 = K_2 + L_3 = H = \partial_t \qquad X_4 = P = \partial_x$$

$$X_2 = 2K_1 = D = 2t\partial_t + x\partial_x + \tfrac{1}{2} \qquad X_5 = B = -t\partial_x + ix/2$$

$$X_3 = -K_2 + L_3 = C = t^2\partial_t + tx\partial_x + t/2 - ix^2/4 \qquad X_6 = E = i. \tag{3.2}$$

Note that {H,D,C} is the algebra of SL(2,R), {P,B,E} that of the Weyl group and that {P,B,E} is an invariant nonabelian subalgebra of LS_1. By construction the generators X_i (i=1,...,6) satisfy

$$[\hat{S},X_i] = \lambda_i(x,t)\hat{S} \tag{3.3}$$

where $\lambda_i(x,t)$ are some functions (actually it turns out that $\lambda_1 = \lambda_4 = \lambda_5 = \lambda_6 = 0$, $\lambda_2 = 2$ and $\lambda_3 = 2t$). Note that H (the Hamiltonian), P (linear momentum), B (the Galilei boost) and E (a constant) generate the Galilei group, D generates dilations and C generates conformal transformations.

Previously developed general methods[2] can again be applied to find representatives of all subalgebras of the Schrödinger algebra LS_1. The result in this case is quite complicated and is reproduced on Figure 1.

<u>Figure 1</u>. Subalgebras of LS_1 classified under S_1 ($a\neq 0, b>0, \varepsilon=\pm 1$).

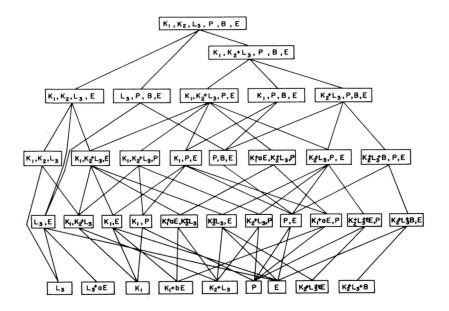

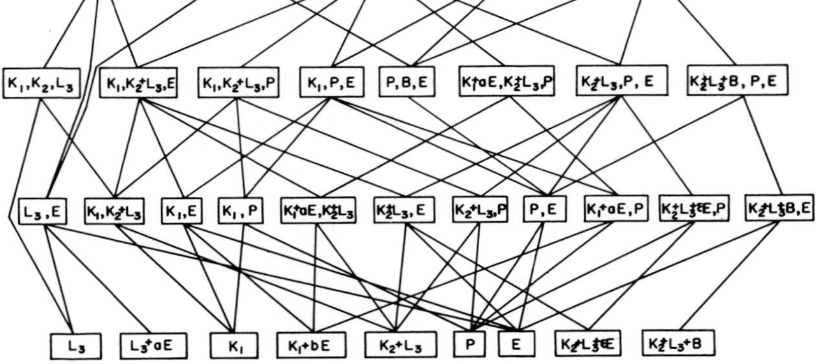

We now consider the Schrödinger equation with a quite general "interaction term", depending on x,t, the wave function ψ and its complex conjugate ψ^*:

$$\frac{\partial^2\psi}{\partial x^2} + i\,\frac{\partial\psi}{\partial t} = F(x,t,\psi,\psi^*). \tag{3.4}$$

Let us require that eq. (3.4) be invariant under a subgroup of the Schrödinger group. Consider a one dimensional subgroup acting on space time as in

$$(x',t') = (x,t)\cdot g \tag{3.5}$$

and on the wave function as

$$[T_g\psi](x,t) = [e^{\alpha X}\psi](x,t) = \mu(g,x',t')\psi(x',t'). \tag{3.6}$$

Here X is the generator of the group (some linear combination of the X_i of (3.2)), α is a real parameter, $\mu(g,x',t')$ a multiplier. Expanding T_g into a Taylor series about $\alpha=0$ we find

$$X\psi = \left\{\frac{dx'}{d\alpha}\,\partial_x + \frac{dt'}{d\alpha}\,\partial_t + \frac{d\mu(g,x',t')}{d\alpha}\right\}\psi$$

$$= \{a(x,t)\partial_x + b(x,t)\partial_t + c(x,t)\}\psi. \tag{3.7}$$

Remembering that

$$X = \sum_{j=1}^{6} \alpha_j X_j$$

with α_j real we can spell out a(x,t), b(x,t) and c(x,t) in each specific case. Notice that a and b are real and c is complex. The requirement that (3.4) be invariant

$$\left(\frac{\partial^2}{\partial x^2} + i\,\frac{\partial}{\partial t}\right)[T_g\psi](x,t) = F(x,t[T_g\psi](x,t),[T_g\psi]^*(x,t)) \tag{3.8}$$

upon differentiation with respect to α implies the relation

$$\left(\frac{\partial^2}{\partial x^2} + i\frac{\partial}{\partial t}\right)(X\psi) = (X\psi)F_\psi + (X\psi)^*F_{\psi^*}. \tag{3.9}$$

Since X is an element of a subalgebra of LS_1 it must also satisfy (3.3) and we obtain a partial differential equation for F:

$$(X+\lambda)F = (X\psi)F_\psi + (X\psi)^*F_{\psi^*}, \tag{3.10}$$

or more explicitly

$$a(x,t)F_t + b(x,t)F_x - c\psi F_\psi - c^*\psi^*F_{\psi^*} = -(c+\lambda)F. \tag{3.11}$$

To find the general solution of (3.11) we must solve a set of ordinary differential equations, namely the subsidiary equations

$$\frac{dt}{a} = \frac{dx}{b} = -\frac{d\psi}{c\psi} = -\frac{d\psi^*}{c^*\psi^*} = -\frac{dF}{(c+\lambda)F}. \tag{3.12}$$

Here ψ and ψ^* together with x and t are independent variables and (3.12) implies that F is some known function of x, t, ψ and ψ^*, times an arbitrary function of three variables obtained by solving the first three equations of (3.12).

For higher dimensional Lie subalgebras we proceed in the same manner for each generator in some chosen basis and obtain a set of equations of the type (3.11) and (3.12).

In this manner we run through all subalgebras of LS_1 and generate an invariant interaction (the most general invariant interaction of the type $F(x,t,\psi,\psi^*)$) for each subalgebra. Invariance under the subalgebra can then be used to generate specific solutions of the Schrödinger equation (3.4) by adding the condition

$$X\psi(x,t) = 0 \tag{3.13}$$

to the Schrödinger equation (3.4). Equation (3.13) is linear; finding its solution reduces to solving the first two of equations (3.12). It leads to a separation of variables in the (generally nonlinear) Schrödinger equation (3.4).

Let us consider one specific subgroup as an example, namely a one dimensional subgroup generated by

$$2K_1 + aE = X_2 + aX_6 = 2t\partial_t + x\partial_x + \tfrac{1}{2} + ia, \quad 0 \leq a < \infty \tag{3.14}$$

(a is a fixed parameter). The equations (3.12) providing a solution of the partial differential equation for the interaction F are in this case

$$\frac{dt}{2t} = \frac{dx}{x} = -\frac{d\psi}{(\frac{1}{2}+ia)\psi} = -\frac{d\psi^*}{(\frac{1}{2}-ia)\psi^*} = -\frac{dF}{(\frac{5}{2}+ia)F}. \tag{3.15}$$

The first equation provides us with the "similarity" variable to be used, namely $\xi = x/t^{\frac{1}{2}}$. Adding the next equation we find that the wave function satisfies

$$\psi(x,t) = t^{-(1+2ia)/4}\exp(\frac{i\xi^2}{8})\phi(\xi), \tag{3.16}$$

where $\phi(\xi)$ is so far an arbitrary function of the similarity variable. Function (3.16) by construction satisfies

$$(2K_1+aE)\psi(x,t) = 0 \tag{3.17}$$

and this equation has led to the separation of variables. Finally, equations (3.15) yield the interaction

$$F(x,t,\psi,\psi^*) = \frac{\psi(x,t)}{t}\, G(\xi, \psi\psi^*\sqrt{t}, \frac{\psi}{\psi^*}\, t^{ia}\exp(-\frac{i\xi^2}{4})), \tag{3.18}$$

where G is an arbitrary function of the three indicated variables. Substituting (3.18) and (3.16) back into the Schrödinger equation (3.4) we obtain an ordinary differential equation for $\phi(\xi)$:

$$\phi''(\xi) + (\frac{\xi^2}{16} + \frac{a}{2})\phi(\xi) = \phi(\xi)G(\xi,|\phi|^2, \frac{\phi}{\phi^*}). \tag{3.19}$$

Notice that the parameter a in (3.17) plays the role of an eigenvalue. If the equation is linear i.e. $G = G(\xi)$ in (3.18), or if it is nonlinear, but G does not depend on the phase of ψ, i.e. $G = G(\xi,\psi\psi^*\sqrt{t})$, then a does not figure in the interaction term. It can hence be considered a free parameter in (3.19) and for each value of a we obtain a different solution. In particular, if (3.19) happens to be linear, then we thus obtain a complete set of solutions.

The one dimensional subalgebras $2L_3 + aE$ $(-\infty<a<\infty)$, $2K_1 + bE$ $(0\le b<\infty)$, $K_2 + L_3 + \kappa E$ $(\kappa=0,\pm1)$, $K_2 + L_3 + B$ and P can be treated in the same manner. We always obtain a wave function of the form

$$\psi(x,t) = \frac{\phi(\xi)}{\rho(t)e^{i\mu(\xi,t)}} \tag{3.20}$$

and an interaction term

$$F = \frac{\psi(x,t)}{\rho^4(t)} G(\xi,\psi\psi^*\rho^2, \frac{\psi}{\psi^*} e^{2i\mu(\xi,t)}), \tag{3.21}$$

where $\phi(\xi)$ satisfies an ordinary differential equation obtained by substituting (3.20) and (3.21) into the Schrödinger equation. The functions $\{\rho(t),\mu(\xi,t)\}$ are known elementary functions (equal to $\{(1+t^2)^{\frac{1}{4}},[(1+it)/(1-it)]^{a/2}\exp(-i\xi^2 t/4)\}$, $\{t^{\frac{1}{4}},t^{ib/2}\exp(-i\xi^2/8)\}$, $\{1,\exp i\kappa t\}$, $\{1,[\exp it(\xi-t^2/6)]/2\}$ and $\{1,1\}$ for the algebras listed above, respectively). The similarity variable ξ is $x/(1+t^2)^{\frac{1}{2}}$, $x/t^{\frac{1}{2}}$, x, $x + \frac{1}{2}t^2$ and t, respectively. The only remaining subgroup of dimension 1 is generated by E. Since E is a constant (E=i) the condition $E\psi = 0$ only leads to a trivial result $\psi = 0$. The requirement that E be a symmetry does however have nontrivial implications, namely that the interaction term satisfies

$$F = \psi(x,t)G(x,t,|\psi|^2). \qquad (3.22)$$

In the case of higher dimensional subalgebras further constraints are imposed upon the form of the interaction F. In some cases it will involve an arbitrary function of two variables, in other cases a function of one variable and in still other cases it will be completely specified. Indeed, the interaction $F = c\psi|\psi|^4$ (c = const.) is invariant under the entire algebra and and it is also the only interaction invariant under the subalgebras $\{K_1, K_2+L_3, P, E\}$ and $\{K_1, K_2+L_3, P, B, E\}$. Invariance of the Schrödinger equation under a higher dimensional Lie algebra makes it possible to obtain several different solutions, or classes of solutions, by imposing the condition $X\psi = 0$ where X is a linear combination of the generators of the subgroup. A systematic approach would require that the operators X be classified into orbits under the action of the subgroup generated by the corresponding Lie algebra. Each class of operators provides a class of solutions.

4. THE HYDROGEN ATOM IN AN EXTERNAL ELECTROMAGNETIC FIELD

The results of this section were obtained in collaboration with E.G. Kalnins and W. Miller Jr.[7]. It is well known[8] that the stationary Schrödinger equation for a hydrogen atom (in three dimensions)

$$H\psi = (-\tfrac{1}{2}\Delta - \frac{\lambda}{r})\psi(x_1,x_2,x_3) = E\psi(x_1,x_2,x_3) \qquad (4.1)$$

is invariant under the group O(4), generated by the components of the angular momentum operator $\vec{L}$ and of the Laplace-Runge-Lenz vector

$$\vec{K} = (-2E)^{-\frac{1}{2}}\{\tfrac{1}{2}(\vec{p}\times\vec{L}-\vec{L}\times\vec{p}) - \frac{\lambda\vec{r}}{r}\} \equiv (-2E)^{-\frac{1}{2}}\vec{A}. \qquad (4.2)$$

Equation (4.1) with a Coulomb potential in Euclidean three space can be rewritten as a free Laplace equation on a four dimensional sphere (in momentum space)[8]. Wave functions for the hydrogen atom can thus be obtained either by solving the Schrödinger equation (4.1) or the Laplace equation on an O(4) sphere. It is hence of physical interest to consider the separation of variables in these two equations. Separation of variables on an O(4) sphere has been considered both geometrically[9] and from the point of view of group theory[7]. Here let us dwell briefly on separation of variables in (4.1) and on its relation to symmetry and symmetry breaking.

Separation of variables in equation (4.1) can be characterized by the fact that a pair of operators {X,Y}, commuting with each other and with the Hamiltonian H, are diagonalized together with H, i.e. we are looking for the simultaneous eigenfunctions of these operators:

$$H\psi = E\psi, \quad X\psi = \mu\psi, \quad Y\psi = \nu\psi \tag{4.3}$$

(μ and ν are constants). The operators X and Y are second order polynomials in the generators of O(4).

Exactly four separable coordinate systems exist[7], namely the following.

1. Spherical coordinates:

$$r = (r \sin\theta \cos\phi, r \sin\theta \sin\phi, r \cos\theta), \quad 0 \le r < \infty, \quad 0 \le \theta \le \pi, \quad 0 \le \phi < 2\pi. \tag{4.4}$$

We have:

$$X = \vec{L}^2, \quad Y = L_3 \tag{4.5}$$

and the separated eigenfunctions are

$$\psi^n_{\ell m}(\vec{r}) = \left(\frac{(n-\ell)!}{n+\ell+1}\right)^{\frac{1}{2}} (2\alpha)^{3/2} e^{-\alpha r} (2\alpha r)^{\ell} L^{2\ell+1}_{n-1}(2\alpha r) Y_{\ell m}(\theta,\phi) \tag{4.6}$$

$$\alpha = \frac{\lambda}{2n+1}, \quad 0 \le \ell \le n, \quad -\ell \le m \le \ell$$

where $L_{n-1}^{2\ell+1}(2\alpha r)$ is a Laguerre polynomial.

These coordinates correspond to the group reduction $O(4) \supset O(3) \supset O(2)$, i.e. the three equations (4.3) are invariant under $O(4)$, $O(3)$ and $O(2)$, respectively.

2. Spheroconical coordinates:

$$\vec{r} = (r \text{ sn } \alpha \text{ dn } \beta, r \text{ cn } \alpha \text{ cn } \beta, r \text{ dn } \alpha \text{ sn } \beta) \tag{4.7}$$

$$0 \le r < \infty, \quad -K \le \alpha \le K, \quad -2K \le \beta \le 2K'$$

where sn α, cn α and dn α are Jacobi elliptic function of modulus k; sn β, cn β and dn β are of modulus k', with $k^2 + k'^2 = 1$, k and k' real. The real and imaginary periods of the Jacobi elliptic functions of modulus k are $4K$ and $4iK'$, respectively.

In this case we have

$$X = \vec{L}^2, \quad Y = L_1^2 + k'^2 L_2^2, \quad 0 < k' < 1 \tag{4.8}$$

and the separated eigenfunctions are

$$\psi_{\ell m}^{npq}(\vec{r}) = \left[\frac{(n-\ell)!}{(n+\ell+1)!}\right]^{\frac{1}{2}} (2\alpha)^{3/2} e^{-\alpha r} (2\alpha r)^{\ell} L_{n-1}^{2\ell+1}(2\alpha r) F_{\ell m}^{pq}(-i\alpha+iK+K',\beta) \tag{4.9}$$

where $F_{\ell m}^{pq}$ are ellipsoidal harmonics[10,11] (products of Lamé polynomials).

These coordinates correspond to the reduction $O(4) \supset O(3) \supset D_2$ where D_2 is a dihedral group (in this case corresponding to rotations through π about the axes x, y and z and the identity transformation). Again, the operator X is invariant under $O(3)$, Y only under its subgroup D_2.

3. Parabolic coordinates:

$$\vec{r} = (\xi\eta \cos \phi, \xi\eta \sin \phi, \tfrac{1}{2}(\xi^2-\eta^2)) \tag{4.10}$$

$$0 \le \xi < \infty, \quad 0 \le \eta < \infty, \quad 0 \le \phi < 2\pi.$$

The diagonalized operators are

$$X = L_3, \qquad Y = A_3 \tag{4.11}$$

and the separated eigenfunctions are

$$\psi_{n_1 n_2 m}(\vec{r}) = \alpha^{|m|+3/2}\left\{\frac{n_1!n_2!}{n\pi(n_1+|m|)(n_2+|m|)}\right\}^{\frac{1}{2}}(\xi\eta)^{|m|/2}$$

$$\times \exp\{-\tfrac{1}{2}\alpha(\xi+\eta)+im\phi\}L_{n_1}^{|m|}(\alpha\xi)L_{n_2}^{|m|}(\alpha\eta) \tag{4.12}$$

$$0 \le |m| < n, \quad 0 \le n_1 < n - |m| + 1,$$

$$0 \le n_2 < n - |m| + 1, \quad n = n_1 + n_2 + |m|.$$

These coordinates correspond to the group reduction $O(4) \supset O(2) \otimes O(2)$.

4. Prolate spheroidal coordinates:

$$\vec{r} = \{f \sinh \xi \sin \eta \cos \phi, f \sinh \xi \sin \eta \sin \phi, f(\cosh \xi \cos \eta + 1)\}$$

$$0 \le \xi < \infty, \quad 0 \le \eta \le \pi, \quad 0 \le \phi < 2\pi. \tag{4.13}$$

The parameter $f(0<f<\infty)$ is a fixed constant ($2f$ is a focal distance). The diagonalized operators are

$$X = L_3^2, \qquad Y = \vec{L}^2 - 2fA_3 \tag{4.14}$$

and the separated eigenfunctions are products of two spheroidal Coulomb functions and an exponential[12].

The coordinates and operators correspond to the group reduction $O(4) \supset D_\infty \supset O(2)$ where D_∞ is the group generated by all rotations about the z axis and rotations through π about an axis perpendicular to z.

Thus, four different separable coordinate systems correspond to four different reductions of O(4) to subgroups (involving both Lie groups and discrete groups).

Let us now relate the above considerations to symmetry breaking. Consider a hydrogen atom in an electromagnetic field, i.e. the Hamiltonian

$$H = \tfrac{1}{2}(\vec{p}-e\vec{A}(\vec{r}))^2 + V(\vec{r}) = -\tfrac{1}{2}\Delta + B_i\partial_{x_i} + W \tag{4.15}$$

where $\vec{A}(\vec{r})$ and $V(\vec{r})$ are vector and scalar potentials and $\vec{B} = ie\vec{A}$, $W = V + \frac{1}{2}\,\mathrm{div}\,\vec{B} - \frac{1}{2}\vec{B}^2$. Clearly the Hamiltonian (4.15) will only be O(4) invariant if $\vec{A}(\vec{r}) = 0$, $V(\vec{r}) = -\frac{\lambda}{r}$. However, some "remnants" of the O(4) symmetry may survive for more general potentials and these may in particular lead to the separation of variables in one of the above coordinate systems. We pose the following somewhat limited question. What are the forms of $\vec{A}$ and V (or $\vec{B}$ and W) for which a pair of operators $\tilde{X}$, $\tilde{Y}$ exists, such that

$$[H,\tilde{X}] = [H,\tilde{Y}] = [\tilde{X},\tilde{Y}] = 0 \tag{4.16}$$

and the pair $\{\tilde{X},\tilde{Y}\}$ is obtained from one of the four above pairs $\{X,Y\}$ by adding terms involving only first derivatives and scalar functions.

The corresponding potentials $\vec{A}$ and V and the corresponding commuting integrals of motion are easily found (using the appropriate coordinates) and in each case we obtain a separable Schrödinger equation. Omitting all details[7] we just list the resulting function $\vec{B}$ and W and the operators $\tilde{X}$ and $\tilde{Y}$.

1. Spherical coordinates:

$$B_1 = (u(r)\sin\theta + \frac{1}{r}v(\theta)\cos\theta)\cos\phi - \frac{w(\phi)\sin\phi}{r\sin\theta},$$

$$B_2 = (u(r)\sin\theta + \frac{1}{r}v(\theta)\cos\theta)\sin\phi + \frac{w(\phi)\cos\phi}{r\sin\theta},$$

$$B_3 = u(r)\cos\theta - \frac{1}{r}v(\theta)\sin\theta,$$

$$W = p(r) + \frac{1}{r^2}q(\theta) + \frac{1}{r^2\sin^2\theta}s(\phi), \tag{4.17}$$

$$\tilde{X} = L_1^2 + L_2^2 + L_3^2 + 2(q(\theta)+v(\theta)\frac{\partial}{\partial\theta}) + \frac{2}{\sin^2\theta}(s(\phi)+w(\phi)\frac{\partial}{\partial\phi}),$$

$$\tilde{Y} = L_3^2 + 2(s(\phi)+w(\phi)\frac{\partial}{\partial\phi}).$$

2. Spheroconical coordinates:

$$B_1 = u(r)\,\mathrm{sn}\,\alpha\,\mathrm{dn}\,\beta + \frac{\mathrm{cn}\,\alpha\,\mathrm{dn}\,\alpha\,\mathrm{dn}\,\beta\,v(\alpha) - k'^2\mathrm{sn}\,\alpha\,\mathrm{cn}\,\beta\,\mathrm{sn}\,\beta\,w(\beta)}{r(k^2\mathrm{cn}^2\alpha + k'^2\mathrm{cn}^2\beta)},$$

$$B_2 = u(r)\,\mathrm{dn}\,\alpha\,\mathrm{sn}\,\beta + \frac{-k'^2\mathrm{sn}\,\alpha\,\mathrm{cn}\,\alpha\,\mathrm{cn}\,\beta\,v(\alpha) + \mathrm{dn}\,\alpha\,\mathrm{cn}\,\beta\,\mathrm{dn}\,\beta\,w(\beta)}{r(k^2\mathrm{cn}^2\alpha + k'^2\mathrm{cn}^2\beta)},$$

$$B_3 = u(r)\,\mathrm{cn}\,\alpha\,\mathrm{cn}\,\beta - \frac{\mathrm{sn}\,\alpha\,\mathrm{dn}\,\alpha\,\mathrm{cn}\,\beta\,v(\alpha) + \mathrm{cn}\,\alpha\,\mathrm{sn}\,\beta\,\mathrm{dn}\,\beta\,w(\beta)}{r(k^2\mathrm{cn}^2\alpha + k'^2\mathrm{cn}^2\beta)},$$

$$W = p(r) + \frac{q(\alpha)+s(\beta)}{r^2(k^2\mathrm{cn}^2\alpha + k'^2\mathrm{cn}^2\beta)}, \tag{4.18}$$

$$\tilde{X} = L_1^2 + L_2^2 + L_3^2 + \frac{2}{(k^2\mathrm{cn}^2\alpha + k'^2\mathrm{cn}^2\beta)}[v(\alpha)\frac{\partial}{\partial\alpha} + w(\beta)\frac{\partial}{\partial\beta} + q(\alpha) + s(\beta)],$$

$$\tilde{Y} = L_1^2 + k'^2L_2^2 + \frac{2}{(k^2\mathrm{cn}^2\alpha + k'^2\mathrm{cn}^2\beta)}$$

$$\times\,[k'^2\mathrm{sn}^2\beta(v(\alpha)\frac{\partial}{\partial\alpha} + q(\alpha)) + \mathrm{dn}^2\alpha(w(\beta)\frac{\partial}{\partial\beta} + h(\beta))].$$

3. Parabolic coordinates:

$$B_1 = -\frac{u(\phi)}{\xi\eta}\sin\phi + \frac{(v(\xi)\eta + w(\eta)\xi)}{\xi^2+\eta^2}\cos\phi,$$

$$B_2 = \frac{u(\phi)}{\xi\eta}\cos\phi + \frac{(v(\xi)\eta + w(\eta)\xi)}{\xi^2+\eta^2}\sin\phi,$$

$$B_3 = \frac{u(\phi)}{\xi^2\eta^2} + \frac{v(\xi)+w(\eta)}{\xi^2+\eta^2},$$

$$W = \frac{p(\xi)+q(\eta)}{\xi^2+\eta^2} + \frac{s(\phi)}{\xi^2\eta^2}, \tag{4.19}$$

$$\tilde{X} = L_3^2 - 2(s(\phi)+u(\phi)\ \frac{\partial}{\partial\phi}),$$

$$\tilde{Y} = A_3^2 - 2\left[\frac{\eta^2(v(\xi)\frac{\partial}{\partial\xi}+p(\xi)) - \xi^2(w(\eta)\frac{\partial}{\partial\eta}+q(\eta))}{(\xi^2+\eta^2)} + \left(\frac{1}{\xi^2} - \frac{1}{\eta^2}\right)(u(\phi)\frac{\partial}{\partial\phi}+s(\phi))\right].$$

4. Prolate spheroidal coordinates:

$$B_1 = \frac{d\cos\phi}{(\mathrm{sh}^2\xi+\sin^2\eta)}(u(\xi)\mathrm{ch}\ \xi\ \sin\eta+v(\eta)\mathrm{sh}\ \xi\ \cos\eta) - \frac{\ell\sin\phi}{\mathrm{sh}\ \xi\ \sin\eta}\ w(\phi),$$

$$B_2 = \frac{d\sin\phi}{(\mathrm{sh}^2\xi+\sin^2\eta)}(u(\xi)\mathrm{ch}\ \xi\ \sin\eta+v(\eta)\mathrm{sh}\ \xi\ \cos\eta) + \frac{\ell\cos\phi}{\mathrm{sh}\ \xi\ \sin\eta}\ w(\phi),$$

$$B_3 = \frac{d}{(\mathrm{sh}^2\xi+\sin^2\eta)}[u(\xi)\mathrm{sh}\ \xi\ \cos\eta - v(\eta)\mathrm{ch}\ \xi\ \sin\eta],$$

$$W = \frac{p(\xi)+q(\eta)}{\mathrm{sh}^2\xi+\sin^2\eta} + \frac{s(\phi)}{\mathrm{sh}^2\xi\ \sin^2\eta}, \tag{4.20}$$

$$X = L_1^2 + L_2^2 + L_3^2 - 2dA_3$$

$$-2[\{\mathrm{ch}^2\xi(u)(\xi)\frac{\partial}{\partial\xi}+p(\xi)) + \cos^2\eta(v(\eta)\frac{\partial}{\partial\eta}+q(\eta))\}/(\mathrm{sh}^2\xi+\sin^2\eta)$$

$$+ \frac{\mathrm{sh}^2\xi+\cos^2\eta}{\mathrm{sh}^2\xi\ \sin^2\eta}(w(\phi)\frac{\partial}{\partial\phi}+s(\phi))],$$

$$Y = L_3^2 - 2(w(\phi)\frac{\partial}{\partial\phi}+s(\phi)).$$

5. DISCRETE SUBGROUPS OF O(3) AND SYMMETRY ADAPTED WAVE FUNCTIONS

The results of this section were obtained in collaboration with J. Patera[13]. Let us consider a situation typical for molecular physics and other fields, namely one in which we have an initial Hamiltonian, invariant under a Lie group, say O(3). To this a further interaction term is added, breaking the symmetry down to a discrete subgroup $\Gamma \subset O(3)$. More generally we could have several perturbing terms added successively, breaking the symmetry along the chain $O(3) \supset \Gamma_1 \supset \Gamma_2 \ldots \supset \Gamma_n$.

In such a situation it is most advantageous to make use of the representation theory of O(3) in a basis corresponding to the reduction $O(3) \supset \Gamma$. The basis functions $|\ell\gamma\rangle$ will be labelled by the angular momentum ℓ and by the irreducible representation γ of Γ that they belong to. However, representations of O(3) are not simply reducible to representations of Γ. In other words the representation γ of Γ can occur in a representation ℓ of O(3) more than once and we are faced with a degeneracy or a "missing label problem". Let us denote a complete and orthogonal set of basis functions $|\ell\kappa\gamma\rangle$, where κ is a label or set of labels to be specified.

We propose to construct the basis functions $|\ell\kappa\gamma\rangle$ and to resolve the missing label problem by providing a certain complete set of commuting hermitian operators and requiring that $|\ell\kappa\gamma\rangle$ be eigenfunctions of this set. The set will include the Casimir operator $\vec{L}^2$ of O(3) and a maximal set of commuting generators of Γ (providing the labels γ). To complete the set we go into the enveloping algebra of O(3) and there construct a complete set of commuting Γ invariant operators.

To do this we must for each subgroup Γ of O(3) construct all such invariants or rather construct an integrity basis (an arbitrary Γ invariant can then be expressed as a polynomial of the invariants in the integrity basis). This task is greatly simplified by the fact that every subgroup of O(3) can be regarded as a transformation of a real three dimensional vector space (the coordinates of a vector $\vec{r}$). Integrity bases for invariants in $\vec{r}$ space are known[14] for all subgroups of O(3). The generators of O(3), i.e. the components of the angular momentum $\vec{L}$, transform like the coordinate vector $\vec{r}$ under proper rotations. When adapting invariants from $\vec{r}$ space to $\vec{L}$ space, two factors must be taken into account: 1. Under reflections $\vec{r}$ is a vector, $\vec{L}$ a pseudovector. 2. The components of $\vec{r}$ commute, those of $\vec{L}$ do not.

Let us now consider the subgroups of O(3) specifically. The groups consisting of rotations only are of types C_n, D_n, T, O, and I, the cyclic, dihedral, tetrahedral, octahedral, and icosahedral groups respectively. The n-fold axis for C_n and D_n is taken to be the third axis, the 2-fold axis of D_n is the first axis. The cube corresponding to O is placed with its centre at the origin and its faces parallel to the coordinate planes. The tetrahedron corresponding to T is placed so that its 3-fold axes coincide with 3-fold axes of O, and the 2-fold axes are the coordinate axes. In the case of the icosahedral group I, the coordinate axes pass through the mid-points of opposite edges.

The groups denoted by C_{ni}, D_{ni}, T_i, O_i, and I_i are obtained from C_n, D_n, T, O, and I respectively by adjoining to them an inversion

$$J = \begin{pmatrix} -1 & 0 & 0 \\ 0 & -1 & 0 \\ 0 & 0 & -1 \end{pmatrix}.$$

Finally, there are groups denoted in general by $G_1[G_2$, where G_2 is a rotational group and G_1 is its subgroup of index 2. A group $G_1[G_2$ consists of all rotations of G_1 and all elements Jr, r being a rotation in G_2 but not in G_1. Thus one has the following groups: $C_n[D_n$, $C_n[C_{2n}$, $D_n[D_{2n}$, and $T[O$.

The $\vec{L}$-space invariants of the subgroups of SO(3), i.e. C_n,

$\mathcal{D}_n$, T, O and I can be obtained from the $\vec{r}$ space ones by first symmetrizing and then replacing $\vec{r} \to \vec{L}$. The groups C_{ni}, $\mathcal{D}_{ni}$, T_i, O_i and I_i will have the same invariants in $\vec{L}$ space as C_n, $\mathcal{D}_n$, T, O and I, respectively, since the inversion J does not act in $\vec{L}$ space (as opposed to $\vec{r}$ space). The invariants of the groups of the type $G_1[G_2$ will for the same reason coincide with those of G_2. Taking these comments into account we can use Meyer's results[14] to write out integrity bases for all the considered invariants. From the obtained list for each algebra we must still eliminate operators that can be written as commutators of other operators in the list. This leads to a very simple result (much simpler than the corresponding $\vec{r}$ space result), summarized in Table 1. The notations used are:

$$C_n = \sum_{k=0}^{[n/2]} (-1)^k \binom{n}{2k} \{L_1^{n-2k} L_2^{2k}\}$$

$$O_3 = \{L_1 L_2 L_3\}$$

$$O_4 = L_1^4 + L_2^4 + L_3^4 \qquad (5.1)$$

$$I_6 = \{(\tau^2 L_1^2 - L_2^2)(\tau^2 L_2^2 - L_3^2)(\tau^2 L_3^2 - L_1^2)\}$$

$$I_{10} = \{(L_1^4 + L_2^4 + L_3^4)(\tau^{-2} L_1^2 - \tau^2 L_2^2)(\tau^{-2} L_2^2 - \tau^2 L_3^2)(\tau^{-2} L_3^2 - \tau^2 L_1^2)\}.$$

Here $[n/2]$ denotes the integer part of $n/2$, $\tau = \frac{1}{2}(1+\sqrt{5})$, and the curly brackets mean complete symmetrization of each term with respect to permutations of L_1, L_2, and L_3. Thus e.g. $\{L_1^3 L_2\} \equiv \frac{1}{4}[L_1^3 L_2 + L_1^2 L_2 L_1 + L_1 L_2 L_1^2 + L_2 L_1^3]$.

Thus, for all subgroups Γ the integrity basis for Γ invariants in the enveloping algebra of O(3) consists of precisely two operators, given in Table 1 (plus the Casimir operator of O(3)). These two operators never commute, hence only one of them (or an arbitrary function of both of them) can be added to $\vec{L}^2$ and the

TABLE 1. Integrity basis for Γ-invariant polynomials. Notations are explained in the text.

No.	Γ	Integrity basis for operator invariants
1	$\mathcal{C}_n$	L_3, C_n
2	$\mathcal{D}_n$	L_3^2, C_n
3	$\mathcal{C}_n[\mathcal{D}_n$	L_3^2, C_n
4	$\mathcal{C}_{ni}$	L_3, C_n
5	$\mathcal{D}_{ni}$	L_3^2, C_n
6	$\mathcal{C}_n[\mathcal{C}_{2n}$	L_3, C_{2n}
7	$\mathcal{D}_n[\mathcal{D}_{2n}$	L_3^2, C_{2n}
8	$\mathcal{T}$	O_3, O_4
9	$\mathcal{O}$	O_3^2, O_4
10	$\mathcal{T}[\mathcal{O}$	O_3^2, O_4
11	$\mathcal{T}_i$	O_3^2, O_4
12	$\mathcal{O}_i$	O_3^2, O_4
13	$\mathcal{I}$	I_6, I_{10}
14	$\mathcal{I}_i$	I_6, I_{10}

generators of Γ to form a complete set of commuting operators.

One particular example deserves a somewhat more complete treatment, namely the group $\Gamma = D_2$. The integrity basis consists of the operators L_3^2 and $C_2 = L_1^2 - L_2^2$. Let us restrict ourselves to linear combinations of these two operators. The complete set of commuting operators then provides us with an eigenvalue problem:

$$
\begin{aligned}
L^2|\ell\kappa pq\rangle &= \ell(\ell+1)|\ell\kappa pq\rangle\,, \\
(L_1^2+rL_2^2)|\ell\kappa pq\rangle &= \kappa|\ell\kappa pq\rangle\,, \\
C_x^\pi|\ell\kappa pq\rangle &= p(-1)^\ell|\ell\kappa pq\rangle\,, \\
C_y^\pi|\ell\kappa pq\rangle &= pq(-1)^\ell|\ell\kappa pq\rangle\,, \\
C_z^\pi|\ell\kappa pq\rangle &= q|\ell\kappa pq\rangle\,,
\end{aligned}
\tag{5.2}
$$

where C_i^π is a rotation through π about the i-th axis. If we realize $|\ell\kappa pq\rangle$ as functions on an O(3) sphere then equations (5.2) allow the separation of variables in elliptic coordinates (on the sphere) and the separated functions will be products of Lamé polynomials[10]. These are also the wave functions for an asymmetrical quantum mechanical top.

6. CONCLUSIONS

We have examined four different problems in which a classification of subgroups of a Lie group makes possible a systematic study of symmetry breaking. Each of the examples can serve as a prototype for a certain class of problems and each of them is currently being pursued. The stationary Schrödinger equation with its E(3) symmetry and the considered symmetry breaking are typical for external field problems. The problem of a relativistic elementary particle in an external field is of considerable interest in a quantum field theory context. It is presently being approached by considering Poincaré invariant equations (e.g. the Bargmann-Wigner equations[15]) and introducing external fields, reducing the symmetry to subgroups of the Poincaré group. The treatment of the time dependent Schrödinger equation with a possibly nonlinear interaction term typifies a possible group theoretical approach to nonlinear differential equations and in particular to soliton like phenomena[16]. The hydrogen atom in an electromagnetic field is on one hand a prototype for the breaking of dynamic (or canonical) symmetries, rather than kinematic ones. On the other hand it exemplifies the usefulness of a systematic group theoretical approach to the separation of variables in a partial differential equation. Indeed, each separable system for the hydrogen atom allows for the incorporation of different types of electromagnetic fields. Finally, the treatment of discrete subgroups of O(3) and the resolution of the missing label problem in this case can be extended to other Lie groups and their discrete subgroups. This is of considerable interest in many body calculations of various types, in particular in molecular physics.

REFERENCES

1. J. PATERA, P. WINTERNITZ and H. ZASSENHAUS, J. Math. Phys. 15, 1378 (1974), 15, 1932 (1974).
2. J. PATERA, P. WINTERNITZ and H. ZASSENHAUS, J. Math. Phys. 16, 1597 (1975), 16, 1615 (1975), 17, 717 (1976).
3. J. PATERA, R.T. SHARP, P. WINTERNITZ and H. ZASSENHAUS, J. Math. Phys. 17, 986 (1976), 17, 977 (1976), Can. J. Phys. 54, 950 (1976).
4. M. PERROUD, J. Math. Phys. 17, 1028 (1976).
5. J. BECKERS, J. PATERA, M. PERROUD and P. WINTERNITZ, J. Math. Phys. 18, 72 (1977).
6. C.P. BOYER, R.T. SHARP and P. WINTERNITZ, J. Math. Phys. 17, 1439 (1976).
7. E.G. KALNINS, W. MILLER, Jr. and P. WINTERNITZ, SIAM J. Appl. Math. 30, 630-664 (1976).

8. V. FOCK, Z. Phys. 98, 145 (1935); V. BARGMANN, Z. Phys. 99, 576 (1936).

9. M.P. OLEVSKIĬ, Mat. Sb. 27, 379 (1950).

10. J. PATERA and P. WINTERNITZ, J. Math. Phys. 14, 1130 (1973).

11. E.G. KALNINS and W. MILLER, Jr., J. Math. Phys. 15, 1263 (1974).

12. C.A. COULSON and P.D. ROBINSON, Proc. Phys. Soc. 71, 815 (1958).

13. J. PATERA and P. WINTERNITZ, J. Chem. Phys. 65, 2725 (1976).

14. B. MEYER, Can. J. Math. 6, 135 (1954).

15. V. BARGMANN and E.P. WIGNER, Proc. Nat. Acad. Sci. 34, 211 (1948).

16. A.C. SCOTT, F.Y.F. CHU and D. McLAUGHLIN, Proc. IEEE 61, 1443 (1973).

CONFORMAL HARMONIC ANALYSIS AND INTERTWINING DIFFERENTIAL OPERATORS

H.P. Jakobsen

INTRODUCTION

Earlier in this conference, we have heard Professor I.E. Segal's lecture "Causality and Symmetry in Cosmology and the Conformal Group". I refer to that for motivation, general background, and references. Here shall be described a few results concerning the development of the theory towards elementary particle physics; in particular some results of a study, jointly with Michele Vergne (1), of the interplay between representations of the conformal group (in the form of SU(2,2)) and classical mass-zero-equations, will be given.

GENERALITIES

For technical reasons it is convenient to work with the group

$$G = \{g \in SL(4,\mathbb{C}) \mid g^*\begin{pmatrix} 0 & i \\ -i & 0 \end{pmatrix} g = \begin{pmatrix} 0 & i \\ -i & 0 \end{pmatrix}\}.$$

This group is isomorphic to the group

$$SU(2,2) = \{g \in SL(4,\mathbb{C}) \mid g^*\begin{pmatrix} I & 0 \\ 0 & -I \end{pmatrix} g = \begin{pmatrix} I & 0 \\ 0 & -I \end{pmatrix}\}$$

which is used in Segal's theory, and it is straightforward to translate results from G to SU(2,2).

G acts on the generalized upper half-plane $\mathcal{D}$ = {2 × 2 complex matrices z | $\frac{z-z^*}{2i}$ is positive definite} by, for G ∋ g = $\begin{pmatrix} a & b \\ c & d \end{pmatrix}$; a, b, c, d, 2 × 2 complex matrices,

$$g\cdot z = \frac{az+b}{cz+d}.$$

One can prove that $(g\cdot z^*)^* = g\cdot z$, and thus G leaves the boundary of $\mathcal{D}$, $\partial\mathcal{D}$ = H(2) = {2 × 2 complex matrices h|h = h*}, invariant, and this gives rise to the well-known local action of SU(2,2) on Minkowski space M, identified with H(2) by

$$M \ni (t,x,y,z) \Longleftrightarrow h = \begin{pmatrix} t+z & x+iy \\ x-iy & t-z \end{pmatrix} \in H(2).$$

Finally, let c: $\begin{pmatrix} \lambda_1 & \lambda_2 \\ \lambda_3 & \lambda_4 \end{pmatrix} \to \begin{pmatrix} \lambda_4 & -\lambda_2 \\ -\lambda_3 & \lambda_1 \end{pmatrix}$ denote the co-factor mapping on the space of 2 × 2 complex matrices and observe that c's restriction to H(2) is the operation of space-reversal.

THE REPRESENTATIONS

Let ψ be a function from H(2) to $\mathbb{C}$, $g^{-1} = \begin{pmatrix} a & b \\ c & d \end{pmatrix} \in G$, and $n \in \mathbb{Z}$ and define

$$(U_n(g)\psi)(h) = \det(ch+d)^{-(n+2)}\psi(g^{-1}h).$$

Moreover, let φ be a function from H(2) to $\mathbb{C}^2$ and define

$$(D_n^+(g)\varphi)(h) = (ch+d)^{-1}\det(ch+d)^{-(n+2)}\varphi(g^{-1}h),$$

$$(D_n^-(g)\varphi)(h) = (hc^*+d^*)\det(hc^*+d^*)^{-(n+3)}\varphi(g^{-1}h),$$

and finally define, on functions from H(2) to $\mathbb{C}^4$,

$$D_n(g) = D_n^+(g) \oplus D_n^-(g).$$

U_n and D_n are then well-defined actions on measurable functions.

It is proved in (1), but has also been observed elsewhere (see References in (1)), that one has

Theorem 1. For $n \geq 0$ there exist Hilbert spaces H_n and K_n, obtained as completions of spaces of positive energy and positive mass functions, such that U_n is unitary and irreducible in H_n, and D_n is the direct dum of two unitary irreducible representations in K_n. Moreover, there exists a Hilbert space H_{-1} (K_{-1}) consisting of solutions to $\Box\psi = 0$ $(\not\nabla\varphi = 0)$ in which U_{-1} (D_{-1}) is unitary.

The significance of the representations U_n and D_n for integers $n \leq -1$ and the relation of these to positive mass functions is illustrated by

THE INTERTWINING RELATIONS

For the sake of being explicit, let

$$\Box = \frac{\partial^2}{\partial t^2} - \frac{\partial^2}{\partial x^2} - \frac{\partial^2}{\partial y^2} - \frac{\partial^2}{\partial z^2},$$

$$\sigma = \begin{pmatrix} \frac{\partial}{\partial t} + \frac{\partial}{\partial z} & \frac{\partial}{\partial x} + i\frac{\partial}{\partial y} \\ \frac{\partial}{\partial x} - i\frac{\partial}{\partial y} & \frac{\partial}{\partial t} - \frac{\partial}{\partial z} \end{pmatrix}, \qquad \tilde{\sigma} = \begin{pmatrix} \frac{\partial}{\partial t} - \frac{\partial}{\partial z} & -\frac{\partial}{\partial x} - i\frac{\partial}{\partial y} \\ -\frac{\partial}{\partial x} + i\frac{\partial}{\partial y} & \frac{\partial}{\partial t} + \frac{\partial}{\partial z} \end{pmatrix},$$

and $\not\nabla = \begin{pmatrix} 0 & \sigma \\ \tilde{\sigma} & 0 \end{pmatrix}$. (Then $\not\nabla^2 = \Box$.)

Theorem 2. (1) $\forall\ g \in G$:

$$\Box^n U_{-n}(g) = U_n(g)\Box^n$$

$$\not\nabla^{2n+1} D_{-1-n}(g) = D_n(g)\not\nabla^{2n+1}$$

whenever both sides make sense.

Let, for $n \in \mathbb{N}$, two functions $\psi_i : H(2) \to \mathbb{C}$ $(i=1,2)$ be equivalent if $\square^n(\psi_1-\psi_2) = 0$ and denote the equivalence classes by $[\cdot]^0_n$. Likewise define, for non-negative integers n, two functions $\varphi_i:H(2) \to \mathbb{C}^4$ (i=1,2) to be equivalent if $\not\nabla^{2n+1}(\varphi_1-\varphi_2) = 0$ and denote the equivalence classes by $[\cdot]^{\frac{1}{2}}_n$.

Theorem 3. (1) It is possible to put a Hilbert space structure $\mathcal{H}_n$ ($\mathcal{K}_n$) on a space of equivalence classes $[\cdot]^0_n([\cdot]^{\frac{1}{2}}_n)$ of positive mass function, such that U_{-n} (D_{-1-n}) is a unitary representation in $\mathcal{H}_n$ ($\mathcal{K}_n$).

Remark. If one restricts the representations U_n and D_n to the Poincaré group together with the scale-transformations, the proofs of the above mentioned facts are quite simple. Also, theorem 3 can then be sharpened, in the sense that for each $U_{-n}(D_{-1-n})$ (n positive (or zero)) there is a natural Hilbert space H_{-n} (K_{-n}) of positive mass functions, on which this restriction is unitary. However, this is not the case for the full group G. In fact, for the one-parameter group

$$\tau(\theta) = \begin{pmatrix} \cos\theta & -\sin\theta \\ \sin\theta & \cos\theta \end{pmatrix} \subseteq G$$

(which is the G-version of the uni-time translation subgroup of SU(2,2)), $U_{-n}(\tau(\theta))$ ($D_{-1-n}(\tau(\theta))$) does not leave H_{-n} (K_{-n}) invariant, as can be seen easily by computations on special functions.

REFERENCE

1. H.P. JAKOBSEN and M. VERGNE, Wave and Dirac operators, and representations of the conformal group. (To appear in J. Functional Analysis.)

GROUP THEORETIC ASPECTS OF CONSERVATION LAWS OF NONLINEAR TIME EVOLUTION EQUATIONS: THE KdV EQUATION AND THE CUBIC SCHRODINGER EQUATION*

Sukeyuki Kumei

In this short communication, we present a few results of a study of group theoretic properties of time evolution equations. The aim of the study is to understand clearly a group theoretic meaning of the well-known conservation laws associated with non-linear equations describing wave propagation. Proofs of the statements made below will appear elsewhere along with some detailed analysis of the groups[1]. The groups we consider in the following discussion are of generalized Lie type[2]. We consider an operator of the form

$$U = \eta\partial_u + \zeta\partial_v + \ldots + \eta_{i\ldots j}\partial_{u_{i\ldots j}} + \zeta_{i\ldots j}\partial_{v_{i\ldots j}} + \ldots \qquad (1)$$

where

$$\eta_{i\ldots j} = D_{i\ldots j}\eta, \qquad \zeta_{i\ldots j} = D_{i\ldots j}\zeta$$

with

* This work has been supported by a Research Corporation grant.

$$D_{i\ldots j} = D_i \ldots D_j,$$

$$D_i = \partial_{x^i} + u_i\partial_u + v_i\partial_v + \ldots + u_{ij\ldots k}\partial_{u_{j\ldots k}} + v_{ij\ldots k}\partial_{v_{j\ldots k}} + \ldots$$

The coefficients η and ζ are functions of the *coordinates* x^i, u, v, u_i, v_i, u_{ij}, v_{ij},... where i = 0,1,...,N. In the following, we write the operator of the form (1) simply as $U = \eta\partial_u + \zeta\partial_v$. It is well-known that the group e^{aU} leaves invariant the differential equations

$$\begin{cases} F^p(x^i,u,v,u_i,v_i,u_{ij},v_{ij},\ldots,c) = 0, \quad p = 1,2 \\ u = f(x),\ v = g(x),\ u_i = f_i(x),\ v_i = g_i(x),\ldots \end{cases}$$

if and only if

$$UF^p|_{\mathcal{F}} = 0, \qquad p = 1,2.$$

Here $x = (x^0,x^1,\ldots,x^N)$, $f_i(x) = \partial_{x^i}f(x)$, $g_i(x) = \partial_{x^i}g(x),\ldots$, and $(\cdot)|_{\mathcal{F}}$ is an instruction to evaluate the quantity under the condition

$$F^p = 0, \quad D_{p_1\ldots p_i}F^p = 0, \quad i = 1,2,\ldots,\infty.$$

In the present paper, we are interested in invariance groups of time evolution equations: x^0 = time coordinate

$$\begin{cases} F^1 = H^1(x^i,u,v,u_i,v_i,\ldots;c) + u_0 = 0 \\ F^2 = H^2(x^i,u,v,u_i,v_i,\ldots;c) + v_0 = 0. \end{cases} \tag{2}$$

Let us consider an operator of the type (1) associated with (2):

$$\mathcal{H} = H^2\partial_u + H^1\partial_v.$$

It is easy to check that the group $e^{a\mathcal{H}}$ leaves the equation (2) invariant. Furthermore,

I. If U is a generator of an invariance group of the equation (2), then

$$[\mathcal{H},U] + \frac{\partial U}{\partial x^0} = 0. \qquad \blacksquare$$

One special case of the equation (2) is when $u = f(x)$ and $v = f(x)^*$. Clearly, we must have $H^2 = H^1(x^i,v,u,v_i,u_i,\ldots;c^*) = (H^1)^\#$ where # represents a conjugation operation to interchange u and v and to replace all the numbers c in the equation by their complex conjugate c*. For this class of equations we define the following quantity:

$$\langle U \rangle = \mathrm{Re} \int (vUu)_{\substack{u=f(x)\\ v=f^*(x)}} dx^1 dx^2 \ldots dx^N, \qquad \mathrm{Re} = \text{real part} \tag{3}$$

where the integration should be taken over the whole space of interest. Obviously $\langle U \rangle$ is a function of x^0 only. The following statement describes how it develops in time for a class of nonlinear systems:

II. If H of the equation $H+u_0 = 0$ satisfies the equation

$$\begin{aligned} H^\# + vH_u + u(H_v)^\# - D_i\{vH_{u_i} + u(H_{v_i})^\#\} + \ldots \\ + (-1)^\pi D_{p_1\ldots p_\pi}\{vH_{u_{p_1\ldots p_\pi}} + u(H_{v_{p_1\ldots p_\pi}})^\#\} = 0 \end{aligned} \tag{4}$$

and if all the boundary integrals

$$\int_S [u\eta_{j\ldots k}H_{u_{ij\ldots k}}]_{\substack{u=f(x)\\ v=f^*(x)}} \nu^i d\Omega$$

and

$$\int_S [vn^{\#}_{j\ldots k} H_{u_{ij\ldots k}}]_{\substack{u=f(x)\\ v=f^*(x)}} \nu^i d\Omega$$

vanish for $\nu = (\nu^1,\ldots,\nu^N)$ = normal vector on the boundary surface, then

$$\frac{d}{dx^0}\langle U \rangle = \langle [U,\mathcal{H}] + \frac{\partial U}{\partial x^0} \rangle . \qquad \blacksquare$$

The combination of the statements I and II leads to a method for associating a constant of motion with an invariance group of the equation:

III. If the operator U defined by (1) is a generator of an invariance group of the equation $H + u_0 = 0$, and if H satisfies all the conditions in II, then the quantity $\langle U \rangle$ defined by (3) is a constant of motion i.e.

$$\frac{d}{dx^0}\langle U \rangle = 0. \qquad \blacksquare$$

The following differential equations which have been attracting considerable attention in the studies of nonlinear wave propagation are found to satisfy the condition (4):

Generalized Körteweg-de Vries equation

$$(\partial_x)^{2n+1}u + u^m\partial_x u + \partial_t u = 0$$

Cubic Schrödinger equation in n-dimensions

$$-\, i\,\{\sum_{k=1}^{n} (\partial_{x^k})^2 + uu^*\}u + \partial_t u = 0$$

Hirota equation

$$a(\partial_x)^3u + ib(\partial_x)^2u + cuu^*\partial_x u + idu^2u^* + \partial_t u = 0,$$

where we used $x^0 = t$.

Now we consider special cases of the equation (2). First, we consider a coupled nonlinear Schrodinger equation: with $x^0=t$, $x^1=x$,

$$\begin{cases} u_{xx} + a(u,v;c) + iu_t = 0, \quad v_{xx} + a(u,v;c)^{\#} - iv_t = 0 \\ u = f(x,t), \quad u_x = f_x(x,t), \quad u_t = f_t(x,t),\ldots \\ v = g(x,t), \quad v_x = g_x(x,t), \quad v_t = g_t(x,t),\ldots \end{cases} \tag{5}$$

where the function $\underline{a}$ is subject to the condition

$$a_u(f,g;c) = [a_u(f,g;c)]^{\#}, \qquad a_u = \partial_u a. \tag{6}$$

The condition (6) amounts to the requirement that the equation can be written as a Hamiltonian system:

$$\frac{\delta H}{\delta g} = -if_t, \qquad \frac{\delta H}{\delta f} = ig_t$$

where $\frac{\delta H}{\delta g}$ and $\frac{\delta H}{\delta f}$ are Frechet derivatives of $H = \int E(f,g)dx$. The equation (5) reduces to the cubic Schrödinger equation for the special case $a = u^2v$ and $g = f^*$.

We assume that an initial value problem is well posed either for a periodic condition $f(x,t) = f(x+x_0,t)$, $g(x,t) = g(x+x_0,t)$ or for a boundary condition $f(\pm\infty,t) = 0$, $g(\pm\infty,t) = 0$. Let us suppose that the system described by (5) has a constant of motion $I(f,g) = \int \rho(f,g)dx$ where the integration is over the period or from $-\infty$ to $+\infty$. The following theorem establishes the relationship between I and an invariance group of the equation. In the following, the quantities $\frac{\delta I}{\delta u}$ and $\frac{\delta I}{\delta v}$ represent

$$\left\{\frac{\delta I}{\delta f}\right\}_{\substack{f=u\\g=v}} \quad \text{and} \quad \left\{\frac{\delta I}{\delta g}\right\}_{\substack{f=u\\g=v}}.$$

Theorem 1. If $\frac{\delta I}{\delta f}$ and $\frac{\delta I}{\delta g}$ are Frechet derivatives of a constant of motion $I(f,g) = \int \rho(f,g)dx$ associated with the equation (5), then the operator $U = i\frac{\delta I}{\delta v}\partial_u - i\frac{\delta I}{\delta u}\partial_v$ is a generator of an invariance group of the equation.

The second case of interest is the Korteweg-de Vries type equation: $F^1 = u_{xxx} + a(u)u_x + u_t$, $F^2 \equiv 0$, i.e.

$$f_{xxx} + a(f)f_x + f_t = 0 \tag{7}$$

where $a(f)$ is a function of f. We assume that an initial value problem for this equation is well posed for a periodic boundary condition $f(x,t) = f(x+x_o,t)$ or for a condition $f(-\infty,t) = f(\infty,t) = 0$. Let us suppose that the system has a constant of motion of integral type $I(f) = \int \rho(f)dx$. The limits of the integration are either over the period or from $-\infty$ to ∞. Then

Theorem 2. If $\Gamma(u)$ is the gradient[3)] of a constant of motion $I(f) = \int \rho(f)dx$ associated with the equation $f_{xxx} + a(f)f_x + f_t = 0$, then the operator $U = \eta\partial_u$ which has $\eta(u) = D_x\Gamma(u)$ is a generator of an invariance group of the equation.

This theorem establishes a relationship between constants of motion and invariance groups of Eq. (7):

$$I(f) = \int \rho(f)dx \leftrightarrow \Gamma(u) \leftrightarrow U = \eta\partial_u,\quad \eta = D_x\Gamma.$$

Now we apply the theorems to the cubic Schrödinger equation $f_{xx} + f^2f^* + if_t = 0$ and to the Korteweg-de Vries equation $f_{xxx} + ff_x + f_t = 0$ to obtain constants of motion associated with these equations. The process of obtaining constants of motion from the generators involves a simple integration process. Some of the generators for the cubic Schrödinger equation are

$$U_1 = (-\tfrac{1}{2}ixu+tu_x)\partial_u,\quad U_2 = (itu_{xx}+itu^2v+\tfrac{1}{2}xu_x+\tfrac{1}{2}u)\partial_u,$$

$$U_3 = iu\partial_u, \quad U_4 = u_x\partial_u, \quad U_5 = i(u_{xx}+u^2v)\partial_u, \quad U_6 = (u_{xxx}+3uvu_x)\partial_u,$$

$$U_7 = i(u_{xxxx}+u^2v_{xx}+4uvu_{xx}+2uu_xv_x+3vu_x^2 + \frac{3}{2}u^3v^2)\partial_u,$$

$$U_8 = \{u_{xxxxx}+5(uvu_{xxx}+uu_xv_{xx}+2vu_xu_{xx}+uv_xu_{xx}+u_x^2v_x) + \frac{15}{2}u^2v^2u_x\}\partial_u.$$

The constants of motion associated with these may be written $I^i = \int \rho^i(u,v,u_x,v_x,\ldots)dx$ with

$$\rho^1 = xuv - it(u_xv-uv_x), \quad \rho^3 = uv, \quad \rho^4 = iu_xv,$$

$$\rho^5 = \tfrac{1}{2}(u_xv_x-\tfrac{1}{2}u^2v^2), \quad \rho^6 = i(u_{xxx}v + \frac{3}{2}uu_xv^2)$$

$$\rho^7 = u_{xx}v_{xx} + \tfrac{1}{2}u^3v^3 - 2(u_xv+uv_x)^2 - 3u_xv_xuv$$

$$\rho^8 = u_{xxxxx}v + 5(uu_{xxx}v+uu_xv_{xx}+2u_xu_{xx}v+uu_{xx}v_x+u_x^2v_x) + \frac{1}{3}u^2u_xv^3.$$

For the KdV equation, some of the generators and their associated constants are the following:

$$U^1 = (tu_x-1)\partial_u, \quad U^2 = \frac{1}{3}\{xu_x-3t(u_{xxx}+uu_x)+2u\}\partial_u, \quad U^3 = u_x\partial_u,$$

$$U^4 = (u_{xxx}+uu_x)\partial_u, \quad U^5 = (\frac{3}{5}u_{xxxxx}+uu_{xxx}+2u_xu_{xx}+\tfrac{1}{2}u^2u_x)\partial_u.$$

$$I^1 = \int(\tfrac{1}{2}tu^2-xu)dx, \quad I^3 = \int \tfrac{1}{2}u^2dx, \quad I^4 = \int(\frac{1}{3}u^3-u_x^2)dx,$$

$$I^5 = \int(\frac{1}{4}u^4 - 3uu_x^2 + \frac{9}{5}u_{xx}^2)dx.$$

The constants obtained here agree with the ones due to Zakharov and Shabat[4)] and Miura, Gardner and Kruskal[5)]. The existence of an infinite number of constants of motion implies that these two equations are invariant under an infinite number of groups.

ACKNOWLEDGMENT

I am grateful to Professor Carl Wulfman for many helpful and stimulating discussions.

REFERENCES

1. S. KUMEI, *Group theoretic aspects of conservation laws of nonlinear time evolution equations: the* KdV *equation and the cubic Schrödinger equation* (accepted for publication in Journal of Mathematical Physics).

2. R.L. ANDERSON, S. KUMEI and C.E. WULFMAN, Phys. Rev. Lett. 28 988 (1972). N.I. IBRAGIMOV and R.L. ANDERSON, J. Math. Anal. (in press).

3. P.D. LAX, Commun. Pure Appl. Math. 21 467 (1968), 28 141 (1975).

4. V.E. ZAKHAROV and A.B. SHABAT, JETP 34 62 (1972).

5. R.M. MIURA, C.S. GARDNER and M.D. KRUSKAL, J. Math. Phys. 9 1204 (1968).

APPLICATION OF GROUP THEORETIC METHODS IN BIFURCATION THEORY

D.H. Sattinger

Suppose $G(\lambda,u)$ is a Frechet differentiable mapping from $K \times \mathcal{E}$ to F, where K denotes the field of real or complex numbers, and $\mathcal{E}$ and F are Banach spaces; and that $G(0,0) = 0$. One is then interested in constructing all solutions of the functional equation

$$G(\lambda,u) = 0 \tag{1}$$

in a neighborhood of $\lambda = 0$, $u = 0$. If the Frechet derivative $G_u(0,0)$ is invertible, that is, if it is an isomorphism from $\mathcal{E}$ to F, then by the implicit function theorem there is a continuously differentiable $\mathcal{E}$-valued function $u(\lambda)$, defined for small λ, such that $G(\lambda,u(\lambda)) \equiv 0$. Furthermore, all solutions of (1) in a sufficiently small neighborhood of $(0,0)$ in $K \times \mathcal{E}$ lie on the curve $(\lambda,u(\lambda))$.

Bifurcation theory concerns itself with the situation where $G_u(0,0)$ is not invertible. The simplest case is that in which $L_0 = G_u(0,0)$ is a Fredholm operator: L_0 has a finite dimensional kernel N and a closed range $\mathcal{R}$, with codim $\mathcal{R}$ = dim N = n. In that case the solution of (1) near (0,0) may be reduced, via an alternative method to a system of n algebraic equations in n unknowns, *viz.*

$$F_i(\lambda, v) = 0, \quad i = 1, \ldots, n$$
$$v = (v_1, \ldots, v_n). \tag{2}$$

The subject of bifurcation is an important topic for applied mathematics inasmuch as it arises naturally in any physical system described by a nonlinear set of equations depending on a set of parameters. Equation (1) may typically represent a nonlinear system of partial differential equations, ordinary differential equations, etc. When discussing applied problems stability considerations are fundamental. In fact, the phenomenon of bifurcation is intimately associated with "loss of stability". A complete resolution of the branching problem therefore requires an analysis of the stability of the bifurcating solutions.

Let us suppose (as is often the case) that (1) is the time independent equation for a dynamical system $\frac{\partial u}{\partial t} = G(\lambda, u)$. We state loosely the *Principle of Linearized Stability. An equilibrium point* u_0 *of* (1) *is stable if the spectrum of the linear operator* $G_u(\lambda, u_0)$ *lies strictly in the left half plane and unstable if points in the spectrum of* $G_u(\lambda, u_0)$ *lie in the right half plane.*

In case (1) is a system of ordinary differential equations the Principle of Linearized Stability is valid and is known as Lyapunov's first theorem. The principle is also valid for a number of important problems in partial differential equations, for example the Navier-Stokes equations and other nonlinear parabolic systems. For a discussion of bifurcation and stability theory, including physical applications, the reader is referred to the research monograph [4].

When the kernel N is one dimensional the bifurcation equations (2) reduce to a single equation $f(\lambda, v) = 0$ (v = scalar), and the situation can be analyzed completely. (See [4].) But when $n > 1$ the analysis of the equations (2) can be extremely complex, especially since it is difficult to compute any but the terms of lowest order in (2). It is at this juncture that group theoretic methods become important.

In physical applications the presence of multiple eigenvalues ($n>1$) is often due to the invariance of the given problem under a transformation group. The phenomenon is well known in quantum mechanics, where invariance of the Hamiltonian under a symmetry group leads to a degeneracy of the energy levels. The same situation obtains in bifurcation theory, and this fact permits a significant simplification of the analysis of the bifurcation equations.

Let T_g be a representation of a group G onto the vector spaces $\mathcal{E}$ and F and suppose that (1) is covariant with respect to T_g: $T_g G(\lambda,u) = G(\lambda,T_g u)$. This situation is typical in physical applications. For example, in classical continuum mechanics, the group G may be the group of rigid motions or one of its subgroups. The following theorem is easily proved [5].

Theorem. The Frechet derivative $G_u(0,0)$ *commutes with* T_g *and the kernel* N *is invariant under* T_g. *Furthermore, the bifurcation equations* (2) *are covariant with respect to the finite dimensional representation:* $T_g F(\lambda,v) = F(\lambda,T_g v)$. (Here $F = (F_1,\ldots,F_n)$.)

Knowing the group G the methods of group representation theory permit one to calculate the general covariant mappings F. If the original operator G is analytic the mapping F is analytic also, and can be expanded in a Taylor series in v:

$$F(\lambda,u) = B_1(\lambda)v + B_2(\lambda;V,V) + B_3(\lambda;V,V,V) + \ldots .$$

where $B_1(\lambda)$ is a linear operator, B_2 is quadratic, etc. If the representation T_g restricted to the kernel N is irreducible, then necessarily $B_1(\lambda)$ is a scalar multiple of the identity, by Schur's theorem. If $B_k(v)$ is a k-linear covariant mapping and if (,) is a scalar product on N, relative to which T_g is unitary, then $F_k(v) = (B_k(v),v)$ is an invariant tensor of rank $k + 1$. Thus the general structure of the covariant mapping F can be computed by group theoretic methods.

Let us close with a description of some physical problems in which these questions arise. Of particular interest in physical problems is the phenomenon of pattern formation, or "symmetry breaking instabilities". The appearance of convection cells in the Bénard problem [7] constitutes a classical example of a symmetry breaking instability. In this problem a layer of fluid is heated from below. The relevant physical parameter is the Rayleigh number $\mathcal{R}$ which is proportional to the temperature drop across the layer. If $\mathcal{R}$ lies below a certain critical value $\mathcal{R}_c$ there is no fluid motion, and the solution is invariant under the entire group of rigid motions. For $\mathcal{R} > \mathcal{R}_c$ convective motion sets in and, under certain circumstances, hexagonal convection cells are formed [7]. This bifurcating solution is invariant only under a crystallographic subgroup of the group of rigid motions. A treatment of the Bénard problem using group theoretic methods is given in [6].

Suppose (1) is covariant with respect to rigid motions. Since cellular patterns are observed in experiments one looks for doubly periodic solutions of (1) in the neighborhood of $\lambda = 0$. (By doubly periodic we mean $u(\underline{x}+\underline{w}) = u(\underline{x})$ for all vectors $\underline{w}$ belonging to a lattice Λ in the plane.) For simplicity assume that u is a scalar function on the plane. Then the kernel N is spanned by Bloch functions

$$e^{i\langle \underline{w}_j,\underline{x}\rangle}, \quad j=1,2,\ldots,$$

where the vectors $\underline{w}_j$ are precisely those vectors in Λ' (dual lattice) which lie on the circle of critical wave vectors. The action of T_g on N can be computed and the form of the lowest order terms of the bifurcation equations can be determined. For example, in the case of hexagonal solutions Λ is the lattice generated by two vectors $\underline{w}_1$ and $\underline{w}_2$ 60° apart. Dim $N = 6$, but using the fact that the original equations are real analytic ($\overline{G(\lambda,u)} = G(\bar{\lambda},\bar{u})$)

we can cut the number of equations from six to three in the non-negative quantities x_1, x_2, x_3. We get $F_1(\lambda, x_1, x_2, x_3) = \lambda x_1 + ax_2x_3 + b(x_3^2+x_2^2)x_1 + cx_1^3$ while the components F_2 and F_3 are obtained by successive cyclic permutation of the arguments x_1, x_2, x_3. The parameters a,b,c depend on the physical parameters of the problem. The stability of the bifurcating solutions depends on the relative sizes of a,b, and c.

Other bifurcation problems which are invariant are the buckling of spherical shells [3] and convection in spherical shells [1]. The relevant group is O(3). One well known area in which bifurcation theory and group theory interact is that of phase transitions in statistical mechanics. See Birman [2] as well as the references in [5] and [6].

REFERENCES

1. F. BUSSEL, "Patterns of convection in spherical shells", Jour. Fluid Mech. 72, 67 (1975).

2. J.L. BIRMAN, "Symmetry changes, phase transitions, and ferroelectricity", Ferroelectricity, Elsevier, Amsterdam, 1967.

3. D. SATHER, "Branching and stability for nonlinear shells", Proceedings of the IUTAM/IMU Symposium Marseille, 1975, Springer Lecture Notes in Mathematics, 503.

4. D. SATTINGER, Topics in stability and bifurcation theory, Springer Lecture Notes in Mathematics No. 309, Springer-Verlag, 1973.

5. D. SATTINGER, "Group representation theory and branch points of nonlinear functional equations", to appear SIAM Jour. Math. Analysis.

6. D. SATTINGER, "Group representation theory, bifurcation theory, and pattern formation", to appear, *Jour. Funct. Anal.*

7. L.A. SEGEL, "Nonlinear problems in hydrodynamic stability", in Nonequilibrium Thermodynamics, Variational Techniques, and Stability, Univ. of Chicago Press, 1965.

HARMONIC ANALYSIS ON DOUBLE CLASSES

T.H. Seligman and K.B. Wolf

I. INTRODUCTION

The concept of double classes has been introduced recently [1] as a generalized classification scheme for group elements. It contains as particular cases the well-known cosets, double cosets, classes and subclasses. Here we give some results on properties of double classes and see their particularization to known properties of these cases. The question of complete and orthogonal sets of functions in terms of which to expand any function on the set of double classes is answered in terms of "partial traces" which particularize to spherical functions and traces in the known cases. Finally, we investigate the circumstances under which the double-class partitioning of the group generates subrings of the full group ring.

II. DEFINITIONS AND NOTATION

We establish an equivalence relation ($\sim$) in a group G stating that two of its elements g, g' are equivalent ($g \sim g'$) iff there exist $h \in H \subset G$ and $k \in K \subset G$ such that $g' = hgk^{-1}$ with the additional requirement that the elements of H and K be "partially correlated". Loosely speaking, this means that for a fixed h, only some k are allowed and vice versa. To make this precise, we establish a *correlation* function φ' from sets in H into sets in K which preserves group multiplication in G for sets. In Ref. [1] it is shown that these ideas lead to an equivalence relation iff there exist normal subgroups $\hat{H} \triangleleft H$ and $\hat{K} \triangleleft K$ such that φ' is an isomorphism between the factor group $H/\hat{H}$ and $K/\hat{K}$. Let $\{h_\alpha\}$ and $\{k_\alpha\}$ be two systems of coset representatives of the factor groups, so that $h = \hat{h}h_\alpha$, $(\hat{h} \in \hat{H})$ and $k = \hat{k}k_\alpha$, $(\hat{k} \in \hat{K})$, and φ the mapping $\varphi(h_\alpha) = k_\alpha$ (i.e. φ' restricted to the coset representatives). Then, the equivalence relation ($\sim$) divides the group G into disjoint equivalence classes:

$$C_i = \{g \in G \,|\, g_i \sim g = \hat{h}h_\alpha g_i \varphi(h_\alpha)^{-1}\hat{k}^{-1}\}, \tag{1}$$

called *double classes*. It is easy to verify the identity, symmetry and transitivity of (1).

Note that a double class (1) is the union of entire double cosets $H\backslash G/K$, so the partition produced by (1) is coarser than the latter. When H and K *split* (i.e. when they are semidirect products $H = \hat{H} \wedge [H/\hat{H}]$, $K = \hat{K} \wedge [K/\hat{K}]$), the systems of representatives can be chosen so that the product law in the factor groups is that in G. In this case, a double class (1) is the union of entire φ-*twisted classes* $\{g \in G \,|\, g = h_\alpha g_i \varphi(h_\alpha)^{-1}\}$. We shall use this expression keeping in mind that, an "untwisted" situation $h_\alpha = k_\alpha$ may be impossible.

As detailed in [1], double cosets arise from "totally uncorrelated" groups $H = \hat{H}$ and $K = \hat{K}$ (so that $H/\hat{H} = \{e\} = K/\hat{K}$ and φ is irrelevant), while subclasses correspond to the "completely correlated" situation when $H = K$, $\hat{H} = \hat{K} = \{e\}$ (φ is the identity mapping).

Let the Unitary Irreducible Representation (UIR) matrix elements of $G \ni g$ be $D^\gamma_{\rho\rho'}(g)$, where γ labels the UIR of dimension $d(\gamma)$, and ρ,ρ' are row and column labels. Let the set of UIRs of of G be $\tilde{G}$. Our construction needs the use of different group chains for row and column labelling: define the *skew* representation matrices [2]

$$\Delta^{\gamma}_{\rho\mu}(g) = \sum_{\rho'} D^{\gamma}_{\rho\rho'}(g) V_{\rho'\mu}, \tag{2}$$

where V is a unitary $d(\gamma) \times d(\gamma)$ matrix transforming in our case an orthonormal basis labelled (as will be detailed below) according to the chain of subgroups $G \supset H \supset \hat{H}$ to one labelled according to $G \supset K \supset \hat{K}$. In order not to involve ourselves with the definition of measures, we restrict ourselves in what follows to finite-order groups (letting $|G|$, $|H|$, etc. be the order of G, H, etc.); yet we wish to point out that if the proper Haar and Plancherel measures for G and $\tilde{G}$ are known, the arguments carry over quite trivially to continuous Lie groups.

Under this restriction, it may easily be seen that the skew UIR matrices (2) are a complete and orthogonal basis for functions over G, i.e.

$$\sum_{g\in G} \Delta^{\gamma}_{\rho\mu}(g)^* \Delta^{\gamma'}_{\rho'\mu'}(g) = \frac{|G|}{d(\gamma)} \delta_{\rho\rho'} \delta_{\mu\mu'} \delta_{\gamma\gamma'} \tag{3a}$$

$$\sum_{\gamma\in\tilde{G}} \frac{d(\gamma)}{|G|} \sum_{\rho,\mu} \Delta^{\gamma}_{\rho\mu}(g_1)^* \Delta^{\gamma}_{\rho\mu}(g_2) = \delta_{g_1,g_2} \tag{3b}$$

so that a function A(g) may be expanded in terms of $\Delta^{\gamma}_{\rho\mu}(g)^*$-functions in a sum as (3b), its harmonic (Fourier) coefficients $A^{\gamma}_{\rho\mu}$ obtained from A(g) though a sum (3a) in the usual fashion.

Regarding the structure of the row and column indices ρ,μ, these are constituted as

$$\rho = (p,\eta,r), \quad \mu = (s,\kappa,u), \tag{4}$$

where, as indicated, η and κ label the UIRs of H and K, whose row- and column-labels are r and u. The possible multiple occurrence of UIRs of H and K in G are distinguished by p and s.

III. INVARIANT FUNCTIONS ON DOUBLE CLASSES

We will consider functions A on G which are constant for all $g \in C_i$, for fixed i, i.e., functions of the set Γ of double classes C_i:

$$A(g_i) = A(\hat{h}h_\alpha g_i \varphi(h_\alpha)^{-1}\hat{k}^{-1}) = A(C_i), \tag{5}$$

where the symbols involved are those of (1), for the equivalence relations defined by H, $\hat{H}$, K, $\hat{K}$ and φ. The property (5) of a function is expected to bring about a corresponding property in the harmonic (Fourier) coefficients. Indeed, we have

$$A^{\gamma}_{p\eta r,s\kappa u} = \sum_{g\in G} A(g)\Delta^{\gamma}_{p\eta r,s\kappa u}(g) \tag{6a}$$

$$= \sum_{g\in G} A(g) \sum_{r'u'} D^{\eta}_{rr'}(h_\alpha^{-1}\hat{h}^{-1})\Delta^{\gamma}_{p\eta r',s\kappa u'}(g)D^{\kappa}_{u'u}(\hat{k}\varphi(h_\alpha)) \tag{6b}$$

$$= \delta_{\eta,\eta_0}\delta_{\kappa,\kappa_0} \sum_{g\in G} A(g) \sum_{\bar{r}'\bar{u}'} D^{\eta_0(\tau)}_{\bar{r}\bar{r}'}(h_\alpha^{-1})$$

$$\times \Delta^{\gamma}_{p\eta_0(\tau)\bar{r}',s\kappa_0(\tau')\bar{u}'}(g)D^{\kappa_0(\tau')}_{\bar{u}'\bar{u}}(\varphi(h_\alpha)) \tag{6c}$$

$$= \delta_{\eta,\eta_0}\delta_{\kappa,\kappa_0}\delta_{\tau\tau'}\delta_{\bar{r}\bar{u}} \frac{1}{d(\tau)} \sum_{g\in G} A(g) \sum_{\bar{r}} \Delta^{\gamma}_{p\eta_0(\tau)\bar{r},s\kappa_0(\tau)\bar{r}}(g). \tag{6d}$$

The equality between (6a) and (6b) is obtained making use of (5), the invariance of the sum over $g \in G$ under right- and left-multiplication, and the decomposition of the UIR $\Delta^{\gamma}(h_\alpha^{-1}\hat{h}^{-1}g\hat{k}\varphi(h_\alpha))$ into three factors, UIRs of H, G and K. To obtain (6c) we sum over $\hat{h} \in \hat{H}$ and $\hat{k} \in \hat{K}$ making use of (3a) for $\hat{H}$ and $\hat{K}$, whereupon the UIRs of H and K, (labelled by η and κ) must be such η_0 and κ_0 that contain the trivial representation of their normal subgroups. Further we may choose the labels r and u such that they define UIR of $\hat{H}$ and $\hat{K}$ and the range of these labels will then be restricted to the ones containing the trivial representation of $\hat{H}$

and $\hat{K}$ respectively. The corresponding basis vectors of the UIR are uniquely characterized by basis vectors for the UIR of the factor groups $H/\hat{H}$ and $K/\hat{K}$ as well as the labels p and s. As the two factor groups are isomorphic we may choose the row labels of the UIR of the two factor groups such that

$$D^{\eta_0(\tau)}_{\bar{r}\bar{r}'}(h_\alpha) = D^{\kappa_0(\tau)}_{\bar{r}\bar{r}'}(\varphi(h_\alpha)) = D^{\tau}_{\bar{r}\bar{r}'}(h_\alpha), \tag{7}$$

where τ labels an UIR of the factor groups and $\bar{r}$ and $\bar{r}'$ are appropriate row labels for the UIR η and κ restricted such that they only contain the trivial representation of the normal subgroups; $\eta_0(\tau)$ indicates a UIR of H which contains the trivial representation of $\hat{H}$ and is characterized by the UIR τ of the factor group and similarly for $\kappa_0(\tau)$. From here, (6d) is obtained summing over the representatives h_α of the factor group $H/\hat{H}$ (under which A(g) is invariant) and using (3a) for the UIRs of the factor group.

It is thus convenient to define the *partial traces*, functions over double classes, which can be written as

$$\begin{aligned}\chi^{\gamma}_{p\tau s}(g_i) = \chi^{\gamma}_{p\tau s}(C_i) &= \sum_{\bar{r}} \Delta^{\gamma}_{p\eta_0(\tau)\bar{r},s\kappa_0(\tau)\bar{r}}(g_i)\\ &= \frac{d(\tau)}{|C_i|}\sum_{g\in C_i} \Delta^{\gamma}_{p\eta_0(\tau)\bar{r},s\kappa_0(\tau)\bar{r}}(g),\end{aligned} \tag{8}$$

where in the last expression $|C_i|$ is the number of elements in C_i; the expression is independent of $\bar{r}$ and can be proven from (6) factorizing the sum over $g \in G$ into a sum over $g \in C_i$ and another sum over $C_i \in \Gamma$.

From (3), following a similar procedure, we can prove the orthogonality and completeness relations for the partial traces (8) as

$$\sum_{C_i \in \Gamma} \frac{|C_i|}{d(\tau)} \chi^{\gamma}_{p\tau s}(C_i)^* \chi^{\gamma'}_{p'\tau' s'}(C_i) = \delta_{pp'}\delta_{ss'}\delta_{\tau\tau'}\delta_{\gamma\gamma'} \frac{|G|}{d(\gamma)}, \tag{9a}$$

$$\sum_{\gamma \in \tilde{G}} \frac{d(\gamma)}{|G|} \sum_{p\tau s} \frac{|C_i|}{d(\tau)} \chi^{\gamma}_{p\tau s}(C_i)^* \chi^{\gamma}_{p\tau s}(C_j) = \delta_{i,j}. \tag{9b}$$

As a consequence, the function (5) over double classes can be expanded as

$$A(C_i) = \sum_{\gamma \in \tilde{G}} \frac{d(\gamma)}{|G|} \sum_{p\tau s} A^{\gamma}_{p\tau s} \chi^{\gamma}_{p\tau s}(C_i)^* \tag{10a}$$

with harmonic (Fourier) coefficients

$$A^{\gamma}_{p\tau s} = \sum_{C_i \in \Gamma} \frac{|C_i|}{d(\tau)} A(C_i) \chi^{\gamma}_{p\tau s}(C_i). \tag{10b}$$

For the same reason that double cosets and conjugation classes are particular cases of double classes, spherical functions and characters as well as the UIR D-functions themselves are special cases of the partial traces.

IV. GROUP RING STRUCTURE

Given a group G, we can construct a |G|-dimensional vector space $\mathcal{G}$ whose elements are the formal sums

$$a = \sum_{g \in G} A(g) g, \tag{11}$$

where A(g) are complex-valued functions on G and the group elements $g \in G$ are basis vectors. The product law in G induces a corresponding product in $\mathcal{G}$ (which does not in general have an inverse), so $\mathcal{G}$ is endowed with a *ring* structure.

If $ab = c$; $ab \in \mathcal{G}$, the corresponding coordinates C(g) of c are the *group convolution* of A(g) and B(g):

$$C(g) = \sum_{g'\in G} A(g')B(g'^{-1}g) = \sum_{g'\in G} A(gg'^{-1})B(g'). \tag{12}$$

We shall call the group ring of $\hat{H}$, $\hat{\mathcal{H}}$ and the one of $\hat{K}$, $\hat{\mathcal{K}}$ and further we define the subsets $\mathcal{F} \subset \mathcal{G}$ and $\varphi(\mathcal{F}) \subset \mathcal{G}$ by their elements $\mathfrak{f}$ and $\varphi(\mathfrak{f})$:

$$\mathfrak{f} = \sum_{\alpha} F(h_\alpha)h_\alpha, \quad \varphi(\mathfrak{f}) = \sum_{\alpha} F(h_\alpha)\varphi(h_\alpha). \tag{13}$$

We also choose the coset representatives such that if $h_\alpha\hat{h} = \varphi(h_\alpha)\hat{k}$ we actually have $h_\alpha = \varphi(h_\alpha)$. We then can state the *theorem*:

The following two statements are equivalent:

1) $A(g)$ is invariant on double classes defined by $(H,\hat{H},K,\hat{K},\varphi)$.

2) All elements of the subset $\mathcal{A} \subset \mathcal{G}$ with coordinates $A(g)$ fulfill the relations:

a) $\hat{h}a = \text{const } a, \quad \forall\ \hat{h} \in \hat{\mathcal{H}},$

b) $a\hat{k} = \text{const } a, \quad \forall\ \hat{k} \in \hat{\mathcal{K}},$

c) $\mathfrak{f}a = a\varphi(\mathfrak{f}), \quad \forall\ \mathfrak{f} \in \mathcal{F}.$

Note that a) and b) indicate that $\mathcal{A}$ is made up of one dimensional ideals with respect to multiplication with $\hat{h} \in \hat{\mathcal{H}}$ from the left and $\hat{k} \in \hat{\mathcal{K}}$ from the right. If φ is the identity then condition c) indicates that $\mathcal{A}$ is in the commutant of $\mathcal{F}$ in $\mathcal{G}$.

We first prove that 2) follows from 1). a) and b) follow from the fact that double classes are unions of double cosets with respect to $\hat{H}$ on the left and $\hat{K}$ on the right. Using (12) and $A(h_\alpha^{-1}g) = A(g\varphi(h_\alpha)^{-1})$ we obtain c) as

$$\mathfrak{f}a = \sum_{g\in G} [\sum_{\alpha} F(h_\alpha)A(h_\alpha^{-1}g)]g = \sum_{g\in G} [\sum_{\alpha} F(h_\alpha)A(g\varphi(h_\alpha)^{-1})]g = a\varphi(\mathfrak{f}). \tag{14}$$

To show that 1) follows from 2) we first note that $A(g) = A(\hat{h}g\hat{k}^{-1})$

is an immediate consequence of a) and b). It remains to show that $A(h_\alpha^{-1}g) = A(g\varphi(h_\alpha)^{-1})$ is a consequence of c). We choose $\mathcal{g} = h_\alpha \in \mathcal{F}$ and obtain

$$h_\alpha \mathcal{A} = \sum_{g\in G} A(h_\alpha^{-1}g)g = \mathcal{A}\varphi(h_\alpha) = \sum_{g\in G} A(g\varphi(h_\alpha)^{-1})g. \qquad (15)$$

As the $g \in G$ form a basis for the vector space $\mathcal{G}$ the two sums can only be identical if their coefficients coincide.

We thus have characterized the elements of $\mathcal{A}$ and we may ask under which conditions $\mathcal{A}$ forms a subring of $\mathcal{G}$. The answer is given by the theorem:

The subset $\mathcal{A} \subset \mathcal{G}$ with coordinates $A(g)$ invariant on double classes is a subring iff φ is the identity mapping.

The "if" part follows immediately from the fact that a,b) and c) define three rings and $\mathcal{A}$ is their intersection. It may also be verified directly from (12). The "only if" part can be proven in the matrix basis.

REFERENCES

1. W. HÄSSELBARTH, E. RUCH, D. KLEIN and T.H. SELIGMAN, (These proceedings).
2. D. KLEIN, in "Group Theory and its Applications III" (ed. E.M. Loebl), Academic Press, New York 1975. T.H. SELIGMAN, "Double Cosets and the Many Body Problem", Burg Verlag, Basel 1975.

Part VI

Representation Theory

INDUCED PROJECTIVE REPRESENTATIONS

R. Dirl

I. INTRODUCTION

A topic of interest in mathematical[1] and physical[2,3] literature is projective representations of finite groups. The present work deals mainly with the problem of constructing by means of induction[1] all projective unitary irreducible representations (projective unirreps) of a finite group G for a given standard factor system[1-4] from the projective unirreps of a normal subgroup N of G. Thereby we prefer to use the corresponding group algebra to aid the induction procedure. This induction procedure has a direct application to the problem of constructing in a systematic way the projective unirreps of the little co-groups (of the groups of the $\vec{q}$-vectors) of non-symmorphic space groups[2-4].

II. PROJECTIVE REPRESENTATIONS

We define by

$$[V(x)f](z) = Q(x,x^{-1}z)f(x^{-1}z)$$

$$\text{for all} \quad x,z \in G \quad \text{and all} \quad f \in L^2(G) \qquad (2.1)$$

the unitary projective left-regular representation

$V(G) = \{V(x): x \in G\}$ of the finite group G (of order $|G|$), where Q is a standard factor system[1-4]. It is well-known[4] that the set of all matrix elements of the projective unirreps for a given factor system Q forms an orthogonal basis of $L^2(G)$. The elements of the set

$$\{\mathbb{D}^{\beta}_{pq} : \beta \in A_{G(Q)}; p,q=1,2,\ldots,n_{\beta}\} \tag{2.2}$$

where $A_{G(Q)}$ denotes the set of all equivalence classes and n_{β} the dimension of the corresponding projective unirrep, must satisfy apart from the multiplication law, the orthogonality and completeness relation.

The definition of the group algebra A(G) of a finite group G and its properties for ordinary vector representations[5-7] can be easily transferred to projective representations. We call the set

$$A(G) = \{\mathbb{F} = |G|^{-1} \sum_{x} F(x)V(x) : F \in L^2(G)\} \tag{2.3}$$

the (left) group algebra A(G) belonging to the standard factor system Q. Analogous to ordinary vector representations A(G) is a semi-simple algebra [5-7] which decomposes into a direct orthogonal sum of n_{β}^2-dimensional simple algebras $A^{\beta}(G)$. For every simple algebra $A^{\beta}(G)$ there exists a basis

$$\{\mathbb{E}^{\beta}_{pq} = n_{\beta}|G|^{-1} \sum_{x} \mathbb{D}^{\beta *}_{pq}(x)V(x) : p,q=1,2,\ldots,n_{\beta}\} \tag{2.4}$$

whose elements are called "units" satisfying

$$\mathbb{E}^{\beta\dagger}_{pq} = \mathbb{E}^{\beta}_{qp}; \qquad \mathbb{E}^{\beta}_{pq}\mathbb{E}^{\xi}_{rs} = \delta_{\beta\xi}\delta_{qr}\mathbb{E}^{\beta}_{ps} \tag{2.5}$$

$$V(x)\mathbb{E}^{\beta}_{pq} = \sum_{r} \mathbb{D}^{\beta}_{rp}(x)\mathbb{E}^{\beta}_{rq}. \tag{2.6}$$

Relation (2.6) is used in the following to calculate actually the matrix elements of the projective unirreps of G.

III. INDUCED PROJECTIVE REPRESENTATIONS

We assume that for a normal subgroup $N = \{e,n,\ldots\}$ of G the projective unirreps are already known. This implies that the elements of the set

$$\{\mathbb{D}^{\mu}_{pq} : \mu \in A_{N(Q)}; p,q=1,2,\ldots,n_{\mu}\} \tag{3.1}$$

can be used to define by means of the rule (2.4) the corresponding units $\mathbb{E}^{\mu}_{pq}$ of the n_{μ}^2-dimensional simple algebras $A^{\mu}(N)$ of $A(N)$ ($\subset A(G)$).

We realize that the automorphism of the projective regular representation of N

$$V(y)^{\dagger}V(n)V(y) = Q(n,y)Q^*(y,y^{-1}ny)V(y^{-1}ny)$$
$$\text{for all} \quad n \in N; \quad y \in G \tag{3.2}$$

implies an automorphism of the group algebra $A(N)$:

$$\mathbb{F} \in A(N) \Longleftrightarrow V(y)^{\dagger}\mathbb{F}\, V(y) \in A(N)$$
$$\text{for all} \quad y \in G; \quad \mathbb{F} \in A(N). \tag{3.3}$$

Consequently we call the set

$$N\{\mu\} = \{x : V(x)^{\dagger}\mathbb{F}^{\mu}V(x) \in A^{\mu}(N), \mathbb{F}^{\mu} \in A^{\mu}(N)\} \tag{3.4}$$

forming a group, the "little group" which belongs to the projective unirrep $\mathbb{D}^{\mu}$ of N. To be more concrete the automorphism of the simple algebras $A^{\mu}(N)$ defined by (3.3) implies that the projective unirrep

$$\mathbb{D}^{\mu}(xnx^{-1})Q(n,x^{-1})Q^*(x^{-1},xnx^{-1}) \sim \mathbb{D}^{\mu}(n) \Longleftrightarrow x \in N\{\mu\} \tag{3.5}$$

must be equivalent to itself, provided $x \in N\{\mu\}$. Because of Schur's Lemma there must exist for every x a unitary matrix $\mathbb{B}^{\mu}(x)$ satisfying

$$\mathbb{D}^{\mu}(xnx^{-1})Q(n,x^{-1})Q^*(x^{-1},xnx^{-1}) = \mathbb{B}^{\mu}(x)\mathbb{D}^{\mu}(n)\mathbb{B}^{\mu}(x)^{\dagger}. \quad (3.6)$$

The set $\mathbb{B}^{\mu} = \{\mathbb{B}^{\mu}(x) : x \in N\{\mu\}\}$ forms a projective unirrep of $N\{\mu\}$ which belongs however to a standard factor system P different from the given one:

$$\mathbb{B}^{\mu}(x_1)\mathbb{B}^{\mu}(x_2) = P(x_1,x_2)\mathbb{B}^{\mu}(x_1x_2), \quad (3.7)$$

$$P(x_1,x_2) = Q(x_1,x_2)K^*(x_1,x_2), \quad (3.8)$$

$$K(x_1,x_2) = K(\underline{x}_1,\underline{x}_2)$$

$$\text{with} \quad x_i = \underline{x}_i n_i; \quad \underline{x}_i \in N\{\mu\}:N, \; n_i \in N. \quad (3.9)$$

This implies that the matrices $\mathbb{B}^{\mu}(x)$ can be chosen in such a way that the standard factor system K is a constant function on the left cosets $(\underline{x}_1N \times \underline{x}_2N)$ of the direct product group $N\{\mu\} \times N\{\mu\}$ where $\underline{x}(\in N\{\mu\}:N)$ denotes a left coset representative. If $\mathbb{D}^{\mu}$ is 1-dimensional relation (3.5) reduces to an equality and the matrices $\mathbb{B}^{\mu}(x)$ can be chosen as

$$\mathbb{B}^{\mu}(x) = \mathbb{B}^{\mu}(\underline{x}n) = Q^*(\underline{x},n)\mathbb{D}^{\mu}(n) \quad (3.10)$$

which implies that K must take the special form

$$K(x_1,x_2) = Q(\underline{x}_1,\underline{x}_2)Q^*(\underline{x}_{12},n_{12})\mathbb{D}^{\mu}(n_{12})$$

$$\text{with} \quad \underline{x}_1\underline{x}_2 = \underline{x}_{12}n_{12}. \quad (3.11)$$

Because of the transformation properties

$$V(x)^{\dagger}\mathbb{E}^{\mu}_{pq}V(x) = \sum_{rs} \mathbb{B}^{\mu}_{rp}(x)^{\dagger}\mathbb{E}^{\mu}_{rs}\mathbb{B}^{\mu}_{qs}(x) \quad (3.12)$$

it is suggestive to introduce in the $|N\{\mu\}:N|n_{\mu}^2$-dimensional subalgebra $A(\mathbb{D}^{\mu} \uparrow N\{\mu\})$ instead of the obvious basis

$\{V(\underline{x})\mathbb{E}^{\mu}_{pq}:\underline{x} \in N\{\mu\}:N;p,q=1,2,\ldots,n_{\mu}\}$ the following new one

$$\{\mathbb{F}^{\mu;\underline{x}}_{pq} = \sum_{k} \mathbb{B}^{\mu}_{kp}(\underline{x})^{\dagger}V(\underline{x})\mathbb{E}^{\mu}_{kq}:\underline{x} \in N\{\mu\}:N;p,q=1,2,\ldots,n_{\mu}\}. \quad (3.13)$$

This new basis implies $|N\{\mu\}:N|n_{\mu}$-dimensional reducible projective representations of $N\{\mu\}$ belonging to the original factor system Q.

$$V(\underline{x}_1 n)\mathbb{F}^{\mu;\underline{x}_2}_{pq} = K(\underline{x}_1,\underline{x}_2) \sum_{k} \mathbb{B}^{\mu}_{kp}(\underline{x}_1 n)\mathbb{F}^{\mu;\underline{x}_{12}}_{kq}. \quad (3.14)$$

In order to obtain the projective unirreps of $N\{\mu\}$ belonging to the factor Q one has to construct the projective unirreps of the factor group $N\{\mu\}/N$ belonging to the factor system K. We denote the set of all projective unirreps of $N\{\mu\}/N$ by

$$\{\mathbb{D}^{\kappa}_{ab}:\kappa \in A_{N\{\mu\}/N(K)};a,b=1,2,\ldots,n_{\kappa}\}. \quad (3.15)$$

If we then identify the elements of $N\{\mu\}/N$ with the corresponding left-coset representatives by means of the mapping $\underline{x}N \to \underline{x}$ we can write for the multiplication law $\mathbb{D}^{\kappa}(\underline{x}_1)\mathbb{D}^{\kappa}(\underline{x}_2) = K(\underline{x}_1,\underline{x}_2)\mathbb{D}^{\kappa}(\underline{x}_{12})$ where $\underline{x}_{12}$ has to be understood as $\underline{x}_1 N\underline{x}_2 N = \underline{x}_{12}N \to \underline{x}_{12}$. Therefore the elements

$$\mathbb{L}^{\kappa,\mu}_{ap,bq} = n_{\kappa}|N\{\mu\}/N|^{-1} \sum_{\underline{x}} \mathbb{D}^{\kappa *}_{ab}(\underline{x})\mathbb{F}^{\mu;\underline{x}}_{pq} \quad (3.16)$$

engender the desired projective unirreps $\mathbb{D}^{\kappa,\mu}$ of $N\{\mu\}$ belonging to the standard factor system Q.

$$\mathbb{D}^{\kappa,\mu} = \{\mathbb{D}^{\kappa,\mu}(x) = \mathbb{D}^{\kappa,\mu}(\underline{x}n) = \mathbb{D}^{\kappa}(\underline{x}) \otimes \mathbb{B}^{\mu}(\underline{x}n):x \in N\{\mu\}\}. \quad (3.17)$$

As the last and simplest step we carry out the induction $\mathbb{D}^{\kappa,\mu} \uparrow G$. For this purpose we introduce in A(G) the following basis[8]

$$\mathbb{P}^{(\kappa,\mu)\uparrow G}_{\underline{z}_1 ap,\underline{z}_2 bq} = V(\underline{z}_1)\mathbb{L}^{\kappa,\mu}_{ap,bq}V(\underline{z}_2)^{\dagger}:\mu \in \Delta A_{N(Q)};\ \kappa \in A_{N\{\mu\}/N(K)}$$

$$p,q=1,2,\ldots,n_\mu;\ a,b=1,2,\ldots,n_\kappa$$

$$\underline{z}_j \in G:N\{\mu\} \tag{3.18}$$

which satisfy the necessary and sufficient conditions (2.5,2.6) in order to be units of A(G). ($\Delta A_{N(Q)}$ denotes the fundamental domain of $A_{N(Q)}$ and $\underline{z}_j$ left-coset representatives.) By means of (2.6) we obtain the $|G:N\{\mu\}|n_\kappa n_\mu$-dimensional projective unirreps $\mathbb{D}^{(\kappa,\mu)\uparrow G}$ of G belonging to the standard factor system Q.

$$\begin{aligned} V(y_j)\mathbb{P}^{(\kappa,\mu)\uparrow G}_{\underline{z}_1 ap,\underline{z}_2 bq} &= V(\underline{z}_j x_j)\mathbb{P}^{(\kappa,\mu)\uparrow G}_{\underline{z}_1 ap,\underline{z}_2 bq} \\ &= Q(y_j,\underline{z}_1)Q^*(\underline{z}_{j,11(j)},x_{j,11(j)}x_{11(j)}x_{j(1)}) \\ &\times V(\underline{z}_{j,11(j)})\sum_{cr}\mathbb{D}^{\kappa,\mu}_{cr,ap}(x_{j,11(j)}x_{11(j)}x_{j(1)}) \\ &\times \mathbb{P}^{(\kappa,\mu)\uparrow G}_{ecr,\underline{z}_2 bq} = \sum_{\underline{z}_3 cr}\mathbb{D}^{(\kappa,\mu)\uparrow G}_{\underline{z}_3 cr,\underline{z}_1 ap}(y_j)\mathbb{P}^{(\kappa,\mu)\uparrow G}_{\underline{z}_3 cr,\underline{z}_2 bq}, \end{aligned} \tag{3.19}$$

$$\begin{aligned} &y = \underline{z}x \quad \text{with} \quad \underline{z} \in G:N\{\mu\} \quad \text{and} \quad x \in N\{\mu\} \\ &\underline{z}_1\underline{z}_2 = \underline{z}_{12}x_{12} \quad \text{with} \quad \underline{z}_{12} \in G:N\{\mu\} \quad \text{and} \quad x_{12} \in N\{\mu\} \\ &\underline{z}_j^{-1}x_k\underline{z}_j = \underline{z}_{j(k)}x_{k(j)} \quad \text{for all} \quad \underline{z}_j \in G:N\{\mu\} \quad \text{and} \quad x_k \in N\{\mu\} \end{aligned} \tag{3.20}$$

such that $\underline{z}_{j(k)} \in G:N\{\mu\}$ and $x_{k(j)} \in N\{\mu\}$.

IV. NON-SYMMORPHIC SPACE GROUPS

The described induction procedure can be applied to construct the projective unirreps of the little co-groups $P^{\vec{q}}$ (being isomorphic to the factor group $G^{\vec{q}}/T$, where T denotes the translation and $G^{\vec{q}}$ the groups of the $\vec{q}$-vectors of a non-symmorphic space group). The little co-groups are defined by

$$P^{\vec{q}} = \{\alpha:\vec{q}(\alpha)=D(\alpha)\vec{q}=\vec{q}+\vec{Q}\{\vec{q}(\alpha)\}\} \tag{4.1}$$

where the $\vec{Q}$'s are (uniquely determined) reciprocal lattice vectors. In order to obtain the ordinary vector unirreps of $G^{\vec{q}}$ one method consists of determining the projective unirreps of $P^{\vec{q}}$ which belong to the standard factor system $R^{\vec{q}}$.

$$R^{\vec{q}}(\alpha,\alpha') = e^{-i\vec{q}\cdot\vec{t}(\alpha,\alpha')};\ \vec{t}(\alpha,\alpha') = \vec{\tau}(\alpha) + D(\alpha)\vec{\tau}(\alpha') - \vec{\tau}(\alpha\alpha'). \tag{4.2}$$

(Thereby $D = \{D(\alpha):\alpha \in P^{\vec{q}}\}$ is a n-dimensional (n = dimension of the lattice) orthogonal representation of $P^{\vec{q}}$ and the symbol $\vec{\tau}(\alpha)$ denotes non-primitive lattice translations.)

In order to be able to apply the induction procedure we start from the "trivial little co-groups" $M^{\vec{q}}$ which are defined by

$$M^{\vec{q}} = \{\sigma:\vec{q}(\sigma)=\vec{q}\}. \tag{4.3}$$

We note that $M^{\vec{q}}$ is a subgroup of $P^{\vec{q}}$ for which the factor system reduces to a trivial one. This implies that the projective unirreps of $M^{\vec{q}}$ are uniquely determined (up to the unimodular factor $e^{-i\vec{q}\cdot\vec{\tau}(\alpha)}$)[9] by the ordinary vector unirreps of $M^{\vec{q}}$. Therefore, if $\vec{q}$ does not lie on the "surface" of the fundamental domain of the Brillouin zone, we are confronted with a trivial case, since $M^{\vec{q}} = P^{\vec{q}}$. However, if $\vec{q}$ lies on the surface of the zone, $M^{\vec{q}}$ is a proper subgroup of $P^{\vec{q}}$ but in general not a normal one. Therefore we introduce the subsidiary group ("trivial extended little co-group")

$$N^{\vec{q}} = \{\delta:\vec{Q}\{\vec{q}(\delta\sigma\delta^{-1})\} = \vec{0};\sigma \in M^{\vec{q}}\} \tag{4.4}$$

containing $M^{\vec{q}}$ as a normal subgroup. Now if $N^{\vec{q}}$ is a normal subgroup of $P^{\vec{q}}$ the induction procedure can be applied twice where the starting point is (up to the already mentioned unimodular factor) the vector unirreps of $M^{\vec{q}}$; otherwise a more general induction procedure must be considered.

ACKNOWLEDGEMENT

A financial grant from the Université de Montréal is gratefully acknowledged.

REFERENCES

1. G.W. MACKEY, Acta Mathematica 99, 265 (1958).
2. J.L. BIRMAN, "Theory of Crystall Space Groups and Infra-red and Raman Processes of Insulated Crystalls" in Handbuch der Physik, ed. S. Flügge (Springer, Berlin, 1974).
3. C.J. BRADLEY and A.P. CRACKNELL, "The Mathematical Theory of Symmetry in Solids" (Clarendon, Oxford, 1972).
4. S.L. ALTMANN, "Induced Representations in Crystalls and Molecules" (Oxford, 1975, unpublished).
5. L. JANSEN and M. BOON, "Theory of Finite Groups. Applications in Physics" (North-Holland, Amsterdam, 1969).
6. R. DIRL and P. KASPERKOVITZ, "Gruppentheorie, Anwendungen in der Atom- und Festkörperphysik" (Vieweg, Braunschweig, 1976, in press).
7. P. KASPERKOVITZ and R. DIRL, J. Math. Phys. 15, 1203 (1974).
8. R. DIRL, "Induced Projective Representations" (submitted for publication).
9. O.V. KOVALEV, "Irreducible Representations of Space Groups" (Gordon and Breach, New York, 1965).

UNE IDENTITÉ DU BINÔME ET LES REPRÉSENTATIONS LINÉAIRES FINIES DE $M_p(K)$ ET $GL(p,K)$

J.P. Gazeau

La littérature sur les fonctions spéciales foisonne de formules souvent surprenantes par leur simplicité. Citons par exemple les suivantes[(1)(2)]:

$$Z_{\ell m}(\vec{x}+\vec{y}) = \sum_{\ell' m'} Z_{\ell' m'}(\vec{x}) Z_{\ell-\ell', m-m'}(\vec{y}), \tag{1.1}$$

$$Z_{\ell m}(\vec{x}+\vec{y}) = \sum_{\ell' m'} Z_{\ell' m'}(\vec{x}) Z_{\ell+\ell', m-m'}(\vec{y}), \qquad (|\vec{x}| < |\vec{y}|) \tag{1.2}$$

où

$$Z_{\ell m}(\vec{x}) = [(\ell+m)!(\ell-m)!]^{-\frac{1}{2}} r^{\ell} Y_{\ell m}(\theta,\varphi),$$

$$Z_{\ell m}(\vec{x}) = [(\ell+m)!(\ell-m)!]^{\frac{1}{2}} r^{-\ell-1} (-1)^{\ell-m} Y_{\ell m}(\theta,\varphi),$$

r,θ,φ: coordonnées sphériques de x.

Si on porte une attention spéciale aux translations sur les indices ℓ, m on ne peut s'empêcher de penser à une sorte de généralisation du développement du binôme. Or les coefficients du binôme obéissent à une équation aux différences finies ("triangle de Pascal"):

$$\binom{n}{m} = \sum_{m'} \binom{n-n'}{m-m'} \binom{n'}{m'} \tag{2}$$

Nous exploitons ici cette voie heuristique, abandonnant le traitement analytique classique(1), ou en termes de théorie des groupes(2)(3)(4), pour replacer ces théorèmes d'addition dans un cadre plus large d'ordre combinatoire.

Mais il nous faut un jeu d'indices adéquat, généralisant les "n,m" de l'équation (2) ou les "ℓ,m" de (1). Ce jeu d'indices nous sera précisément fourni par les tableaux de Gelfand(5) auxquels on imposera d'obéir à une équation aux différences finies du type (2).

A. TABLEAU T DE GELFAND

Considérons le tableau ordonné T:

$T = \{m_i^j ; 1 \leq i,j \leq p\} =$

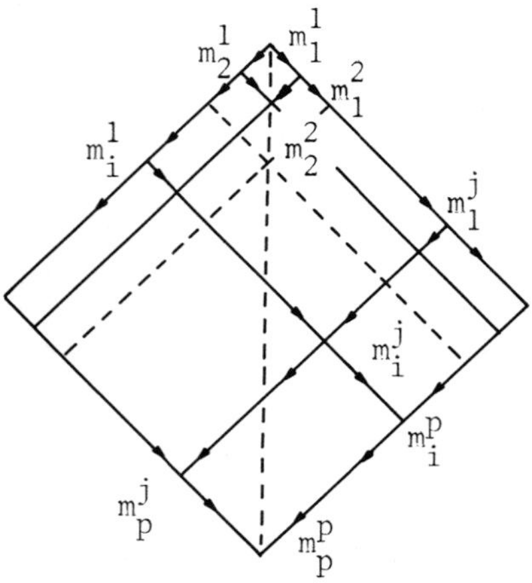

$; \quad m_i^j - m_p^p \in \mathbb{N} .$ (3)

Les flèches indiquent les relations d'ordre:

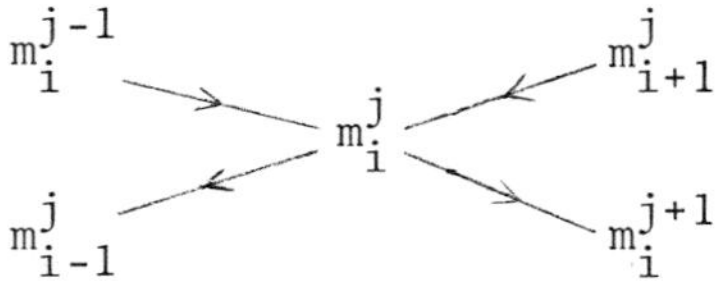

$, \quad \longrightarrow \; \equiv \; \geq$ (4)

la décroissance allant vers le bas selon les obliques. Introduisons les sous-tableaux suivants:

$$P = \{m_i^i; 1 \le i \le p\}; \qquad M = \{m_i^j; j<i\}; \qquad N = \{m_i^j; i<j\}.$$

On notera:

$$T = M^P N$$

T n'est rien d'autre que deux bases de Gelfand[5] accolées l'une à l'autre par leur ligne supérieure P (où l'indice m_i^i est exactement le m_{ip} de Gelfand).

B. UNE EQUATION AUX DIFFERENCES FINIES POUR T

T prenant ses valeurs dans K, corps commutatif de caractéristique nulle, considérons l'équation de récurrence:

$$T = \sum_{M',N'} T\text{-}T' \cdot T', \tag{5}$$

où $T' = M'^{P'}N'$ est composée d'entiers naturels m'^j_i et où $T\text{-}T' = \{m_i^j - m'^j_i;\ m_i^j \in T,\ m'^j_i \in T'\}$.

Adjoignons à (5) l'ensemble x des "conditions initiales" prises dans K:

$\mathbb{0} \overset{\text{def}}{=}$ 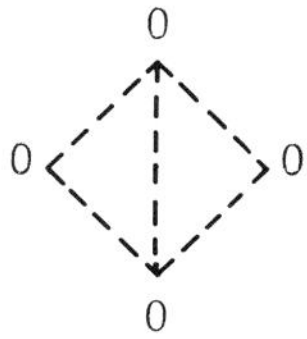 $= 1,$

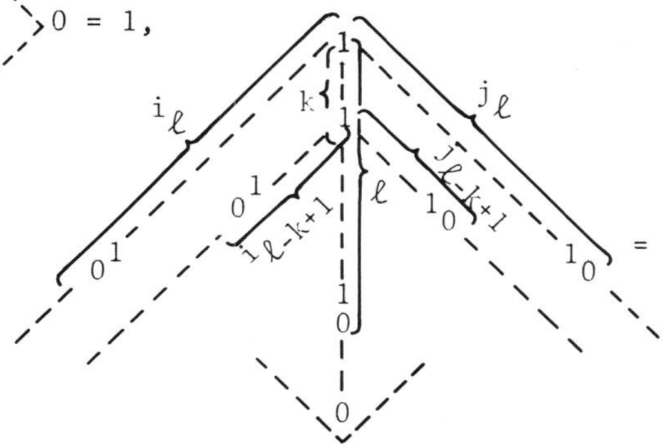

$= x^{j_1,j_2,\ldots,j_\ell}_{i_1,i_2,\ldots,i_\ell}$

(6)

$$\mathbb{1} \overset{\text{def}}{=}$$ 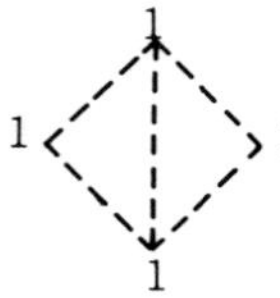 $$= x^{1,2,\dots,p}_{1,2,\dots,p} \qquad (6)$$

avec:

$$1 \le i_1 < i_2 < \dots < i_1 \le p$$

$$1 \le j_1 < j_2 < \dots < j_1 \le p.$$

Nous avons alors le théorème:

Théorème 1. Pour une condition initiale donnée, x, avec $\mathbb{1} = x^{1,2,\dots,p}_{1,2,\dots,p} \neq 0$ il existe une et une seule solution non nulle à (5).

Cette solution sera notée $T = x^T$ et dite "K-puissance" de x, plus précisément la "$T^{\text{ème}}$ K-puissance" de x. Elle est donnée par la formule:

$$T = x^T = (\mathbb{1})^{m^p_p} \sum_{\substack{M^{(\ell)} \\ 2\le\ell\le p-1}} \prod_{\ell=1}^{p-1} (M^{(\ell)} - M^{(\ell+1)}; m^\ell_\ell - m^{\ell+1}_{\ell+1}; N^{(\ell)} - N^{(\ell+1)})_\ell \qquad (7)$$

où:

$$M^{(1)} = M;\ N^{(1)} = N;\ M^{(p)} = N^{(p)} = \{m^j_i; m^j_i = m^p_p\};$$

$$(M;n;N)_\ell \overset{\text{def}}{=}$$ 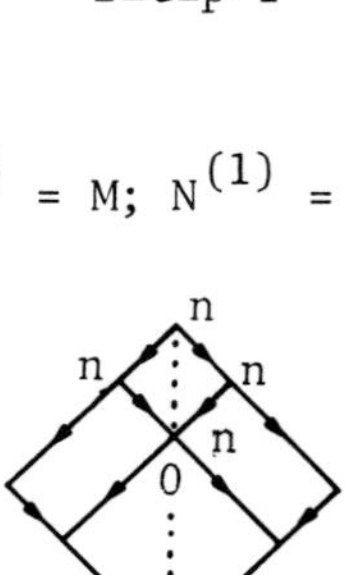 $$(8)$$

$$= n! \sum_{d^{j_1,j_2,\dots,j_\ell}_{i_1,i_2,\dots,i_\ell}} \prod_{\substack{1\le i_1<i_2<\dots<i_\ell\le p \\ j_1,j_2,\dots,j_\ell}} \frac{\left(x^{j_1,j_2,\dots,j_\ell}_{i_1,i_2,\dots,i_\ell}\right)^{d^{j_1,j_2,\dots,j_\ell}_{i_1,i_2,\dots,i_\ell}}}{\left(d^{j_1,j_2,\dots,j_\ell}_{i_1,i_2,\dots,i_\ell}\right)!}$$

où la sommation s'effectue en tenant compte des contraintes:

$$m_i^j = \sum_{\substack{i_1,i_2,\ldots,i_\ell \\ j_1,j_2,\ldots,j_\ell}} d_{i_1,i_2,\ldots,i_\ell}^{j_1,j_2,\ldots,j_\ell}$$

où:

$$i_{\ell-j+1} \geq i-j+1 \text{ pour } i \geq \ell \geq j \geq 1$$

$$j_{\ell-i+1} \geq j-i+1 \text{ pour } j \geq \ell \geq i \geq 1.$$

C. CAS PARTICULIER DIT "SYMETRIQUE 1er"

Définition. On appellera tableau symétrique 1er celui où tous les éléments de la colonne centrale P sont nuls sauf le premier, $m_1^1 = n$. Dans ce cas T se réduit aux deux obliques supérieures, et la condition initiale x, jointe à (5), à l'ensemble des x_i^j, $1 \leq i, j \leq p$, c'est-à-dire à un élément de $M_p(K)$, l'algèbre des matrices p×p à coefficients dans K.

Théorème 2. La solution x^T de (5) pour tout T symétrique 1er et tout $x \in M_p(K)$ est un polynôme homogène de degré $m_1^1 = n$ en les éléments de matrice x_i^j de x. Explicitement:

$$x^T = n! \sum_{\substack{M^{(j)} \\ 1 \leq j \leq p}} \prod_{i,j=1}^{p} \frac{(x_i^j)^{d_i^j}}{d_i^j!} \tag{9}$$

où les $M^{(j)} = \{m_{(i)}^{(j)};\ 1 \leq i \leq p\}$ sont des multi-indices ordonnées de sommation, obéissant à:

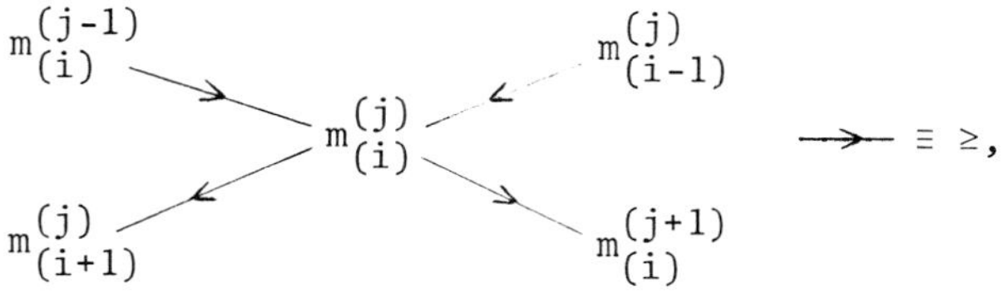

$$m_{(i)}^{(1)} = m_i^1; \quad m_{(1)}^{(j)} = m_1^j; \quad m_{(i)}^{(p+1)} = m_{(p+1)}^{(j)} = 0;$$

$$d_i^j = m_{(i)}^{(j)} - m_{(i)}^{(j+1)} - m_{(i+1)}^{(j)} + m_{(i+1)}^{(j+1)}.$$

Théorème 3 ("théorème d'addition"). On a le "développement du binôme": $x,y \in M_p(K)$

$$(x+y)^T = \sum_{T'} \binom{n}{n'} x^{T-T'} y^{T'}. \tag{10}$$

Cette formule est illustrée moyennant quelques considérations techniques, par (1.1).

Théorème 4 ("théorème de multiplication"). Pour n donné, la matrice carrée $M_n(x)$, d'ordre $\binom{n+p-1}{p-1}$, dont les éléments sont les

$$\frac{x^T}{\prod_{j=1}^{p} \begin{pmatrix} m_1^j \\ m_1^{j+1} \end{pmatrix}}$$

est une représentation linéaire finie du monoïde multiplicatif $M_p(K)$:

$$\frac{(xy)^T}{\prod_{j=1}^{p} \begin{pmatrix} m_1^j \\ m_1^{j+1} \end{pmatrix}} = \sum_{N'} \frac{x^{M^P N'}}{\prod_{j=1}^{p} \begin{pmatrix} m_1'^j \\ m_1'^{j+1} \end{pmatrix}} \frac{y^{N'^P N}}{\prod_{j=1}^{p} \begin{pmatrix} m_1^j \\ m_1^{j+1} \end{pmatrix}} \tag{11}$$

$$N' = \{m_1'^j, \ 1 \le j \le p\}.$$

Posons K = C. Soit $U_n(x)$ la matrice d'ordre $\binom{n+p-1}{p-1}$ dont les éléments sont:

$$\left[\prod_{i=1}^{p} (m_i^1 - m_{i+1}^1)! \prod_{j=1}^{p} (m_1^j - m_1^{j+1})!\right]^{\frac{1}{2}} \frac{x^T}{n!}.$$

Corollaire 4. $U_n(x)$ est une représentation unitaire de dimension $\binom{n+p-1}{p-1}$ de $U(p)$.

Enfin, on a les deux théorèmes de développement, le dernier étant une généralisation de (1.2).

Théorème 5. Pour tout $x \in GL(p,C)$ on désigne par $\rho(x)$ son rayon spectral. Alors pour tout $\lambda \in C$ tel que $|\lambda|\rho(x) < 1$ on a le développement:

$$\det^{-1}(I-\lambda x) = \sum_{n=0}^{+\infty} \lambda^n \mathrm{Tr} M_n(x). \tag{12}$$

Pour tous $x,y \in GL(p,\mathbb{C})$ tels que $x+y \in GL(p,\mathbb{C})$ et $\|xy^{-1}\| < 1$, où $\|\ \ \|$ est la norme de Hilbert Schmidt, on a:

$$\det^{-1}(x+y)(x+y)^{-T} = \sum_{T'} (-1)^{n'} \binom{n+n'}{n'} x^{T'} \times \det^{-1}y\ y^{-(T+{}^tT')}, \tag{13}$$

où:

$$x^{-T} \stackrel{def}{=} (x^{-1})^T, \quad {}^tT = N^P M \text{ (tableau transposé de T)}.$$

Ayant établi un certain nombre de propriétés dans le cas symétrique 1er il est tentant de rechercher celles restant vraies pour la solution générale (7), en particulier de se poser le problème important. Avons-nous obtenu là les éléments de matrice de toutes les représentations linéaires finies de $GL(p,K)$ lorsque les conditions initiales

$$x^{j_1,j_2,\ldots,j_\ell}_{i_1,i_2,\ldots,i_\ell}$$

représentent les sous-déterminants de la matrice (x_i^j)? La méthode présentée ici aurait alors l'avantage de sa rapidité et de son efficacité par rapport aux tentatives variées pour résoudre ce problème[6][7][8].

REFERENCES

1. M. DANOS et L.X. MAXIMON, J. Math. Phys. 6, 766 (1965).
2. J. TALMAN, "Special functions", Benjamin Inc., p. 228 et 232.
3. N. VILENKIN, "Special functions and the theory of group representations", Providence, R.I. (1968).
4. W. MILLER, "Lie theory and special functions", Ac. Press (1968).
5. I.M. GELFAND et M.I. GRAEV, Izv. Akad. Nauk. SSSR. Ser. Math. 29, 1329 (1965) (Amer. Math. Soc. Transl. Ser. 2, 64, 116 (1967)).
6. G.E. BAIRD et L.C. BIEDENHARN, J. Math. Phys. 4, 1449 (1963).
7. J.D. LOUCK et L.C. BIEDENHARN, J. Math. Phys. 14, 1336 (1973).
8. C.J. HENRICH, J. Math. Phys. 16, 2271 (1975) et bibliographie.

DOUBLE CLASSES: A NEW CLASSIFICATION SCHEME FOR GROUP ELEMENTS

W. Hässelbarth, E. Ruch,
D.J. Klein and T.H. Seligman

Classification is one of the main goals of science, and the increasing importance of group theory makes a unified classification scheme for group elements desirable. This scheme should contain the well-known cosets, double cosets, classes of conjugacy and subclasses as special cases. Nevertheless it should still relate explicitly to the fact that the set whose elements are to be classified is a group. This goal will be achieved by the double classes introduced in this paper.

First we shall briefly recall some useful facts about orbits: if G is a set and Π a group of operators acting on G (i.e. $\pi \circ g \in G$ and $(\pi\pi') \circ g = \pi \circ (\pi' \circ g)$, for $\pi, \pi' \in \Pi$, $g \in G$), then G may be rewritten as a union of disjoint subsets $\{\pi \circ g | \pi \in \Pi\}$ called orbits. Orbits may be shown to form equivalence classes with the equivalence relation $g \sim g' \Leftrightarrow g = \pi \circ g'$ for some $\pi \in \Pi$. Further the stabilizers of all elements in one orbit are conjugate to each other. If the stabilizers for an orbit α are finite we may define a repetition frequency m_α giving the number of elements $\pi \in \Pi$ that fulfill $g = \pi \circ g'$ for fixed g and g'. This number is constant for any pair g, g' in one orbit and equal to the order of the stabilizer of g.

We now proceed to define double classes: Choose the set G to be a group and consider the direct product $G \times G$ with elements (a,b); $a,b \in G$. Next we define an action of the elements of $G \times G$ on G by

$$(a,b)\circ g = agb^{-1}. \tag{1}$$

If P is any subgroup of $G \times G$ it may be readily checked that, with the action (1), P may be considered a group of operators on G with the properties discussed above. Therefore

$$\{agb^{-1};(a,b) \in P\}, \qquad g \in G \text{ fixed} \tag{2}$$

forms an orbit we call a P-*double class*, or if no confusion can arise simply a double class. The equivalence relation for a given group P is given by

$$g \sim g' \Leftrightarrow g = ag'b^{-1} \text{ for some } (a,b) \in P. \tag{3}$$

In the table I we give the groups P and their action on $g \in G$ for the well-known classification schemes with the notation that $H,K \subseteq G$ are subgroups; $h \in H$, $k \in K$; I is the identity group; and $(H \times H)_D$ is the diagonal part of $H \times H$ with elements (h,h).

Classification Scheme	P	Action
Double Coset	$H \times K$	hgk^{-1}
Right Coset	$H \times I$	hg
Left Coset	$I \times H$	gh^{-1}
Subclass	$(H \times H)_D$	hgh^{-1}
Conjugacy Class	$(G \times G)_D$	$g'gg'^{-1}$

Table I

Note that while right and left cosets are special cases of double cosets and conjugacy classes are a special case of subclasses, we need double classes to get a unified view of all the schemes of Table I.

Next we wish to investigate what new equivalence relations we can expect in the framework of double classes. As the relation is defined by the group P we are interested in learning the structure of all subgroups of G×G assuming that we know the subgroup structure of G. This problem was solved by Goursat (1,2) back in 1889, and we shall state his result without proof: If $H,K \subseteq G$ have normal subgroups $\hat{H} \lhd H$ and $\hat{K} \lhd K$ such that $H/\hat{H}$ is isomorphic to $K/\hat{K}$ we may choose any isomorphism φ of the factor groups and map the coset generators h_α of H with respect to $\hat{H}$ on the coset generators k_α of K with respect to $\hat{K}$ writing $k_\alpha = \varphi(h_\alpha)$. Then the elements $(\hat{h}h_\alpha, \hat{k}k_\alpha)$ with $\hat{h} \in \hat{H}$ and $\hat{k} \in \hat{K}$ form a subgroup of G×G. Further, every subgroup of G×G can be so expressed.

Comparing with Table I we find for double cosets $\hat{H}=H$, $\hat{K}=K$. The factor groups are trivial and so is by consequence the isomorphism φ. For subclasses we find $H=K$, $\hat{H}=\hat{K}=I$. Again φ is the trivial mapping but in this case nontrivial mappings are possible giving rise to the twisted subclasses mentioned in (3). The related class concepts mentioned in (4) are also special cases of double classes. Note that the general form of P given above implies that a double class is always a union of complete double cosets with respect to $\hat{H}$ and $\hat{K}$.

As double classes are orbits their elements have conjugate stabilizers and, if the latter are finite, constant repetition frequency. To illustrate the importance of this fact consider a function f(g) over a finite group G with elements g, and take this function to be constant over a P-double class, i.e. $f(g) = f(agb^{-1})$, $(a,b) \in P$. If we wish to evaluate a sum over all elements of G we have

$$\sum_{g\in G} f(g) = \sum_\alpha \frac{1}{m_\alpha} \sum_{(a,b)\in P} f(ag_\alpha b^{-1}) = \frac{1}{|P|} \sum_\alpha \frac{1}{m_\alpha} f(g_\alpha) \qquad (4)$$

where g_α is a generator for the double class α and $|P|$ indicates the order of P. We thus reduce a sum over G to a sum over double classes using the constant repetition frequency.

The importance of double classes appears naturally if we discuss classifications of bijections $\psi: B \to A$ between the sets A and B. Using a single reference bijection ψ_0 from B to A, we may write an arbitrary bijection ψ in terms of ψ_0 and permutations p of S_A acting on A or $\bar{p}$ of S_B acting on B, thusly $\psi = p^{-1}\psi_0 = \psi_0\bar{p}$. Hence ψ_0 fixes an isomorphism between S_A and S_B. If we have a symmetry for the range B and image A of the bijections, different bijections become physically indistinguishable and we consider them as being equivalent. Obviously any symmetry group for A and B must be a subgroup of $S_A \times S_B$. Choosing any such subgroup $\bar{P}$ with elements $(a,\bar{b}) \in \bar{P}$, $a \in S_A$, $\bar{b} \in S_B$ we have

$$\psi = p\psi_0 \sim \psi' = p'\psi_0 \Leftrightarrow \psi = a\psi'\bar{b} \text{ for some } (a,\bar{b}) \in \bar{P}$$

but as $p\psi_0 = a\psi'\bar{b} = ap'\psi_0\bar{b} = ap'b^{-1}\psi_0$ we have $\psi \sim \psi'$ if $p = ap'b^{-1}$ with $(a,b) \in P \approx \bar{P}$ if $(a,\bar{b}) \in \bar{P}$. Thus $\psi \sim \psi'$ if p and p' are in the same P-double class.

The well-known classification schemes lead to two extreme cases: For double cosets we have two completely uncorrelated symmetry groups in range and image spaces and P is a direct product of these two groups. For subclasses on the contrary we have a totally correlated symmetry; i.e., the same transformation has to be carried out on range and image. This is reasonable e.g. if A=B, i.e., if we map a space onto itself. In general double classes fill the gap between these two extremes as the normal subgroups give independent symmetries while the mapping φ of the factor groups establishes a correlation.

Practically, a nontrivial double class is found in a problem of molecular physics, namely that of classification of rearrangements between different permutational isomers. Assume a molecular skeleton $\mathcal{L}$ with a point symmetry group $S = R \wedge J$ where R is the normal subgroup of proper rotations, J is the inversion group (with the inversion being with respect to either a point or a plane), and $\wedge$ indicates a semidirect product. Permutational isomers are molecules, that by definition only differ by a permutation of the ligands (e.g. atoms) sitting at the n sites of the skeleton. Therefore any rearrangement is identified with a permutation of S_n. We thus have to decide under what conditions we wish to consider two permutations $p, p' \in S_n$ as equivalent. Keeping in mind that $S_n \supseteq S \triangleright R$ and using the notation $s \in S$, $r, r', r'' \in R$, and $\sigma \in J$ we can enounce two physical criteria sufficient for equivalence:

a) the result of p and p' are distinguished only by a proper rotation (note that mirror images can be distinguished) i.e. $p \sim p'$ if $p = p'r''$.

b) p and p' affect the skeleton in a symmetric fashion i.e. $p \sim p'$ if $p = sp's^{-1}$.

Combining a and b we have that $p \sim p'$ if $p = sp's^{-1}r''$. But as $s = r\sigma$ we may rewrite this as

$$p = r\sigma p'\sigma^{-1}r^{-1}r'' = r\sigma p'\sigma^{-1}r'. \tag{5}$$

This is exactly the equivalence relation for a P-double class with

$$P = (R\times R) \wedge (J\times J)_D. \tag{6}$$

We thus have a nontrivial example where double classes are important. The equivalence relation 5 was first given in (5) where it was discussed as a union of double cosets. Note that if the molecule has no improper rotations (J=I) the double classes reduce to double cosets and if it has no proper rotations they reduce to subclasses.

REFERENCES

1. E. GOURSAT, Annales Scientifiques de l'Ecole Normale Supérieure, Paris (3) 6, 9 (1889).
2. J. LAMBEK, Can. J. Math. 10, 45 (1958).
3. T.H. SELIGMAN and K.B. WOLF, these proceedings.
4. R. REE, Ill. J. Math. 3, 440 (1959).
5. W. HÄSSELBARTH and E. RUCH, Theor. Chim. Acta 29, 259 (1973).

A CHARACTERIZATION OF THE STANDARD POLYNOMIAL OF EVEN DEGREE

Yehiel Ilamed

I. INTRODUCTION

Let F be a field of characteristic zero, $M_n(F)$ be the algebra over F of the n×n matrices with entries in F, and $F[x] = F[x_1,x_2,\ldots]$ be the algebra over F generated by the non-commutative indeterminates $x_1,x_2,\ldots$. Let $s_k(x_1,\ldots,x_k)$ denote the *standard polynomial* of degree k, i.e.

$$s_k(x_1,\ldots,x_k) = \Sigma(\mathrm{sg}\pi)x_{\pi 1} \ldots x_{\pi k} \tag{1}$$

where Σ means summation over the permutations π of $1,\ldots,k$ and $\mathrm{sg}\pi$ means the sign of π.

Using the notations, $p(x_1,\ldots,x_i)$, $p_1(x_1,\ldots,x_j)$, $p_2(x_1,\ldots,x_k)$ are polynomials in F[x] and $A_1^{(n)}$, $A_2^{(n)},\ldots$ are arbitrary matrices in $M_n(F)$, we define:

(i) $p(x_1,\ldots,x_i)$ to be a *trace zero polynomial* and denote this by $\mathrm{tr}(p(x_1,\ldots,x_i)) = 0$, if $\mathrm{tr}(p(A_1^{(n)},\ldots,A_i^{(n)})) = 0$ for all n.

(ii) $p_1 = p_1(x_1,\ldots,x_j)$ to be *orthogonal to* $p_2 = p_2(x_1,\ldots x_k)$ and denote this by $p_1 \perp p_2$, if $\mathrm{tr}(p_1(x_1,\ldots,x_j)p_2(x_1,\ldots,x_k)) = 0$.

The standard polynomial of degree 2k may be characterized as the only multilinear polynomial of minimal degree which is (up to a multiplicative element in F) an identity for $M_k(F)$, [1], [2].

In this note we obtain another characterization of the standard polynomial of even degree, namely: *between the polynomials* $p(x_1,\ldots,x_{2k})$ *of degree* 2k, *which are linear in* x_i, $i=1,\ldots,2k$, *the standard polynomial of degree* 2k *is* (up to a factor in F) *the only polynomial which is orthogonal to* x_i, $i=1,\ldots,2k$. As a result of this characterization we construct, using two arbitrary polynomials as generators, a list $q_1,q_2,\ldots$ of polynomials which are mutually orthogonal, i.e. $tr(q_iq_j) = 0$ for $i \neq j$, $i,j = 1,2,\ldots$.

II. THE MAIN THEOREMS

Theorem 1. Let $p(x_1,\ldots,x_k) \in F[x]$ *be linear in each variable separately. If* $x_i \perp p(x_1,\ldots,x_k)$, $i=1,\ldots,k$, *then:*

(1) $k = 2n$ *implies* $p(x_1,\ldots,x_{2n}) = \alpha s_{2n}(x_1,\ldots,x_{2n})$, $\alpha \in F$

(2) $k = 2n+1$ *implies* $p(x_1,\ldots,x_{2n+1}) = 0$.

Proof. By definition if $x_i \perp p(x_1,\ldots,x_k)$ then

$$tr(x_ip(x_1,\ldots,x_i,\ldots,x_k)) = 0. \tag{2}$$

By linearization $(x_i \to x_i+x_{k+1})$ we obtain from Eq. (2)

$$\begin{aligned} &tr(x_ip(x_1,\ldots,x_{i-1},x_{k+1},x_{i+1},\ldots,x_k)) \\ &\quad + tr(x_{k+1}p(x_1,\ldots,x_i,\ldots,x_k)) = 0. \end{aligned} \tag{3}$$

The polynomial $q(x_1,\ldots,x_k)$ in

$$tr(x_{k+1}(q(x_1,\ldots,x_k)) = tr(x_ip(x_1,\ldots,x_{i-1},x_{k+1},x_{i+1},\ldots,x_k)) \tag{4}$$

is determined, by the right side of Eq. (4), using the commutative property of the trace to move x_{k+1} as first factor in each term of the sum representing the right side of Eq. (4).

From Eq. (3) and Eq. (4)

$$\operatorname{tr}(x_{k+1}(p(x_1,\ldots,x_k) + q(x_1,\ldots,x_k))) = 0. \tag{5}$$

By assumption Eq. (5) is satisfied by $M_h(F)$ for all h. Since the trace bilinear form is nondegenerate

$$p(x_1,\ldots,x_k) + q(x_1,\ldots,x_k) \tag{6}$$

is an identity for $M_h(F)$ for all h. It follows

$$p(x_1,\ldots,x_k) + q(x_1,\ldots,x_k) = 0 \tag{7}$$

identically, since the degree of p+q is k (given) and since the minimal degree of the polynomial identity for $M_h(F)$ is 2h, [1]. From Eq. (7) we obtain k! equations for the k! coefficients, $\alpha_\pi \in F$, of $p(x_1,\ldots,x_k) = \Sigma\alpha_\pi x_{\pi 1} \ldots x_{\pi k}$; the solutions of these equations are $\alpha_\pi = \alpha \operatorname{sg}\pi$ for k even and $\alpha_\pi = 0$ for k odd.

Theorem 2. $s_{2n}(x_1,\ldots,x_{2n}) \perp x_i$, i=1,...,2n.

Proof. To prove this theorem we use

$$\operatorname{tr}(s_{2n+1}(x_1,\ldots,x_{2n+1})) = (2n+1)\operatorname{tr}(x_{2n+1}s_{2n}(x_1,\ldots,x_{2n})) \tag{8}$$

that was proved by Kostant [3]. Substituting x_i for x_{2n+1} in Eq. (8)

$$\operatorname{tr}(x_{2n+1}(x_1,\ldots,x_{2n},x_i)) = (2n+1)\operatorname{tr}(x_i s_{2n}(x_1,\ldots,x_{2n})). \tag{9}$$

For i=1,...,2n the left side of the Eq. (9) is zero since the standard polynomial is alternating. Hence in zero characteristic

$$\operatorname{tr}(x_i s_{2n}(x_1,\ldots,x_{2n})) = 0, \qquad i=1,\ldots,2n. \tag{10}$$

III. AN APPLICATION

Let p_1, p_2 be two polynomials in F[x]. The list $q_1, q_2, \ldots$ defined by

$$q_1 = p_1, \quad q_{2k} = s_{2k}(q_1, \ldots, q_{2k-1}, p_2),$$

$$q_{2k+1} = s_{2k}(q_1, \ldots, q_{2k}), \quad k=1,2,\ldots \tag{11}$$

is a list of mutually orthogonal polynomials i.e. $tr(q_i q_j) = 0$, $i \neq j$, $i,j = 1,2,\ldots$.

REFERENCES

1. J. LEVITZKI, A theorem on polynomial identities, Proc. Amer. Math. Soc. 1 (1950), 334-341.

2. S.A. AMITSUR and J. LEVITZKI, Minimal identities for algebras, Proc. Amer. Math. Soc. 1 (1950), 449-463.

3. B. KOSTANT, A theorem of Frobenius, a theorem of Amitsur-Levitzki and cohomology theory, J. of Mathematics and Mechanics 7 (1958), 237-264.

A GENERALIZATION OF THE "ASSOCIATIVE" PROPERTY OF THE KILLING FORM

Yehiel Ilamed

I. INTRODUCTION

Let F be a field of characteristic zero, $M_n(F)$ the algebra of the n×n matrices over F and, $F[x] = F[x_1,x_2,\ldots]$ the algebra of polynomials over F generated by the noncommutative indeterminates $x_1,x_2,\ldots$.

Let $\tau(A^{(n)}) : F[x] \to M_n(F)$ be the mapping defined by

$$\tau(A^{(n)})p(x_1,\ldots,x_k) = p(A_1,\ldots,A_k) \tag{1}$$

where $p(x_1,\ldots,x_k) \in F[x]$ and $A^{(n)}$ is a given list, $A_1,A_2,\ldots$, of matrices in $M_n(F)$.

Let $B_{\tau(A^{(n)})} : F[x] \times F[x] \to F$ be the mapping defined by

$$B_{\tau(A^{(n)})}(p_1,p_2) = tr(\tau(A^{(n)})(p_1p_2)) \tag{2}$$

where $p_1 = p_1(x_1,\ldots,x_i)$ and $p_2 = p_2(x_1,\ldots,x_j)$ are elements of F[x] and tr means trace. We write

$$B(p_1,p_2) = tr(p_1p_2) \tag{2'}$$

to denote the definition given by Eq. (2) in short.

If $p_1 = x_1$, $p_2 = x_2$ are elements of a Lie algebra and $\tau(A^{(n)})$ is the adjoint representation of this Lie algebra, then the above bilinear form, Eq. (2), is the usual Killing form $B(x_1,x_2) = tr(x_1x_2)$.

It is known that

$$B(x_1,[x_2,x_3]) = B([x_1,x_2],x_3) \tag{3}$$

where $[x_1,x_2] = x_1x_2-x_2x_1$. The triple products appearing in Eq. (3), $tr(x_1[x_2,x_3]) = tr([x_1,x_2]x_3)$, are of a mixed type (an associative product and a Lie product); this is the reason for the quotation marks in the title.

It is not difficult to show that

$$tr((x_1 \circ x_2) \circ x_3) = tr(x_1 \circ (x_2 \circ x_3)) \tag{4}$$

where $x_1 \circ x_2$ is the symmetric (Jordan) product defined by $x_1 \circ x_2 = (x_1x_2+x_2x_1)/2$. Using the commutative property of the trace, we obtain from Eq. (4)

$$B(x_1 \circ x_2,x_3) = B(x_1,x_2 \circ x_3). \tag{5}$$

In this case the trace bilinear form, $B(x_1,x_2)$, has an associative property expressed by Eq. (4); the Jordan product is a non-associative product.

Let $s_k(x_1,\ldots,x_k)$ and $\sigma_k(x_1,\ldots,x_k)$ denote respectively the alternating polynomial of degree k (the standard polynomial of degree k) and the symmetric polynomial of degree k, i.e.

$$s_k(x_1,\ldots,x_k) = \Sigma(sg\pi)x_{\pi 1} \ldots x_{\pi k} \tag{6}$$

$$\sigma_k(x_1,\ldots,x_k) = \Sigma x_{\pi 1} \ldots x_{\pi k} \tag{7}$$

where Σ means summation over the permutations π of $1,\ldots,k$ and $sg\pi$ means the sign of π.

In this note we show that the properties of $B(p_1,p_2)$ expressed by Eq. (3) and Eq. (5) remain valid if we substitute $s_{2k}(x_1,\ldots,x_{2k})$ for the commutator in Eq. (3) and $\sigma_k(x_1,\ldots,x_k)$ for the Jordan product in Eq. (5). A generalization of the symmetry properties of $B([x_1,x_2],x_3)$ and $B(x_1\circ x_2,x_3)$ is also given.

II. THE MAIN THEOREMS

Theorem 1.

$$B(s_{2k}(x_1,\ldots,x_{2k}),x_{2k+1}) = B(x_1,s_{2k}(x_2,\ldots,x_{2k+1})) \tag{8}$$

$$B(\sigma_k(x_1,\ldots,x_k),x_{k+1}) = B(x_1,\sigma_k(x_2,\ldots,x_{k+1})). \tag{9}$$

To prove this theorem we need to prove the following two equations

$$tr(s_{2k+1}(x_1,\ldots,x_{2k+1})) = (2k+1)tr(x_{2k+1}s_{2k}(x_1,\ldots,x_{2k})) \tag{10}$$

$$tr(\sigma_k(x_1,\ldots,x_{k+1})) = (k+1)tr(x_{k+1}\sigma_k(x_1,\ldots,x_k)). \tag{11}$$

The Eq. (10) was proved by Kostant [1, lemma 3.4] using the following known property of the trace $tr(x_1 \ldots x_i x_{i+1} \ldots x_k) = tr(x_{i+1} \ldots x_k x_1 \ldots x_i)$. In a similar way it is easy to prove Eq. (11).

Proof of theorem 1.

From Eq. (10) we obtain

$$tr(x_{2k+1}s_{2k}(x_1,\ldots,x_{2k})) = tr(x_1 s_{2k}(x_2,\ldots,x_{2k+1})) \tag{12}$$

since $s_{2k+1}(x_1,\ldots,x_{2k+1}) = s_{2k+1}(x_2,\ldots,x_{2k+1},x_1)$ and since the characteristic of F is zero. We obtain Eq. (8) from Eq. (12) using Eq. (2). The proof of Eq. (9) is obtained in a similar way using Eq. (11).

Theorem 2. Let $f(x_1,\dots,x_{2k+1})$ and $g(x_1,\dots,x_{k+2})$, $k = 1,2,\dots,$ be defined by

$$f(x_1,\dots,x_{2k+1}) = B(s_{2k}(x_1,\dots,x_{2k}),x_{2k+1}) \tag{13}$$

$$g(x_1,\dots,x_{k+2}) = B(\sigma_{k+1}(x_1,\dots,x_{k+1}),x_{k+2}). \tag{14}$$

Then for all permutations π of $1,\dots,2k+1$ and all permutations φ of $1,\dots,k+2$

$$f(x_1,\dots,x_{2k+1}) = (sg\pi)f(x_{\pi 1},\dots,x_{\pi(2k+1)}) \tag{15}$$

$$g(x_1,\dots,x_{k+2}) = g(x_{\varphi 1},\dots,x_{\varphi(k+2)}). \tag{16}$$

Proof. From Eq. (8) we obtain that $f(x_1,\dots,x_{2k+1})$ is alternating, Eq. (15), since $s_{2k}(x_1,\dots,x_{2k})$ is alternating. From Eq. (9) we obtain that $g(x_1,\dots,x_{k+2})$ is symmetric, Eq. (16), since $\sigma_{k+1}(x_1,\dots,x_{k+1})$ is symmetric.

III. CONCLUDING REMARK

If we define $\sigma_k(x_1,\dots,x_k)$ as a (k-1)-ary operator (by holding k-1 arguments of σ_k fixed and acting on the remaining argument), then Eq. (9) states that this (k-1)-ary operator is self-adjoint. In the same way we can define a (2k-1)-ary operator using $s_{2k}(x_1,\dots,x_{2k})$ and learn the properties of this operator from Eq. (8).

REFERENCES

1. B. KOSTANT, A theorem of Frobenius, a theorem of Amitsur-Levitzki and cohomology theory, J. of Mathematics and Mechanics 7 (1958), 237-264.

HIGHEST WEIGHTS OF SEMISIMPLE LIE ALGEBRAS

W. Laskar

INTRODUCTION

This paper is concerned with semisimple Lie algebras defined over an algebraically closed field of characteristic zero only (in brief s.L.a.), i.e. with the type of algebras widely used by physicists. Calculations of highest wieght vectors in particular cases [4, 11-13] have of course been done already. However here the use of a general procedure yields general formulas which give a very simple proof that no other s.L.a. than the well-known ones do exist.

To make the paper relatively self contained and to define notations we first recall the usual definitions of roots of an algebra, the Dynkin diagram and the highest weight vector (in brief h.w.v.) of a given representation of that algebra [1-14].

In the second part the calculation of the h.w.v. is performed firstly when all the roots have the same length and secondly when the roots have two different lengths of ratio equal to $\sqrt{c}$; these two cases correspond respectively to the two classes of s.L.a. $W_{\ell pc=1}$ and $W_{\ell zc\neq 1}$ (c = 2 or 3).

The third part is devoted to the interpretation of the results obtained in the second part; in particular it is very simply shown that no other simple Lie algebras (defined over an algebraically closed field of characteristic zero) than the ones already known to exist, namely the four infinite series A_ℓ, B_ℓ, C_ℓ, D_ℓ and the five "exceptional" Lie algebras $\{E_6, E_7, E_8, F_4, G_2\}$ which we reclassify according to our scheme as

$$W_{\ell pc=1} = \{A_\ell, D_\ell, E_6, E_7, E_8\}$$

$$W_{\ell zc\neq 1} = \{B_\ell, C_\ell, F_4, G_2\}.$$

The time and space allowed prevent us from giving here any uses and extensions of the present results; a forthcoming publication [17] will deal with them.

I. ROOTS, DYNKIN DIAGRAM AND HIGHEST WEIGHT

The following fundamental facts are well-known:

§1. If $\Sigma = \{\alpha_1, \ldots, \alpha_i, \ldots, \alpha_j, \ldots, \alpha_\ell\}$ is an irreducible fundamental system of simple roots we have

i) $\alpha_1, \ldots, \alpha_\ell$ are linearly independent;

ii) $$\frac{2\langle \alpha_i, \alpha_j \rangle}{\langle \alpha_i, \alpha_i \rangle} = -m, \qquad \frac{2\langle \alpha_i, \alpha_j \rangle}{\langle \alpha_j, \alpha_j \rangle} = -c \qquad (m, c \in \mathbb{Z} > 0); \tag{1}$$

iii) Σ is not decomposable into two mutually orthogonal subsets. Consequently

$$\frac{[2\langle \alpha_i, \alpha_j \rangle]^2}{\langle \alpha_i, \alpha_i \rangle \langle \alpha_j, \alpha_j \rangle} = 4\cos^2\theta = mc \leq 4 \tag{2}$$

and for m=1 one only gets:

c=0, ($\theta = 90°$);	c=1, ($\theta = 120°$);	c=2, ($\theta = 135°$);	c=3, ($\theta = 150°$);
0 line	1 line	2 lines	3 lines

$$c=4, \begin{cases} \theta = 0 & \alpha_j = \alpha_i \\ \theta = \pi & \alpha_j = -\alpha_i \end{cases} . \tag{3}$$

Also

$$\frac{\dfrac{2\langle \alpha_i,\alpha_j\rangle}{\langle \alpha_j,\alpha_j\rangle}}{\dfrac{2\langle \alpha_i,\alpha_j\rangle}{\langle \alpha_i,\alpha_i\rangle}} = \frac{\langle \alpha_i,\alpha_i\rangle}{\langle \alpha_j,\alpha_j\rangle} = c \qquad (4)$$

i.e. the roots have only two possible lengths. Hence

(5-a)

$$\langle \alpha_j,\alpha_j\rangle = \lambda, \quad \langle \alpha_i,\alpha_i\rangle = c\lambda, \quad \langle \alpha_i,\alpha_j\rangle = \begin{cases} -\dfrac{c\lambda}{2} & \text{if } \alpha_i,\alpha_j \text{ are connected roots;} \\ 0 & \text{if } \alpha_i,\alpha_j \text{ are not connected roots.} \end{cases}$$

Normalizing α_j so that $\lambda = \frac{2}{c}$ yields the following relations:

(5-b)

$$\langle \alpha_j,\alpha_j\rangle = \lambda = \frac{2}{c}, \quad \langle \alpha_i,\alpha_i\rangle = 2, \quad \langle \alpha_i,\alpha_j\rangle = \begin{cases} -1 & \text{if } \alpha_i,\alpha_j \text{ are connected roots} \\ 0 & \text{if } \alpha_i,\alpha_j \text{ are not connected roots.} \end{cases}$$

§2. To every given irreducible representation (IR) corresponds a unique vector L (in the idempotent $\mathcal{D}$) called the highest weight vector (denoted h.w.v.) of the given IR. From this h.w.v. L all the properties of the IR can be deduced; for instance the H. Weyl formula giving the dimension N is well-known:

$$N = \prod_{\mu\in\Sigma_+} \frac{(L+R,\mu)}{(R,\mu)} = \prod_{\mu\in\Sigma_+} \left[\frac{(L,\mu)}{(R,\mu)} + 1\right] \qquad (6)$$

Σ_+ being the subset of positive roots and

$$R = \frac{1}{2} \prod_{\mu\in\Sigma_+} \mu. \qquad (7)$$

From the h.w.v. L, a set of N ordinary weight vectors

$\{\lambda_1,\ldots,\lambda_r,\ldots,\lambda_N\}$ can be deduced (all distinct if there is no degeneracy) and used in turn to calculate the diagonal matrices of the IR

$$(F_\mu)^r_r = (\mu,\lambda_r), \qquad \mu \in \Sigma_+ \tag{8}$$

and the non diagonal ones

$$(E_\alpha)^t_r = \pm\sqrt{(F_\alpha)^r_r + [(E_\alpha)^r_s]^2} \text{ where } (E_\alpha)^r_s \neq 0 \text{ if } \lambda_s = \lambda_r + \alpha \tag{9}$$

using

$$(E_{-\alpha})^r_s = -(E_\alpha)^s_r. \tag{10}$$

II. CALCULATION OF THE HIGHEST WEIGHT VECTOR

Having emphasized the importance of the h.w.v., it seems natural to calculate it for each of the two type of IR given by the following Dynkin diagrams:

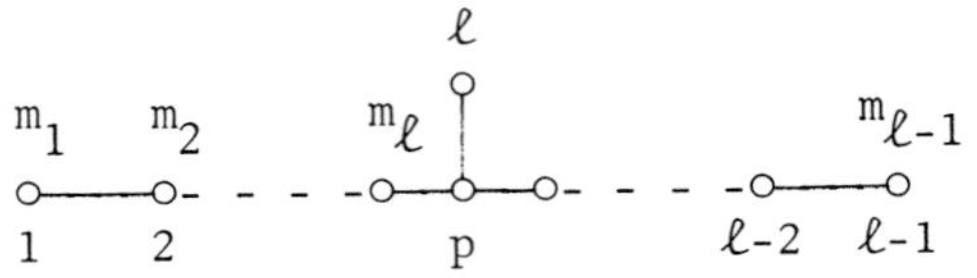

(Type I)

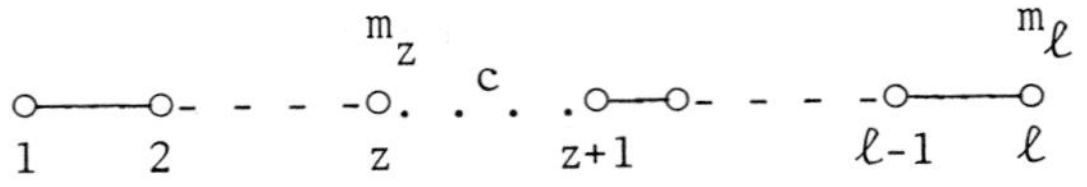

(Type II)

where

$$m_i = L_{\alpha_i} = \frac{2\langle L,\alpha_i\rangle}{\langle \alpha_i,\alpha_i\rangle} \quad (m_i \in \mathbb{Z} > 0;\ i = 1,2,\ldots,\ell-1,\ell). \tag{11}$$

Writing

$$L = \sum_{k=1}^{\ell} a_k\alpha_k$$

and using (5) we get the a_k's as the solution of a system of ℓ

linear equations:

$$m_i = \frac{2}{\langle \alpha_i, \alpha_i \rangle} [a_{i-1} \langle \alpha_{i-1}, \alpha_i \rangle + a_i \langle \alpha_i, \alpha_i \rangle + a_{i+1} \langle \alpha_{i+1}, \alpha_i \rangle + a_\ell \langle \alpha_\ell, \alpha_i \rangle \delta_{ip}] \tag{12}$$

the last term occuring only for diagrams of type (I) when i=p. The system (12) has been solved for each of the two types of diagrams (I) and (II). The corresponding results are given in Tables I and II for diagrams (I) and (II) respectively. If one writes

$$a_k = \frac{1}{\Delta} \sum_{i=1}^{\ell} \xi_k^i m_i ,$$

then one gets two different expressions of Δ according to the type of diagram, say Δ_p for (I) and Δ_z for (II). These expressions will be analyzed in §3 to give the reason for the limitation of the number of simple Lie algebras. As a consequence of Chevalley's theorem the classification of Dynkin diagrams is equivalent to that of simple algebraic groups over algebraically closed fields of zero characteristic.

TABLE I: ξ_k^i for Type I (algebras $W_{\ell pc=1}$)

$$\Delta_p \equiv \Delta = p^2+(2-p)\ell \qquad \delta = \ell-p-2 \qquad \Delta+p\delta = 2(\ell-p)$$

ξ_k^i	$1 \le i \le p-1$	$p \le i \le \ell-1$	ℓ
$1 \le k \le p-1$	$(\Delta+i\delta)k$ $(\Delta+k\delta)i$	$2k(\ell-i)$	$k(\ell-p)$
$p \le k \le \ell-1$	$(\ell-k)2i$	$(\ell-i)[p^2+(2-p)k]$ $[p^2+(2-p)i](\ell-k)$	$(\ell-k)p$
ℓ	$(\ell-p)i$	$p(\ell-i)$	ℓ

TABLE II: ξ_k^i for Type II (algebras $W_{\ell zc}$)

$$\Delta_z(j) = \ell+1-j + (1-c)(\ell-z)(z-j) \qquad j = 1,k,z \text{ or } 0$$

ξ_k^i	$1 \le i \le z-1$	$z \le i \le \ell$
$1 \le k \le z-1$	$k\Delta_z(i)$ (above diagonal); $i\Delta_z(k)$ (below diagonal)	$k(\ell+1-i)$
z	$i(\ell+1-z)$	$z(\ell+1-i)$
$z+1 \le k \le \ell$	$ic(\ell+1-k)$	$(\ell+1-i)[k+(1-c)z(k-z-1)]$ (above diagonal); $(\ell+1-k)[i+(1-c)z(i-z-1)]$ (below diagonal)

III. ANALYSIS OF RESULTS

As the h.w.v. has been written

$$L = \sum_{k=1}^{\ell} a_k \alpha_k$$

with

$$a_k = \frac{1}{\Delta} \sum_{i=1}^{\ell} \xi_k^i m_i$$

we must have $\Delta \neq 0$ and $\Delta > 0$.

In the case of diagrams of Type I, i.e. of $W_{\ell pc=1}$, we have

$$\Delta \equiv \Delta_p = p^2 + (2-p)\ell = \ell+1 + (p-1)(p-\ell+1) > 0$$

$p = \ell-1$	(or 1)	$\Delta = \ell+1 > 0$	for all ℓ	A_ℓ
$p = 2$	(or $\ell-2$)	$\Delta = p^2 = 4 > 0$	for all ℓ	D_ℓ
$p = 3$	(or $\ell-3$)	$\Delta = 9-\ell > 0$	for $\ell = 6,7,8$	E_6, E_7, E_8
p big		$\Delta \sim p(p-\ell) > 0$	for $p > \ell$	nonsense.

In the case of diagrams of Type II, i.e. of $W_{\ell zc}$, we have

$$\Delta \equiv \Delta_z = \ell+1 + (1-c)(\ell-z)z$$

$c = 1$: We come back to the previous case where all the roots have the same length with a linear diagram ($\ell = p-1$), i.e. to A_ℓ

$c = 2$: $\Delta = \ell+1 - (\ell-z)z = 2 + (z-1)(z-\ell+1) > 0$

$z = \ell-1$	$\Delta = 2 > 0$	for all ℓ	B_ℓ
$z = 1$	$\Delta = 2 > 0$	for all ℓ	C_ℓ
$z = 2$	$\Delta = 5-\ell > 0$	for $\ell = 4$	F_4

$c = 3$: $\Delta = \ell+1 - 2z(\ell-z)$

$z = 1$ $\Delta = 3-\ell > 0$ for $\ell = 2$ G_2

$c > 1$:

z big $\Delta \sim z(z-\ell) > 0$ for $z > \ell$ nonsense.

When one writes for instance $9-\ell > 0$, of course one can take $\ell=5$ (or 4) which gives D_5 (or A_4) already seen; similarly for $5-\ell > 0$ $\ell=3$ gives B_3 already seen.

As all other diagrams lead to a null h.w.v. one is left with only the nine s.L.a. already known and widely used by physicists.

Within the time and space allowed it is not possible to do more here. More applications will be given in a further publication [17].

CONCLUSION

As stated by Chevalley's theorem [7,14] the classification of Dynkin diagrams is equivalent to that of simple algebraic groups over algebraic closed fields of characteristic zero. Let us stress the fact that this classification is based on equation (1); once this relation is established it follows that $c = 1,2,3$ and that only two classes of algebra can exist, namely $W_{\ell pc=1}$ (Type I: all the roots have the same length) and $W_{\ell zc\neq 1}$ (Type II: the roots have only two different lengths). In contrast to the point of view (recently discussed in [15,16]) which consists in breaking a given algebra into subalgebras, we have considered here the building of two classes of algebras out of known algebras

$$W_{\ell pc=1} = \{A_\ell, D_\ell, E_6, E_7, E_8\},$$

$$W_{\ell zc\neq 1} = \{B_\ell, C_\ell, F_4, G_2\}.$$

ACKNOWLEDGEMENTS

It is a pleasure to express my gratitude to the board of the Summer School and Colloquium on Group Theory for physicists of the Université de Montréal, particularly to its Director, Professor A. Daigneault, for his kind hospitality at the Département de Mathématiques of the Université de Montréal as well as to Professor Hans Zassenhaus for his enlightening lectures on Lie groups.

REFERENCES

N.B. The 7 first references are given in order to justify the initials used to designate the two classes of algebras with possibilities of alternative interpretation.

1. H. WEYL, The classical groups (Princeton Univ. Press, 1946).
2. S. LIE and F. ENGEL, Theorie der Transformationsgruppen (Leipzig, 1893).
3. L.S. PONTRJAGIN, Topological groups (Princeton Univ. Press, 1939).
4. J. PATERA and D. SANKOFF, Tables of branching rules (Presses de l'Université de Montréal, 1973).
5. H. ZASSENHAUS, Lecture Notes of the Summer School (Université de Montréal, juin 1976) on Lie Groups, Lie algebras and Representation Theory.
6. E. CARTAN, Thèse, Paris 1894, 2nd ed. Vuibert, Paris 1933.
7. C. CHEVALLEY, Théorie des groupes de Lie (Hermann, Paris 1968).
8. H. BACRY, Leçons sur la théorie des groupes (Gordon & Breach 1968).
9. E.B. DYNKIN, Am. Math. Soc. translations, no. 17 (1950).
10. H. FREUDENTHAL & DE VRIES, Linear Lie Groups (Academic Press 1969) and Serie 2, 6 (1957), p. 111-244.
11. N. JACOBSON, Lie algebras (John Wiley, Interscience, 1962).
12. J.E. HUMPHREYS, Introduction to Lie algebras, etc. (Springer, G.T.M. 1972).
13. P.A. ROWLATT, Group Theory and Elementary Particles (Longmans & Co. 1966).

14. I. SATAKE, Classification theory of semi-simple algebraic groups (M. Dekker, New-York 1971).

15. H. BACRY, Ph. Combe and P. Sorba, Rept. Math. Phys. 5, 2, 145 (1974); 5, 3, 361 (1974).

16. J. PATERA, R. SHARP, P. WINTERNITZ and H. ZASSENHAUS, Subgroups of the Poincaré group and their invariants (J.M.P. 17, 6, 1976).

17. W. LASKAR, Highest weights of semi-simple Lie algebras (to be published in J.M.P. 1977).

PROJECTIVE IRREDUCIBLE UNITARY REPRESENTATIONS OF THE SCHRODINGER GROUP WITH A NONTRIVIAL FACTOR

M. Perroud

We will give a classification and a realization of the projective irreducible unitary representations of the Schrödinger group which have a nontrivial factor, i.e., representations which are not equivalent to true representations. The connection between some of these representatives and the realizations found by U. Niederer on spaces of solutions of Schrödinger equations will then be established.

The Schrödinger group S(3) can be defined as the group of the following 5×5 matrices [1]

$$g \equiv (\vec{a},\vec{v},R,S) = \begin{pmatrix} R & (\vec{v}\vec{a}) \\ 0 & S \end{pmatrix},$$

$$R \in SO(3), \quad S \in SL(2,\mathbb{R}). \quad (\vec{v}\vec{a}) \in M_{3\times 2}(\mathbb{R}).$$

Referring to V. Bargmann [2], we know that we can look for the projective irreducible unitary representations (PIUR's) of S(3) among the irreducible unitary representations (IUR's) of the central extensions of its universal covering group S(3)* by a one dimensional Abelian group. It is always possible to restrict oneself to nontrivial extensions. Only one class of such extensions exists:

$$\tilde{S}(3) = \tilde{G}'(3) \mathbin{\square} SL(2,\mathbb{R})^*$$

where

$$\tilde{G}'(3) = (\mathbb{R} \mathbin{\square} T_6) \mathbin{\square} SU(2)$$

is the isochronous subgroup of the extended Galilei group [3]. Denoting by $g = (\xi,\vec{a},\vec{v},s,\sigma)$ an element of $\tilde{S}(3)$, we have the canonical projection

$$(\xi,\vec{a},\vec{v},s,\sigma) \to (\vec{a},\vec{v},R_s,S_\sigma)$$

where $s \to R_s$ and $\sigma \to S_\sigma$ are the canonical projections $SU(2) \to SO(3)$ and $SL(2,\mathbb{R})^* \to SL(2,\mathbb{R})$.

We take advantage of the semidirect product structure of $\tilde{S}(3)$ to use the induction method of G.W. Mackey. The PIUR's of $S(3)$ for which the factor is nontrivial are among the IUR's of $\tilde{S}(3)$, the kernels of which do not contain the Abelian group of the extension.

Firstly we look for the IUR's of $\tilde{G}'(3)$ which satisfy this condition: they belong to the classes (m,j), $m \in \mathbb{R}_*$, $2j \in \mathbb{N}$ and can be realized on $L^2(\mathbb{R}^3,\mathbb{C}^{2j+1})$:

$$(L_{m,j}(\xi,\vec{a},\vec{v},s)f)(\vec{p}) = e^{i(m\xi+\vec{p}\cdot\vec{a})} D_j(s) f(R_s^{-1}(\vec{p}+m\vec{v}))$$

where D_j is a spin j representation of $SU(2)$.

Secondly we determine the orbits of $\tilde{S}(3)$ in the classes $\{(m,j)\}$; the action of $\tilde{S}(3)$ on this set, $g : (m,j) \to (m',j')$ is defined by

$$L_{m',j'}(h) = L_{m,j}(ghg^{-1}), \quad \forall\, L_{m,j} \in (m,j), \quad \forall\, h \in \tilde{G}'(3).$$

We find that these orbits are reduced to a point, their stabilizer is therefore all of $\tilde{S}(3)$. The next step consists of "extending" $L_{m,j}$ to a unitary representation $\mathcal{L}_{m,j}$ of the stabilizer $(\tilde{S}(3))$ of its class. This "extension" can be constructed in the form

$$\mathcal{L}_{m,j}(h,\sigma) = T(\sigma)L_{m,j}(h)$$

where

$$T(\sigma) \quad (\sigma \equiv (0,\bar{0},\bar{0},1,\sigma) \in SL(2,\mathbb{R})^*)$$

is defined up to a phase ($L_{m,j}$ is irreducible) by

$$L_{m,j}(\sigma h \sigma^{-1}) = T(\sigma)L_{m,j}(h)T(\sigma)^{-1}.$$

Using a parametrization of $SL(2,\mathbb{R})^*$ related to the Iwasawa decomposition of $SL(2,\mathbb{R})$ by the projection

$$\sigma(\theta,u,v) \to \begin{pmatrix} \cos\theta & -\sin\theta \\ \sin\theta & \cos\theta \end{pmatrix} \begin{pmatrix} u & 0 \\ 0 & u^{-1} \end{pmatrix} \begin{pmatrix} 1 & v \\ 0 & 1 \end{pmatrix} \equiv \begin{pmatrix} a & b \\ c & d \end{pmatrix},$$

$$\theta \in \mathbb{R}, \quad u \in \mathbb{R}^+, \quad v \in \mathbb{R},$$

we get

$$(T(\sigma)f)(\bar{p}) = \frac{e^{i\Lambda(\sigma)}}{|2\pi mc|^{3/2}} \int_{\mathbb{R}} d^3q \; e^{\frac{i}{2mc}(a|\bar{p}|^2 - 2\bar{p}\cdot\bar{q} + d|\bar{q}|^2)} f(\bar{q})$$

with

$$\Lambda(\sigma) = \frac{\pi}{2}\,\mathrm{sig}(m)(\mathrm{sig}(\sin\frac{\theta}{2}) + \frac{3}{2}\,\mathrm{sig}(\sin\theta)).$$

This is a metaplectic representation of $SL(2,\mathbb{R})$; the factor $\Lambda(\sigma)$ contains the Maslov index of $T(\sigma)$.

Since the orbits of $\tilde{S}(3)$ in $\{(m,j)\}$ are points, the induced IUR's of $\tilde{S}(3)$ are simply

$$U_{m,j,q,h} = \mathcal{L}_{m,j} \otimes \mathcal{D}_{q,h}$$

where $\mathcal{D}_{q,h}$ is an IUR of $SL(2,\mathbb{R})^*$; these latter are known [5]. We verify at last the one-to-one correspondence between the equivalence classes of these IUR's of $\tilde{S}(3)$ and the projective equivalence classes of the PIUR's of $S(3)$ obtained by putting $\xi = 0$ in $U_{m,j,q,h}$.

In this realization, the representations act on $L^2(\mathbb{R}^3,\mathbb{C}^{2j+1} \otimes \mathcal{H})$ where $\mathcal{H}$ is the Hilbert space of the representation $\mathcal{D}_{q,h}$; $\dim \mathcal{H} = 1$ or $\dim \mathcal{H} = \infty$ depending on whether $\mathcal{D}_{q,h}$ is trivial or not.

The restriction of $U_{m,j,q,h}$ to the Galilei subgroup $G(3) = \{(\bar{a},\bar{v},R,\begin{pmatrix}1 & \tau\\ 0 & 1\end{pmatrix})\}$ of $S(3)$ is completely reducible into the direct integral

$$U_{m,j,q,h} \downarrow G(3) = \int^{\otimes} d\mu\, U_{m,j,\mu}.$$

Here $U_{m,j,\mu} \in (m,j)$, i.e. belongs to the class of PIUR's of $G(3)$ which traditionally are interpreted as elementary localizable systems with mass m and spin j. The generalization of this interpretation to the classes (m,j,q,h) is rather artificial and its interest is dubious when $\dim \mathcal{H} = \infty$. When $\dim \mathcal{H} = 1$, we return to the traditional interpretation. Let us consider this case and let us introduce a dynamical postulate for the system by choosing as dynamical group a one parameter subgroup of $S(3)$; for example

$$t \to U_t = U(\vec{0},\bar{0},1,\begin{pmatrix}1 & -t\\ 0 & 1\end{pmatrix})$$

or

$$t \to U_t = U(\bar{0},\bar{0},1,\begin{pmatrix}\cos\omega t & \sin\omega t\\ -\sin\omega t & \cos\omega t\end{pmatrix}), \qquad \omega > 0$$

which define respectively a free particle and a harmonic oscillator. The choice of such an evolution operator makes it possible to realize the PIUR's of $(m,j,0,0)$ on the space $\mathcal{E}_H$ of the solutions of the Schrödinger equation associated with U_t.

By a Fourier transformation

$$\tilde{U}_{m,j,0,0}(\ldots) = F\, U_{m,j,0,0}(\ldots)F^{-1}$$

we obtain a representation acting on the space $L^2(\mathbb{R}^3,\mathbb{C}^{2j+1})$ of state functions defined on the spectrum of the position operators

$(Q_k f)(\bar{x}) = x_k f(\bar{x})$. The wave functions are given by

$$\psi(\bar{x},t) = (\tilde{U}_t f)(\bar{x})$$

and this formula defines an application

$$T : L^2(\mathbb{R}^3, \mathbb{C}^{2j+1}) \to \mathcal{E}_H$$

which turns out to be an isometry if $\mathcal{E}_H$ is provided with the Hilbert space structure $(\psi|\psi)_{\mathcal{E}_H} = (f|f)_{L^2}$.

The PIUR's

$$\hat{U}_{m,j,0,0}(\ldots) = T\tilde{U}_{m,j,0,0}(\ldots)T^{-1}$$

act on $\mathcal{E}_H$ and, for the two given examples, are precisely the realizations given by U. Niederer [6] (j=0).

REFERENCES

1. M. PERROUD, Helv. Phys. Acta, 50, XXX (1977).
2. V. BARGMANN, Ann. Math. 59, 1 (1954).
3. J.-M. LEVY-LEBLOND, in "*Group Theory and its Applications*", Vol. 2 (E.M. Loebl, ed.), Acad. Press, New York and London (1971).
4. G.W. MACKEY, "*Theory of Group Representations*", Lecture Notes, Univ. of Chicago, Summer 1955.
5. L. PUKANSKY, Math. Annalen 156, 36 (1964).
6. U. NIEDERER, Helv. Phys. Acta 45, 802 (1972). Helv. Phys. Acta 46, 191 (1973).

REPRESENTATION OF THE RACAH COEFFICIENT AS A GENERALIZED HYPERGEOMETRIC FUNCTION

K. Srinivasa Rao and K. Venkatesh

In an earlier article[1], a set of three series representations for the Racah coefficient[2] were obtained and these were rearranged into a set of generalized hypergeometric functions of unit argument, viz. ${}_4F_3$(ABCD;EFG;1). It was shown that the set of three ${}_4F_3$(1)s (hereafter referred to as "set I") is necessary and sufficient to account for the known[3] 144 symmetries of the Racah coefficient.

Here we show that it is possible to obtain an equivalent set of four series representations. Observations on the symmetries of the Racah coefficient are made in terms of these. We rearrange these four allowed series representations into a new set of four ${}_4F_3$(1)s (hereafter referred to as "set II") and discuss their domains of validity.

Finally, we obtain binomial expansions for the Racah coefficient, following the procedure of Sato[4], and we point out a connection between the two equivalent sets of ${}_4F_3$(1)s and the twelve binomial expansions, through the number of terms in a Racah coefficient.

The Racah coefficient[2] W(abcd;ef) is defined by the series:

$$\begin{Bmatrix} a & b & e \\ d & c & f \end{Bmatrix} = (-1)^{a+b+c+d} W(abcd;ef)$$
$$= N(-1)^{\beta_1} \sum_P (-1)^P (P+1)! \times \left[\prod_{i=1}^{4} (P-\alpha_i)! \prod_{j=1}^{3} (\beta_j - P)! \right]^{-1} \qquad (1)$$

where the range of P is restricted to non-negative integral values of the factorials:

$$\max \begin{bmatrix} \alpha_1 = a+b+e, & \alpha_2 = c+d+e \\ \alpha_3 = a+c+f, & \alpha_4 = b+d+f \end{bmatrix} \le P \le \min \begin{bmatrix} \beta_1 = a+b+c+d \\ \beta_2 = a+d+e+f \\ \beta_3 = b+c+e+f \end{bmatrix}, \qquad (2)$$

$$N = \Delta(abe)\Delta(cde)\Delta(acf)\Delta(bdf) \qquad (3)$$

and

$$\Delta(xyz) = \{(x+y-z)!(x-y+z)!(-x+y+z)![(x+y+z+1)!]^{-1}\}^{1/2} \qquad (4)$$

which vanishes unless the usual triangular condition in x, y and z, viz. $|x-y| \le z \le (x+y)$, is satisfied.

By setting in (1), $n = P-\alpha_k$, $k = 1,2,3,4$, in succession, we get the following set of four series representations:

$$\begin{Bmatrix} a & b & e \\ d & c & f \end{Bmatrix} = N(-1)^{\alpha_k+\beta_1} \sum_n (-1)^n (n+\alpha_k+1)!$$
$$\times \left[\prod_{i=1}^{4} (n+\alpha_k-\alpha_i)! \prod_{j=1}^{3} (\beta_j-\alpha_k-n)! \right]^{-1}. \qquad (5)$$

If we denote these by (I), (II), (III) and (IV), then:

(i) the column permutations of $\{\begin{smallmatrix} a & b & e \\ d & c & f \end{smallmatrix}\}$ leave a given series representation unaltered;

(ii) the permutation of any two elements in a row with the corresponding elements in the other now take a given series representation into any one of the other three series representations; and

(iii) the effect of the Regge[3] symmetries (which we refer to as RI, RII, RIII, RIV and RV, as per their order in ref. 3) is as follows: RI leaves (I) and (III) invariant and interchanges (II) and (IV); RII and RIII leave each one of the four series invariant; RIV and RV both leave (I) invariant but take (IV) → (III) → (II) → (IV) and (II) → (III) → (IV) → (II), respectively.

The set of four series representations (5) can be rearranged[5)] into a set of generalized hypergeometric functions (defined in the usual manner[6)]) of unit argument as:

$$\begin{Bmatrix} a & b & e \\ d & c & f \end{Bmatrix} = N(-1)^{A-2}\Gamma(1-E)[\Gamma(1-A)\Gamma(1-B)\Gamma(1-C)\Gamma(1-D)$$

$$\Gamma(F)\Gamma(G)]^{-1}{}_4F_3(ABCD;EFG;1) \qquad (6)$$

where the parameters of the set II of ${}_4F_3(1)$s are given by:

$$A = a+b+e+2, \quad B = a-c-f, \quad C = b-d-f, \quad D = e-c-d,$$

$$E = a+b-c-d+1, \quad F = a+e-d-f+1, \quad G = b+e-c-f+1; \qquad (7)$$

$$A = c+d+e+2, \quad B = c-a-f, \quad C = d-b-f, \quad D = e-a-b,$$

$$E = c+d-a-b+1, \quad F = c+e-b-f+1, \quad G = d+e-a-f+1; \qquad (8)$$

$$A = a+c+f+2, \quad B = c-d-e, \quad C = a-b-e, \quad D = f-b-d,$$

$$E = a+c-b-d+1, \quad F = a+f-d-e+1, \quad G = c+f-b-e+1; \qquad (9)$$

$$A = b+d+f+2, \quad B = b-a-e, \quad C = d-c-e, \quad D = f-a-c,$$

$$E = b+d-a-c+1, \quad F = b+f-c-e+1, \quad G = d+f-a-e+1. \qquad (10)$$

For all the allowed values of a,b,c,d,e and f, three of the four numerator parameters, B, C and D, are non-positive, while A is positive and A+B+C+D+1 = E+F+G. Thus, the ${}_4F_3(1)$s are of the terminating Saalschutzian type[6)], provided the denominator parameters (E, F and G) are positive. If E, F and G are negative, then the ${}_4F_3(1)$ is convergent, only if the condition[5)]

$$(B,\ C \text{ or } D) \geq (E,\ F \text{ or } G), \tag{11}$$

is satisfied.

Whereas the denominator parameters of the ${}_4F_3(1)$s belonging to set I consulted the relative magnitudes of the column sums of the Racah coefficient $\left\{\begin{smallmatrix} a & b & e \\ d & c & f \end{smallmatrix}\right\}$, the four ${}_4F_3(1)$s belonging to set II consult the relative magnitudes of two of the sums of the remaining 12 possible pairs of the arguments of the Racah coefficient. By making a comparison of the denominator parameters with the numerator parameters and using the triangular conditions, we deduce that

$$(E,\ F \text{ or } G) \geq (B,\ C \text{ or } D) \tag{12}$$

in all the four cases. From (11) and (12) it follows that E, F and G must all be positive for the ${}_4F_3(1)$ belonging to set II to be convergent.

We find that except when a=b=c=d=e=f and a=d, b=c, e=f, when all the four ${}_4F_3(1)$s are convergent, there are 4 cases when only one of the four, 6 cases when two of the four and 4 cases when three of the four ${}_4F_3(1)$s are convergent. For example, when $a+c > b+d$, $a+f > d+e$, $c+f > b+e$, only (9) is convergent; when $a+b > c+d$, $a+e > d+f$, $b+e > c+f$, (7) and (9) are convergent; when $a+c = b+d$, $a+f = d+e$, $c+f = b+e$, (8), (9) and (10) are convergent.

Choosing the ${}_4F_3(1)$ whose parameters are given by (7), we find that:

(i) the column permutations of $\left\{\begin{smallmatrix} a & b & e \\ d & c & f \end{smallmatrix}\right\}$ give rise to the same ${}_4F_3(1)$ but with the parameters permuted as in: ${}_4F_3(ACDB;GEF;1)$, ${}_4F_3(ADBC;FGE;1)$, ${}_4F_3(ACBD;EGF;1)$, ${}_4F_3(ADCB;GFE;1)$ and ${}_4F_3(ABDC;FEG;1)$;

(ii) the symmetries due to the interchange of a pair of row elements with the corresponding elements in the other row of a Racah coefficient, when imposed on a given ${}_4F_3(1)$ give rise to one of the other three ${}_4F_3(1)$s belonging to set II;

(iii) as in the case of set I, the allowed numerator and denominator parameter permutations, other than those explicitly mentioned in (i) above, lead to Regge symmetries on (most of) which the tetrahedral symmetries are super-imposed.

Thus, it is straightforward to show that any one of the ${}_4F_3(1)$s belonging to set II accounts for only 36 of the known 144 symmetries of the Racah coefficient. For, while all the 3! denominator parameter permutations lead to symmetries of the Racah coefficient, only 3! of the 4! numerator parameter permutations result in meaningful symmetries. This is due to one of the numerator parameters (A) being positive while the other three (B, C and D) are negative. Permutations of A with B, C or D lead to relationships of the type (see eq. 7) given in ref. 1.

Jahn and Howell[7)] have obtained this set of ${}_4F_3(1)$s but they have not discussed either their domains of validity or the symmetries of the Racah coefficient in terms of the parameter permutations of the arguments of the ${}_4F_3(1)$s. Our considerations above, lead us to the conclusion that the set (I or II) of ${}_4F_3(1)$s is necessary and sufficient to account for all the known 144 symmetries of the Racah coefficient. From the sets of ${}_4F_3(1)$ representations for the Racah coefficient, one can readily arrive at special formulae for W(abcd;ef) with any one, two or three stretched angular momenta (e.g. $e = a+b$).

The Racah coefficient can also be expressed symbolically[4)] as a binomial expansion. Following Sato's notation and making a different set of substitutions, corresponding to those made in ref. 1, given below:

$$Z = \beta_0 - k, \quad 0 \le k \le n, \quad n = \beta_0 - \alpha_0, \quad A_i = \beta_0 - \alpha_i \quad (i = p,q,r)$$

and

$$B_j = \beta_j - \beta_0 \quad (j = u,v) \tag{13}$$

where the indices p,q,r and u,v are used for those values of the lower limits (α's) and upper limits (β's) of P (2) which are other than $\alpha_0 = \alpha_{max}$ and $\beta_0 = \beta_{min}$, we obtain the binomial expansion:

$$W(abcd;ef) = N(-1)^{\beta_1-\beta_0}(\beta_0+1)!\,[n!A_p!A_q!A_r!(B_u+n)!(B_v+n)!]^{-1}$$

$$\cdot\ \{(B_u+n)(B_v+n)-A_pA_qA_r(\beta_0+1)^{(-1)}\}^{(n)} \tag{14}$$

where we have used the notation:

$$p^{(\sigma)} = p!/(p-\sigma)! \quad \text{and} \quad p^{(-\sigma)} = 1/p^{(\sigma)}. \tag{15}$$

Though this binomial expansion appears to be different in form from that given by Sato, (eq. 8 of ref. 4), it can be shown that these two binomial expansions are strictly equivalent.

Obviously, there are 12 possible values for $n = \beta_{min}-\alpha_{max}$ and correspondingly there are 12 possible binomial expansions for the Racah coefficient and the number of terms in an expansion is n+1. A given binomial expansion is left invariant only by two column permutations of $\{\begin{smallmatrix} a & b & e \\ d & c & f \end{smallmatrix}\}$ and two Regge symmetries. Other symmetries lead to different (one of the eleven other) binomial expansions. Thus, a discussion of the symmetries of the Racah coefficient in terms of its 12 binomial expansions is tedious.

The methods adopted for rearranging the single sum series of Racah to arrive at the binomial expansions or sets of ${}_4F_3(1)$s are different. It should be noted that while it is necessary to fix both the upper and lower limits of P to arrive at a binomial expansion, it is sufficient to choose either of these limits to arrive at a ${}_4F_3(1)$. In set I of ${}_4F_3(1)$s, all the four numerator parameters are negative and the number of terms (n+1) in the ${}_4F_3(1)$ is determined by n = min(-A,-B,-C,-D). In set II of ${}_4F_3(1)$s, only three of the four numerator parameters are negative, and the number of terms (n+1) in a ${}_4F_3(1)$ belonging to set II is determined by n = min(-B,-C,-D). Taking these facts into account, the twelve possible values of n+1 (= $\beta_{min}-\alpha_{max}+1$) can be written down as a block with three columns and four rows. Corresponding to each one of the twelve entries in Table 1 we have a binomial expansion and corresponding to the columns and rows, we have set I or set II of ${}_4F_3(1)$s, respectively.

TABLE 1. Possible number of terms, in the binomial series representation and the ${}_4F_3(1)$ representations for the Racah coefficient.

	Numerator parameters of set I of ${}_4F_3(1)$s (column-wise)		
Numerator parameters of set II of ${}_4F_3(1)$s (row-wise)	c+d-e+1	c+f-a+1	d+f-b+1
	b+d-f+1	b+e-a+1	d+e-c+1
	a+c-f+1	c+e-d+1	a+e-b+1
	a+b-e+1	b+f-d+1	a+f-c+1

Therefore, if we wish to rearrange the single sum series of Racah into generalized hypergeometric functions of unit argument, we necessarily end up with either a set I of three, or an equivalent set II of four, ${}_4F_3(1)$s. Since all the symmetries do not follow from the one ${}_4F_3(1)$ given in refs. 5 and 8, as has been envisaged by Smorodinskii and Shelepin[8], any one of these sets is necessary and sufficient to account for all the known (144) symmetries of the Racah coefficient. The parameter permutations of a given ${}_4F_3(1)$ for the Racah coefficient represent only 48 (if the ${}_4F_3(1)$ belongs to set I) or 36 (if the ${}_4F_3(1)$ belongs to set II) symmetries of the Racah coefficient. The other parameter permutations of a given ${}_4F_3(1)$ represent Racah coefficients for unphysical values (negative j-values) of angular momentum and Racah coefficients of a group other than O(3).

It is a pleasure to thank Professor Alladi Ramakrishnan for his interest in this work. One of us (K.S.R.) wishes to thank

Professor Marcos Moshinsky, Instituto de Fisica, UNAM, Mexico, D.F., for interesting discussions and Mrs. H. Kennelly of McMaster University for readily typing the paper.

REFERENCES

1. K. SRINIVASA RAO, T.S. SANTHANAM and K. VENKATESH, Jour. Math. Phys. 16, 1528 (1975).
2. G. RACAH, Phys. Rev. 12, 438 (1942).
3. T. REGGE, Nuo. Cim. 11, 116 (1959); also, Proc. Int. School of Phys., "Enrico Fermi", course LIV (1972), p. 11.
4. M. SATO, Prog. Theor. Phys. 13, 405 (1955); also, Y. LEHRER, private communication, quoted in "Nuclear Shell Theory" by A. de-Shalit and I. Talmi (Acad. Press, N.Y. 1963), p. 139.
5. M.E. ROSE, "Multipole Fields", (John Wiley and Sons, 1955), Appendix B, p. 92.
6. L.J. SLATER, "Generalized Hypergeometric Functions", (Cambridge Univ. Press, Cambridge, 1965), Chapter 2.
7. H.A. JAHN and K.M. HOWELL, Proc. Camb. Phil. Soc. 55, 338 (1959).
8. Ya. A. SMORODINSKII and L.A. SHELEPIN, Sov. Phys. Uspekhi 15, 1 (1972).

POLYNÔMES INVARIANTS D'UN CERTAIN GROUPE DE TRANSFORMATIONS LINÉAIRES

Ghislain Roy

RAPPELS, DEFINITIONS ET NOTATIONS

Si G est un groupe de transformations d'un espace E, on dira que la fonction $f(X_1,\ldots,X_\ell)$, à ℓ arguments en E, *est un invariant à* ℓ *points du groupe* G si

$$f(X_1,X_2,\ldots,X_\ell) = f(X_1',X_2',\ldots,X_\ell') \quad \forall\ X_i \in E; \quad \forall\ g \in G$$

où l'on note $X_i' = g(X_i)$.

Un ensemble: $f_1,f_2,\ldots,f_m$ (m minimum) de fonctions à plusieurs arguments en E est dit *système complet d'invariants de* G si tout invariant F de G peut s'écrire en fonction des $f_1,f_2,\ldots,f_m$: $F(X_1,\ldots,X_\ell) = \phi(f_1,\ldots,f_m)$.

Supposons que E est un espace vectoriel réel de dimension finie n, dont les points sont notés

$$X = \begin{pmatrix} x^1 \\ x^2 \\ \vdots \\ x^n \end{pmatrix}$$

et que G est un sous-groupe différentiable de $GL(n,\mathbb{R})$ dont la paramétrisation a été choisie de façon à ce que pour tout $g(t^1,\ldots,t^p) \in G$ le développement en série soit, dans un voisinage de $(t^1,\ldots,t^p) = 0$, $g(t^1,t^2,\ldots,t^p) = I + t^h K_h + \ldots$ où l'on note

$$\left.\frac{\partial g(t^1,\ldots,t^p)}{\partial t^h}\right|_{(t^1,\ldots,t^p)=0} = K_h = (K^i_{hj}).$$

Si $f(X_1,\ldots,X_\ell)$ est une fonction à ℓ argument en E qui est un polynôme homogène de degré un dans les coordonnées: $x^1_i, x^2_i, \ldots, x^n_i$ de chacun de ses arguments $X_1, X_2, \ldots, X_\ell$ cette fonction peut s'écrire:

$$f(X_1,\ldots,X_\ell) = \sum_{k=1}^{\ell} \sum_{i_k=1}^{n} a_{i_1 i_2,\ldots,i_\ell} x_1^{i_1} x_2^{i_2},\ldots,x_\ell^{i_\ell}.$$

Dans un article en cours de publication dans la Revue Roumaine de Mathématiques Pures et Appliquées on a démontré que si f est un invariant à ℓ points du groupe G, ses coefficients $a_{i_1 i_2,\ldots,i_\ell}$ sont solution du système linéaire

$$\sum_{j=1} (a_{j i_2 i_3,\ldots,i_\ell} K^j_{h i_1} + a_{i_1 j i_3,\ldots,i_\ell} K^j_{h i_2} + \ldots + a_{i_1 i_2,\ldots,j} K^j_{h i_\ell}) = 0$$

$$\forall\ h = 1,2,\ldots,p; \quad \forall\ i_1, i_2,\ldots,i_\ell = 1,2,\ldots,n.$$

EXEMPLE D'UTILISATION DE CE RESULTAT

Soit E l'ensemble des matrices réelles M, de dimensions n×n et G le groupe des transformations de E de la forme

$$M \to {}^T PMP, \quad \text{où} \quad P \in O(n,\mathbb{R}).$$

Si n=2, on note

$$M_k = \begin{pmatrix} a_k & b_k \\ c_k & d_k \end{pmatrix} \qquad (k = 1,2);$$

en résolvant le système plus haut, on trouve que les trois fonctions suivantes constituent un système complet d'invariants de ce groupe

$$F_1 = a_1+d_1 \qquad \left(\text{ou} \quad F_1 = a_2+d_2 \right)$$

$$F_2 = b_1-c_1 \qquad \left(\phantom{\text{ou}} \quad F_2 = b_2-c_2 \right)$$

$$F_3 = \begin{vmatrix} a_1 & b_1 \\ c_2 & d_2 \end{vmatrix} + \begin{vmatrix} a_2 & b_2 \\ c_1 & d_1 \end{vmatrix} .$$

Si l'on prend $M_1 = M_2 = M$, les fonctions F_1 et F_3 sont des invariants bien connus des transformations considérées de M, ce sont les coefficients de son polynôme caractéristique, multipliés par 1! et 2!, l'autre fonction: F_2 est un invariant moins connu, c'est le coefficient non nul du polynôme minimal de $M-{}^{T}M$.

Semblablement si n=3, on note

$$M_k = \begin{pmatrix} a_k & b_k & c_k \\ d_k & e_k & f_k \\ g_k & h_k & i_k \end{pmatrix} \qquad (k = 1,2,3)$$

et l'on trouve en résolvant le système linéaire donné plus haut que les 6 fonctions suivantes constituent un système complet d'invariants de ce groupe.

$$F_1 = a_k + e_k + i_k \qquad (k = 1,\ 2 \text{ ou } 3)$$

$$F_2 = \begin{vmatrix} a_k & b_k \\ d_\ell & e_\ell \end{vmatrix} + \begin{vmatrix} a_\ell & b_\ell \\ d_k & e_k \end{vmatrix} + \begin{vmatrix} a_k & c_k \\ g_\ell & i_\ell \end{vmatrix} + \begin{vmatrix} a_\ell & c_\ell \\ g_k & i_k \end{vmatrix}$$

$$+ \begin{vmatrix} e_k & f_k \\ h_\ell & i_\ell \end{vmatrix} + \begin{vmatrix} e_\ell & f_\ell \\ h_k & i_k \end{vmatrix}$$

(où $(k,\ell) = (1,2), (1,3)$ ou $(2,3)$)

$$F_3 = \begin{vmatrix} 0 & b_k-d_k \\ b_\ell-d_\ell & 0 \end{vmatrix} + \begin{vmatrix} 0 & c_k-g_k \\ c_\ell-g_\ell & 0 \end{vmatrix} + \begin{vmatrix} 0 & f_k-h_k \\ f_\ell-h_\ell & 0 \end{vmatrix}$$

où (k,ℓ) prend les mêmes valeurs que pour F_2, de même que pour la fonction suivante.

$$F_4 = \begin{vmatrix} 2a_k & b_k+d_k \\ b_\ell+d_\ell & 2e_\ell \end{vmatrix} + \begin{vmatrix} 2a_\ell & b_\ell+d_\ell \\ b_k+d_k & 2e_k \end{vmatrix} + \begin{vmatrix} 2a_k & c_k+g_k \\ c_\ell+g_\ell & 2i_\ell \end{vmatrix}$$

$$+ \begin{vmatrix} 2a_\ell & c_\ell+g_\ell \\ c_k+g_k & 2i_k \end{vmatrix} + \begin{vmatrix} 2e_k & f_k+h_k \\ f_\ell+h_\ell & 2i_\ell \end{vmatrix} + \begin{vmatrix} 2e_\ell & f_\ell+h_\ell \\ f_k+h_k & 2i_k \end{vmatrix}$$

$$F_5 = \sum \begin{vmatrix} a_k & b_k & c_k \\ d_\ell & e_\ell & f_\ell \\ g_m & h_m & i_m \end{vmatrix}$$

$$F_6 = \sum \begin{vmatrix} 2a_k & b_k+d_k & c_k+g_k \\ b_\ell+d_\ell & 2e_\ell & f_\ell+h_\ell \\ c_m+g_m & f_m+h_m & 2i_m \end{vmatrix}.$$

Pour les deux fonctions précédentes (k,ℓ,m) parcourt les six permutations de $\{1,2,3\}$.

Si l'on prend $M_1 = M_2 = M_3 = M$, les six invariants précédents sont les coefficients des polynômes caractéristiques des matrices M, $M+{}^TM$ ou $M-{}^TM$, multipliés, suivant le cas, par 1! 2! ou 3!.

THE CLEBSCH-GORDAN DECOMPOSITION AND THE COEFFICIENTS FOR THE SYMMETRIC GROUP

Susan Schindler and R. Mirman

The decomposition of the tensor product of two irreducible representations of the symmetric group into a direct sum of irreducible representations has been studied, providing a formalism for the calculation of the Clebsch-Gordan coefficients. We have written a computer program using this formalism and have generated tables of these coefficients.

The defining space for the left regular representation of the symmetric group is the group ring. The group ring elements are operators, each given by a linear combination of permutations. There are several standard constructions for a basis of the group ring. The coefficients in the expansion of an element of the group in terms of the basis vectors give the entries of the matrices of the left regular representation of the symmetric group.

The irreducible representations of the group are found by the reduction of the left regular representation. In this reduction, the irreducible representations fall into classes of equivalent representations. The number of representations in a class equals the dimension of any of its members. To label basis vectors for an irreducible representation we need three indices, one for the equivalence class, and one each for the representation and vector. Representation matrices for equivalent, irreducible representations are equal, not merely equivalent. This approach is standard for finite groups.

The irreducible representations for the symmetric group can also be realized on the space of polynomials in a set of non-commuting variables. The group elements are viewed as operators on the subscripts of the variables and the group ring elements as linear combinations of these operators. The action of the basis vectors of the group ring, on a single monomial, gives a basis for this space, and the action of a group element on this space is induced by its action on the group ring.

We can consider a second, distinct, but completely equivalent, polynomial space by starting with a different set of non-commuting variables. By using polynomials in products of variables from each of the two non-commuting sets, we obtain a third polynomial representation space. This space can be reduced into a direct sum of subspaces, each spanned by products of pairs of polynomials, one from each of the previous two spaces.

One view of the decomposition problem comes from the expansion of polynomials in products, in terms of the above products of pairs of polynomials. This gives a relationship involving three-indexed basis vectors. The expansion coefficients, the tensor coupling coefficients, are nine-index symbols. We give, below, their relations to the Clebsch-Gordan coefficients.

Another expansion, that of products of polynomials in terms of polynomials in products, can also be considered. This gives a reduction of the tensor product of the left regular representation with itself. To do this requires the construction of additional polynomial spaces. The expansion coefficients are related to the tensor coupling coefficients.

An alternate approach to the decomposition is to study the representation matrices. We start with the matrix for the tensor product of two irreducible representations, and we seek a similarity transformation reducing it, that is, transforming the matrix for the tensor product to block diagonal form, where each block is a matrix for some irreducible representation. The entries of the matrix for the similarity transformation are the Clebsch-Gordan coefficients.

In the direct sum (that is, in the block diagonal matrix), a representation matrix may occur more than once. Thus the Clebsch-Gordan coefficients bear an extra index to label the occurrence, the multiplicity index.

The tensor coupling coefficient is an entry of the representation matrix of the tensor product of three representations, multiplied by a function of the dimension (and so is completely defined). The Clebsch-Gordan coefficients arise in the reduction of a matrix for the tensor product of two representations. For the unitary case, we may rewrite the equation for this reduction. We get an equality between a sum of products of three matrix elements, which gives a tensor coupling coefficient, and a sum of products of pairs of Clebsch-Gordan coefficients. This relation gives the tensor coupling coefficients in terms of the Clebsch-Gordan coefficients.

Going the other way is not quite so easy. For multiplicity one, however, Clebsch-Gordan coefficients can be obtained simply by dividing the appropriate tensor coupling coefficient by the square root of a fixed tensor coupling coefficient. In the general procedure (which works for multiplicity one, also) we use some of the linear equations between sums of products of Clebsch-Gordan coefficients and tensor coupling coefficients which we discussed above. These equations state that the product of two matrices (formed from the Clebsch-Gordan coefficients) equals a matrix formed from the tensor coupling coefficients. If we know one of the matrices in the product, then, by matrix inversion and multiplication, we could solve for the other, giving the Clebsch-Gordan coefficients.

In fact, the columns of one of these matrices consist of eigenvectors of a matrix formed from the tensor coupling coefficients. This latter matrix defines an orthogonal projection, and so its range is spanned by its independent columns. Thus we find one of the matrices in the product by taking a particular matrix formed by the tensor coupling coefficients, searching for its independent columns, and replacing the resulting set by an orthonormal one, using the Gram-Schmidt process. The new set gives the required matrix.

This procedure generates the Clebsch-Gordan coefficients; our construction guarantees they satisfy appropriate orthonormality conditions.

We can find the tensor coupling coefficients from their expression in terms of the representation matrices. Direct calculation is impractical and we have developed an alternate approach.

Before discussing this, we have to consider how to label basis vectors. This can be done in several ways. One is to use diagrams, called frames (for equivalence classes) and tableaux (for representations and basis vectors). These are the standard labels for the symmetric group, S_n. A frame is a set of n boxes arranged in rows of non-increasing length, and a tableau for a particular frame is an arrangement of the numbers 1,...,n in these boxes, such that the numbers increase moving to the right, in a row, and down, in a column.

A tensor coupling coefficient is labeled by the indices of the basis vectors it links; for these we use frames and tableaux.

For any S_n tableau there is a corresponding S_{n-1} frame and tableau, found by deleting, from the tableau, the box containing the number n. We use the term "starring" for this deletion. To any tensor coupling coefficient for S_n there corresponds a set of tensor coupling coefficients for S_{n-1}, obtained by starring the S_n frames and tableaux labeling the given coefficient.

To calculate the tensor coupling coefficients we have developed an iterative formula, which expresses an S_n coefficient in terms of S_{n-1} coefficients and representation matrices for that neighboring transposition which interchanges the numbers n-1 and n. The formula gives the coefficient as a sum of two terms. The first term is an S_{n-1} coefficient, which is obtained by starring, multiplied by a function of the dimensions of the various representations (starred and unstarred) involved. The second term is given by a sum of products of pairs of S_{n-1} coefficients, obtained by starring, multiplied by the product of three matrix entries for the neighboring transposition (n-1,n). This sum is multiplied by another function of the dimensions of the representations involved.

For this procedure we need the coefficients for S_{n-1}, and the matrix of only one element of the group at a time, and overall, only the matrices of the neighboring transpositions.

Our procedure, of course, also provides an iterative method for generating the Clebsch-Gordan coefficients. The S_{n-1} tensor coupling coefficients in the formula can be replaced by sums of products of Clebsch-Gordan coefficients for S_{n-1}. Then, by methods discussed previously, from the tensor coupling coefficients for S_n, we can find Clebsch-Gordan coefficients for S_n, which can be used for the next stage of the iteration.

It is not actually necessary to compute all the coefficients. There are symmetry relations, suggesting the definition of equivalence classes of triplets of representations. We pick one triplet in an equivalence class, which we call the "working triplet", and we find its coefficients in the manner described previously. The coefficients of any other member of this class can be found, simply, from those of the working triplet. Explicit rules for doing this have been developed. The coefficients so generated depend on the rules given, as we would expect, since the Clebsch-Gordan coefficients are not unique (although the tensor coupling coefficients are).

We use two types of symmetry: interchange and conjugation. The first relates the coefficients for one triplet of representations to the coefficients for a second triplet, of the same three representations, taken in a different order. The latter symmetry relates the coefficients for one triplet to those of the triplet obtained from it by conjugating the last two of its frames (two frames are conjugate if they are obtained from one another by interchanging rows and columns. Conjugation changes the ordering we use for the tableaux.)

There is a difference between symmetry operations on the tensor coupling coefficients and on the Clebsch-Gordan coefficients. The former are unique so the symmetry expresses a well determined relation between two coefficients. The latter are not unique, but must obey certain conditions (orthonormality and the relationship with the tensor coupling coefficients). Thus a symmetry of the Clebsch-Gordan coefficients is a relationship arising from options we choose, within limits, by defining the coefficients so they possess certain properties. There are several different types of symmetries we can impose. However, we cannot construct coefficients satisfying, simultaneously, all of them. They depend on the symmetries we choose to build into them.

However all the symmetry constructions are used simultaneously to define the equivalence classes of triplets, and to set up chains of operations leading from the coefficients of one member of the class to those of any other member. We do not use this method to impose any of the above symmetry types. Rather there is a simple procedure for getting one coefficient from another, allowing a reduction in the amount of computation required.

The tensor coupling coefficients, and the Clebsch-Gordan coefficients satisfy certain linear homogeneous equations. In these equations, the coefficients are multiplied by entries of the representation matrices. This system appears useful mainly for checking computations.

In order to carry out the necessary computations, we must specify ordering schemes for the frames and tableaux. The ones we use are those of last letter ordering. Here the earlier frame has the longer last row. If the last rows of two frames have the same length, the relation between them is determined by their next to last rows, and so on. For tableaux, the one with n in the lower row comes first. If two tableaux have n in the same row, their order is determined by the row of n-1, and so on. Thus frames and tableaux have ordinals, providing another way of labeling them. Of course, in the computations we actually work with the indices. However, in setting up the computations, we work with the diagrams, the frames and tableaux.

This ordering scheme, applied to tableaux for the same frame, is preserved under starring, allowing it to be introduced into the iterative formula in a consistent manner.

Under conjugation the ordering for tableaux is reversed. Thus, the conjugate of the first tableau of a frame is the last tableau of the conjugate frame, and so on. Thus, we must, in certain cases, conjugate a self-conjugate frame. Under this operation the frame remains the same but the order of the tableaux is reversed.

Construction of coefficients by conjugation requires a sign function. When the relation between two coefficients is obtained by the conjugation of two of the frames (even self-conjugate ones) labeling them, the sign function must be included.

This function, defined on tableaux, is the sign of the permutation which produces a given tableau from the first tableau of a frame. The definition does not provide a reasonable way of calculating the function. However, we have found a formula for it, involving the positions occupied by the numbers 1,...,n in a tableau, which is simple to use.

This completes the theoretical machinery necessary for the

calculation of the Clebsch-Gordan coefficients. The formulas, algorithms, and procedures are readily translated into computer language. The number of coefficients increases quite rapidly with n. Thus computer calculation is feasible, within the limits set by storage and running time.

Although this discussion has been presented in the terminology of the symmetric group, large parts of it carry over with little, if any, changes, to any other finite group. In fact, we would expect that it would also carry over with some fairly simple modifications to the case of a compact group.

The computational methods, in particular the iterative formula, do seem closely tied to the symmetric group, and are probably not easily generalized. However, the general method of finding the Clebsch-Gordan coefficients from the tensor coupling coefficients, for any multiplicity, holds for any finite group.

Further details of this work, and the coefficients, will be published in the Journal of Mathematical Physics.